The Principles of
Nonlinear Optics

The Principles of
Nonlinear Optics

Y. R. SHEN

Wiley Classics Library Edition Published 2003

A JOHN WILEY & SONS, INC., PUBLICATION

Published by John Wiley & Sons, Inc., Hoboken, New Jersey.
Published simultaneously in Canada

For general information on our other products and services please contact our Customer Care Department within the U.S. at 877-762-2974, outside the U.S at 317-572-3993 or fax 317-572-4002.

Wiley also publishes its books in a variety of electronic formats. Some content that appears in print, however, may not be available in electronic format.

Library of Congress Cataloging-in-Publication Data is available.

ISBN 0-471-43080-3

10 9 8 7 6 5 4 3 2 1

To
Hsiao-Lin,
Kai, and Hao

Preface

The laser is certainly one of the greatest inventions in the history of science. Its arrival, a quarter of a century ago, created many fascinating new fields, among which nonlinear optics undoubtedly has the broadest scope and the most influential proponents. The field originated with the experimental work of P. A. Franken and co-workers on optical second-harmonic generation in 1961 and the theoretical work of N. Bloembergen and co-workers on optical wave mixing in 1962. Since that time the field has grown at such a prodigious rate that today it has already found applications in nearly all areas of science.

The very broad expanse of nonlinear optics certainly is most exciting, but it also makes the field difficult to comprehend. The vast amount of knowledge generated over the years is scattered everywhere in the literature. Beginners in nonlinear optics often have a hard time acquainting themselves with the many facets of the field. Even workers in nonlinear optics may sometimes have difficulty finding some rudimentary information about a subarea of the field they are not familiar with. A book on nonlinear optics offering a fair introduction to all branches of the field is clearly needed.

Actually, there already exist a number of books on the subject of nonlinear optics. The most authoritative, by N. Bloembergen, lays the foundation for nonlinear optics. However, since the book was written in 1965, it is clearly outdated, as is the 1964 book by S. A. Akhmanov and R. V. Khokhlov (English translation 1972). Among the remaining books found in most academic libraries, some are elementary or narrow in scope and others tend to concentrate on special topics of nonlinear optics. Conference proceedings may provide a broader perspective but usually are very advanced and lack continuity. What one would like to have is a book that not only logically presents the basic principles of nonlinear optics but also systematically describes the subareas of the field. This book is meant to fulfill that need.

To write a book covering the entire field of nonlinear optics in depth is an impossible task for a single author. This book falls short of the full details in the description of some subject matter. Furthermore, to limit the size of the book several areas were omitted. These include collision-induced nonlinear optical excitations, optical multistabilities, bifurcations and chaos, quantum statistics of nonlinear optics, and many highly nonlinear optical effects. In writing the book, I chose to emphasize the fundamentals as well as the interplay between theory and experiment. Physical concepts are stressed in the

vii

theoretical presentation, although equations are usually unavoidable for a careful account. In the illustration of a particular process, a brief description of the experimental situation is given to provide the readers with a realistic picture. References at the end of each chapter supplement the omitted details in the text, but each list is purposely brief.

This book grows out of a graduate physics course on modern optics I taught several times at Berkeley. The frustrating experience of selecting appropriate materials for the course led me to write this book. It was therefore prepared at a level intended for a physics graduate student. With some effort, students of chemistry and engineering who are serious in learning about nonlinear optics should also be able to appreciate the material without much difficulty. And the book should be a useful reference for professionals in their review of the field.

The book begins with a general introduction, followed by a description of the fundamentals in Chapters 2 and 3. Electrooptical and magnetooptical effects are considered in Chapter 4 as special nonlinear optical phenomena, and their inverse effects are discussed in Chapter 5. The more familiar second-order nonlinear optical effects are discussed in Chapters 6 to 9, and the third-order effects are considered in Chapters 10 to 17. In the discussion, parametric conversion in Chapter 9 is treated as the inverse of a mixing process. Stimulated light scattering is shown in Chapters 10 to 11 to behave like a parametric process from the general coupled-wave point of view, although it is often understood as a two-photon process resulting in material excitation. While the first half of the book deals with the traditional type of nonlinear optics, the second half is on special topics. Chapters 13, 15, and 18 to 28 are devoted to the discussion of various nonlinear optical effects and applications which have fascinated researchers in recent years. In many of these areas, new results and discoveries are still being reported frequently at meetings and in journals. Some parts of the text are bound to become obsolete sooner or later, but it is hoped that the principles should always remain unchanged.

As a tribute to the twenty-fifth anniversary of the invention of lasers, this book is written to manifest a part of the intellectual wealth lasers have created. I am deeply indebted to Professor Bloembergen for introducing nonlinear optics to me during its early development stage. His teaching and guidance have led to the great satisfaction I have experienced in research in the past 20 years. I should like to express my gratitude to all my friends and colleagues who have supported my effort in writing this book. Special thanks are due to S. J. Gu, whose critical reading of the manuscript resulted in many changes and corrections. I also thank T. F. Heinz, X. D. Zhu, M. Mate, Y. Twu, and many others for their contribution in proofreading and improving the manuscript. With respect to the preparation of the manuscript, I am most grateful to Rita Jones, who not only typed the entire manuscript but also helped and supported the project in all possible ways. Without her devoted effort, completion of this book would not have been possible. Finally, my wife,

Hsiao-Lin, deserves my warmest appreciation. It is her patience, understanding, encouragement, and help in many details that gave me faith and strength in the course of writing this book.

Y. R. SHEN

Berkeley, California
April 1984

Contents

1	Introduction	1
2	Nonlinear Optical Susceptibilities	13
3	General Description of Wave Propagation in Nonlinear Media	42
4	Electrooptical and Magnetooptical Effects	53
5	Optical Rectification and Optical Field-Induced Magnetization	57
6	Sum-Frequency Generation	67
7	Harmonic Generation	86
8	Difference-Frequency Generation	108
9	Parametric Amplification and Oscillation	117
10	Stimulated Raman Scattering	141
11	Stimulated Light Scattering	187
12	Two-Photon Absorption	202
13	High-Resolution Nonlinear Optical Spectroscopy	211
14	Four-Wave Mixing	242
15	Four-Wave Mixing Spectroscopy	266
16	Optical-Field-Induced Birefringence	286
17	Self-Focusing	303
18	Multiphoton Spectroscopy	334

19 **Detection of Rare Atoms and Molecules**   349

20 **Laser Manipulation of Particles**   366

21 **Transient Coherent Optical Effects**   379

22 **Strong Interaction of Light with Atoms**   413

23 **Infrared Multiphoton Excitation and Dissociation of Molecules**   437

24 **Laser Isotope Separation**   466

25 **Surface Nonlinear Optics**   479

26 **Nonlinear Optics in Optical Waveguides**   505

27 **Optical Breakdown**   528

28 **Nonlinear Optical Effects in Plasmas**   541

Index   555

1

Introduction

Physics would be dull and life most unfulfilling if all physical phenomena around us were linear. Fortunately, we are living in a nonlinear world. While linearization beautifies physics, nonlinearity provides excitement in physics. This book is devoted to the study of nonlinear electromagnetic phenomena in the optical region which normally occur with high-intensity laser beams. Nonlinear effects in electricity and magnetism have been known since Maxwell's time. Saturation of magnetization in a ferromagnet, electrical gas discharge, rectification of radio waves, and electrical characteristics of p–n junctions are just a few of the familiar examples. In the optical region, however, nonlinear optics became a subject of great common interest only after the laser was invented. It has since contributed a great deal to the rejuvenation of the old science of optics.

1.1 HISTORICAL BACKGROUND

The second harmonic generation experiment of Franken et al.[1] marked the birth of the field of nonlinear optics. They propagated a ruby laser beam at 6942 Å through a quartz crystal and observed ultraviolet radiation from the crystal at 3471 Å. Franken's idea was simple. Harmonic generation of electromagnetic waves at low frequencies had been known for a long time. Harmonic generation of optical waves follows the same principle and should also be observable. Yet an ordinary light source is much too weak for such an experiment. It generally takes a field of about 1 kV/cm to induce a nonlinear response in a medium. This corresponds to a beam intensity of about 2.5 kW/cm². A laser beam is therefore needed in the observation of optical harmonic generation.

Second harmonic generation is the first nonlinear optical effect ever observed in which a coherent input generates a coherent output. But nonlinear optics covers a much broader scope. It deals in general with nonlinear

1

interaction of light with matter and includes such problems as light-induced changes of the optical properties of a medium. Second harmonic generation is then not the first nonlinear optical effect ever observed. Optical pumping is certainly a nonlinear optical phenomenon well known before the advent of lasers. The resonant excitation of optical pumping induces a redistribution of populations and changes the properties of the medium. Because of resonant enhancement, even a weak light is sufficient to perturb the material system strongly to make the effect easily detectable. Low-power CW atomic lamps were used in the earlier optical pumping experiments on atomic systems. Optical pumping is also one of the effective schemes for creating an inverted population in a laser system.

In general, however, observation of nonlinear optical effects requires the application of lasers. Numerous nonlinear optical phenomena have been discovered since 1961. They have not only greatly enhanced our knowledge about interaction of light with matter, but also created a revolutionary change in optics technology. Each nonlinear optical process may consist of two parts. The intense light first induces a nonlinear response in a medium, and then the medium in reacting modifies the optical fields in a nonlinear way. The former is governed by the constitutive equations, and the latter by the Maxwell's equations.

At this point, one may raise a question: Are all media basically nonlinear? The answer is yes. Even in the case of a vacuum, photons can interact through vacuum polarization. The nonlinearity is, however, so small that with currently available light sources, photon-photon scattering and other nonlinear effects in vacuum are still difficult to observe.[2] So, in a practical sense, a vacuum can be regarded as linear. In the presence of a medium, the nonlinearity is greatly enhanced through interaction of light with matter. Photons can now interact much more effectively through polarization of the medium.

1.2 MAXWELL'S EQUATIONS IN NONLINEAR MEDIA

All electromagnetic phenomena are governed by the Maxwell's equations for the electric and magnetic fields $\mathbf{E}(\mathbf{r}, t)$ and $\mathbf{B}(\mathbf{r}, t)$:

$$\nabla \times \mathbf{E} = -\frac{1}{c}\frac{\partial \mathbf{B}}{\partial t},$$

$$\nabla \times \mathbf{B} = \frac{1}{c}\frac{\partial \mathbf{E}}{\partial t} + \frac{4\pi}{c}\mathbf{J}, \qquad (1.1)$$

$$\nabla \cdot \mathbf{E} = 4\pi\rho,$$

$$\nabla \cdot \mathbf{B} = 0$$

where $\mathbf{J}(\mathbf{r}, t)$ and $\rho(\mathbf{r}, t)$ are the current and charge densities, respectively. They are related by the charge conservation law

$$\nabla \cdot \mathbf{J} + \frac{\partial \rho}{\partial t} = 0 \tag{1.2}$$

We can often expand $\mathbf{J}$ and ρ into series of multipoles:[3]

$$\mathbf{J} = \mathbf{J}_0 + \frac{\partial \mathbf{P}}{\partial t} + c\nabla \times \mathbf{M} + \frac{\partial}{\partial t}(\nabla \cdot \mathbf{Q}) + \cdots,$$
$$\rho = \rho_0 - \nabla \cdot \mathbf{P} - \nabla(\nabla \cdot \mathbf{Q}) + \cdots. \tag{1.3}$$

Here, $\mathbf{P}$, $\mathbf{M}$, $\mathbf{Q}, \ldots$, are respectively the electric polarization, the magnetization, the electric quadrupole polarization, and so on. However, as pointed out by Landau and Lifshitz,[4] it is not really meaningful in the optical region to express $\mathbf{J}$ and ρ in terms of multipoles because the usual definitions of multipoles are unphysical. In many cases, for example in metals and semiconductors, it is more convenient to use $\mathbf{J}$ and ρ directly as the source terms in the Maxwell's equations, or to use a generalized electric polarization $\mathbb{P}$ defined by

$$\mathbf{J} = \mathbf{J}_{dc} + \frac{\partial \mathbb{P}}{\partial t} \tag{1.4}$$

where $\mathbf{J}_{dc}$ is the dc current density. In other cases, the magnetic dipole and higher-order multipoles can be neglected. Then, the generalized $\mathbb{P}$ reduces to the electric-dipole polarization $\mathbf{P}$. The difference between $\mathbb{P}$ and $\mathbf{P}$ is that $\mathbb{P}$ is a nonlocal function of the field and $\mathbf{P}$ is local. In this book, we assume electric dipole approximation, $\mathbb{P} = \mathbf{P}$, unless specified.

With (1.2) and (1.4), the Maxwell's equations appear in the form

$$\nabla \times \mathbf{E} = -\frac{1}{c}\frac{\partial \mathbf{B}}{\partial t},$$
$$\nabla \times \mathbf{B} = \frac{1}{c}\frac{\partial}{\partial t}(\mathbf{E} + 4\pi\mathbb{P}) + \frac{4\pi}{c}\mathbf{J}_{dc}, \tag{1.5}$$
$$\nabla \cdot (\mathbf{E} + 4\pi\mathbb{P}) = 0,$$
$$\nabla \cdot \mathbf{B} = 0$$

where $\mathbb{P}$ is now the only time-varying source term. In general, $\mathbb{P}$ is a function of $\mathbf{E}$ that describes fully the response of the medium to the field, and it is often known as the constitutive equation. If we could just write the constitutive equation and find the solution for the resulting set of Maxwell's equations with appropriate boundary conditions, then all optical phenomena would be predictable and easily understood. Unfortunately, this seldom is possible. Physically reasonable approximations must be resorted to in order to make the mathematical solution of the equations feasible. This is where physics comes into play.

The polarization $\mathbb{P}$ is usually a complicated nonlinear function of $\mathbf{E}$. In the linear case, however, $\mathbb{P}$ takes a simple linearized form

$$\mathbb{P}(\mathbf{r}, t) = \int_{-\infty}^{\infty} \chi^{(1)}(\mathbf{r} - \mathbf{r}', t - t') \cdot \mathbf{E}(\mathbf{r}', t') \, d\mathbf{r}' \, dt' \qquad (1.6)$$

where $\chi^{(1)}$ is the linear susceptibility. If $\mathbf{E}$ is a monochromatic plane wave with $\mathbf{E}(\mathbf{r}, t) = \mathbf{E}(\mathbf{k}, \omega) = \mathscr{E}(\mathbf{k}, \omega)\exp(i\mathbf{k} \cdot \mathbf{r} - i\omega t)$, then Fourier transformation of (1.6) yields the familiar relation

$$\begin{aligned} \mathbb{P}(\mathbf{r}, t) &= \mathbb{P}(\mathbf{k}, \omega) \\ &= \chi^{(1)}(\mathbf{k}, \omega) \cdot \mathbf{E}(\mathbf{k}, \omega) \end{aligned} \qquad (1.7)$$

with

$$\chi^{(1)}(\mathbf{k}, \omega) = \int_{-\infty}^{\infty} \chi^{(1)}(\mathbf{r}, t)\exp(-i\mathbf{k} \cdot \mathbf{r} + i\omega t) \, d\mathbf{r} \, dt. \qquad (1.8)$$

The linear dielectric constant $\varepsilon(\mathbf{k}, \omega)$ is related to $\chi^{(1)}(\mathbf{k}, \omega)$ by

$$\varepsilon(\mathbf{k}, \omega) = 1 + 4\pi\chi^{(1)}(\mathbf{k}, \omega). \qquad (1.9)$$

In the electric dipole approximation, $\chi^{(1)}(\mathbf{r}, t)$ is independent of $\mathbf{r}$, and hence both $\chi^{(1)}(\mathbf{k}, \omega)$ and $\varepsilon(\mathbf{k}, \omega)$ are independent of $\mathbf{k}$.

In the nonlinear case, when $\mathbf{E}$ is sufficiently weak, the polarization $\mathbb{P}$ as a function of $\mathbf{E}$ can be expanded into a power series of $\mathbf{E}$:

$$\begin{aligned} \mathbb{P}(\mathbf{r}, t) = {} & \int_{-\infty}^{\infty} \chi^{(1)}(\mathbf{r} - \mathbf{r}', t - t') \cdot \mathbf{E}(\mathbf{r}, t') \, d\mathbf{r}' \, dt \\ & + \int_{-\infty}^{\infty} \chi^{(2)}(\mathbf{r} - \mathbf{r}_1, t - t_1; \mathbf{r} - \mathbf{r}_2, t - t_2) : \mathbf{E}(\mathbf{r}_1, t_1) \\ & \times \mathbf{E}(\mathbf{r}_2, t_2) \, d\mathbf{r}_1 \, dt_1 \, d\mathbf{r}_2 \, dt_2 \\ & + \int_{-\infty}^{\infty} \chi^{(3)}(\mathbf{r} - \mathbf{r}_1, t - t_1; \mathbf{r} - \mathbf{r}_2, t - t_2; \\ & \qquad \mathbf{r} - \mathbf{r}_3, t - t_3) : \mathbf{E}(\mathbf{r}_1, t_1) \\ & \times \mathbf{E}(\mathbf{r}_2, t_2)\mathbf{E}(\mathbf{r}_3, t_3) \, d\mathbf{r}_1 \, dt_1 \, d\mathbf{r}_2 \, dt_2 \, d\mathbf{r}_3 \, dt_3 + \cdots \end{aligned} \qquad (1.10)$$

where $\chi^{(n)}$ is the nth-order nonlinear susceptibility. If $\mathbf{E}$ can be expressed as a group of monochromatic plane waves

$$\mathbf{E}(\mathbf{r}, t) = \sum_l \mathbf{E}(\mathbf{k}_l, \omega_l), \qquad (1.11)$$

then, as in the linear case, Fourier transform of (1.10) gives

$$\mathbb{P}(\mathbf{k}, \omega) = \mathbb{P}^{(1)}(\mathbf{k}, \omega) + \mathbb{P}^{(2)}(\mathbf{k}, \omega) + \mathbb{P}^{(3)}(\mathbf{k}, \omega) + \cdots \qquad (1.12)$$

with

$$\begin{aligned}
\mathbb{P}^{(1)}(\mathbf{k}, \omega) &= \chi^{(1)}(\mathbf{k}, \omega) \cdot \mathbf{E}(\mathbf{k}, \omega), \\
\mathbb{P}^{(2)}(\mathbf{k}, \omega) &= \chi^{(2)}(\mathbf{k} = \mathbf{k}_i + \mathbf{k}_j, \omega = \omega_i + \omega_j) : \mathbf{E}(\mathbf{k}_i, \omega_i)\mathbf{E}(\mathbf{k}_j, \omega_j), \\
\mathbb{P}^{(3)}(\mathbf{k}, \omega) &= \chi^{(3)}(\mathbf{k} = \mathbf{k}_i + \mathbf{k}_j + \mathbf{k}_l, \omega = \omega_i + \omega_j + \omega_l) \\
&\qquad \vdots \ \mathbf{E}(\mathbf{k}_i, \omega_i)\mathbf{E}(\mathbf{k}_j, \omega_j)\mathbf{E}(\mathbf{k}_l, \omega_l),
\end{aligned} \qquad (1.13)$$

and

$$\begin{aligned}
\chi^{(n)}(\mathbf{k} &= \mathbf{k}_1 + \mathbf{k}_2 + \cdots + \mathbf{k}_n, \omega = \omega_1 + \omega_2 + \cdots \omega_n) \\
&= \int_{-\infty}^{\infty} \chi^{(n)}(\mathbf{r} - \mathbf{r}_1, t - t_1; \cdots ; \mathbf{r} - \mathbf{r}_n, t - t_n) \\
&\qquad \times e^{-i[\mathbf{k}_1(\mathbf{r} - \mathbf{r}_1) - \omega_1(t - t_1) + \cdots + \mathbf{k}_n(\mathbf{r} \cdots \mathbf{r}_n) - \omega_n(t - t_n)]} d\mathbf{r}_1\, dt_1 \cdots d\mathbf{r}_n\, dt_n
\end{aligned} \qquad (1.14)$$

Again, in the electric dipole approximation, $\chi^{(n)}(\mathbf{r}, t)$ is independent of $\mathbf{r}$, or $\chi^{(n)}(\mathbf{k}, \omega)$ independent of $\mathbf{k}$.

The linear and nonlinear susceptibilities characterize the optical properties of a medium. If $\chi^{(n)}$ is known for a given medium, then at least in principle, the nth-order nonlinear optical effects in the medium can be predicted from the Maxwell's equations in (1.5). Physically, $\chi^{(n)}$ is related to the microscopic structure of the medium and can be properly evaluated only with a full quantum-mechanical calculation. Simple models are, however, often used to illustrate the origin of optical nonlinearity and some characteristic features of $\chi^{(n)}$. We consider here the anharmonic oscillator model and the free electron gas model.

1.3   ANHARMONIC OSCILLATOR MODEL

In this model, a medium is composed of a set of N classical anharmonic oscillators per unit volume. The oscillator describes physically an electron bound to a core or an infrared-active molecular vibration. Its equation of motion in the presence of a driving force is

$$\frac{d^2x}{dt^2} + \Gamma\frac{dx}{dt} + \omega_0^2 x + ax^2 = F. \qquad (1.15)$$

We consider here the response of the oscillator to an applied field with Fourier components at frequencies $\pm\omega_1$ and $\pm\omega_2$:

$$F = \frac{q}{m}\left[E_1(e^{-i\omega_1 t} + e^{i\omega_1 t}) + E_2(e^{-i\omega_2 t} + e^{i\omega_2 t})\right]. \qquad (1.16)$$

The anharmonic term ax^2 in (1.15) is assumed to be small so that it can be treated as a perturbation in the successive approximation of finding a solution:

$$x = x^{(1)} + x^{(2)} + x^{(3)} + \cdots . \qquad (1.17)$$

The induced electric polarization is simply

$$\mathbf{P} = Nqx. \qquad (1.18)$$

The first-order solution is obtained from the linearized equation of (1.15):

$$x^{(1)} = x^{(1)}(\omega_1) + x^{(1)}(\omega_2) + \text{c.c.}$$

$$x^{(1)}(\omega_i) = \frac{(q/m)E_i}{(\omega_0^2 - \omega_i^2 - i\omega_1\Gamma)}e^{-i\omega_i t} \qquad (1.19)$$

where c.c. is a complex conjugate. Then, the second-order solution is obtained from (1.15) by approximating ax^2 by $ax^{(1)2}$:

$$x^{(2)} = x^{(2)}(\omega_1 + \omega_2) + x^{(2)}(\omega_1 - \omega_2) + x^{(2)}(2\omega_1) + x^{(2)}(2\omega_2) + x^{(2)}(0) + \text{c.c}$$

$$x^{(2)}(\omega_1 \pm \omega_2) = \frac{-2a(q/m)^2 E_1 E_2}{(\omega_0^2 - \omega_1^2 - i\omega_1\Gamma)(\omega_0^2 - \omega_2^2 \mp i\omega_2\Gamma)}$$

$$\times \frac{1}{\left[\omega_0^2 - (\omega_1 \pm \omega_2)^2 - i(\omega_1 \pm \omega_2)\Gamma\right]}e^{-i(\omega_1 \pm \omega_2)t}$$

$$(1.20)$$

$$x^{(2)}(2\omega_i) = \frac{-a(q/m)^2 E_i^2}{(\omega_0^2 - \omega_i^2 - i\omega_i\Gamma)^2(\omega_0^2 - 4\omega_i^2 - i2\omega_i\Gamma)}e^{-i2\omega_i t}$$

$$x^{(2)}(0) = -a\left(\frac{q}{m}\right)^2 \frac{1}{\omega_0^2}\left(\frac{1}{\omega_0^2 - \omega_1^2 - i\omega_1\Gamma} + \frac{1}{\omega_0^2 - \omega_2^2 - i\omega_2\Gamma}\right)$$

By successive iteration, higher-order solutions can also be obtained. As seen in the second-order solution, new frequency components of the polarization at $\omega_1 \pm \omega_2$, $2\omega_1$, $2\omega_2$, and 0 have appeared through quadratic interaction of the field with the oscillator via the anharmonic term. The oscillating polarization components will radiate and generate new em waves at $\omega_1 \pm \omega_2$, $2\omega_1$, and $2\omega_2$.

Thus, sum- and difference-frequency generation and second-harmonic generation are readily explained. The appearance of the zero-frequency polarization component is known as optical rectification. More generally, frequency components at $\omega = n_1\omega_1 \pm n_2\omega_2$, with n_1 and n_2 being integers, are expected in the higher-order solutions. In this model, the anharmonicity a determines the strength of the nonlinear interaction.

Treating ax^2 as a small perturbation in the foregoing calculation is equivalent to the assumption that E is small and P can be expanded into power series of E. We can give a rough estimate on how the nonlinear polarization should diminish with increasing order. Assuming the nonresonant case with $\omega_0 \gg \omega_1$ and ω_2, we find from (1.19) and (1.20),

$$\left| \frac{P^{(2)}}{P^{(1)}} \right| \sim \left| \frac{qaE}{m\omega_0^4} \right|. \tag{1.21}$$

For an electron bound to a core, if x is so large that the harmonic force $m\omega_0^2 x$ and the anharmonic force max^2 are of the same order of magnitude, then both will be of the same order of magnitudes as the total binding force on the electron $|qE_{at}|$:

$$|qE_{at}| \sim m\omega_0^2 x \sim max^2$$

or

$$|qE_{at}| \sim \frac{m\omega_0^4}{a}. \tag{1.22}$$

Equation (1.21) then becomes

$$\left| \frac{P^{(2)}}{P^{(1)}} \right| \sim \left| \frac{E}{E_{at}} \right|. \tag{1.23}$$

In fact, one can show in general

$$\left| \frac{P^{(n+1)}}{P^{(n)}} \right| \sim \left| \frac{E}{E_{at}} \right| \tag{1.24}$$

such that $|E/E_{at}|$ acts as an expansion parameter in the perturbation calculation. Typically, $E_{at} \sim 3 \times 10^8$ V/cm. The E field for a 2.5-W/cm^2 laser beam is only 30 V/cm with $|E/E_{at}| \sim 10^{-7}$. The nonlinear polarization is much weaker than the linear polarization. This suggests that the observation of nonlinear optical effects requires high-intensity laser beams.

Relation (1.24), however, is true only for optical frequencies away from resonance. Near resonance, the resonant denominators may drastically enhance the ratio $|P^{(n+1)}/P^{(n)}|$. Consequently, the nonlinear effects can be detected with much weaker light intensity. Optical pumping is an example. With resonant enhancement, it may even happen that $|P^{(n+1)}/P^{(n)}| > 1$. When this is the case, the perturbation expansion is no longer valid, and the full nonlinear expression of P as a function of E must be included in the calculation. The problem then falls into the domain of strong interaction of light with matter.

1.4 FREE ELECTRON GAS

A simple but realistic model to illustrate optical nonlinearity in a medium is the free electron gas model. It properly describes the optical properties of an electron plasma. The simplified version of the model starts with the equation of motion for an electron

$$\frac{d^2\mathbf{r}}{dt^2} = -\frac{e}{m}\left(\mathbf{E} + \frac{1}{c}\mathbf{v} \times \mathbf{B}\right). \tag{1.25}$$

Damping is neglected here for simplicity. Clearly, the only nonlinear term in this equation is the Lorentz force term. Since $v \ll c$ in a plasma, the Lorentz force is much weaker than the Coulomb force, and then $(e/mc)\mathbf{v} \times \mathbf{B}$ in (1.25) can be treated as a perturbation in the successive approximation of the solution. For $\mathbf{E} = \mathscr{E}_1 e^{i\mathbf{k}_1 \cdot \mathbf{r} - i\omega_1 t} + \mathscr{E}_2 e^{i\mathbf{k}_2 \cdot \mathbf{r} - i\omega_2 t} + \text{c.c.}$, we obtain

$$\mathbf{r}^{(1)}(\omega_i) = \frac{e}{m\omega_i^2}\mathscr{E}_i e^{i\mathbf{k}_i \cdot \mathbf{r}^{(0)} - i\omega_i t} + \text{c.c,}$$

$$\mathbf{r}^{(2)}(\omega_1 + \omega_2) = \frac{-ie^2}{m^2\omega_1\omega_2(\omega_1 + \omega_2)^2}$$

$$\times \left[\mathscr{E}_1 \times (\mathbf{k}_2 \times \mathscr{E}_2) + \mathscr{E}_2 \times (\mathbf{k}_1 \times \mathscr{E}_1)\right] \tag{1.26}$$

$$\times e^{i(\mathbf{k}_1 + \mathbf{k}_2) \cdot \mathbf{r}^{(0)} - i(\omega_1 + \omega_2)t} + \text{c.c,}$$

and so on. For a uniform plasma with an electronic charge density ρ, the current density is given by

$$\mathbf{J} = \mathbf{J}^{(1)} + \mathbf{J}^{(2)} + \cdots$$

$$= \rho\frac{\partial}{\partial t}\left(\mathbf{r}^{(1)} + \mathbf{r}^{(2)} + \cdots\right) \tag{1.27}$$

with, for example,

$$\mathbf{J}^{(2)}(\omega_1 + \omega_2) = \rho\frac{\partial}{\partial t}\mathbf{r}^{(2)}(\omega_1 + \omega_2),$$

and so on. This shows explicitly how an electron gas can respond nonlinearly to the incoming light through the Lorentz term.

In a more rigorous treatment of an electron gas, we must also take into account the spatial variations of the electron density ρ and velocity $\mathbf{v}$. Two equations, the equation of motion and the continuity equation,[5] are now necessary to describe the electron plasma:

$$\frac{\partial \mathbf{v}}{\partial t} + (\mathbf{v} \cdot \nabla)\mathbf{v} = -\frac{\nabla p}{m\rho} - \frac{e}{m}\left(\mathbf{E} + \frac{1}{c}\mathbf{v} \times \mathbf{B}\right)$$

and   (1.28)

$$\frac{\partial \rho}{\partial t} + \nabla \cdot (\rho \mathbf{v}) = 0$$

where p is the pressure and m is the electron mass. The pressure gradient term in the equation of motion is responsible for the dispersion of plasma resonance, but in the following calculation we assume $\nabla p = 0$ for simplicity. Then, coupled with (1.28), is the set of Maxwell's equations

$$\nabla \times \mathbf{E} = -\frac{1}{c}\frac{\partial \mathbf{B}}{\partial t},$$

$$\nabla \times \mathbf{B} - \frac{1}{c}\frac{\partial \mathbf{E}}{\partial t} = \frac{4\pi \mathbf{J}}{c} = \frac{4\pi \rho \mathbf{v}}{c},$$ (1.29)

$$\nabla \cdot \mathbf{E} = 4\pi(\rho - \rho^{(0)}),$$

and

$$\nabla \cdot \mathbf{B} = 0$$

We assume here that there is a fixed positive charge background in the plasma to assure charge neutrality in the absence of external perturbation. Successive approximation can be used to find $\mathbf{J}$ as a function of E from (1.28) and (1.29). Let[6]

$$\rho = \rho^{(0)} + \rho^{(1)} + \rho^{(2)} + \cdots,$$

$$\mathbf{v} = \mathbf{v}^{(1)} + \mathbf{v}^{(2)} + \cdots,$$

and

$$\mathbf{j} = \mathbf{j}^{(1)} + \mathbf{j}^{(2)} + \cdots$$ (1.30)

with

$$\mathbf{j}^{(1)} = \rho^{(0)}\mathbf{v}^{(1)}$$

and   (1.31)

$$\mathbf{j}^{(2)} = \rho^{(0)}\mathbf{v}^{(2)} + \rho^{(1)}\mathbf{v}^{(1)}.$$

We shall find the expression for $\mathbf{j}^{(2)}(2\omega)$ as an example assuming $\mathbf{E} = \mathscr{E} \times \exp(i\mathbf{k} \cdot \mathbf{r} - i\omega t)$. Substitution of (1.30) into (1.28) and (1.29) yields

$$\frac{\partial \mathbf{v}^{(1)}(\omega)}{\partial t} = -i\omega \mathbf{v}^{(1)} = -\frac{e}{m}\mathbf{E},$$

$$\frac{\partial \rho^{(1)}(\omega)}{\partial t} = -i\omega \rho^{(1)} = -\nabla \cdot \left(\rho^{(0)}\mathbf{v}^{(1)}\right)$$

$$\nabla \cdot \mathbf{E} = 4\pi\rho^{(1)}(\omega), \tag{1.32}$$

$$\frac{\partial \mathbf{v}^{(2)}(2\omega)}{\partial t} = -i2\omega \mathbf{v}^{(2)} = -\left(\mathbf{v}^{(1)} \cdot \nabla\right)\mathbf{v}^{(1)} - \frac{e}{mc}\mathbf{v}^{(1)} \times \mathbf{B}.$$

The second-order current density is then given by

$$\mathbf{J}^{(2)}(2\omega) = \frac{i\rho^{(0)}}{2\omega}\left[\frac{e^2}{m^2\omega^2}(\mathbf{E} \cdot \nabla)\mathbf{E} + \frac{ie^2}{m^2\omega c}\mathbf{E} \times \mathbf{B}\right]$$

$$+ \frac{e}{i4\pi m\omega}(\nabla \cdot \mathbf{E})\mathbf{E}. \tag{1.33}$$

The last term in (1.33) has the following equalities:

$$\frac{e}{i4\pi m\omega}(\nabla \cdot \mathbf{E})\mathbf{E} = -\frac{e}{m\omega^2}\left[\nabla \cdot \left(\rho^{(0)}\mathbf{v}^{(1)}\right)\right]\mathbf{E}$$

$$= \frac{ie^2}{m^2\omega^3}\left[\nabla \cdot \left(\rho^{(0)}\mathbf{E}\right)\right]\mathbf{E} \tag{1.34}$$

$$= \frac{ie^2}{m^2\omega^3}\left[\frac{\nabla\rho^{(0)} \cdot \mathbf{E}}{1 - \omega_p^2/\omega^2}\right]\mathbf{E}$$

where $\omega_p = (4\pi\rho^{(0)}e/m)^{1/2}$ is the plasma resonance frequency. With (1.34), $\mathbf{B} = (c/i\omega)\nabla \times \mathbf{E}$, and the vector relation $\mathbf{E} \times (\nabla \times \mathbf{E}) + (\mathbf{E} \cdot \nabla)\mathbf{E} = \frac{1}{2}\nabla(\mathbf{E} \cdot \mathbf{E})$, the current density in (1.33) can be written as

$$\mathbf{J}^{(2)}(2\omega) = \frac{i\rho^{(0)}e^2}{4m^2\omega^3}\nabla(\mathbf{E} \cdot \mathbf{E}) + \frac{ie^2}{m^2\omega^3}\left[\nabla \cdot \left(\rho^{(0)}\mathbf{E}\right)\right]\mathbf{E}$$

$$= \frac{i\rho^{(0)}e^2}{4m^2\omega^3}\nabla(\mathbf{E} \cdot \mathbf{E}) + \frac{ie^2}{m^2\omega^3}\left[\frac{\nabla\rho^{(0)} \cdot \mathbf{E}}{1 - \omega_p^2/\omega^2}\right]\mathbf{E}. \tag{1.35}$$

Equation (1.33) shows explicitly that aside from the Lorentz term, there are also terms related to the spatial variation of $\mathbf{E}$. They actually arise from the nonuniformity of the plasma. In a uniform plasma, $\nabla\rho^{(0)} = 0$ and hence $\nabla \cdot \mathbf{E} = 0$ from (1.34). This means that $\mathbf{k}$ is perpendicular to $\mathbf{E}$ and therefore $(\mathbf{E} \cdot \nabla)\mathbf{E}$ also vanishes. The Lorentz term is then the only term in $\mathbf{J}^{(2)}(2\omega)$.

The induced current density $\mathbf{J}^{(2)}(2\omega)$ should now act as a source for second harmonic generation in the plasma. In a uniform plasma with a single pump beam, $\mathbf{J}^{(2)}(2\omega) \propto \mathbf{E} \times \mathbf{B}$ is only along the direction of beam propagation. Since an oscillating current cannot radiate longitudinally, no coherent second harmonic generation along the axis of beam propagation is expected from the bulk of a uniform plasma. In the bulk of a nonuniform plasma or at the boundary surface of a uniform plasma, however, it is possible to find second harmonic generation through the nonvanishing $\nabla\rho^{(0)}$.

Equation (1.35) shows that when $\nabla\rho^{(0)} \neq 0$, the nonlinear response of the medium $\mathbf{J}^{(2)}(2\omega)$ is greatly enhanced if ω is near the plasma resonance. From the general principle, the nonlinear response of a medium is resonantly enhanced when the incoming field hits the resonance of the medium. One should, of course, also expect a resonant enhancement in the second harmonic generation when 2ω is at resonance. This actually does come in through the response of the second harmonic field to $\mathbf{J}^{(2)}(2\omega)$. As $2\omega \to \omega_p$, the current density will excite the longitudinal field at 2ω resonantly.

As shown in (1.33) or (1.35), the current density $\mathbf{J}^{(2)}(2\omega)$ depends exclusively on the spatial variation of $\mathbf{E}$. In fact, using vector identities, the expression of $\mathbf{J}^{(2)}(2\omega)$ in (1.33) or (1.35) can be put into the form[7] $\mathbf{J}^{(2)}(2\omega) = c\nabla \times (\ \ ) - i2\omega\nabla \cdot (\ \ )$. Comparing it with (1.3), we recognize that the two terms in $\mathbf{J}^{(2)}(2\omega)$ represent the magnetic dipole and electric quadrupole contributions, respectively. No induced electric dipole polarization exists in a plasma. Also, the induced electric quadrupole polarization depends on the gradient of the electric field, and therefore cannot show up in the bulk of a uniform plasma.

The free electron model here is applicable to a number of real problems. First, it can be used to describe the optical nonlinearities due to plasmas in metals and semiconductors. Second harmonic generation from metal surfaces is readily observable.[6] Then, with some modification to take into account the net charge distribution, the nonvanishing ∇p, and so on, it can also be used to describe the optical nonlinearities of a gas plasma. Various nonlinear optical effects in gas plasmas have been observed. They will be discussed in some detail in Chapter 28. The model has also been used to describe the observation of nonlinear effect in a crystal in the X-ray region.[8] The electron binding energy is much weaker than the X-ray photon energy, and therefore the electrons in the crystal will respond to the X-ray as if they were free.

REFERENCES

1   P. A. Franken, A. E. Hill, C. W. Peters, and G. Weinreich, *Phys. Rev. Lett.* **7**, 118 (1961).

2   See, for example, G. Rosen and F. C Whitmore, *Phys. Rev.* **137B**, 1357 (1965).

3   J. D. Jackson, *Classical Electrodynamics* (McGraw-Hill, New York 1975), 2nd ed , p. 739; W. K. H. Panofsky and M. Phillips, *Classical Electricity and Magnetism* (Addison-Wesley, Reading, Mass., 1962), p. 131.

4 L. D. Landau and E. M. Lifshitz, *Electrodynamics in Continuous Media* (Pergamon Press, New York, 1960), p. 252.

5 J D. Jackson, *Classical Electrodynamics* (McGraw-Hill, New York, 1975), 2nd ed , p. 469.

6 N. Bloembergen, R. K. Chang, S. S. Jha, and C. H. Lee, *Phys. Rev* **174**, 813 (1968).

7 P. S. Pershan, *Phys. Rev.* **130**, 919 (1963).

8 P. E. Eisenberger and S. L. McCall, *Phys. Rev. Lett.* **26**, 684 (1971).

BIBLIOGRAPHY

Bloembergen, N., *Nonlinear Optics* (Benjamin, New York, 1965).

Bloembergen, N., *Rev. Mod. Phys.* **54**, 685 (1982).

2

Nonlinear Optical Susceptibilities

For lower-order nonlinear optical effects, nonlinear polarizations and nonlinear susceptibilities characterize the steady-state nonlinear optical response of a medium and govern the nonlinear wave propagation in the medium. Chapter 1 showed how the nonlinear optical response can be calculated for two model systems. Chapter 2 gives a more general discussion of nonlinear susceptibilities starting from the microscopic theory.

2.1 DENSITY MATRIX FORMALISM

Nonlinear optical susceptibilities are characteristic properties of a medium and depend on the detailed electronic and molecular structure of the medium. Quantum mechanical calculation is needed to find the microscopic expressions for nonlinear susceptibilities.[1] Density matrix formalism is probably most convenient for such calculation and is certainly more correct when relaxations of excitations have to be dealt with.[2]

Let ψ be the wave function of the material system under the influence of the electromagnetic field. Then the density matrix operator is defined as the ensemble average over the product of the ket and bra state vectors

$$\rho = \overline{|\psi\rangle\langle\psi|} \tag{2.1}$$

and the ensemble average of a physical quantity $\mathbf{P}$ is given by

$$\begin{aligned}
\langle \mathbf{P} \rangle &= \overline{\langle\psi|\mathbf{P}|\psi\rangle} \\
&= \mathrm{Tr}(\rho\mathbf{P})
\end{aligned} \tag{2.2}$$

In our calculation here, $\mathbf{P}$ corresponds to the electric polarization. From the

13

definition of ρ in (2.1) and from the Schrödinger equation for $|\psi\rangle$, we can readily obtain the equation for motion for ρ,

$$\frac{\partial \rho}{\partial t} = \frac{1}{i\hbar}[\mathcal{H}, \rho], \tag{2.3}$$

known as the Liouville equation. The Hamiltonian $\mathcal{H}$ is composed of three parts,

$$\mathcal{H} = \mathcal{H}_0 + \mathcal{H}_{int} + \mathcal{H}_{random}. \tag{2.4}$$

In the semiclassical approach, $\mathcal{H}_0$ is the Hamiltonian of the unperturbed material system with eigenstates $|n\rangle$ and eigenenergies $\mathbb{E}_n$ so that $\mathcal{H}_0|n\rangle = \mathbb{E}_n|n\rangle$, $\mathcal{H}_{int}$ is the interaction Hamiltonian describing the interaction of light with matter, and $\mathcal{H}_{random}$ is a Hamiltonian describing the random perturbation on the system by the thermal reservoir around the system. The interaction Hamiltonian in the electric dipole approximation is given by

$$\mathcal{H}_{int} = e\mathbf{r} \cdot \mathbf{E}. \tag{2.5}$$

We consider here only the electronic contribution to the susceptibilities. For the ionic contribution, we would have to replace $e\mathbf{r} \cdot \mathbf{E}$ by $-\Sigma_i q_i \mathbf{R}_i \cdot \mathbf{E}$ with q_i and $\mathbf{R}_i$ being the charge and position of the ith ion, respectively. The Hamiltonian $\mathcal{H}_{random}$ is responsible for the relaxations of material excitations, or in other words, the relaxation of the perturbed ρ back to thermal equilibrium. We can then express (2.3) as[3,4]

$$\frac{\partial \rho}{\partial t} = \frac{1}{i\hbar}[\mathcal{H}_0 + \mathcal{H}_{int}, \rho] + \left(\frac{\partial \rho}{\partial t}\right)_{relax} \tag{2.6}$$

with

$$\left(\frac{\partial \rho}{\partial t}\right)_{relax} \equiv \frac{1}{i\hbar}[\mathcal{H}_{random}, \rho]. \tag{2.7}$$

If the eigenstates $|n\rangle$ are now used as base vectors in the calculation, and $|\psi\rangle$ is written as a linear combination of $|n\rangle$, that is, $|\psi\rangle = \Sigma_n a_n|n\rangle$, then the physical meaning of the matrix elements of ρ is clear. The diagonal matrix element $\rho_{nn} \equiv \langle n|\rho|n\rangle = \overline{|a_n|^2}$ represents the population of the system in state $|n\rangle$, while the off-diagonal matrix element $\rho_{nn'} \equiv \langle n|\rho|n'\rangle = \overline{a_n a_{n'}^*}$ indicates that the state of the system has a coherent admixture of $|n\rangle$ and $|n'\rangle$. In the latter case, if the relative phase of a_n and $a_{n'}$ is random (or incoherent), then $\rho_{nn'} = 0$ through the ensemble average. Thus at thermal equilibrium $\rho_{nn}^{(0)}$ is given by the thermal population distribution, for example, the Boltzmann distribution in the case of atoms or molecules, and $\rho_{nn'}^{(0)} = 0$ for $n \neq n'$.

We can use a simple physical argument to find a more explicit expression for $(\partial\rho/\partial t)_{\text{relax}}$. The population relaxation is a result of transitions between states induced by interaction with the thermal reservoir. Let $W_{n\to n'}$ be the thermally induced transition rate from $|n\rangle$ to $|n'\rangle$. Then the relaxation rate of an excess population in $|n\rangle$ should be

$$\left(\frac{\partial\rho_{nn}}{\partial t}\right)_{\text{relax}} = \sum_{n'}[W_{n'\to n}\rho_{n'n'} - W_{n\to n'}\rho_{nn}]. \tag{2.8}$$

At thermal equilibrium we have

$$\frac{\partial\rho_{nn}^{(0)}}{\partial t} = \sum_{n'}\left[W_{n'\to n}\rho_{n'n'}^{(0)} - W_{n\to n'}\rho_{nn}^{(0)}\right] = 0. \tag{2.9}$$

Therefore, (2.8) can also be written as

$$\frac{\partial}{\partial t}\left[(\rho_{nn})_{\text{relax}} - \rho_{nn}^{(0)}\right] = \sum_{n'}\left[W_{n'\to n}\left(\rho_{n'n'} - \rho_{n'n'}^{(0)}\right) - W_{n\to n'}\left(\rho_{nn} - \rho_{nn}^{(0)}\right)\right]. \tag{2.10}$$

The relaxation of the off-diagonal elements is more complicated.[2] In simple cases, however, we expect the phase coherence to decay exponentially to zero. Then, we have, for $n \neq n'$,

$$\left(\frac{\partial\rho_{nn'}}{\partial t}\right)_{\text{relax}} = -\Gamma_{nn'}\rho_{nn'} \tag{2.11}$$

with $\Gamma_{nn'}^{-1} = \Gamma_{n'n}^{-1} = (T_2)_{nn'}$ being a characteristic relaxation time between the states $|n\rangle$ and $|n'\rangle$. In magnetic resonance, the population relaxation is known as the longitudinal relaxation, and the relaxation of the off-diagonal matrix elements is known as the transverse relaxation. In some cases, the longitudinal relaxation of a state can be approximated by

$$\frac{\partial}{\partial t}\left(\rho_{nn} - \rho_{nn}^{(0)}\right)_{\text{relax}} = -(T_1)_n^{-1}\left(\rho_{nn} - \rho_{nn}^{(0)}\right). \tag{2.12}$$

Then T_1 is called the longitudinal relaxation time. Correspondingly, T_2 is called the transverse relaxation time.

Thus, at least in principle, if $\mathcal{H}_0$, $\mathcal{H}_{\text{int}}$, and $(\partial\rho/\partial t)_{\text{relax}}$ are known, the Liouville equations in (2.6) together with (2.2) fully describe the response of the medium to the incoming field. It is, however, not possible in general to combine (2.6) and (2.2) into a single equation of motion for $\langle\mathbf{P}\rangle$. Only in special cases can this be done. In this chapter we consider only the case of steady-state response with $\langle\mathbf{P}\rangle$ expandable into power series of $\mathbf{E}$. The transient response is discussed in Chapter 21.

To find nonlinear polarizations and nonlinear susceptibilities of various orders, we use perturbation expansion in the calculation. Let

$$\rho = \rho^{(0)} + \rho^{(1)} + \rho^{(2)} + \cdots$$

and

$$\langle \mathbf{P} \rangle = \langle \mathbf{P}^{(1)} \rangle + \langle \mathbf{P}^{(2)} \rangle + \cdots \tag{2.13}$$

with

$$\langle \mathbf{P}^{(n)} \rangle = \text{Tr}(\rho^{(n)} \mathbf{P}) \tag{2.14}$$

where $\rho^{(0)}$ is the density matrix operator for the system at thermal equilibrium, and we assume no permanent polarization in the medium so that $\langle \mathbf{P}^{(0)} \rangle = 0$. Inserting the series expansion of ρ into (2.6) and collecting terms of the same order with $\mathcal{H}_{\text{int}}$ treated as a first-order perturbation, we should obtain

$$\frac{\partial \rho^{(1)}}{\partial t} = \frac{1}{i\hbar} \left[\left(\mathcal{H}_0, \rho^{(1)} \right) + \left(\mathcal{H}_{\text{int}}, \rho^{(0)} \right) \right] + \left(\frac{\partial \rho^{(1)}}{\partial t} \right)_{\text{relax}},$$

$$\frac{\partial \rho^{(2)}}{\partial t} = \frac{1}{i\hbar} \left[\left(\mathcal{H}_0, \rho^{(2)} \right) + \left(\mathcal{H}_{\text{int}}, \rho^{(1)} \right) \right] + \left(\frac{\partial \rho^{(2)}}{\partial t} \right)_{\text{relax}}, \tag{2.15}$$

and so on. We are interested here in the response to a field that can be decomposed into Fourier components, $\mathbf{E} = \sum_l \boldsymbol{\mathscr{E}}_l \exp(i\mathbf{k}_l \cdot \mathbf{r} - i\omega_l t)$. Then, since $\mathcal{H}_{\text{int}} = \sum_l \mathcal{H}_{\text{int}}(\omega_l)$ and $\mathcal{H}_{\text{int}}(\omega_l) \propto \boldsymbol{\mathscr{E}}_l \exp(-i\omega_l t)$, the operator $\rho^{(n)}$ can also be expanded into a Fourier series

$$\rho^{(n)} = \sum_j \rho^{(n)}(\omega_j).$$

With $\partial \rho^{(n)}(\omega_j)/\partial t = -i\omega_j \rho^{(n)}(\omega_j)$, (2.15) can now be solved explicitly for $\rho^{(n)}(\omega_j)$ in successive orders. The first- and second-order solutions are

$$\rho_{nn'}^{(1)}(\omega_j) = \frac{\left[\mathcal{H}_{\text{int}}(\omega_j) \right]_{nn'}}{\hbar(\omega_j - \omega_{nn'} + i\Gamma_{nn'})} \left(\rho_{n'n'}^{(0)} - \rho_{nn}^{(0)} \right)$$

$$\rho_{nn'}^{(2)}(\omega_j + \omega_k) = \frac{\left[\mathcal{H}_{\text{int}}(\omega_j), \rho^{(1)}(\omega_k) \right]_{nn'} + \left[\mathcal{H}_{\text{int}}(\omega_k), \rho^{(1)}(\omega_j) \right]_{nn'}}{\hbar(\omega_j + \omega_k - \omega_{nn'} + i\Gamma_{nn'})}$$

$$= \frac{1}{\hbar(\omega_j + \omega_k - \omega_{nn'} + i\Gamma_{nn'})} \tag{2.16}$$

$$\times \sum_{n''} \left\{ \left[\mathcal{H}_{\text{int}}(\omega_j) \right]_{nn''} \rho_{n''n'}^{(1)}(\omega_k) \right.$$

$$- \rho_{nn''}^{(1)}(\omega_k) \left[\mathcal{H}_{\text{int}}(\omega_j) \right]_{n''n'}$$

$$+ \left[\mathcal{H}_{\text{int}}(\omega_k) \right]_{nn''} \rho_{n''n'}^{(1)}(\omega_j)$$

$$\left. - \rho_{nn''}^{(1)}(\omega_j) \left[\mathcal{H}_{\text{int}}(\omega_k) \right]_{n''n'} \right\}.$$

We use here the notation $A_{nn'} = \langle n|A|n' \rangle$. Higher-order solutions can be obtained readily, although the derivation is long and tedious. Whenever diagonal elements $\rho_{mm}^{(n)}(0)$ appear in the derivation, further approximation on $(\partial\rho_{mm}/\partial t)_{\text{relax}}$ in (2.8) is often necessary in order to find a closed-form solution. We also note that the expression for $\rho_{nn'}^{(2)}(\omega_j + \omega_k)$ in (2.16) is valid even for $n = n'$ as long as $\omega_j + \omega_k \neq 0$ since the term $(\partial\rho_{nn}^{(2)}/\partial t)_{\text{relax}}$ can then be neglected in the calculation.

2.2 MICROSCOPIC EXPRESSIONS FOR NONLINEAR SUSCEPTIBILITIES

The full microscopic expressions for the nonlinear polarizations $\langle \mathbf{P}^{(n)} \rangle$ and the nonlinear susceptibilities $\langle \mathbf{\chi}^{(n)} \rangle$ follow immediately from the expressions of $\rho^{(n)}$. With $\mathcal{H}_{\text{int}} = e\mathbf{r} \cdot \mathbf{E}$ and $\mathbf{P} = -Ne\mathbf{r}$ in (2.14) and (2.16), the first- and second-order susceptibilities due to electronic contribution are readily obtained. They are given here in explicit Cartesian tensor notation:

$$\chi_{ij}^{(1)}(\omega) = \frac{P_i^{(1)}(\omega)}{E_j(\omega)}$$

$$= N\frac{e^2}{\hbar}\sum_{gn}\left[\frac{(r_i)_{ng}(r_j)_{gn}}{\omega + \omega_{ng} + i\Gamma_{ng}} - \frac{(r_j)_{ng}(r_i)_{gn}}{\omega - \omega_{ng} + i\Gamma_{ng}}\right]\rho_g^{(0)},$$

$$\chi_{ijk}^{(2)}(\omega = \omega_1 + \omega_2)$$

$$= \frac{P_i^{(2)}(\omega)}{E_j(\omega_1)E_k(\omega_2)}$$

$$= -N\frac{e^3}{\hbar^2}\sum_{g,n,n'}\left[\frac{(r_i)_{gn}(r_j)_{nn'}(r_k)_{n'g}}{(\omega - \omega_{ng} + i\Gamma_{ng})(\omega_2 - \omega_{n'g} + i\Gamma_{n'g})}\right.$$

$$+ \frac{(r_i)_{gn}(r_k)_{nn'}(r_j)_{n'g}}{(\omega - \omega_{ng} + i\Gamma_{ng})(\omega_1 - \omega_{n'g} + i\Gamma_{n'g})}$$

$$+ \frac{(r_k)_{gn'}(r_j)_{n'n}(r_i)_{ng}}{(\omega + \omega_{ng} + i\Gamma_{ng})(\omega_2 + \omega_{n'g} + i\Gamma_{n'g})} \tag{2.17}$$

$$+ \frac{(r_j)_{gn'}(r_k)_{n'n}(r_i)_{ng}}{(\omega + \omega_{ng} + i\Gamma_{ng})(\omega_1 + \omega_{n'g} + i\Gamma_{n'g})}$$

$$- \frac{(r_j)_{ng}(r_i)_{n'n}(r_k)_{gn'}}{(\omega - \omega_{nn'} + i\Gamma_{nn'})}\left(\frac{1}{\omega_2 + \omega_{n'g} + i\Gamma_{n'g}} + \frac{1}{\omega_1 - \omega_{ng} + i\Gamma_{ng}}\right)$$

$$\left.- \frac{(r_k)_{ng}(r_i)_{n'n}(r_j)_{gn'}}{(\omega - \omega_{nn'} + i\Gamma_{nn'})}\left(\frac{1}{\omega_2 - \omega_{ng} + i\Gamma_{ng}} + \frac{1}{\omega_1 + \omega_{n'g} + i\Gamma_{n'g}}\right)\right]\rho_g^{(0)}.$$

There are two terms in $\chi_{ij}^{(1)}$ and eight terms in $\chi_{ijk}^{(2)}$. The calculation can be extended to third order to find $\chi_{ijkl}^{(3)}$ ($\omega = \omega_1 + \omega_2 + \omega_3$), which will have 48 terms. The complete expression for $\chi_{ijkl}^{(3)}$ is given in the literature[5] and is not reproduced here. The resonant structure of $\chi_{ijkl}^{(3)}$, however, is discussed in Chapter 14. In nonresonant cases, the damping constants in the denominators in (2.17) can be neglected. The second-order susceptibility can then be reduced to a form with six terms, noting that the last two terms in the expression for $\chi_{ijk}^{(2)}$ in (2.17) become

$$\frac{(r_j)_{gn}(r_i)_{nn'}(r_k)_{n'g}}{(\omega_1 - \omega_{ng})(\omega_2 + \omega_{n'g})} + \frac{(r_k)_{gn}(r_i)_{nn'}(r_j)_{n'g}}{(\omega_1 + \omega_{n'g})(\omega_2 - \omega_{ng})}.$$

With N denoting the number of atoms or molecules per unit volume, the expressions in (2.17) are actually more appropriate for gases or molecular liquids or solids, and $\rho_g^{(0)}$ is given by the Boltzmann distribution. For solids whose electronic properties are described by band structure, the eigenstates are the Bloch states, and $\rho_g^{(0)}$ corresponds to the Fermi distribution. The expression for $\chi_{ij}^{(1)}$ and $\chi_{ijk}^{(2)}$ should then be properly modified. Since the band states form essentially a continuum, the damping constants in the resonant denominators can be ignored. In the electric dipole approximation with the photon wavevector dependence neglected, $\chi_{ijk}^{(2)}$ for such solids has the form[3]

$$\chi_{ijk}^{(2)}(\omega = \omega_1 + \omega_2)$$

$$= -\frac{e^3}{\hbar^2}\int d\mathbf{q} \sum_{v,c,c'} \left\{ \frac{\langle v,\mathbf{q}|r_i|c,\mathbf{q}\rangle\langle c,\mathbf{q}|r_j|c',\mathbf{q}\rangle\langle c',\mathbf{q}|r_k|v,\mathbf{q}\rangle}{[\omega - \omega_{cv}(\mathbf{q})][\omega_2 - \omega_{c'v}(\mathbf{q})]} \right.$$

$$+ \frac{\langle v,\mathbf{q}|r_i|c,\mathbf{q}\rangle\langle c,\mathbf{q}|r_k|c',\mathbf{q}\rangle\langle c',\mathbf{q}|r_j|v,\mathbf{q}\rangle}{[\omega - \omega_{cv}(\mathbf{q})][\omega_1 - \omega_{c'v}(\mathbf{q})]}$$

$$+ \frac{\langle v,\mathbf{q}|r_k|c,\mathbf{q}\rangle\langle c,\mathbf{q}|r_j|c',\mathbf{q}\rangle\langle c',\mathbf{q}|r_i|v,\mathbf{q}\rangle}{[\omega + \omega_{c'v}(\mathbf{q})][\omega_2 + \omega_{cv}(\mathbf{q})]} \qquad (2.18)$$

$$+ \frac{\langle v,\mathbf{q}|r_j|c,\mathbf{q}\rangle\langle c,\mathbf{q}|r_k|c',\mathbf{q}\rangle\langle c',\mathbf{q}|r_i|v,\mathbf{q}\rangle}{[\omega + \omega_{c'v}(\mathbf{q})][\omega_1 + \omega_{cv}(\mathbf{q})]}$$

$$+ \frac{\langle v,\mathbf{q}|r_j|c,\mathbf{q}\rangle\langle c,\mathbf{q}|r_i|c',\mathbf{q}\rangle\langle c',\mathbf{q}|r_k|v,\mathbf{q}\rangle}{[\omega_1 - \omega_{cv}(\mathbf{q})][\omega_2 + \omega_{c'v}(\mathbf{q})]}$$

$$\left. + \frac{\langle v,\mathbf{q}|r_k|c,\mathbf{q}\rangle\langle c,\mathbf{q}|r_i|c',\mathbf{q}\rangle\langle c',\mathbf{q}|r_j|v,\mathbf{q}\rangle}{[\omega_1 + \omega_{c'v}(\mathbf{q})][\omega_2 - \omega_{cv}(\mathbf{q})]} \right\} f_v(\mathbf{q})$$

where $\mathbf{q}$ denotes the electron wavevector, v, c, and c' are the band indices, and $f_v(\mathbf{q})$ is the Fermi distribution factor for the state $|v,\mathbf{q}\rangle$.

For condensed matter, there should be a local field arising from the induced dipole–dipole interaction. A local field correction factor $L^{(n)}$ should then appear as a multiplication factor in $\chi^{(n)}$. We discuss the local field correction in more detail in Section 2.4. For Bloch (band-state) electrons in solids with wavefunctions extended over many unit cells, the local field tends to get averaged out, and $\mathbf{L}^{(n)}$ may approach 1.

2.3 DIAGRAMMATIC TECHNIQUE

Perturbation calculations can be facilitated with the help of diagrams. Feynman diagrams have been used in perturbation calculations on wavefunctions. Here, since the density matrices involve products of two wavefunctions, perturbation calculations require a kind of double-Feynman diagram. We introduce in this section a technique devised by Yee and Gustafson.[6] Only the steady-state response is considered here.

The important aspects of any diagrammatic technique are that the diagrams provide a simple picture to the corresponding physical process as well as allowing one to write down immediately the corresponding mathematical expression. It is essential to find the complete set of diagrams for a perturbation process of a given order. The scheme we adopt for calculating $\rho^{(n)}$ involves in each diagram a pair of Feynman diagrams with two lines of propagation, one for the $|\psi\rangle$ side of ρ and the other for the $\langle\psi|$ side. Figure 2.1 shows one of

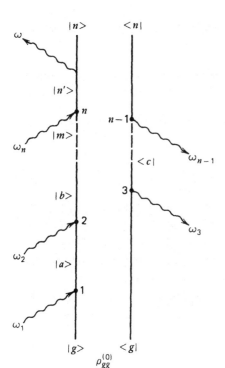

Fig. 2.1 A representative double-Feynman diagram describing one of the many terms in $\rho^{(n)}(\omega = \omega_1 + \omega_2 + \cdots + \omega_n)$.

the many diagrams describing the various terms in $\rho^{(n)}(\omega = \omega_1 + \omega_2 + \cdots + \omega_n)$. The system starts initially from $|g\rangle\langle g|$ with a population $\rho_{gg}^{(0)}$. The ket state propagates from $|g\rangle$ to $|n'\rangle$ through interaction with the radiation field at $\omega_1, \omega_2, \ldots, \omega_n$, and the bra state propagates from $\langle g|$ to $\langle n|$ through interaction with the field at $\omega_3, \ldots, \omega_{n-1}$. Then, the final interaction with the output field at ω puts the system in $|n\rangle\langle n|$. Through permutation of the interaction vertices and rearrangement of the positions of the vertices on the lines of propagation, the other diagrams for $\rho^{(n)}$ can also be drawn.

The microscopic expression for a given diagram can now be obtained using the following general rules describing the various multiplication factors:

1   The system starts with $|g\rangle\rho_{gg}^{(0)}\langle g|$.

2   The propagation of the ket state appears as multiplication factors on the left, and that of the bra state on the right.

3   A vertex bringing $|a\rangle$ to $|b\rangle$ through absorption at ω_i on the left (ket) side of the diagram is described by the matrix element $(1/i\hbar)\langle b|\mathcal{H}_{\text{int}}(\omega_i)|a\rangle$

with $\mathcal{H}_{\text{int}}(\omega_i) \propto e^{-i\omega_i t}$ $\left(\text{denoted by} \quad \overset{|b\rangle}{\underset{\omega_i}{\nearrow}} \quad \text{in Fig. 2.1}\right)$. If it is emission

$\left(\begin{array}{c} \nwarrow \quad |b\rangle \\ \omega_i \quad |a\rangle \end{array}\right)$ instead of absorption, the vertex should be described by

$(1/i\hbar)\langle b| \mathcal{H}_{\text{int}}^{\dagger}(\omega_i)|a\rangle$. Because of the adjoint nature between the bra and ket sides, an absorption process on the ket side appears as an emission process on the bra side, and vice versa.* Therefore, on the right (bra) side

of the diagram, the vertices for emission $\left(\begin{array}{c} \langle b| \\ \langle a| \quad \omega_i \end{array}\right)$ and absorption

$\left(\begin{array}{c} \langle b| \\ \langle a| \quad \searrow_{\omega_i} \end{array}\right)$ are described by $-(1/i\hbar)\langle a|\mathcal{H}_{\text{int}}(\omega_i)|b\rangle$ and $-(1/i\hbar)\langle a|$

$\mathcal{H}_{\text{int}}^{\dagger}(\omega_i)|b\rangle$, respectively.

4   Propagation from the jth vertex to the $(j+1)$th vertex along the $|l\rangle\langle k|$ double lines is described by the propagator $\Pi_j = \pm[i(\Sigma_{i=1}^{j}\omega_i - \omega_{lk} + i\Gamma_{lk})]^{-1}$ The frequency ω_i is taken as positive if absorption of ω_i at the ith vertex occurs on the left or emission of ω_i on the right; it is taken as negative if absorption of ω_i occurs on the right or emission on the left.

5   The final state of the system is described by the product of the final ket and bra states, for example, $|n'\rangle\langle n|$ after the nth vertex in Fig. 2.1 for $\rho^{(n)}$.

6   The product of all factors describes the propagation from $|g\rangle\langle g|$ to $|n'\rangle\langle n|$ through a particular set of states in the diagram. Summation of these

---

*If the field is also quantized, $\mathcal{H}_{\text{int}}(\omega_i)$ operating on a ket state will annihilate a photon at ω_i, while if operating on a bra state it will create a photon.

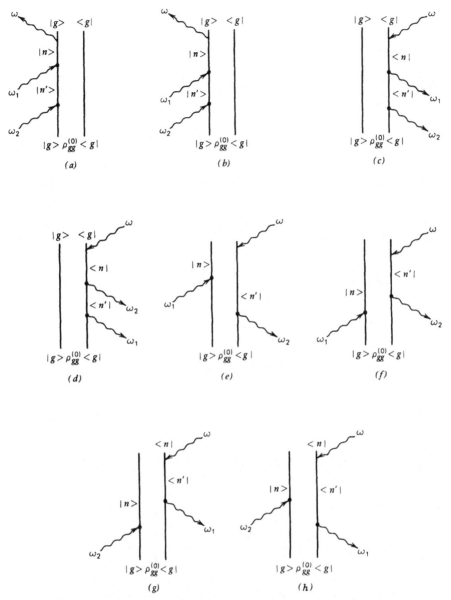

Fig. 2.2 The complete set of eight diagrams for the eight terms in $\rho^{(2)}(\omega = \omega_1 + \omega_2)$.

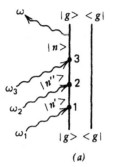
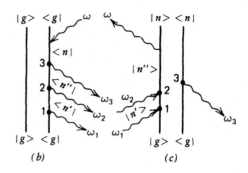

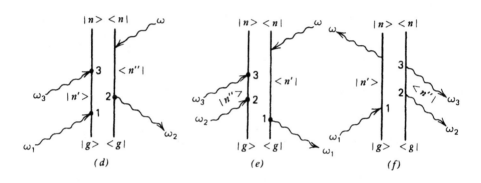

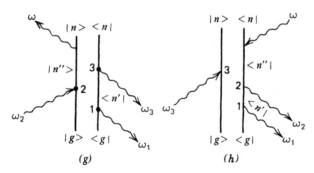

Fig. 2.3 The eight basic diagrams for $\rho^{(3)}(\omega = \omega_1 + \omega_2 + \omega_3)$.

products over all possible sets of states yields the final result with contributions from all states.

By using these rules, the diagram in Fig. 2.1 leads to the expression

$$
\sum_{g, a, \ldots, n} \left\{ \frac{|n'\rangle\langle n'|\mathcal{H}(\omega_n)|m\rangle \cdots \langle b|\mathcal{H}(\omega_2)|a\rangle\langle a|\mathcal{H}(\omega_1)|g\rangle}{\hbar^n\left(\sum_{i=1}^{n}\omega_i - \omega_{n'n} + i\Gamma_{n'n}\right)\left(\sum_{i=1}^{n-1}\omega_i - \omega_{mn} + i\Gamma_{mn}\right) \cdots}\right.
$$

$$
\left. \times \frac{\rho_{gg}^{(0)}\langle g|\mathcal{H}(\omega_3)|c\rangle \cdots \langle|\mathcal{H}(\omega_{n-1})|n\rangle\langle n|}{(\omega_1 + \omega_2 + \omega_3 - \omega_{bc} + i\Gamma_{bc})(\omega_1 + \omega_2 - \omega_{bg} + i\Gamma_{bg})(\omega_1 - \omega_{ag} + i\Gamma_{ag})} \right\}
$$

$$
\propto e^{-i(\omega_1 + \omega_2 + \cdots + \omega_n)t},
$$

(2.19)

which is just one term in the full expression for $\rho^{(n)}$ ($\omega = \omega_1 + \omega_2 + \cdots + \omega_n$).

As a more concrete example, Fig. 2.2 gives the complete set of diagrams for $\rho^{(2)}$ ($\omega = \omega_1 + \omega_2$) that leads to $\chi_{ijk}^{(2)}$ ($\omega = \omega_1 + \omega_2$) in (2.17). The eight diagrams (a)–(h) correspond in successive order to the eight terms in (2.17). Note that $\chi_{ijk}^{(2)}$ ($\omega = \omega_1 + \omega_2$) is derived from $\text{Tr}(\rho^{(2)}P_i)/E_j(\omega_1)E_k(\omega_2)$. There are in fact only four basic diagrams, (a), (c), (e), and (g), in Fig. 2.2. The others can be obtained by permutation of the ω_1 and ω_2 vortices.

As another example, Fig. 2.3 presents eight basic diagrams for $\rho^{(3)}$ ($\omega = \omega_1 + \omega_2 + \omega_3$) that lead to $\chi_{ijkl}^{(3)}$ ($\omega = \omega_1 + \omega_2 + \omega_3$). There should be 48 diagrams in the complete set corresponding to the 48 terms in $\chi_{ijkl}^{(3)}$. The other 40 diagrams are obtained from permutations of the three vertices $(1, 2, 3)$ in the eight basic diagrams in Fig. 2.3. The full expression of $\chi_{ijkl}^{(3)}$ can then be written down from the diagrams according to the rules.

What happens if identical photons appear at a number of vertices? Diagrams obtained from permutations of these vertices in a given diagram yield identical terms in $\rho^{(n)}$. They should not be discarded, and should be taken into account by a degeneracy factor attached to the terms in $\rho^{(n)}$. For example, $\chi_{iiii}^{(3)}(3\omega = \omega + \omega + \omega)$ has 48 diagrams, but 40 of them yield terms identical to others. Thus $\chi_{iiii}^{(3)}(3\omega = \omega + \omega + \omega)$ has only eight terms, each having a degeneracy factor of 6. It reduces further to four terms when the damping constants in the denominators of the expression can be neglected.

2.4   LOCAL FIELD CORRECTION TO $\chi^{(n)}$

The expressions for $\chi^{(n)}$ in the previous sections are strictly correct only for dilute media. They can be written as $\chi^{(n)} = N\alpha^{(n)}$ with N being the number of atoms or molecules per unit volume and $\alpha^{(n)}$ the nth-order nonlinear polarizabilities. In condensed matter, however, the induced dipole–dipole interaction becomes important and leads to the so-called local field correction. The susceptibilities $\chi^{(n)}$ are no longer simply proportional to $\alpha^{(n)}$. The usual derivation of local field correction applies to isotropic or cubic media with well-localized bound electrons. The general theory applicable to media with any symmetry or with more freely moving electrons is not yet available.

The local field at a local spatial point is the sum of the applied field $\mathbf{E}$ and the field due to neighboring dipoles $\mathbf{E}_{\text{dip}}$,

$$\mathbf{E}_{\text{loc}} = \mathbf{E} + \mathbf{E}_{\text{dip}}. \tag{2.20}$$

In the Lorentz model, $\mathbf{E}_{\text{dip}}$ is proportional to the polarization; for isotropic or cubic media, it is given by[7]

$$\mathbf{E}_{\text{dip}} = \frac{4\pi}{3}\mathbf{P}. \tag{2.21}$$

The polarization can be expressed in terms of either microscopic polarizabilities and local fields or macroscopic susceptibilities and applied fields:

$$\begin{aligned}
P_i(\omega) &= N\left\{ \alpha^{(1)}\left[E_{\text{loc}}(\omega) \right]_i \right\} + \alpha^{(2)}_{ijk}\left[E_{\text{loc}}(\omega_1) \right]_j\left[E_{\text{loc}}(\omega_2) \right]_k + \cdots \\
&= \chi^{(1)}E_i(\omega) + \chi^{(2)}_{ijk}E_j(\omega_1)E_k(\omega_2) + \cdots .
\end{aligned} \tag{2.22}$$

With (2.20) and (2.21), the first expression in (2.22) becomes

$$\begin{aligned}
P_i(\omega) &= N\left[1 - \frac{4\pi}{3}N\alpha^{(1)}(\omega) \right]^{-1} \\
&\times \left\{ \alpha^{(1)}E_i(\omega) + \alpha^{(2)}_{ijk}\left[E_{\text{loc}}(\omega_1) \right]_j\left[E_{\text{loc}}(\omega_2) \right]_k + \cdots \right\}.
\end{aligned} \tag{2.23}$$

If the contribution of $\mathbf{P}^{(n)}$ to $\mathbf{E}_{\text{loc}}$ with $n > 1$ is neglected [which is usually an excellent approximation since $|P^{(n)}|_{n>1} \ll |P^{(1)}|$], then the local field can be written as

$$\mathbf{E}_{\text{loc}}(\omega_i) = \left[1 - \frac{4\pi}{3}N\alpha^{(1)}(\omega_i) \right]^{-1}\mathbf{E}(\omega_i). \tag{2.24}$$

Then, from (2.22) and (2.23), we find

$$\chi^{(1)}(\omega) = \frac{N\alpha^{(1)}(\omega)}{1 - (4\pi/3)N\alpha^{(1)}(\omega)}$$

$$\chi^{(2)}_{ijk}(\omega = \omega_1 + \omega_2) \tag{2.25}$$

$$= \frac{N\alpha^{(2)}_{ijk}(\omega)}{\left[1 - (4\pi/3)N\alpha^{(1)}(\omega) \right]\left[1 - (4\pi/3)N\alpha^{(1)}(\omega_1) \right]\left[1 - (4\pi/3)N\alpha^{(1)}(\omega_2) \right]}$$

and more generally

$$\chi^{(n)}(\omega = \omega_1 + \omega_2 + \cdots + \omega_n)$$

$$= \frac{N\alpha^{(n)}(\omega = \omega_1 + \omega_2 + \cdots + \omega_n)}{\left[1 - (4\pi/3)N\alpha^{(1)}(\omega)\right]\left[1 - (4\pi/3)N\alpha^{(1)}(\omega_1)\right] \cdots \left[1 - (4\pi/3)N\alpha^{(1)}(\omega_n)\right]} \qquad (2.26)$$

Since the linear dielectric constant $\varepsilon^{(1)}$ is related to $\chi^{(1)}$ by

$$\varepsilon^{(1)} = 1 + 4\pi\chi^{(1)} = \frac{1 + (8\pi/3)N\alpha^{(1)}}{1 - (4\pi/3)N\alpha^{(1)}}$$

we can write

$$\left[1 - \frac{4\pi}{3}N\alpha^{(1)}\right]^{-1} = \frac{\varepsilon^{(1)} + 2}{3},$$

and (2.26) becomes[3]

$$\chi^{(n)}(\omega = \omega_1 + \omega_2 + \cdots + \omega_n) = NL^{(n)}\alpha^{(n)}(\omega = \omega_1 + \omega_2 + \cdots + \omega_n) \tag{2.27}$$

with

$$L^{(n)} = \left[\frac{\varepsilon^{(1)}(\omega) + 2}{3}\right]\left[\frac{\varepsilon^{(1)}(\omega_1) + 2}{3}\right] \cdots \left[\frac{\varepsilon^{(1)}(\omega_n) + 2}{3}\right] \tag{2.28}$$

being the local field correction factor for the nth-order nonlinear susceptibilities. In media with other symmetry, the expression (2.27) is still valid, but $L^{(n)}$ will be a complicated tensorial function of $\varepsilon^{(1)}(\omega)$, $\varepsilon^{(1)}(\omega_1),\ldots,$ and $\varepsilon^{(1)}(\omega_n)$.[8]

2.5 PERMUTATION SYMMETRY OF NONLINEAR SUSCEPTIBILITIES

There is inherent symmetry in the microscopic expressions of susceptibilities. As can be readily seen from (2.17), the linear susceptibility $\chi_{ij}^{(1)}$ has the symmetry

$$\chi_{ij}^{(1)}(\omega) = \chi_{ij}^{(1)*}(-\omega), \tag{2.29}$$

which is actually a special case of the Onsager relation. Similarly, the nonlinear susceptibility $\chi_{ijk}^{(2)}(\omega = \omega_1 + \omega_2)$ in (2.17) or a similar expression for $\chi_{ijk}^{(2)}(2\omega = \omega + \omega)$ has the following permutation symmetry when the damping constants in the frequency denominators can be neglected (i.e., the nonresonant cases):[1, 9]

$$\chi_{ijk}^{(2)*}(\omega = \omega_1 + \omega_2) = \chi_{jki}^{(2)}(\omega_1 = -\omega_2 + \omega)$$

$$= \chi_{kij}^{(2)}(\omega_2 = \omega - \omega_1), \tag{2.30}$$

$$\chi_{ijj}^{(2)*}(2\omega = \omega + \omega) = \tfrac{1}{2}\chi_{jij}^{(2)}(\omega = 2\omega - \omega) = \tfrac{1}{2}\chi_{jji}^{(2)}(\omega = -\omega + 2\omega).$$

In the permutation operation, the Cartesian indices are permutated together with the frequencies with their signs properly chosen. More generally, one can show that the nth-order nonlinear susceptibility also has the permutation symmetry[9]

$$\chi_{il_1l_2\cdots l_n}^{(n)\,*}(\omega = \omega_1 + \omega_2 + \cdots \omega_n) = \chi_{l_1l_2\cdots l_n i}^{(n)}(\omega_1 = -\omega_2 \cdots -\omega_n + \omega)$$

$$= \cdots \tag{2.31}$$

$$= \chi_{l_n il_1 \cdots l_{n-1}}^{(n)}(\omega_n = \omega - \omega_1 \cdots -\omega_{n-1}).$$

If the dispersion of $\chi^{(n)}$ can also be neglected, then the permutation symmetry in (2.31) becomes independent of the frequencies. Consequently, a symmetry relation now exists between different elements of the same $\chi^{(n)}$ tensor, that is, $\chi_{i,l_1,\cdots l_n}^{(n)}$ remains unchanged when the Cartesian indices are permuted. This is known as Kleinman's conjecture,[10] with which the number of independent elements of $\chi^{(n)}$ can be greatly reduced. For example, it reduces 27 elements of $\chi^{(2)}$ to only 10 independent elements. We should, however, note that since all media are dispersive, Kleinman's conjecture is good approximation only when all frequencies involved are far from resonances such that dispersion of $\chi^{(n)}$ is relatively unimportant.

2.6 STRUCTURAL SYMMETRY OF NONLINEAR SUSCEPTIBILITIES

As optical properties of a medium, the nonlinear susceptibility tensors should have certain forms of symmetry that reflect the structural symmetry of the medium. Accordingly, some tensor elements are zero and others are related to each other, greatly reducing the total number of independent elements. As an illustration, we consider here the second-order nonlinear susceptibility tensor $\chi^{(2)}$.

Each medium has a certain point symmetry with a group of symmetry operations $\{S\}$, under which the medium is invariant, and therefore $\chi^{(n)}$ remains unchanged. In real manipulation, S is a second-rank three-dimensional tensor S_{lm}. Then, invariance of $\chi^{(2)}$ under a symmetry operation is

Table 2.1
Independent Nonvanishing Elements of $\chi^{(2)}(\omega = \omega_1 + \omega_2)$ for Crystals of Certain Symmetry Classes

Symmetry Class	Independent Nonvanishing Elements
Triclinic	
1	All elements are independent and nonzero
Monoclinic	
2	$xyz, xzy, xxy, xyx, yxx, yyy, yzz, yzx, yxz, zyz,$ zzy, zxy, zyx (two fold axis parallel to $\hat{y}$)
m	$xxx, xyy, xzz, xzx, xxz, yyz, yzy, yxy, yyx, zxx,$ zyy, zzz, zzx, zxz (mirror plane perpendicular to $\hat{y}$)
Orthorhombic	
222	$xyz, xzy, yzx, yxz, zxy, zyx$
$mm2$	$xzx, xxz, yyz, yzy, zxx, zyy, zzz$
Tetragonal	
4	$xyz = -yxz, xzy = -yzx, xzx = yzy, xxz = yyz,$ $zxx = zyy, zzz, zxy = -zyx$
$\bar{4}$	$xyz = yxz, xzy = yzx, xzx = -yzy, xxz = -yyz,$ $zxx = -zyy, zxy = zyx$
422	$xyz = -yxz, xzy = -yzx, zxy = -zyx$
$4mm$	$xzx = yzy, xxz = yyz, zxx = zyy, zzz$
$\bar{4}2m$	$xyz = yxz, xzy = yzx, zxy = zyx$
Cubic	
432	$xyz = -xzy = yzx = -yxz = zxy = -zyx$
$\bar{4}3m$	$xyz = xzy = yzx$
23	$= yxz = zxy = zyx$
Trigonal	
3	$xxx = -xyy = -yyz = -yxy, xyz = -yxz, xzy = -yzx,$ $xzx = yzy, xxz = yyz, yyy = -yxx = -xxy = -xyx,$ $zxx = zyy, zzz, zxy = -zyx$
32	$xxx = -xyy = -yyx = -yxy, xyz = -yxz, xzy = -yzx,$ $zxy = -zyx$
$3m$	$xzx = yzy, xxz = yyz, zxx = zyy, zzz, yyy = -yxx =$ $-xxy = -xyx$ (mirror plane perpendicular to $\hat{x}$)
Hexagonal	
6	$xyz = -yxz, xzy = -yzx, xzx = yxy, xxz = yyz,$ $zxx = zyy, zzz, zxy = -zyx$
$\bar{6}$	$xxx = -xyy = -yxy = -yyx, yyy = -yxx = -xyx =$ $-xxy$
622	$xyz = -yxz, xzy = -yxz, zxy = -zyx$
$6mm$	$xzx = yzy, xxz = yyz, zxx = zyy, zzz$
$\bar{6}m2$	$yyy = -yxx = -xxy = -xyx$

explicitly described by

$$(\hat{\imath} \cdot \mathbf{S}^\dagger) \cdot \chi^{(2)} : (\mathbf{S} \cdot \hat{\jmath})(\mathbf{S} \cdot \hat{k}) = \chi^{(2)}_{ijk}. \tag{2.32}$$

For a medium with a symmetry group that consists of n symmetry operations, n such equations should exist. They yield many relations between various elements of $\chi^{(2)}$, although often only a few are independent. These relations can then be used to reduce the 27 elements of $\chi^{(2)}$ to a small number of independent ones.

An immediate consequence of (2.32) is that $\chi^{(2)} = 0$ in the electric dipole approximation for a medium with inversion symmetry: with $\mathbf{S}$ being the inversion operation, $\mathbf{S} \cdot \hat{e} = -\hat{e}$, (2.32) yields $\chi^{(2)}_{ijk} = -\chi^{(2)}_{ijk} = 0$. This explains why $\chi^{(2)}$ for a free electron gas does not have an electric dipole contribution as shown in Chapter 1. Among crystals without inversion symmetry, those with the zincblende structure such as the III-V semiconductors have the simplest form of $\chi^{(2)}$. They belong to the class of $T_d(\bar{4}3m)$ cubic point symmetry.

Table 2.2

Independent Nonvanishing Elements of $\chi^{(3)}(\omega = \omega_1 + \omega_2 + \omega_3)$ for Crystals of Certain Symmetry Classes

Symmetry Class	Independent Nonvanishing Elements
Triclinic	All 81 elements are independent and nonzero
Tetragonal	$xxxx = yyyy, zzzz,$
422, 4mm,	$yyzz = zzyy, zzxx = xxzz, xxyy = yyxx, yzyz = zyzy,$
$4/mmm, \bar{4}2m$	$zxzx = xzxz, xyxy = yxyx, yzzy = zyyz, zxxz = xzzx,$
	$xyyx = yxxy$
Cubic	$xxxx = yyyy = zzzz, yyzz = zzxx = xxyy,$
23, $m3$	$zzyy = yyxx = xxzz, zyzy = xzxz = yxyx,$
	$yzyz = zxzx = xyxy, zzyz = xzzx = yxxy,$
	$yzzy = zxxz = xyyx$
$432, \bar{4}3m, m3m$	$xxxx = yyyy = zzzz$
	$yyzz = zzyy = zzxx = xxzz = xxyy = yyxx$
	$yzyz = zyzy = zxzx = xzxz = yxyx = xyxy$
	$yzzy = zyyz = zxxz = xzzx = xyyx = yxxy$
Hexagonal	$zzzz, xxxx = yyyy = xxyy + xyyx + xyxy$
622, 6mm,	$xxyy = yyxx, xyyx = yxxy, xyxy = yxyx,$
$6/mmm, \bar{6}m2$	$yyzz = xxzz, zzyy = zzxx, zyyz = zxxz,$
	$yzzy = xzzx, yzyz = xzxz, zyzy = zxzx$
Isotropic	$xxxx = yyyy = zzzz,$
	$yyzz = zzyy = zzxx = xxzz = xxyy = yyxx,$
	$yzyz = zyzy = zxzx = xzxz = xyxy = yxyx,$
	$yzzy = zyyz = zxxz = xzzx = xyyz = yxxy,$
	$xxxx = xxyy + xyxy + xyyx$

Although many symmetry operations are associated with $T_d(\bar{4}3m)$, only the 180° rotations about the three four-fold axes and the mirror reflections about the diagonal planes are needed to reduce $\chi^{(2)}$. The 180° rotations make $\chi^{(2)}_{\hat{\imath}\hat{\imath}\hat{\imath}} = -\chi^{(2)}_{\hat{\imath}\hat{\imath}\hat{\imath}} = 0$, $\chi^{(2)}_{\hat{\imath}\hat{\imath}\hat{\jmath}} = -\chi^{(2)}_{\hat{\imath}\hat{\imath}\hat{\jmath}} = 0$, and $\chi^{(2)}_{\hat{\imath}\hat{\jmath}\hat{\jmath}} = -\chi^{(2)}_{\hat{\imath}\hat{\jmath}\hat{\jmath}} = 0$, where $\hat{\imath}$, $\hat{\jmath}$, and $\hat{k}$ refer to the three principal axes of the crystal. The mirror reflections lead to the invariance of $\chi^{(2)}_{\hat{\imath}\hat{\jmath}\hat{k}}(i \neq j \neq k)$ under permutation of the Cartesian indices. Consequently, $\chi^{(2)}_{\hat{\imath}\hat{\jmath}\hat{k}}(i \neq j \neq k)$ is the only independent element in $\chi^{(2)}$ for the zincblende crystals.

For other classes of crystals, the forms of $\chi^{(2)}$ can be similarly derived through the corresponding symmetry operations. The symmetry consideration here is the same as the one used to derive the electrooptical tensor [which is actually a special case of $\chi^{(2)}(\omega = \omega_1 + \omega_2)$ with $\omega_2 \simeq 0$] and the piezoelectric tensor.[11] The forms of $\chi^{(2)}$ for second-harmonic generation are in fact identical to the latter.[12] We reproduce a part of $\chi^{(2)}(\omega = \omega_1 + \omega_2)$ for various classes of crystals in Table 2.1.

The above symmetry consideration for $\chi^{(2)}$ can of course be extended to higher-order nonlinear susceptibilities. In particular, symmetry forms for $\chi^{(3)}$ are most important in view of the many interesting third-order nonlinear optical effects that can be observed readily in almost all media. Table 2.2 lists the $\chi^{(3)}$ tensors for the more commonly encountered classes of media.[12]

2.7 PRACTICAL CALCULATIONS OF NONLINEAR SUSCEPTIBILITIES

Symmetry operations drastically reduce the number of independent elements in a nonlinear susceptibility tensor, but then for a given medium, we would also like to know the values of these independent elements. While they can often be measured (see, for example, Section 7.5), it is also important that they can be calculated from theory. A successful theoretical calculation can help in predicting $\chi^{(n)}$ for media not easily subject to measurements or for the design of new nonlinear crystals. In principle, the microscopic expressions, such as the one for $\chi^{(2)}_{ijk}$ in (2.17), with appropriate local-field correction, can be used for such calculations. However, in most practical cases these expressions are useless because neither the transition frequencies nor the wavefunctions for the material are sufficiently well known. This is especially true for large molecules or solids. Simplifying models or approximations often are needed. If all frequencies involved are far from resonances, one simplifying assumption often used is to replace each frequency denominator in the microscopic expression of $\chi^{(n)}$ by an average one and bring all frequency denominators out of the summation [see, for example, $\chi^{(2)}$ in (2.17)]. Then the summation over matrix elements can be greatly simplified through the closure property of the eigenstates and can be expressed in terms of moments of the ground-state charge distribution. The problem reduces to finding the ground-state wavefunction of the system.[13]

The foregoing approximation, however, is too drastic to yield good results. A more successful calculation of $\chi^{(n)}$ can be done by the bond model. Such a model was used in the early 1930s to calculate the linear polarizability of a molecule or the linear dielectric constant of a crystal.[14] The bond additivity rule was assumed: the induced polarizations on a molecule (or a crystal) is the vector sum of the polarizations induced on all bonds between atoms. In other words, the bond–bond interaction is neglected. The same rule can be used in the calculations of $\chi^{(n)}$. We can write

$$\chi^{(n)} = \sum_K \alpha_K^{(n)} \tag{2.33}$$

where $\alpha_K^{(n)}$ is the nth-order nonlinear polarizability of the Kth bond in the crystal (or medium), and the summation is over all the bonds in a unit volume. Thus, with known crystal structure, the calculation of $\chi^{(n)}$ reduces to the calculation of $\alpha_K^{(n)}$ for different types of bonds.

We discuss here only the calculations of $\chi^{(2)}$, using the zincblende crystals as an example. The general procedure is as follows. The linear bond polarizability $\alpha_K^{(1)}$ is first calculated as a function of the applied field using the recently well developed bond theory.[15] The second-order nonlinear bond polarizability $\alpha_K^{(2)}$ is then obtained from the first derivative of $\alpha_K^{(1)}$ with respect to the applied field. Finally, the summation of (2.33) over the bonds is performed to find $\chi^{(2)}$. We assume here that a simple crystal can be constructed entirely out of the same type of bonds, and the bonds are cylindrically symmetric. The linear susceptibility $\chi_{11}^{(1)}$ of the crystal can then be written as

$$\begin{aligned}
\chi_{11}^{(1)} &= \left(\sum_K \alpha_K^{(1)} \right)_{11} \\
&= G_{\parallel}^{(1)} \alpha_{\parallel}^{(1)} + G_{\perp}^{(1)} \alpha_{\perp}^{(1)} \\
&= \left(G_{\parallel}^{(1)} + \mu G_{\perp}^{(1)} \right) \alpha_{\parallel}^{(1)}
\end{aligned} \tag{2.34}$$

where $\alpha_{\parallel}^{(1)}$ and $\alpha_{\perp}^{(1)}$ are the polarizabilities parallel and perpendicular to the bond, $\mu = \alpha_{\perp}^{(1)}/\alpha_{\parallel}^{(1)}$, and $G_{\parallel}^{(1)}$ and $G_{\perp}^{(1)}$ are the respective geometric factors arising from the vectorial summation over the bonds. Both $G_{\parallel}^{(1)}$ and $G_{\perp}^{(1)}$ are proportional to the number of unit cells per unit volume. For the zincblende structure, $G_{\parallel}^{(1)} = \frac{1}{2} G_{\perp}^{(1)} = 4N/3$, and (2.34) becomes

$$\chi_{11}^{(1)} = \frac{4N}{3} (1 + 2\mu) \alpha_{\parallel}^{(1)}. \tag{2.35}$$

The next step is to find an approximate expression for $\alpha_{\parallel}^{(1)}$ through $\chi_{11}^{(1)}$. The microscopic expression of $\chi_{11}^{(1)}$ in (2.17) away from resonance has the form

$$\chi_{11}^{(1)} = \frac{Ne^2}{\hbar} \sum_{g,n} \frac{|r_i|_{ng}^2}{\omega_{ng}^2 - \omega^2} 2\omega_{ng} \rho_g^{(0)}. \tag{2.36}$$

In the low-temperature limit, $\rho_g^{(0)} = 0$ for all states except the ground state. Then, following the approximation of replacing ω_{ng} in the denominator by an average $\bar{\omega}_{ng}$, and the sum rule[16]

$$2\sum_n \omega_{ng}|r_i|^2_{ng} = \frac{\hbar}{m},$$ (2.37)

(2.36) reduces to

$$\chi_{ii}^{(1)} = \frac{\Omega_p^2}{4\pi\left(\bar{\omega}_{ng}^2 - \omega^2\right)}$$ (2.38)

with $\Omega_p^2 = 4\pi Ne^2/m$ being the electron plasma frequency. This simplified expression for $\chi_{ii}^{(1)}$ has actually been shown more rigorously by Penn for solids in the limit of zero frequency.[17] From (2.34), we now have

$$G_{\parallel}^{(1)}\alpha_{\parallel}^{(1)} + G_{\perp}^{(1)}\alpha_{\perp}^{(1)} = \left(G_{\parallel}^{(1)} + \mu G_{\perp}^{(1)}\right)\alpha_{\parallel}^{(1)} = \frac{\Omega_p^2}{4\pi\left(\bar{\omega}_{ng}^2 - \omega^2\right)}.$$ (2.39)

We are, however, interested in $\alpha_{\parallel}^{(1)}$ as a function of applied field. The polarizability should depend on the field through the field perturbation on the transition frequencies and matrix elements. However, in the approximate form of (2.39), $\alpha_{\parallel}^{(1)}$ can depend on the field only through $\bar{\omega}_{ng}^2$. To find an expression for $\bar{\omega}_{ng}^2$ is where the bond theory comes in. Physically, $\hbar\bar{\omega}_{ng} \equiv \mathbb{E}_g$ can be regarded as an average energy gap between the filled and unfilled states. It can be written as[15]

$$\mathbb{E}_g = \left[\mathbb{E}_h^2 + C^2\right]^{1/2}$$ (2.40)

where $\mathbb{E}_h$ and C are known as the homopolar and heteropolar gaps, respectively, and, in the bond theory, have the expressions

$$\mathbb{E}_h^{-2} \cong ad^{2s}$$

and (2.41)

$$C \cong b\left(\frac{Z_A}{r_A} - \frac{Z_B}{r_B}\right)e^{-k_s d/2}.$$

In these expressions, a, b, and s are constant coefficients, Z_A and Z_B are the valences, and r_A and r_B are the covalent radii of the A and B atoms forming the bond, $d = r_A + r_B$ is the bond length, and $\exp(-k_s d/2)$ is the Thomas–Fermi screening factor. If A and B are identical atoms, then $C = 0$. Equation (2.40) can be derived easily from molecular orbital theory.[18] The bond electrons have two eigenstates, a bonding state and an antibonding state. The energy difference between the two states is $\bar{\mathbb{E}}_g$. For a homopolar bond ($A = B$), the bond electrons see a symmetric potential with respect to the bond center and $\bar{\mathbb{E}}_g = \mathbb{E}_h$. For a heteropolar bond ($A \neq B$), the bond electrons see an antisymmetric potential, and $\bar{E}_g^2 = E_h^2 + C^2$ with C proportional to the asymmetric part of the potential. The wavefunctions of the bonding and antibonding states along the bond are shown in Fig. 2.4. It is seen that in the heteropolar case, there is a charge transfer from the side of the less electronegative atom to the side of the more electronegative atom. According to the molecular orbital theory, the amount of transferred charge Q is related to the heteropolar gap C by

$$Q = -\frac{eC}{\bar{\mathbb{E}}_g}. \tag{2.42}$$

Figure 2.4 also shows that there is a bond charge cloud between the two atoms. The magnitude of the bond charge derived from the bond theory is

$$q = -\frac{2e\bar{\mathbb{E}}_g^2}{\bar{\mathbb{E}}_g^2 + \hbar^2\Omega_p^2}. \tag{2.43}$$

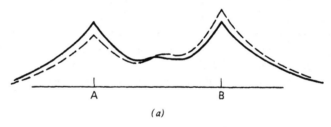

(a)

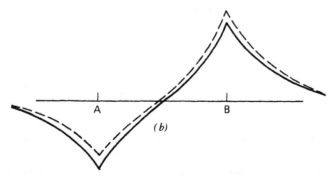

(b)

Fig. 2.4 Sketches of electronic wavefunctions of (a) the bonding state and (b) the antibonding state along the bond connecting the atoms A and B. The solid curves are for the homopolar case and the dashed curves for the heteropolar case.

Levine[19] suggests that the bond charge may be considered as a point charge sitting at distances r_A and r_B, respectively, from atoms A and B.

We can now discuss how the bond polarizability changes when the bond is subject to an external field. The change occurs through the field perturbation on the charge distribution. In our description here, $\alpha_\parallel^{(1)}$ depends on the applied field E through the dependence of $\bar{\mathbb{E}}_g$ on E

$$\frac{\partial \bar{\mathbb{E}}_g}{\partial E_t} = \frac{1}{\mathbb{E}_g}\left(\mathbb{E}_h \frac{\partial \mathbb{E}_h}{\partial E_t} + C\frac{\partial C}{\partial E_t}\right) \tag{2.44}$$

while $\mathbb{E}_h$ and C depend on E through field-induced changes in the charge transfer and bond charge. However, since the applied field is not expected to change the bond length, we have $\partial E_h/\partial E_t = 0$ from (2.41). The second-order nonlinear bond polarizability $\alpha_{ijk}^{(2)}$ is obtained from $\partial\alpha_{ij}^{(1)}/\partial E_k$. If $\hat{\xi}$ and $\hat{\eta}$ denote the two directions parallel and perpendicular to the bond, respectively, then from the symmetry argument, only $\alpha_{\xi\xi\xi}^{(2)}$ and $\alpha_{\eta\eta\eta}^{(2)}$ are nonvanishing. We also neglect $\alpha_{\eta\eta\eta}^{(2)}$ by assuming that a field transverse to the bond will not significantly perturb the charge distribution. Thus $\alpha_{\xi\xi\xi}^{(2)}$ is the only nonvanishing element of $\alpha^{(2)}$. Using (2.39) and (2.44), we find

$$\alpha_{\xi\xi\xi}^{(2)} = \frac{\partial\alpha_\parallel^{(1)}}{\partial E_\xi}$$

$$= \frac{-2\hbar^2\Omega_p^2 C}{4\pi\left(G_\parallel^{(1)} + \mu G_\perp^{(1)}\right)\left(\bar{\mathbb{E}}_g^2 - \hbar^2\omega^2\right)^2}\frac{\partial C}{\partial E_\xi}. \tag{2.45}$$

Now, either (2.41) or (2.42) can be used to calculate $\partial C/\partial E_\xi$. The two, however, correspond to two different physical pictures. In (2.41), the applied field changes r_A and r_B, but keeps $r_A + r_B = d$. In terms of the simple model where the bond charge can be treated as a point charge sitting at distances r_A and r_B away from the atoms A and B, the field then simply shifts the position of the bond charge along the bond. This is known as the bond-charge model.[19] In (2.42), on the other hand, it is the field perturbation on the charge transfer Q that relates C to the field. This is the charge-transfer model.[20]

The bond-charge model involves, with $\Delta r \equiv \Delta r_A = -\Delta r_B$,

$$\frac{\partial C}{\partial E_\xi} = \left(\frac{\partial C}{\partial r_A} - \frac{\partial C}{\partial r_B}\right)\frac{\partial r}{\partial E_\xi} \tag{2.46}$$

and since $q\Delta r = \alpha_\parallel^{(1)}(\omega')\Delta E_\xi(\omega')$ for $\omega' \to 0$, we find from (2.41), (2.45),

(2.46), (2.34), and (2.38)

$$
\begin{aligned}
\left(\alpha^{(2)}_{\xi\xi\xi}\right)_{\mathrm{B\,C}} &= \frac{\hbar^2 \Omega_p^2 \alpha^{(1)}_{\parallel}(\omega')C}{2\pi q\left(G^{(1)}_{\parallel} + \mu G^{(1)}_{\perp}\right)\left(\bar{E}_g^2 - \hbar^2\omega^2\right)^2}\left[\frac{Z_A}{r_A^2} + \frac{Z_B}{r_B^2}\right] be^{-k_s d/2} \\
&= \frac{8\pi C\left[\chi^{(1)}_{ii}(\omega)\right]^2 \chi^{(1)}_{ii}(\omega')}{\left(G^{(1)}_{\parallel} + \mu G^{(1)}_{\perp}\right)^2 q\hbar^2 \Omega_p^2}\left[\frac{Z_A}{r_A^2} + \frac{Z_B}{r_B^2}\right] be^{-k_s d/2}.
\end{aligned}
\tag{2.47}
$$

The charge-transfer model following (2.42) gives

$$
\frac{\partial C}{\partial E_\xi} = -\frac{1}{e}\frac{\bar{E}_g^3}{E_h^2}\frac{\partial Q}{\partial E_\xi}.
\tag{2.48}
$$

It is assumed in this model that the field-induced charge transfer is from atom B to atom A, treating the atoms as points. Since $\alpha^{(1)}_{\parallel}(\omega')\Delta E_\xi(\omega') = \Delta Q d$, we have from (2.45) and (2.48)

$$
\begin{aligned}
\left(\alpha^{(2)}_{\xi\xi\xi}\right)_{\mathrm{C.T.}} &= \frac{\hbar^2 \Omega_p^2 C\bar{E}_g^3 \alpha^{(1)}_{\parallel}(\omega')}{2\pi\left(G^{(1)}_{\parallel} + \mu G^{(1)}_{\perp}\right)\left(\bar{E}_g^2 - \hbar^2\omega^2\right)^2 ed E_h^2} \\
&= \frac{8\pi C\bar{E}_g^3\left[\chi^{(1)}_{ii}(\omega)\right]^2 \chi^{(1)}_{ii}(\omega')}{\left(G^{(1)}_{\parallel} + \mu G^{(1)}_{\perp}\right)^2 ed E_h^2 \hbar^2 \Omega_p^2}.
\end{aligned}
\tag{2.49}
$$

We should, however, keep in mind that the description of how an applied field modifies the charge distribution in both models is still fairly crude. In reality, the electronic charges are broadly distributed in the region between the two atoms. The peak of the distribution is near the center of the bond. As an example, a contour map of the valence electron distribution around a Ga–As bond obtained by empirical pseudopotential calculation is shown in Fig. 2.5.[21] In the presence of a dc external field along the bond, the charge distribution becomes only slightly more asymmetric with its peak essentially unshifted. This is shown in Fig. 2.6 for the charge distributions along the Si–Si and Ga–As bonds.[22] The field-induced shift of the bond charge in the bond-charge model actually refers to the shift of the center of gravity of the valence electron distribution, while the field-induced charge transfer in the charge-transfer model refers to the redistribution of the valence charges around the bond from one side of the bond center to the other.

Finally, we can obtain $\chi^{(2)}_{ijk}$ of a given medium from $\alpha^{(2)}_{\xi\xi\xi}$ for various bonds, where $\hat{\imath}$, $\hat{\jmath}$, and $\hat{k}$ denote the three orthogonal symmetry axes in the crystal:

$$
\begin{aligned}
\chi^{(2)}_{ijk} &= \sum_K \left(\alpha^{(2)}_K\right)_{ijk} \\
&= \sum_\Lambda \left(G^{(2)}_\Lambda\right)_{ijk}\left(\alpha^{(2)}_{\xi\xi\xi}\right)_\Lambda.
\end{aligned}
\tag{2.50}
$$

Fig. 2.5 Contour map of valence electron density distribution (in units of e per primitive cell) for GaAs in the $(1, -1, 0)$ plane. (From Ref. 21.)

where $(G_\Lambda^{(2)})_{ijk}$ is a geometric factor for the Λ-type bonds reflecting the structure of the medium. We note that with $(\alpha_{\xi\xi\xi}^{(2)})_\Lambda$ expressed in terms of $\chi_{ii}^{(1)}$ rather than $\alpha_\parallel^{(1)}$ in (2.47) and (2.49), even the total field correction has been somehow taken into account in the above derivation.

We now use InSb as an example to illustrate the calculation of $\chi_{ijk}^{(2)}$. The crystal has a zincblende structure; therefore, the only nonvanishing elements of $\chi^{(2)}$ are $\chi_{ijk}^{(2)}$ with $i \neq j \neq k$. There is only one type of bond in the crystal: those connecting In and Sb. The geometric factor $G_{xyz}^{(2)}$ is then given by $4N/3\sqrt{3}$ and the density of unit cells N is related to the bond length d by $N = 3\sqrt{3}/16d^3$. We also have $G_\parallel^{(1)} = \frac{1}{2}G_\perp^{(1)} = 4N/3$. From (2.47), (2.49), and (2.50), the bond-

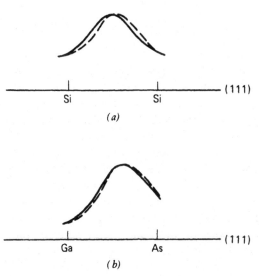

Fig. 2.6 Sketches of the charge distribution along a bond in (a) Si and (b) GaAs. Solid and dashed curves refer to cases with and without an external field along the bond, respectively. (Courtesy of S. Louie.)

charge model gives

$$\left(\chi_{xyz}^{(2)}\right)_{\text{B.C}} = \frac{32\pi d^3 C\left[\chi^{(1)}(\omega)\right]^2\left[\chi^{(1)}(\omega')\right]}{3(1+2\mu)^2 q\hbar^2\Omega_p^2}\left[\frac{Z_A}{r_A^2}+\frac{Z_B}{r_B^2}\right]be^{-k_sd/2} \quad (2.51)$$

and the charge-transfer model gives

$$\left(\chi_{xyz}^{(2)}\right)_{\text{C T}} = \frac{32\pi d^2 C\bar{\mathbb{E}}_g^3\left[\chi^{(1)}(\omega)\right]^2\left[\chi^{(1)}(\omega')\right]}{3(1+2\mu)^2 e\mathbb{E}_h^2\hbar^2\Omega_p^2}. \quad (2.52)$$

We calculate here $\chi_{xyz}^{(2)}$ in the low-frequency limit $\omega \sim \omega' \sim 0$. For InSb, $d = 2.5$ Å, $\bar{\mathbb{E}}_g = 3.7$ eV, $\mathbb{E}_h = 3.1$ eV, $C = 2.1$ eV, $\chi^{(1)} = 1.17$ esu, $\hbar\Omega_p = 13$ eV, $Z_A = 3$, $Z_B = 5$, $r_A \simeq r_B = d/2$, $b\exp(-k_sd/2) \simeq 0.12$ e^2, $\mu \simeq \frac{1}{2}$, and $q \simeq 0.6\,e$,[23] we obtain $(\chi_{xyz}^{(2)})_{\text{B.C.}} \simeq 1.6 \times 10^{-6}$ esu and $(\chi_{xyz}^{(2)})_{\text{C T}} \simeq 2.3 \times 10^{-6}$ esu. The results of both models are in fair agreement with the experimental value of $\chi_{xyz}^{(2)} = (3.3 \pm 0.7) \times 10^{-6}$ esu. This should be considered satisfactory in view of the crude approximations in the models.

The calculations can also be extended to higher-order nonlinear susceptibilities. However, because of the crude approximations involved, they become much less reliable. Also, since we use the covalent bonding picture in the models, the calculations are less suitable for ionic crystals. In nonlinear optics, we are often interested in materials with high nonlinearity. This discussion suggests that the materials should have high nonlinearity in bond polarizabilities. For large $\chi^{(2)}$, the crystal structure should also be as asymmetric as possible so that there is a minimum of vectorial cancellation in summing over $\alpha_K^{(2)}$ of all bonds.

The calculations here are good only in the low-frequency limit. The approximations in the models break down when the optical frequencies are close to the absorption bands. Because of resonant enhancement, the transitions with transition frequencies closer to the optical frequencies contribute much more to the susceptibilities. In order to calculate $\chi^{(n)}$ and its dispersion in these cases, we must use the full microscopic expression of $\chi^{(n)}$ such as those derived in Section 2.2. Then detailed information about the transition matrix elements and frequencies of the material is necessary. Such calculations have been carried out by several authors on $\chi^{(2)}(2\omega)$ of zincblende semiconductors with various degrees of approximation. In most cases, constant matrix elements are assumed. The more accurate calculations, however, are those with wavefunctions and energies of the band states derived from the empirical pseudopotential method,[24] which has been extremely successful in reproducing $\chi^{(1)}(\omega)$ for zincblende semiconductors; it should therefore also yield accurate results for $\chi^{(2)}(2\omega)$. An example is shown in Fig. 2.7 for InSb. The peaks and shoulders in the spectrum generally correspond to resonances of ω or 2ω with the critical point transitions. The results also show that it is important to

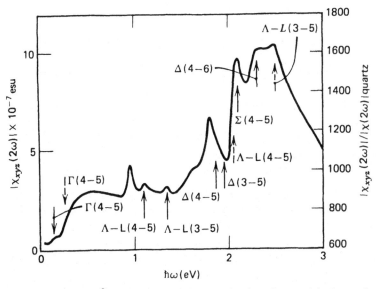

Fig. 2.7 Dispersion of $\chi^{(2)}_{xyz}(2\omega)$ of InSb calculated using the empirical pseudopotential method. The peaks arise from interband transitions in the regions indicated. (From Ref. 24.)

include the dispersive effects of both the matrix elements and the density of states for transitions in the calculations.

Full quantum mechanical calculations of $\chi^{(2)}$ of (2.17) for molecular crystals have also been carried out using semiempirical Hartree–Fock LCAO (linear combination of atomic orbitals) methods by many researchers.[25] They were able to predict quite satisfactorily the measured values of $\chi^{(2)}$. Highly asymmetric molecules with strong charge-transfer bands appear to yield large $|\chi^{(2)}|$ if the crystal structure is also highly asymmetric.

2.8 MILLER'S COEFFICIENT

Miller defined a coefficient[26]

$$\Delta_{ijk} = \frac{\chi^{(2)}_{ijk}(\omega_3 = \omega_1 + \omega_2)}{\chi^{(1)}_{ii}(\omega_3)\chi^{(1)}_{jj}(\omega_1)\chi^{(1)}_{kk}(\omega_2)} \tag{2.53}$$

and found empirically that Δ_{ijk} has only weak dispersion and is almost a constant for a wide range of crystals. This is known as Miller's rule. It suggests that high refractory materials should have large nonlinear susceptibilities. The weak dispersion of Δ_{ijk} can be seen from either the bond-charge or the charge-transfer model. Equations (2.51) and (2.52) show that for $\omega' \to 0$,

$$\Delta_{ijk} \simeq \text{constant independent of frequencies.}$$

The constant is, however, proportional to the heteropolar gap C, and does change, although only mildly, from crystal to crystal. That the measured Δ_{ijk} is indeed proportional to C for a large number of semiconductors has been demonstrated by Levine.[19] For a crystal with several different types of bonds, a weighted average C must be used. The values of Δ_{ijk} for most nonlinear crystals are around few times 10^{-6} esu.

2.9   CONVENTIONS ON NONLINEAR SUSCEPTIBILITIES

The definitions of nonlinear susceptibilities in the literature vary and have caused some confusion. This section clarifies the conventions used in this book. The definition of nonlinear susceptibilities is governed by the following relation between the nonlinear polarization $\mathbf{P}^{(n)}$ and the electric fields $\mathbf{E}_i$:

$$\mathbf{P}^{(n)}(\omega) = \chi^{(n)}(\omega = \omega_1 + \omega_2 + \cdots + \omega_n) : \mathbf{E}_1(\omega_1)\mathbf{E}_2(\omega_2) \cdots \mathbf{E}_n(\omega_n) \tag{2.54}$$

with $\mathbf{E}_i$ and $\mathbf{P}^{(n)}$ expressed as complex quantities:

$$\mathbf{E}_i = \mathscr{E}_i \exp(i\mathbf{k}_i \cdot \mathbf{r} - i\omega_i t)$$
$$\mathbf{P}^{(n)}(\omega) = \mathscr{P}^{(n)} \exp(i\mathbf{k} \cdot \mathbf{r} - i\omega t). \tag{2.55}$$

assuming ω_i and ω are both nonzero. Many authors have written the amplitudes of $\mathbf{E}_i$ and $\mathbf{P}^{(n)}$ in somewhat different forms with

$$\mathbf{E}_i = \tfrac{1}{2}\mathscr{E}_i' \exp(i\mathbf{k}_i \cdot \mathbf{r} - i\omega_i t)$$
$$\mathbf{P}^{(n)}(\omega) = \tfrac{1}{2}\mathscr{P}'^{(n)} \exp(i\mathbf{k} \cdot \mathbf{r} - i\omega t) \tag{2.56}$$

and defined a nonlinear coefficient $\mathbf{d}^{(n)}$ to connect the amplitudes

$$\mathscr{P}'^{(n)} = \mathbf{d}^{(n)} : \mathscr{E}_1' \mathscr{E}_2' \cdots \mathscr{E}_n' \tag{2.57}$$

or

$$\mathbf{P}^{(n)} = (2)^{n-1} \mathbf{d}^{(n)} : \mathbf{E}_1 \mathbf{E}_2 \cdots \mathbf{E}_n. \tag{2.58}$$

Comparison of (2.54) and (2.58) gives

$$\mathbf{d}^{(n)} = (2)^{-n+1} \chi^{(n)} \tag{2.59}$$

and in particular $d_{ijk}^{(2)} = \frac{1}{2}\chi_{ijk}^{(2)}$. Equation (2.59), however, needs modification when there are dc fields present. For $\omega_t = 0$, the corresponding dc field $\mathbf{E}_t$ should be related to $\mathscr{E}_t$ and $\mathscr{E}_t'$ by $\mathbf{E}_t = 2\mathscr{E}_t = \mathscr{E}_t'$. Then, if s of the n fields, namely, $\mathbf{E}_1, \ldots, \mathbf{E}_s$, are dc, we have, following (2.54) and (2.57) as definitions for $\chi^{(n)}$ and $\mathbf{d}^{(n)}$,

$$
\begin{aligned}
\mathbf{P}^{(n)} = \chi^{(n)} : \mathbf{E}_1 \; \cdots \; \mathbf{E}_s \left(\tfrac{1}{2}\right)^{n-s} \mathscr{E}_{s+1}' \; \cdots \; \mathscr{E}_n' \exp\big[i(\mathbf{k}_{s+1} + \cdots + \mathbf{k}_n)\cdot\mathbf{r} \\
- i(\omega_{s+1} + \cdots + \omega_n)t\big] \\
= \tfrac{1}{2}\mathbf{d}^{(n)} : \mathbf{E}_1 \; \cdots \; \mathbf{E}_s \, \mathscr{E}_{s+1}' \; \cdots \; \mathscr{E}_n' \exp\big[i(\mathbf{k}_{s+1} + \cdots + \mathbf{k}_n)\cdot\mathbf{r} \\
- i(\omega_{s+1} + \cdots + \omega_n)t\big],
\end{aligned}
\tag{2.60}
$$

and hence

$$
\mathbf{d}^{(n)} = (2)^{-n+1+s}\chi^{(n)}.
\tag{2.61}
$$

More explicitly, (2.54) takes the form

$$
\begin{aligned}
P_l^{(n)}(\omega = \omega_1 + \omega_2 + \cdots + \omega_n) \\
= \sum_{l_1, l_2 \cdots, l_n} \chi_{l l_1 l_2 \cdots l_n}^{(n)}(\omega = \omega_1 + \omega_2 + \cdots + \omega_n) E_{l_1}(\omega_1) E_{l_2}(\omega_2) \cdots E_{l_n}(\omega_n).
\end{aligned}
\tag{2.62}
$$

Our convention is that the term $\chi_{l l_1 l_2 \cdots l_n}^{(n)}(\omega = \omega_1 + \omega_2 + \cdots + \omega_n)$ $E_{l_1}(\omega_1) E_{l_2}(\omega_2) \cdots E_{l_n}(\omega_n)$ can be written with the fields arranged in any order as long as the subindices of $\chi^{(n)}$ are arranged in the same order, but no additional contribution to $P_l^{(n)}$ should arise from permutation of the fields in (2.62). The conventional notation demands that the field arrangement should always follow the ordering of the frequencies in the argument of $\chi^{(n)}$. This leads to the question of what happens if two or more fields involved have the same frequency. In our convention, permutation of the fields with the same frequency should yield no additional contribution to $P_l^{(n)}$. For example, we have for second-harmonic generation,

$$
\begin{aligned}
P_x^{(2)}(2\omega) = \chi_{xyz}^{(2)}(2\omega = \omega + \omega) E_y(\omega) E_z(\omega) \\
\neq \chi_{xyz}^{(2)} E_y(\omega) E_z(\omega) + \chi_{xzy}^{(2)} E_z(\omega) E_y(\omega).
\end{aligned}
\tag{2.63}
$$

In the convention using the d coefficients, however, all terms derived from permutation of the fields with the same frequency must be included in the expression of the nonlinear polarization. For example,

$$
\begin{aligned}
\tfrac{1}{2}P_x^{(2)}(2\omega) = d_{xyz}^{(2)}(2\omega) E_y(\omega) E_z(\omega) + d_{xzy}^{(2)}(2\omega) E_z(\omega) E_y(\omega) \\
= 2d_{xyz}^{(2)}(2\omega) E_y(\omega) E_z(\omega).
\end{aligned}
\tag{2.64}
$$

In comparison with

$$\tfrac{1}{2}P_x^{(2)}(\omega_3 = \omega_1 + \omega_2) = d_{xyz}^{(2)}(\omega_3 = \omega_1 + \omega_2)E_y(\omega_1)E_z(\omega_2) \quad (2.65)$$

we notice that since the nonlinear response of the medium is not expected to have a sudden change as ω_1 approaches ω_2, the coefficient $d_{xyz}^{(2)}(\omega_3 = \omega_1 + \omega_2)$ should change smoothly to $2d_{xyz}^{(2)}(\omega_3 = 2\omega_1)$. The result that $[d_{ijk}^{(2)}(\omega_3 = \omega_1 + \omega_2)]_{\omega_1 = \omega_2} = 2d_{ijk}^{(2)}(2\omega_1)$ with $j \neq k$ has caused a great deal of confusion. A similar situation occurs when one or more fields have their frequencies approach zero, as we discussed earlier. Our convention here avoids such difficulty: $\chi_{ijk}^{(2)}(\omega_3 = \omega_1 + \omega_2)$ changes continuously to $\chi_{ijk}^{(2)}(\omega_3 = 2\omega_1)$ as ω_1 approaches ω_2, or to $\chi_{ijk}^{(2)}(\omega_1 = 0 + \omega_1)$ as ω_2 approaches zero. The continuous variation of $\chi_{ijk}^{(2)}$ with frequencies can be explicitly seen in the microscopic expression of $\chi_{ijk}^{(2)}$ in (2.17).

Another convention proposed by Maker and Terhune[27] and often used for third-order nonlinearity is to indicate explicitly the number of terms one can obtain by permutation of different field components in the expression of the nonlinear polarization. For example, we write

$$
\begin{aligned}
P_i^{(3)}(\omega &= \omega_1 + \omega_2 + \omega_3) \\
&= \sum_{j,k,l} D_{jkl}C_{ijkl}^{(3)}(\omega = \omega_1 + \omega_2 + \omega_3)E_j(\omega_1)E_k(\omega_2)E_l(\omega_3)
\end{aligned}
\quad (2.66)
$$

where D_{jkl} is the degeneracy factor for the particular terms. If $E_j(\omega_1) \neq E_k(\omega_2) \neq E_l(\omega_3)$, then $D_{jkl} = 6$, indicating that six terms can be obtained by permuting the three fields. For $E_j(\omega_1) = E_k(\omega_2) \neq E_l(\omega_3)$, we have $D_{jkl} = 3$, and for $E_j(\omega_1) = E_k(\omega_2) = E_l(\omega_3)$, we have $D_{jkl} = 1$. This convention also has the difficulty that the nonlinear coefficients $C_{ijkl}^{(3)}(\omega = \omega_1 + \omega_2 + \omega_3)$ vary discontinuously as the frequencies become degenerate.

Further discussions of nonlinear optical susceptibilities appear in later chapters in connection with the specific nonlinear optical problems discussed.

REFERENCES

1   J. A. Armstrong, N. Bloembergen, J. Ducuing, and P. S. Pershan, *Phys. Rev.* **127**, 1918 (1962).

2   N. Bloembergen and Y. R. Shen, *Phys. Rev.* **133**, A37 (1964).

3   N. Bloembergen, *Nonlinear Optics* (Benjamin, New York, 1965).

4   C. P. Slichter, *Principles of Magnetic Resonance*, 2nd ed. (Springer-Verlag, Berlin, 1978), Chapter 5.

5   N. Bloembergen, H. Lotem, and R. T. Lunch, *Indian J. Pure Appl. Phys.* **16**, 151 (1978).

6   T. K. Yee and T. K. Gustafson, *Phys. Rev.* **A18**, 1597 (1978).

7   See, for example, C. Kittel, *Introduction to Solid State Physics*, 5th ed. (Wiley, New York, 1976), p. 406.

8   D. Bedeaux and N. Bloembergen, *Physica* (Amsterdam) **69**, 67 (1973).

9 Y. R. Shen, *Phys. Rev.* **167**, 818 (1968).

10 D. A. Kleinman, *Phys. Rev.* **126**, 1977 (1962).

11 See, for example, J. F. Nye, *Physical Properties of Crystals* (Oxford University Press, London, 1957).

12 P. N. Butcher, *Nonlinear Optical Phenomena* (Ohio State University Press, Columbus, 1965), pp 43–50.

13 F. N. H. Robinson, *Bell Syst. Tech. J.* **46**, 913 (1967); *J. Phys. C* **1**, 286 (1968); S. S. Jha and N. Bloembergen, *Phys. Rev.* **171**, 891 (1968); C. Flytzanis and J. Ducuing, *Phys. Rev.* **178**, 1218 (1969)

14 K. G. Denbigh, *Trans. Faraday Soc.* **36**, 936 (1940).

15 See, for example, J. C. Phillips, *Covalent Bonding in Crystals, Molecules, and Polymers* (University of Chicago Press, Chicago, 1969); *Bonds and Bands in Semiconductors* (Academic Press, New York, 1973).

16 The relation is known as the Thomas–Reiche–Kuhn sum rule in solid state physics. See, for example, J. Ziman, *Principles of the Theory of Solids* (Cambridge University Press, Cambridge, 1965), p. 224.

17 D. R. Penn, *Phys. Rev.* **128**, 2093 (1962).

18 See, for example, C. A. Coulson, *Valences* (Oxford University Press, London, 1961).

19 B. F. Levine, *Phys. Rev Lett.* **22**, 787 (1969); *Phys. Rev.* **B7**, 2591 (1973); 2600 (1973).

20 C. L. Tang and C. Flytzanis, *Phys. Rev.* **B4**, 2520 (1971); C. L. Tang, *IEEE J. Quant. Electron.* **QE-9**, 755 (1973); F. Scholl and C. L. Tang, *Phys. Rev.* **B8**, 4607 (1973).

21 J. P. Walter and M. L. Cohen, *Phys. Rev. Lett.* **26**, 17 (1971).

22 S. Louie and M. L. Cohen, personal communication.

23 Values of the various quantities are obtained from J. C. Phillips and J. A. Van Vechten, *Phys. Rev.* **183**, 709 (1969).

24 C. Y. Fond and Y. R. Shen, *Phys. Rev.* **B12**, 2325 (1975).

25 See, for example, J. L. Oudar and J. Zyss, *Phys. Rev.* **A26**, 2106 (1982); J Zyss and J. L. Oudar, *Phys. Rev.* **A26**, 2028 (1982); C. C. Teng and A. F. Garito, *Phys. Rev. Lett.* **50**, 350 (1983); and references therein.

26 R. C. Miller, *Appl. Phys. Lett.* **5**, 17 (1964).

27 P. D. Maker and R. W. Terhune, *Phys. Rev.* **137**, A801 (1965).

BIBLIOGRAPHY

Bloembergen, N., *Nonlinear Optics* (Benjamin, New York, 1965).

Butcher, P. N., *Nonlinear Optical Phenomenon* (Ohio State University Press, Columbus, 1965).

Ducuing, J., and C. Flytzanis, in F. Abelès, ed., *Optical Properties of Solids* (North-Holland Publishing Co., Amsterdam, 1972), p. 859.

Flytzanis, C., in H. Rabin and C. L. Tang, eds., *Quantum Electronics* (Academic Press, New York, 1975).

Shen, Y. R., in N. Bloembergen, ed., *Nonlinear Spectroscopy, Proceedings of the International School of Physics, Enrico Fermi, Course LXIV* (North-Holland Publishing Co., Amsterdam, 1977), p. 170

3

General Description of
Wave Propagation in
Nonlinear Media

Waves can interact through nonlinear polarization in a medium. Propagation of waves in the presence of wave interaction leads to various nonlinear optical phenomena. The quantitative description of a lower-order nonlinear optical effect usually starts with a set of coupled wave equations with the nonlinear susceptibilities acting as the coupling coefficients. This coupled-wave approach can also be generalized to include waves other than electromagnetic. This chapter is devoted to a general discussion of coupled electromagnetic waves in a medium and the solution of the coupled wave equations under certain approximations. Applications of the analysis to specific nonlinear optical phenomena appear in later chapters.

3.1 COUPLED WAVES IN A NONLINEAR MEDIUM

The wave equation that governs optical wave propagation in a medium is

$$\left[\nabla \times (\nabla \times) + \frac{1}{c^2} \frac{\partial^2}{\partial t^2} \right] \mathbf{E}(\mathbf{r}, t) = - \frac{4\pi}{c^2} \frac{\partial^2}{\partial t^2} \mathbf{P}(\mathbf{r}, t), \tag{3.1}$$

which follows directly from the Maxwell equations (1.5). Wave interaction gives rise to the nonlinear terms in $\mathbf{P}$. We assume that both $\mathbf{E}(\mathbf{r}, t)$ and $\mathbf{P}(\mathbf{r}, t)$

can be decomposed into a set of infinite plane waves:

$$\mathbf{E}(\mathbf{r}, t) = \sum_l \mathbf{E}_l(\mathbf{k}_l, \omega_l)$$

$$= \sum_l \mathscr{E}_l e^{i(\mathbf{k}_l \cdot \mathbf{r} - \omega_l t)},$$

$$\mathbf{P}(\mathbf{r}, t) = \mathbf{P}^{(1)}(\mathbf{r}, t) + \mathbf{P}^{NL}(\mathbf{r}, t),$$

$$\mathbf{P}^{(1)}(\mathbf{r}, t) = \sum_l \mathbf{P}_l^{(1)}(\mathbf{k}_l, \omega_l)$$

$$= \sum_l \chi^{(1)}(\omega_l) \cdot \mathbf{E}_l(\mathbf{k}_l, \omega_l), \qquad (3.2)$$

and

$$\mathbf{P}^{NL}(\mathbf{r}, t) = \sum_{n \geq 2} \mathbf{P}^{(n)}(\mathbf{r}, t)$$

$$= \sum_m \mathbf{P}^{NL}(\mathbf{k}_m, \omega_m)$$

$$= \sum_m \mathscr{P}^{NL} e^{i \mathbf{k}_m \cdot \mathbf{r} - i \omega_m t}$$

where $\mathscr{E}_l$ is taken as essentially independent of time. With $\varepsilon(\omega_l) \equiv 1 + 4\pi\chi^{(1)}(\omega_l)$, (3.1) becomes

$$\left[\nabla \times (\nabla \times) - \frac{\omega^2}{c^2} \varepsilon \cdot \right] \mathbf{E}(\mathbf{k}, \omega) = \frac{4\pi\omega^2}{c^2} \mathbf{P}^{NL}(\mathbf{k}_m, \omega_m = \omega). \qquad (3.3)$$

Suppose $\mathbf{P}^{NL}(\mathbf{k}_m, \omega) = \mathbf{P}^{(n)}(\mathbf{k}_m, \omega)$ is a nonlinear polarization from the product of $\mathbf{E}_1(\mathbf{k}_1, \omega_1) \cdots \mathbf{E}_n(\mathbf{k}_n, \omega_n)$. Then, for the n fields $\mathbf{E}_l(\mathbf{k}_l, \omega_l)$, there should be n corresponding wave equations similar to (3.3). Together with (3.3), they form a set of $(n + 1)$ coupled wave equations. Note that while ω_m should be equal to ω in $\mathbf{P}^{NL}(\mathbf{k}_m, \omega_m)$ because of photon energy conservation in the steady-state case, $\mathbf{k}_m$ need not be exactly equal to $\mathbf{k}$ since wave momentum conservation is not strictly required in a finite medium. Equation (3.3) clearly indicates that the various waves $\mathbf{E}_l(\mathbf{k}_l, \omega_l)$ are nonlinearly coupled through the nonlinear polarization $\mathbf{P}^{NL}$, and their propagations in the medium will consequently be very different from the linear case where $\mathbf{P}^{NL} = 0$. Through nonlinear coupling, energy can now be transferred back and forth between waves, and the larger $\mathbf{P}^{NL}$ is, the more pronounced the effect should be. The coupled wave approach was first used in the description of microwave parametric amplification[1] and later adopted by Armstrong et al.[2] for describing wave interaction in nonlinear optics.

The simplest case of optical wave interaction deals with second-order nonlinear optical effects. We use it here as an example to illustrate the coupled wave formalism. Consider three waves $E(k_1, \omega_1)$, $E(k_2, \omega_2)$, and $E(k, \omega = \omega_1 + \omega_2)$ interacting in a medium by second-order nonlinear polarization. Then, the coupled wave equations from (3.3) are

$$\left[\nabla \times (\nabla \times) - \frac{\omega_1^2}{c^2} \epsilon_1 \cdot \right] E_1(k_1, \omega_1) = \frac{4\pi\omega_1^2}{c^2} P^{(2)}(\omega_1)$$

$$= \frac{4\pi\omega_1^2}{c^2} \chi^{(2)}(\omega_1 = -\omega_2 + \omega) : E_2^*(k_2, \omega_2) E(k, \omega),$$

$$\left[\nabla \times (\nabla \times) - \frac{\omega_2^2}{c^2} \epsilon_2 \cdot \right] E_2(k_2, \omega_2) = \frac{4\pi\omega_2^2}{c^2} P^{(2)}(\omega_2)$$

$$= \frac{4\pi\omega_2^2}{c^2} \chi^{(2)}(\omega_2 = \omega - \omega_1) : E(k, \omega) E_1^*(k_1, \omega_1),$$

$\qquad\qquad\qquad\qquad\qquad\qquad\qquad\qquad\qquad\qquad\qquad (3.4)$

and

$$\left[\nabla \times (\nabla \times) - \frac{\omega^2}{c^2} \epsilon \cdot \right] E(k, \omega) = \frac{4\pi\omega^2}{c^2} P^{(2)}(\omega)$$

$$= \frac{4\pi\omega^2}{c^2} \chi^{(2)}(\omega = \omega_1 + \omega_2) : E(k_1, \omega_1) E(k_2, \omega_2).$$

As seen here, the nonlinear susceptibilities appear explicitly as the coupling coefficients. They determine the rate of energy transfer among the three waves. In the case of a dissipationless medium, the permutation relation $\chi_{ijk}^{(2)*}(\omega = \omega_1 + \omega_2) = \chi_{jki}^{(2)}(\omega_1 = -\omega_2 + \omega) = \chi_{kij}^{(2)}(\omega_2 = \omega - \omega_1)$ exists (see Section 2.5). This is actually a necessary condition for the coupled wave equations to satisfy the requirement that the total energy in the three waves is a constant, as we shall see in Section 3.2. The photon energy and momentum conservations in the present case are $\omega = \omega_1 + \omega_2$ and $k = k_1 + k_2$, respectively. For most effective energy transfer among the waves, one naturally expects that both photon energy and momentum conservations should be satisfied in the wave interaction. Therefore, even though $k = k_1 + k_2$ is not required, as mentioned earlier, satisfaction of the relation is preferred for the maximization of the wave coupling. This photon momentum matching condition is known in nonlinear optics as the phase matching condition, and it is one of the most important considerations in many nonlinear optical processes. Detailed solutions of (3.4) appear in later chapters.

3.2   FIELD ENERGY IN A NONLINEAR MEDIUM

The Maxwell equations (1.5) lead to the following familiar energy relation for the fields:

$$\frac{c}{4\pi} \nabla \cdot (E \times B) = -\frac{1}{8\pi} \frac{\partial}{\partial t}(E^2 + B^2) - E \cdot \frac{\partial P}{\partial t}. \qquad (3.5)$$

With $\mathbf{E} \times \mathbf{B}$ being the Poynting vector, it shows that the rate of electromagnetic energy flowing out of a unit volume is equal to the reduction rate of the stored electromagnetic energy density. If the dispersion of the medium can be neglected, then the polarization $\mathbf{P}$ can be written as

$$\mathbf{P}(\mathbf{r}, t) = \chi^{(1)} \cdot \mathbf{E}(\mathbf{r}, t) + \chi^{(2)} : \mathbf{EE} + \cdots \tag{3.6}$$

and (3.5) reduces to the form

$$\frac{c}{4\pi} \nabla \cdot (\mathbf{E} \times \mathbf{B}) = -\frac{\partial}{\partial t} U(\mathbf{r}, t) \tag{3.7}$$

with

$$U(\mathbf{r}, t) = \frac{1}{8\pi}(E^2 + B^2) + \frac{1}{2}\mathbf{E} \cdot \chi^{(1)} \cdot \mathbf{E} + \frac{2}{3}\mathbf{E} \cdot \chi^{(2)} : \mathbf{EE} + \cdots \tag{3.8}$$

being the instantaneous electromagnetic energy density. This is clearly not valid in a medium with dispersion, since the susceptibilities are defined only in terms of the Fourier components of fields and polarizations. In reality, it is more meaningful to consider the time-averaged energy relation. Let us illustrate the problem by first assuming a linear medium.

We begin by assuming a quasi-monochromatic field

$$\mathbf{E}(\mathbf{r}, t) = \mathscr{E}(t)e^{i(\mathbf{k}\cdot\mathbf{r} - \omega t)} + \mathscr{E}*(t)e^{-i(\mathbf{k}\cdot\mathbf{r} - \omega t)} \tag{3.9}$$

where $\mathscr{E}(t)$ is a slowly varying amplitude. Expressing $\mathscr{E}(t)$ as a Fourier integral, we have

$$\mathbf{E}(\mathbf{r}, t) = \int d\eta \mathscr{E}(\omega + \eta)e^{i(\mathbf{k}\cdot\mathbf{r} - \omega t) - i\eta t} + \text{c.c.} \tag{3.10}$$

Then the linear polarization takes the form

$$\mathbf{P}^{(1)}(\mathbf{r}, t) = \int d\eta \chi^{(1)}(\omega + \eta) \cdot \mathscr{E}(\omega + \eta)e^{i(\mathbf{k}\cdot\mathbf{r} - \omega t) - i\eta t} + \text{c.c.} \tag{3.11}$$

This leads to

$$\frac{\partial}{\partial t}\mathbf{P}^{(1)}(\mathbf{r}, t) = \int d\eta(-i)(\omega + \eta)\left[\chi^{(1)}(\omega) + \frac{\partial \chi^{(1)}}{\partial \omega}\eta + \cdots\right] \cdot \mathscr{E}(\omega + \eta)$$

$$\times e^{i(\mathbf{k}\cdot\mathbf{r} - \omega t) - i\eta t} + \text{c.c.} \tag{3.12}$$

$$\simeq \left[-i\omega\chi^{(1)}(\omega) \cdot \mathscr{E}(t) + \frac{\partial(\omega\chi^{(1)})}{\partial \omega} \cdot \frac{\partial \mathscr{E}(t)}{\partial t}\right]e^{i(\mathbf{k}\cdot\mathbf{r} - \omega t)} + \text{c.c.}$$

With $\chi_{ij}^{(1)}(\omega) = \chi_{ji}^{(1)}(-\omega)^*$, the time average of (3.5) yields[3]

$$\left\langle \frac{c}{4\pi} \nabla \cdot (\mathbf{E} \times \mathbf{B}) \right\rangle = -\frac{\partial}{\partial t} \langle U^{(1)} \rangle - Q \qquad (3.13)$$

where

$$\langle U^{(1)} \rangle = \frac{1}{4\pi} \left[\mathscr{E}^*(t) \cdot \frac{\partial(\omega \varepsilon')}{\partial \omega} \cdot \mathscr{E}(t) + |\mathbf{B}(t)|^2 \right],$$

$$Q = \frac{\omega}{2\pi} \mathscr{E}^*(t) \cdot \varepsilon'' \cdot \mathscr{E}(t), \qquad (3.14)$$

and

$$\varepsilon = \varepsilon' + i\varepsilon'' = 1 + 4\pi\chi^{(1)}.$$

As may be inferred from the preceding equations, $\langle U^{(1)} \rangle$ is the average field energy density stored in the linear medium and Q is the average power density dissipated into heat through absorption because $\varepsilon'' \neq 0$. Equation (3.13) is therefore an energy conservation relation one should expect physically.

The above calculation can be extended to the nonlinear case.[4] Additional terms in the energy relation are expected from wave coupling. Consider, for example, three waves with $\omega_3 = \omega_1 + \omega_2$ and $\mathbf{k}_3 = \mathbf{k}_1 + \mathbf{k}_2$ interacting in a nonlinear medium via $\chi^{(2)}$. One finds, with the help of the permutation symmetry relation of $\chi^{(2)}$, that the average field energy density has an additional term

$$\langle U^{(2)} \rangle = 2\mathscr{E}_1^*(t) \cdot \chi^{(2)}(\omega_1 = -\omega_2 + \omega_3) : \mathscr{E}_2^*(t)\mathscr{E}_3(t)$$

$$+ \mathscr{E}_1^*(t) \cdot \left[\omega_1 \frac{\partial \chi^{(2)}(\omega_1)}{\partial \omega_1} + \omega_2 \frac{\partial \chi^{(2)}(\omega_1)}{\partial \omega_2} + \omega_3 \frac{\partial \chi^{(2)}(\omega_1)}{\partial \omega_3} \right] : \qquad (3.15)$$

$$\times \mathscr{E}_2^*(t)\mathscr{E}_3(t) + \text{cc.}$$

obtained from

$$\frac{\partial \langle U^{(2)} \rangle}{\partial t} = \left\langle \sum_{l=1}^{3} \mathbf{E}_l \cdot \frac{\partial \mathbf{P}_l^{(2)}}{\partial t} \right\rangle.$$

Note that only when $|\omega_m \partial \chi^{(2)}/\partial \omega_m| \ll |\chi^{(2)}|$ can we write

$$\langle U^{(2)} \rangle = 2\mathscr{E}_1^* \cdot \chi^{(2)}(\omega_1 = -\omega_2 + \omega_3) : \mathscr{E}_2^*\mathscr{E}_3 + \text{c.c.} \qquad (3.16)$$

The last equation, however, is frequently used in the literature[5] to describe the interaction free energy density for the wave coupling. One would write the free energy density as $F^{(2)} = -\langle U^{(2)} \rangle$ using (3.16) and derive the nonlinear polarization from $\mathbf{P}^{(2)}(\omega_m) = -\frac{1}{2}\partial F^{(2)}/\partial \mathscr{E}_m^*$. From $\partial^2 \mathbf{P}^{(2)}(\omega_1)/\partial \mathscr{E}_2^* \, \partial \mathscr{E}_3 =$

$\partial^2 \mathbf{P}^{(2)}(\omega_2)/\partial\mathscr{E}_1^* \partial\mathscr{E}_3 = \partial \mathbf{P}^{(2)*}(\omega_3)/\partial\mathscr{E}_1^* \partial\mathscr{E}_2^*$, the permutation symmetry relation of $\chi^{(2)}$ immediately results. Although this is a practice one can indeed follow, we must realize that it is actually the permutation symmetry of $\chi^{(2)}$ that leads to the expression of $\langle U^{(2)} \rangle$ in (3.15). Conversely, it is the existence of $\langle U^{(2)} \rangle$ that physically justifies the permutation symmetry of $\chi^{(2)}$. Near resonances, when dissipation in the medium becomes important, the permutation symmetry relation of $\chi^{(2)}$ breaks down, and accordingly, (3.15) is no longer valid.

More generally, in a nonabsorbing medium, the time-averaged field energy density should be[4]

$$\langle U \rangle = \sum_{n=1}^{\infty} \langle U^{(n)} \rangle \tag{3.17}$$

where $\langle U^{(n)} \rangle$ arises from nonlinear coupling of $(n + 1)$ waves via the nth-order nonlinearity in the medium, and is given by

$$\langle U^{(n)} \rangle = n\mathscr{E}_{n+1}^* \cdot \chi^{(n)}(\omega_{n+1} = \omega_1 + \omega_2 + \cdots + \omega_n) : \mathscr{E}_1\mathscr{E}_2 \cdots \mathscr{E}_n$$
$$+ \mathscr{E}_{n+1}^* \cdot \left[\sum_{l=1}^{n+1} \omega_l \frac{\partial\chi^{(n)}}{\partial\omega_l} \right] : \mathscr{E}_1\mathscr{E}_2 \cdots \mathscr{E}_n + \text{c.c.} \tag{3.18}$$

The time-averaged energy conservation relation takes the form

$$\left\langle \frac{c}{4\pi} \nabla \cdot \mathbf{E} \times \mathbf{B} \right\rangle = -\frac{\partial}{\partial t} \langle U \rangle. \tag{3.19}$$

3.3 SLOWLY VARYING AMPLITUDE APPROXIMATION

In actually solving the coupled wave equations, several simplifying approximations are often made.[2, 6] Among them are the slowly varying amplitude approximation, the infinite plane-wave approximation, and constant pump intensity approximation. We discuss here only the slowly varying amplitude approximation and leave the others to later chapters.

As mentioned earlier, wave coupling in a nonlinear medium results in energy transfer among waves. Therefore, the wave amplitudes are expected to change in propagation. We assume for illustration a plane wave propagating along $\hat{z}$:

$$\mathbf{E}(\omega, z) = \mathscr{E}(z)e^{i(kz-\omega t)}.$$

Since the energy transfer among waves is usually significant only after the waves travel over a distance much longer than their wavelengths, we expect

$$\left| \frac{\partial^2 \mathscr{E}(z)}{\partial z^2} \right| \ll \left| k \frac{\partial\mathscr{E}}{\partial z} \right|. \tag{3.20}$$

The field $\mathbf{E}$ can generally be decomposed into a longitudinal component $\mathbf{E}_\parallel$ parallel to $\mathbf{k}$ and a transverse component $\mathbf{E}_\perp$ perpendicular to $\mathbf{k}$. The wave equation for $\mathbf{E}$, following (3.3), can similarly be split into two equations:

$$\nabla^2 \mathbf{E}_\perp + \frac{\omega^2}{c^2}(\boldsymbol{\varepsilon}\cdot\mathbf{E})_\perp = -\frac{4\pi\omega^2}{c^2}\mathbf{P}_\perp^{\mathrm{NL}}(\omega, z) \tag{3.21a}$$

and

$$\nabla\cdot\left[(\boldsymbol{\varepsilon}\cdot\mathbf{E})_\parallel + 4\pi\mathbf{P}_\parallel^{\mathrm{NL}}\right] = 0. \tag{3.21b}$$

We now have

$$\nabla^2 \mathbf{E}_\perp = \frac{\partial^2}{\partial z^2}\mathbf{E}_\perp = e^{\iota(kz-\omega t)}\left[\frac{\partial^2}{\partial z^2} + \iota 2k\frac{\partial}{\partial z} - k^2\right]\boldsymbol{\mathscr{E}}_\perp(z) \tag{3.22}$$

and

$$-k^2\mathbf{E}_\perp + \frac{\omega^2}{c^2}(\boldsymbol{\varepsilon}\cdot\mathbf{E})_\perp = 0. \tag{3.23}$$

Then the approximation of (3.20) reduces the second-order differential equation (3.21a) to a simple first-order differential equation

$$\frac{\partial\boldsymbol{\mathscr{E}}_\perp}{\partial z} = \frac{\iota 2\pi\omega^2}{kc^2}\mathbf{P}_\perp^{\mathrm{NL}}(\omega, z)e^{-\iota(kz-\omega t)}. \tag{3.24}$$

This is known as the slowly varying amplitude approximation.

The preceding description of the slowly varying amplitude approximation is what is usually found in the literature. However, the real physical implication of the approximation is in neglecting the oppositely propagating field component generated by $\mathbf{P}^{\mathrm{NL}}$. Consider, for example, the wave propagation in an isotropic medium,

$$\left(\frac{\partial^2}{\partial z^2} + \frac{\omega^2}{c^2}\varepsilon\right)\mathbf{E}(\omega, z) = -\frac{4\pi\omega^2}{c^2}\mathbf{P}^{\mathrm{NL}}(\omega, z) \tag{3.25}$$

with plane boundaries at $z = 0$ and l. The equation can be solved by the Green function method. Let $G(z, z')$ be the Green function, which obeys the equation

$$\left(\frac{\partial^2}{\partial z^2} + \frac{\omega^2}{c^2}\varepsilon\right)G(z, z') = -\delta(z, z'). \tag{3.26}$$

Then we find

$$\begin{aligned} G(z, z') &= \frac{-1}{\iota 2k}e^{\iota k(z-z')} \qquad \text{for } z > z' \\ &= \frac{-1}{\iota 2k}e^{-\iota k(z-z')} \qquad \text{for } z < z' \end{aligned} \tag{3.27}$$

with $k = \omega\sqrt{\varepsilon}/c$. The solution of (3.25) is given by

$$\mathbf{E}(\omega, z) = \int_0^l \frac{4\pi\omega^2}{c^2} \mathbf{P}^{\mathrm{NL}}(z')G(z, z')dz' + \left[G\frac{\partial \mathbf{E}}{\partial z'} - \mathbf{E}\frac{\partial G}{\partial z'}\right]_{z'}^l = 0$$

$$= -\left(\frac{2\pi\omega^2}{ikc^2}\right)\left[\int_0^z \mathbf{P}^{\mathrm{NL}}(z')e^{ik(z-z')}dz' + \int_z^l \mathbf{P}^{\mathrm{NL}}(z')e^{-ik(z-z')}dz\right]$$

$$+ \left[G\frac{\partial \mathbf{E}}{\partial z'} - \mathbf{E}\frac{\partial G}{\partial z'}\right]_{z'=0}^l . \tag{3.28}$$

If we write

$$\mathbf{E}(\omega, z) = \mathscr{E}_F(z)e^{i(kz-\omega t)} + \mathscr{E}_B(z)e^{i(-kz-\omega t)} \tag{3.29}$$

and impose the boundary conditions $\partial\mathscr{E}_F/\partial z' = 0$ and $\partial\mathscr{E}_B/\partial z' = 0$ at $z = 0^-$ and l^+, indicating that no amplitude change should occur outside the medium, then we have

$$\left[G\frac{\partial \mathbf{E}}{\partial z'} - \mathbf{E}\frac{\partial G}{\partial z'}\right]_{z'=0}^l = \left[\mathscr{E}_F(0)e^{ikz} + \mathscr{E}_B(l)e^{-ikz}\right]e^{-i\omega t} \tag{3.30}$$

Comparison of (3.28) and (3.29) yields

$$\mathscr{E}_F(z) = \mathscr{E}_F(0) + i\frac{2\pi\omega^2}{kc^2}\int_0^z \mathbf{P}^{\mathrm{NL}}(z')e^{-i(kz'-\omega t)}dz'$$

and

$$\mathscr{E}_B(z) = \mathscr{E}_B(l) + i\frac{2\pi\omega^2}{kc^2}\int_z^l \mathbf{P}^{\mathrm{NL}}(z')e^{i(kz'+\omega t)}dz' . \tag{3.31}$$

The corresponding differential equations for $\mathscr{E}_F$ and $\mathscr{E}_B$ are

$$\frac{\partial\mathscr{E}_F}{\partial z} = i\frac{2\pi\omega^2}{kc^2}\mathbf{P}^{\mathrm{NL}}(\omega, z)e^{-i(kz-\omega t)}$$

and

$$\frac{\partial\mathscr{E}_B}{\partial z} = -i\frac{2\pi\omega^2}{kc^2}\mathbf{P}^{\mathrm{NL}}(\omega, z)e^{i(kz+\omega t)} . \tag{3.32}$$

Comparing (3.24) with (3.32), we recognize that (3.24) can be obtained by neglecting $\mathscr{E}_B$ in $\mathscr{E}$ (or neglecting $\mathscr{E}_F$ if $\mathscr{E}_\perp$ is propagating with wavevector $-\mathbf{k}$).

3.4 BOUNDARY CONDITIONS

The usual boundary conditions for electromagnetic waves should be valid here; for example, the tangential components of $\mathbf{E}$ and $\mathbf{B}$ at a boundary surface must be continuous for each Fourier component. In general, the solution of the wave equation (3.3) for $\mathbf{E}(\mathbf{k}_l, \omega_l)$ driven by $\mathbf{P}^{NL}(\mathbf{k}_m, \omega_m = \omega_l) \propto \exp(i\mathbf{k}_m \cdot \mathbf{r} - i\omega_l t)$ has the form

$$\mathbf{E}(\mathbf{k}_l, \mathbf{k}_m, \omega_l) = \left(\mathscr{E}_H e^{i\mathbf{k}_l \cdot \mathbf{r}} + \mathscr{E}_p e^{i\mathbf{k}_m \cdot \mathbf{r}} \right) e^{-i\omega_l t} \tag{3.33}$$

where the $\mathscr{E}_H$ and $\mathscr{E}_P$ terms correspond to the homogeneous and particular solutions, respectively. At the boundary, an incoming wave $\mathbf{E}_I(\mathbf{k}_{lI}, \mathbf{k}_{mI}, \omega_l)$ splits into a reflected wave $\mathbf{E}_R(\mathbf{k}_{lR}, \mathbf{k}_{mR}, \omega_l)$ and a transmitted wave $\mathbf{E}_T(\mathbf{k}_{lT}, \mathbf{k}_{mT}, \omega_l)$. Let $z = 0$ be the boundary plane, and $\hat{x}$–$\hat{z}$ be the plane of incidence. Then it is easily seen that the continuity of the tangential components of $\mathbf{E}$ and $\mathbf{B}$ leads to the following relations:[7]

$$\mathscr{E}_{HI,x} + \mathscr{E}_{PI,x} + \mathscr{E}_{HR,x} + \mathscr{E}_{PR,x} = \mathscr{E}_{HT,x} + \mathscr{E}_{PT,x}$$

and

$$\begin{aligned}
(\mathbf{k}_{lI} \times \mathscr{E}_{HI})_x &+ (\mathbf{k}_{mI} \times \mathscr{E}_{PI})_x + (\mathbf{k}_{lR} \times \mathscr{E}_{HR})_x + (\mathbf{k}_{mR} \times \mathscr{E}_{PR})_x \\
&= (\mathbf{k}_{LT} \times \mathscr{E}_{HT})_x + (\mathbf{k}_{mT} \times \mathscr{E}_{PT})_x
\end{aligned} \tag{3.34}$$

with two similar equations for the y components, and

$$k_{lI,x} = k_{mI,x} = k_{lR,x} = k_{mR,x} = k_{lT,x} = k_{mT,x}. \tag{3.35}$$

The last equation relating the various tangential wavevectors is most interesting. It prescribes the directions of propagation for all waves (homogeneous and particular) in the media when one of them is given. This is therefore equivalent to Snell's law in linear optics.

3.5 TIME-DEPENDENT WAVE PROPAGATION

Propagating waves with time-varying amplitudes should of course obey the time-dependent wave equation in (3.1). Here again, the slowly varying amplitude approximation is usually valid. We expect that both the second-order time derivative and the second-order spatial derivative of the field amplitude can be neglected in the wave equation. This is illustrated in the following by assuming a quasi-monochromatic plane wave propagating along a symmetry axis, $\hat{z}$, of the medium. The wave equation takes the form

$$\frac{\partial^2}{\partial z^2} \mathbf{E}(z,t) - \frac{1}{c^2} \frac{\partial^2}{\partial t^2} \mathbf{D}(z,t) = \frac{4\pi}{c^2} \frac{\partial^2}{\partial t^2} \mathbf{P}^{NL}(z,t) \tag{3.36}$$

with $D(z, t) \equiv E(z, t) + 4\pi P^{(1)}(z, t)$, and $E(z, t) = \mathcal{E}(z, t)\exp(ikz - i\omega t)$. Then, as shown in Section 3.3, the slowly varying amplitude approximation gives

$$\frac{\partial^2}{\partial z^2}E(z, t) \cong \left(i2k\frac{\partial}{\partial z}\mathcal{E} - k^2\mathcal{E}\right)e^{i(kz - \omega t)}. \qquad (3.37)$$

If $E(z, t)$ is expressed in terms of the Fourier integral

$$E(z, t) = \int \mathcal{E}(\omega + \eta)e^{ikz - i(\omega + \eta)t} \, d\eta,$$

then we have

$$D(z, t) = \int \varepsilon(\omega + \eta)\mathcal{E}(\omega + \eta)e^{ikz - i(\omega + \eta)t} \, d\eta$$

$$\frac{1}{c^2}\frac{\partial^2}{\partial t^2}D(z, t) = \int \frac{-(\omega + \eta)^2}{c^2}\varepsilon(\omega + \eta)\mathcal{E}(\omega + \eta)e^{ikz - i(\omega + \eta)t} \, d\eta$$

$$\cong -\frac{1}{c^2}\int \left[\omega^2\varepsilon(\omega) + 2\omega\eta\varepsilon(\omega) + \omega^2\eta\frac{\partial\varepsilon}{\partial\omega}\right]\mathcal{E}(\omega + \eta) \qquad (3.38)$$

$$\times e^{ikz - i(\omega + \eta)t} \, d\eta$$

$$= \left[-\frac{\omega^2}{c^2}\varepsilon(\omega)\mathcal{E}(z, t) - i2k\frac{1}{v_g}\frac{\partial}{\partial t}\mathcal{E}(z, t)\right]e^{i(kz - \omega t)}$$

where $v_g = (dk/d\omega)^{-1}$ is the group velocity. Insertion of (3.37) and (3.38) in (3.36) with the approximation of $\partial^2 P^{NL}/\partial t^2 \cong -\omega^2 P^{NL}$ yields[8]

$$\left(\frac{\partial}{\partial z} + \frac{1}{v_g}\frac{\partial}{\partial t}\right)\mathcal{E}(z, t) = -\frac{2\pi\omega^2}{ikc^2}P^{NL}(z, t)e^{i(kz - \omega t)}. \qquad (3.39)$$

In fact, as we have shown in the time-independent case in Section 3.3, the field amplitude $\mathcal{E}$ in (3.39) actually corresponds to $\mathcal{E}_F$ for the forward propagating wave. For the backward propagating wave, the corresponding equation is

$$\left(\frac{\partial}{\partial z} - \frac{1}{v_g}\frac{\partial}{\partial t}\right)\mathcal{E}_B(z, t) = \frac{2\pi\omega^2}{ikc^2}P^{NL}(z, t)e^{i(kz + \omega t)}. \qquad (3.40)$$

Equations (3.39) and (3.40) should be used for short pulse propagation in a nonlinear medium. The time-derivative term in the equations is negligible only if the amplitude variation is insignificant during the time $T = l\sqrt{\varepsilon}/c$ it takes for light to traverse the medium. We use (3.39) and (3.40) later in the discussion of nonlinear optical effects with ultrashort pulses.

REFERENCES

1   See, for example, W. H. Louisell, *Coupled Mode and Parametric Electronics* (Wiley, New York, 1960).

2   J. A. Armstrong, N. Bloembergen, J. Ducuing, and P. S. Pershan, *Phys. Rev.* **127**, 1918 (1962).

3   L. Landau and E. M. Lifshitz, *Electrodynamics in Continuous Media* (Addison-Wesley, Reading, Mass., 1959), p. 253.

4   Y. R. Shen, *Phys. Rev.* **167**, 818 (1968).

5   P. S. Pershan, *Phys. Rev.* **130**, 919 (1963), and many books and review articles on nonlinear optics.

6   See, for example, N. Bloembergen, *Nonlinear Optics* (Benjamin, New York, 1965).

7   N. Bloembergen and P. S. Pershan, *Phys. Rev.* **128**, 606 (1962).

8   S. A. Akhmanov, A. S. Chirkin, K. N. Drabovich, A. I. Kovrigin, R. V. Khokhlov, and A. P. Sukhorukov, *IEEE J. Quant. Electron.* **QE-4**, 598 (1968).

BIBLIOGRAPHY

Akhmanov, S. A., and R. V. Khokhlov, *Problems of Nonlinear Optics* (Gordon and Breach, New York, 1972).

Bloembergen, N., *Nonlinear Optics* (Benjamin, New York, 1965).

Ducuing, in R. Glauber, ed., *Quantum Optics, Proceedings of the International School of Physics Enrico Fermi Course XLII* (Academic Press, New York, 1969), p. 421.

4

Electrooptical and Magnetooptical Effects

Optical properties of a material can be modified by an applied electric or magnetic field. The refractive index changes as functions of the applied electric and magnetic fields are responsible for many electrooptical and magnetooptical effects. Although these effects were well known long before the advent of lasers, they can be regarded as nonlinear optical mixing effects in the limit where one of the field components is of zero or nearly zero frequency. This chapter therefore offers a brief discussion on these effects.

4.1 ELECTROOPTICAL EFFECTS

In the presence of an applied dc or low-frequency field $\mathbf{E}_0(\Omega \sim 0)$, the optical dielectric constant $\varepsilon(\omega, \mathbf{E}_0)$ of a medium is a function of $\mathbf{E}_0$. For sufficiently small $\mathbf{E}_0$, $\varepsilon(\omega, \mathbf{E}_0)$ can be expanded into a power series of $\mathbf{E}_0$:

$$\varepsilon(\omega, \mathbf{E}_0) = \varepsilon^{(1)}(\omega) + \varepsilon^{(2)}(\omega + \Omega) \cdot \mathbf{E}_0 + \varepsilon^{(3)}(\omega + 2\Omega) : \mathbf{E}_0 \mathbf{E}_0 + \cdots .$$

$$(4.1)$$

Since $\mathbf{E} + 4\pi \mathbf{P} = \varepsilon \cdot \mathbf{E}$, and $\mathbf{P} = \chi^{(1)} \cdot \mathbf{E} + \chi^{(2)} : \mathbf{EE} + \cdots$, we recognize that

$$\varepsilon^{(2)}(\omega + \Omega) = 4\pi \chi^{(2)}(\omega + \Omega)$$

and

$$\varepsilon^{(3)}(\omega + 2\Omega) = 4\pi \chi^{(3)}(\omega + 2\Omega).$$

$$(4.2)$$

Then, in a medium with no inversion symmetry, the electrooptical effect is dominated by the $\varepsilon^{(2)}$ term linear in $\mathbf{E}_0$. This is known as Pockel's effect. The symmetry forms of the nonvanishing $\varepsilon^{(2)}$ or $\chi^{(2)}$ for the 20 classes of crystals are already given in Table 2.1 with, in addition, $\chi_{ijk}^{(2)}(\omega = \omega + 0) = \chi_{jik}^{(2)}(\omega = \omega + 0)$. In a medium with or without inversion symmetry, the quadratic field-dependent term (4.1) always exists and is known as the dc Kerr effect. The symmetry forms of $\varepsilon^{(3)}$ or $\chi^{(3)}$ for some classes of crystals are given in Table 2.2 with, in addition, $\chi_{ijkl}^{(3)}(\omega = \omega + 0 + 0) = \chi_{jikl}^{(3)}(\omega = \omega + 0 + 0)$ and $\chi_{ijkl}^{(3)} = \chi_{ijlk}^{(3)}$.

The field-induced refractive indices give rise to linear birefringence or double refraction. Traditionally, the electrooptical effect is defined through the idex ellipsoid[1]

$$\frac{x^2}{n_{xx}^2} + \frac{y^2}{n_{yy}^2} + \frac{z^2}{n_{zz}^2} + \frac{2\,yz}{n_{yz}^2} + \frac{2zx}{n_{zx}^2} + \frac{2xy}{n_{xy}^2} = 1 \qquad (4.3)$$

with $n_{ij}^{-1} = (\varepsilon^{-1})_{ij}^{1/2}$ being the refractive index tensor. The power series expansion is carried out for all the coefficients $n_{ij}^{-2}(\mathbf{E}_0)$ of the index ellipsoid

$$\frac{1}{n_{ij}^2(\mathbf{E}_0)} = \left(\frac{1}{n_{ij}^2}\right)_0 + \sum_k r_{ijk} E_{0k} + \sum_{k,l} p_{ijkl} E_{0k} E_{0l} + \cdots. \qquad (4.4)$$

The coefficient r_{ijk} often is called the linear electrooptical tensor, and p_{ijkl} is the quadratic electrooptical tensor. The values of r_{ijk} for many crystals are tabulated in the literature.[2]

Physically, electrooptical effects result from both ionic or molecular movement and distortion of electronic cloud induced by the applied electric field. Even if the induced refractive index change is only around 10^{-5} (typical values of r_{ijk} are around 10^{-10} to 10^{-8} cm/v), a medium 1 cm long can already impose on a visible beam a phase retardation of more than $\pi/2$. Therefore, electrooptical effects have been widely used as optical modulators.

4.2 MAGNETOOPTICAL EFFECTS

The optical dielectric tensor ε of a medium is also a function of an applied dc magnetic field, $\mathbf{H}_0$. It has the symmetry relation [3]

$$\varepsilon_{ij}(\mathbf{H}_0) = \varepsilon_{ji}(-\mathbf{H}_0). \qquad (4.5)$$

Here, even in the absence of dissipation, ε_{ij} is a complex quantity but it has the property of being Hermitian:

$$\varepsilon_{ij}(\mathbf{H}_0) = \varepsilon_{ij}'(\mathbf{H}_0) + i\varepsilon_{ij}''(\mathbf{H}_0) = \varepsilon_{ji}^*(\mathbf{H}_0). \qquad (4.6)$$

Then, in a nondissipative medium, we have

$$\varepsilon'_{ij}(\mathbf{H}_0) = \varepsilon'_{ji}(\mathbf{H}_0) = \varepsilon'_{ij}(-\mathbf{H}_0)$$

$$\varepsilon''_{ij}(\mathbf{H}_0) = -\varepsilon''_{ji}(\mathbf{H}_0) = -\varepsilon''_{ij}(-\mathbf{H}_0). \tag{4.7}$$

Thus the real part of the tensor is symmetric and is an even function of $\mathbf{H}_0$, and the imaginary part is antisymmetric and odd in $\mathbf{H}_0$. The dependence of ε''_{ij} on $\mathbf{H}_0$ leads to circular birefringence or the Faraday effect, while the dependence of ε'_{ij} on $\mathbf{H}_0$ leads to linear birefringence or the Cotton–Mouton effect.[3] This can be illustrated with a medium of uniaxial symmetry, having $\mathbf{H}_0$ parallel to the axis. In this case, the only nonvanishing elements of ε are $\varepsilon'_{xx} = \varepsilon'_{yy}$ and ε'_{zz} even in $\mathbf{H}_0$, and $\varepsilon''_{xy} = -\varepsilon''_{yx}$ odd in $\mathbf{H}_0$. Diagonalization of ε in the coordinate system with orthogonal unit vectors $e_\pm = (\hat{x} \pm i\hat{y})/\sqrt{2}$ and $\hat{z}$ yields the three diagonal elements $\varepsilon_\pm$ and ε_{zz}, where $\varepsilon_\pm = \varepsilon'_{xx} \pm \varepsilon''_{xy}$ are the susceptibilities for right and left circularly polarized waves, respectively. Since $\varepsilon''_{xy} \ll \varepsilon'_{xx}$, the wavevectors of the two circularly polarized waves can be written as

$$k_\pm = \frac{\omega\sqrt{\varepsilon_\pm}}{c} \simeq \left(\frac{\omega\sqrt{\varepsilon'_{xx}}}{c}\right)\left(\frac{1 \pm \frac{1}{2}\varepsilon''_{xy}}{\varepsilon'_{xx}}\right) \tag{4.8}$$

and the circular birefringence in a medium of length l is

$$(k_+ - k_-)l = \left(\frac{\omega\varepsilon''_{xy}}{c\sqrt{\varepsilon'_{xx}}}\right)l. \tag{4.9}$$

A linearly polarized beam propagating along $\hat{z}$ will have its polarization rotated by an angle

$$\phi = \frac{(k_+ - k_-)l}{2} \tag{4.10}$$

which is known as the Faraday rotation. On the other hand, since $\varepsilon'_{xx}(\mathbf{H}_0) - \varepsilon'_{xx}(0)$ is generally different from $\varepsilon'_{zz}(\mathbf{H}_0) - \varepsilon'_{zz}(0)$, the linear birefringence in the $\hat{x}$–$\hat{z}$ plane is also altered by the presence of $\mathbf{H}_0$, known as the Cotton–Mouton effect.

For sufficiently weak $\mathbf{H}_0$, the power series expansion of $\varepsilon(\mathbf{H}_0)$ yields

$$\varepsilon'(\omega, \mathbf{H}_0) = \varepsilon'^{(1)}(\omega) + \varepsilon'^{(3)}(\omega + 2\Omega)\cdot\mathbf{H}_0\mathbf{H}_0 + \cdots$$

and

$$\varepsilon''(\omega\,\mathbf{H}_0) = \varepsilon''^{(2)}(\omega + \Omega)\cdot\mathbf{H}_0 + \cdots. \tag{4.11}$$

Again, $\varepsilon^{(2)}/4\pi$, $\varepsilon^{(3)}/4\pi$, and so on, can be regarded as nonlinear susceptibilities, although they now arise from the magnetic contribution.

Analogous to the electrooptical effects, the magnetooptical effects can also be used for optical modulation. The Faraday effect or circular birefringence causes the linear polarization of a beam traversing the medium to rotate. In the low-field limit, the rotation is proportional to the applied magnetic field. Again, the induced change in the dielectric constant or refractive index is usually small ($\sim 10^{-9}$/gauss for glass doped with a few percent of rare earth ions), but rotation resulting form the relative phase shift between the two circular polarizations can be few tens of degrees in a 1-cm-long medium with a field of several thousand gausses. The Cotton–Mouton effect, however, is much weaker and has limited applications.

REFERENCES

1 See, for example, A. Yariv, *Quantum Electronics*, 2nd ed. (Wiley, New York, 1975), p. 327.

2 R. J. Pressley, ed., *CRC Handbook of Lasers*, (Chemical Rubber Co., Cleveland, Ohio, 1971), p. 447

3 See, for example, J. van den Handel, *Encyclopedia of Physics*, vol. 15, S. Flügge, ed. (Springer-Verlag, Berlin, 1956), p. 15.

BIBLIOGRAPHY

Landau, L. D., and E. M. Lifshitz, *Electrodynamics in Continuous Media* (Pergamon Press, Oxford, 1960).

Nye, J. F., *Physical Properties of Solids* (Oxford University Press, London, 1964).

Pressley, R. J., ed., *Handbook of Lasers* (Chemical Rubber Co., Cleveland, Ohio, 1971).

Wemple, S. H., in F. T. Arecchi and E. O. Schulz-Dubois, eds., *Laser Handbook* (North-Holland Publishing Co., Amsterdam, 1972), p. 975.

5

Optical Rectification and Optical Field-Induced Magnetization

Modulation and demodulation are commonly known processes at radio wave and microwave frequencies. They form the basis of telecommunications. It is natural to believe that these processes should also exist in the optical region. Light modulation by electrooptical and magnetooptical effects has already been discussed in Chapter 4. In this chapter, optical rectification producing dc electric polarization and magnetization is considered.

5.1 OPTICAL RECTIFICATION

In the literature, optical rectification, which was among the first nonlinear optical effects discovered,[1] usually refers to the generation of a dc electric polarization by an intense optical beam in a nonlinear medium. The effect can be seen directly from the nonlinear polarization

$$\mathbf{P}^{(2)}(0) = \chi^{(2)}(0 = \omega - \omega) : \mathbf{E}(\omega)\mathbf{E}^*(\omega) \qquad (5.1)$$

with $\mathbf{E}(\omega) = \mathscr{E}\exp(i\mathbf{k}\cdot\mathbf{r} - i\omega t)$. The nonlinear susceptibility $\chi^{(2)}(0 = \omega - \omega)$ here governs the magnitude of the effect. In a nonabsorbing medium, the permutation relation of $\chi^{(2)}$ relates $\chi^{(2)}(0 = \omega - \omega)$ to the electrooptical coefficients

$$\chi^{(2)}_{ijk}(0 = \omega - \omega) = \chi^{(2)}_{jki}(\omega = \omega + 0) = \chi^{(2)}_{kij}(\omega = 0 + \omega)$$

$$= -\frac{\epsilon^{(1)}_{ii}\epsilon^{(1)}_{jj}}{4\pi}r_{ijk} \qquad (5.2)$$

in a principal-axis system. Thus, from the electrooptical coefficient r_{ijk}, the polarization generated in optical rectification can be predicted. In actual experiments, the induced dc field or voltage, instead of $\mathbf{P}^{(2)}$, is measured, but the two are linearly related. Bass et al.[1] and Ward[2] performed optical rectification experiments and measured $\chi_{ijk}^{(2)}(0 = \omega - \omega)$ for a number of crystals. The experimental arrangement can be simple. A slab of crystal is oriented with its $\hat{\imath}$ axis perpendicular to the two parallel faces of the slab. The faces are coated with silver to form a set of condenser plates. An intense light beam is then directed through the crystal in a direction perpendicular to $\hat{\imath}$ to generate $\mathbf{P}_i^{(2)}(0)$ according to (5.1), and the induced dc voltage across the condenser plates is measured. Let the dc dielectric constant of the crystal along $\hat{\imath}$ be ε_0 and assume that the beam intensity can be approximated as uniform over a rectangular cross section $s \times t$ in the crystal, as shown in Fig. 5.1. Then, following the infinite plane approximation for condensers, the equations governing the dc fields are

$$E_0 d = E_a(d - t) + E_b t$$
$$\varepsilon_0 E_a = \varepsilon_0 E_b + 4\pi P_i^{(2)}$$

(5.3)

and since there is no net charge on the plates

$$-E_0(w - s) = E_a s.$$

(5.4)

The solution of these equations yields a dc voltage across the plates as

$$V = -4\pi\left(\frac{st}{\varepsilon_0 w}\right) P_i^{(2)}$$

$$= -4\pi\left(\frac{st}{\varepsilon_0 w}\right) \sum_{j,k} \chi_{ijk}^{(2)} E_j(\omega) E_k^*(\omega).$$

(5.5)

In the experiments to find $\chi_{ijk}^{(2)}(0 = \omega - \omega)$, both the voltage V (in the

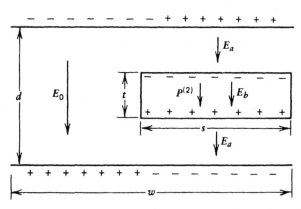

Fig. 5.1 Experimental geometry for measurement of optical rectification.

mv/MW range) and the laser intensity must be accurately measured. The results of Bass et al.[1] and Ward[2] on optical rectification show that the identity of (5.2) indeed holds within experimental error.

5.2 EFFECTIVE FREE ENERGY DENSITY

The time-averaged field energy density $\langle U \rangle$ in a nonlinear medium was derived in Section 3.2, where we saw that we can obtain the polarization $P(\omega)$ in the medium from the derivative of the effective free energy density $F - F^{(0)} = -(\langle U \rangle - \langle U^{(0)} \rangle)$

$$P(\omega) = -\frac{\partial(F - F^{(0)})}{\partial E^*(\omega)} \tag{5.6}$$

if the dispersions of susceptibilities are neglected. Thus the effective free energy density corresponding to a not very intense quasi-monochromatic wave in a nonabsorbing medium subject to a dc electric field E_0 is*

$$F[E(\omega), E_0] \cong F^{(0)}(E_0) - \sum_{j,k} \left[\chi_{jk}(E_0) E_j^*(\omega) E_k(\omega) + \cdots \right]. \tag{5.7a}$$

If $\chi_{jk}(E_0)$ can also be expanded into power series of E_0, it becomes

$$F = F^{(0)}(E_0) + F^{(1)} + F^{(2)} + \cdots$$
$$F^{(1)} = -\sum_{j,k} \chi_{jk}(0) E_j^*(\omega) E_k(\omega), \tag{5.7b}$$
$$F^{(2)} = -\sum_{l,j,k} \chi_{ljk}^{(2)} E_{0l} E_j(\omega)^* E_k(\omega).$$

The free energy density here governs both the electrooptical effect and the optical rectification. From (5.6), the polarization induced in optical rectification is given by

$$P_l^{(2)}(0) = -\frac{\partial F^{(2)}}{\partial E_{0l}} = \sum_{j,k} \chi_{ljk}^{(2)} E_j^*(\omega) E_k(\omega) \tag{5.8}$$

as is expected. The same $\chi_{ljk}^{(2)}$ in (5.7) and (5.8) is clearly responsible for the linear electrooptical effect.

The above description can be extended to the magnetic case.[3, 4] For a not too intense light beam propagating in a nonabsorbing medium in the presence of a magnetic field, the effective free energy density can be written as a power

---

*Note that $E_0 = \lim_{\Omega \to 0} \frac{1}{2}[E_0(\Omega) + E_0^*(\Omega)]$.

series of the optical field

$$F = F_0(\mathbf{H}_0) - \sum_{i,j} \chi_{i,j}(\mathbf{H}_0) E_i^*(\omega) E_j(\omega) + \cdots . \qquad (5.9)$$

In analogy to the electric case, $[\chi_{i,j}(\mathbf{H}_0) - \chi_{i,j}(0)]$ here governs the magneto-optical effect. Then one also expects from $\mathbf{M}(0) = -\partial F/\partial \mathbf{H}_0$ that there should be a dc magnetization induced in the medium by the incoming light. Indeed, this has been observed and is known as the inverse magnetooptical effect.[5]

For illustration, we assume a medium of uniaxial symmetry with light propagating along the axis, say, the $\hat{z}$-axis. It is well known that if the dc magnetic field is also along $\hat{z}$, the two circular polarizations are the eigenmodes of propagation. The effective free energy density can therefore be written as

$$F = F_0(\mathbf{H}_0) - \chi_+(\mathbf{H}_0)|E_+(\omega)|^2 - \chi_-(\mathbf{H}_0)|E_-(\omega)|^2 + \cdots \qquad (5.10)$$

where $\chi_+ = \chi_{xx} + i\chi_{xy}$ and $\chi_- = \chi_{xx} - i\chi_{xy}$ are, respectively, the linear susceptibilities for the right and left circularly polarized waves. Equation (5.10) can be rearranged into

$$
\begin{aligned}
F = F_0(H_0) &- \tfrac{1}{2}[\chi_+(H_0) - \chi_-(H_0)](|E_+|^2 - |E_-|^2) \\
&- \tfrac{1}{2}[\chi_+(H_0) + \chi_-(H_0)](|E_+|^2 + |E_-|^2) + \cdots
\end{aligned}
\qquad (5.11)
$$

with $[\chi_+(H_0) - \chi_-(H_0)]$ and $[\chi_+(H_0) + \chi_-(H_0)]$ being odd and even functions of H_0, respectively. The Faraday effect is now proportional to

$$
\begin{aligned}
-\frac{\partial F}{\partial |E_+|^2} + \frac{\partial F}{\partial |E_-|^2} &= \chi_+(H_0) - \chi_-(H_0) \\
&= \frac{\varepsilon_+(H_0) - \varepsilon_-(H_0)}{4\pi}
\end{aligned}
\qquad (5.12)
$$

as discussed in Section 4.2. We can also show

$$
\begin{aligned}
\chi_{xx}(H_0) &= \chi_{yy}(H_0) \\
\chi_{xx}(H_0) - \chi_{xx}(0) &= \tfrac{1}{2}[\chi_+(H_0) + \chi_-(H_0) - \chi_+(0) - \chi_-(0)].
\end{aligned}
\qquad (5.13)
$$

This magnetic field–induced susceptibility change is connected to the Cotton–Mouton effect. On the other hand, the optical field–induced magnetization can also be derived from (5.11). With the optical field on, the induced magnetization change along $\hat{z}$ is given by

$$\Delta M = -\frac{\partial}{\partial H_0}(F - F_0)$$

$$= \Delta M_F + \Delta M_{CM}, \tag{5.14}$$

$$\Delta M_F = \frac{1}{2}\frac{\partial}{\partial H_0}(\chi_+ - \chi_-)(|E_+|^2 - |E_-|^2),$$

and

$$\Delta M_{CM} = \frac{1}{2}\frac{\partial}{\partial H_0}(\chi_+ + \chi_-)(|E_+|^2 + |E_-|^2).$$

The term ΔM_F even in H_0 comes from the term responsible for the Faraday effect in F, and ΔM_{CM} odd in H_0 from the Cotton–Mouton term in F. The phenomena are therefore called the inverse Faraday effect and the inverse Cotton–Mouton effect, respectively. As seen from (5.14), even if $H_0 = 0$, the inverse Faraday effect is nonvanishing as long as $|E_+| \neq |E_-|$, and is a maximum for circular polarization. The light beam with $|E_+|^2 \neq |E_-|^2$ here plays the role of a dc magnetic field and breaks the time-reversal symmetry of the medium. The inverse Cotton–Mouton effect can, however, exist even with linearly polarized light but vanishes for $H_0 = 0$ since $\partial(\chi_+ + \chi_-)/\partial H_0$ is odd in H_0.

The effective free energy density then allows us to predict the magnitudes of the inverse Faraday and Cotton–Mouton effects from the measured Faraday rotation and Cotton–Mouton effect in the medium. Physically, the Faraday and Cotton–Mouton effects originate from circular and linear dichroism induced by the dc magnetic field, but how do we describe the inverse effects? This is the subject of the following section.

5.3 INVERSE FARADAY AND COTTON–MOUTON EFFECTS

Microscopically, the light-induced dc magnetization arises because the optical field shifts the different magnetic states of the ground manifold differently (known as optical Stark shifts), and mixes into these ground states different amounts of excited states. Let the interaction Hamiltonian be

$$\mathcal{H}_{int} = \mathcal{H}_1 + \mathcal{H}_1^\dagger$$
$$\mathcal{H}_1 = -e[r_+ E_-(\omega) + r_- E_+(\omega)] \tag{5.15}$$

From the time-dependent perturbation calculation, we find the perturbed eigenstate as

$$|n\rangle = |n\rangle_0 + \Delta|n\rangle$$

$$\Delta|n\rangle = \sum_{n' \neq n} |n'\rangle_0 \left[\frac{\langle n'|\mathcal{H}_1|n\rangle_0}{\hbar(\omega - \omega_{n'n})} - \frac{\langle n'|\mathcal{H}_1^\dagger|n\rangle_0}{\hbar(\omega + \omega_{n'n})} \right] \qquad (5.16)$$

and the optical Stark shift for $|n\rangle$ as

$$\Delta E_n = \sum_{n' \neq n} \left[\frac{|\langle n'|\mathcal{H}_1|n\rangle_0|^2}{\hbar(\omega - \omega_{n'n})} - \frac{|\langle n'|\mathcal{H}_1^\dagger|n\rangle_0|^2}{\hbar(\omega + \omega_{n'n})} \right] \qquad (5.17)$$

where $\hbar\omega_{n'n} \equiv E_{n'} - E_n$. To demonstrate the inverse magnetooptical effect, we assume here a simple paramagnetic ion with only two pairs of states as shown in Fig. 5.2. The ground $|+m\rangle$ state is connected to the excited $|-m'\rangle$ state only by the matrix element $\langle -m'|r_-|m\rangle$ and the $|-m\rangle$ state connected to the $|+m'\rangle$ state by $\langle +m'|r_+|-m\rangle$, with $r_\pm \equiv (x \pm iy)/\sqrt{2}$. In an applied magnetic field along $\hat{z}$, the Zeeman splittings for the two pairs are respectively $2g\beta m H_0$ and $2g'\beta m' H_0$, where β is the Bohr magneton. The energy separation between the pairs of states is $\hbar\omega_0 \gg kT$ at $H_0 = 0$. The dc magnetization of a system of N ions per unit volume along $\hat{z}$ is given by

$$\begin{aligned} M &= -Ng\beta\langle J_z\rangle \\ &= -Ng\beta[\langle m|J_z|m\rangle\rho_m + \langle -m|J_z|-m\rangle\rho_{-m}]. \end{aligned} \qquad (5.18)$$

Here J_z is the angular momentum operator and $\rho_{\pm m}$ are the thermal popula-

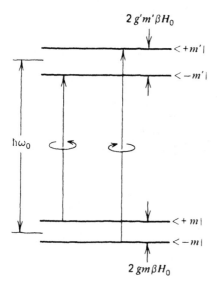

Fig. 5.2 Energy level diagram of an ideal paramagnetic system with only two pairs of states connected by circularly polarized optical fields.

tions in $|\pm m\rangle$ described by the Boltzmann distribution

$$\rho_{\pm m} = \frac{e^{-\mathbb{E}_{\pm m}/kT}}{e^{-\mathbb{E}_m/kT} + e^{-\mathbb{E}_{-m}/kT}} \tag{5.19}$$

with $\mathbb{E}_{\pm m} = \pm g\beta mH_0 + \Delta\mathbb{E}_{\pm m}$. Through its perturbation on $\mathbb{E}_{\pm m}$ and $|m\rangle$, the optical field induces a dc magnetization

$$\Delta M = M(\mathscr{H}_1) - M(\mathscr{H}_1 = 0)$$
$$= \Delta M^P + \Delta M^D, \tag{5.20}$$
$$\Delta M^P = -Ng\beta m\big[(\rho_m - \rho_{-m}) - (\rho_m^0 - \rho_{-m}^0)\big],$$

and

$$\Delta M^D = -Ng\beta\big[(\Delta\langle m|)J_z(\Delta|m\rangle)\rho_m^0 + (\Delta\langle -m|)J_z(\Delta|-m\rangle)\rho_{-m}^0\big]$$

where $\rho_{\pm m}^0 = (\rho_{\pm m})_{\Delta\mathbb{E}_{\pm m}=0}$, and ΔM^P and ΔM^D come from induced changes in the populations and matrix elements, respectively. From (5.15) to (5.20) with $|\Delta\mathbb{E}_{\pm m}| \ll kT$ and $|\langle m'|r_+|-m\rangle|^2 = |\langle -m'|r_-|m\rangle|^2$, we readily find

$$\Delta M^P = \frac{-2Ng\beta m(\Delta\mathbb{E}_m - \Delta\mathbb{E}_{-m})\rho_m^0\rho_{-m}^0}{kT}$$

$$= -\frac{2Ng\beta m\rho_m^0\rho_{-m}^0}{kT}|\langle m'|er_+|-m\rangle|^2$$

$$\times\Bigg\{\Bigg[\frac{\hbar(\omega - \omega_0)}{\hbar^2(\omega - \omega_0)^2 - (gm + g'm')^2\beta^2H_0^2}$$

$$+ \frac{\hbar(\omega + \omega_0)}{\hbar^2(\omega + \omega_0)^2 - (gm + g'm')^2\beta^2H_0^2}\Bigg] \tag{5.21}$$

$$\times\big(|E_+|^2 - |E_-|^2\big)$$

$$+ \Bigg[\frac{-(gm + g'm')\beta H_0}{\hbar^2(\omega - \omega_0)^2 - (gm + g'm')^2\beta^2H_0^2}$$

$$+ \frac{(gm + g'm')\beta H_0}{\hbar^2(\omega + \omega_0)^2 - (gm + g'm')^2\beta^2H_0^2}\Bigg]\big(|E_+|^2 + |E_-|^2\big)\Bigg\}$$

and

$$\Delta M^D = -\tfrac{1}{2} N g \beta m' |\langle m'|er_+|-m\rangle|^2$$

$$\times \left(\left\{ \frac{-\rho_m^0}{\left[\hbar(\omega - \omega_0) + (gm + g'm')\beta H_0\right]^2} \right.\right.$$

$$+ \frac{\rho_m^0}{\left[\hbar(\omega + \omega_0) - (gm + g'm')\beta H_0\right]^2}$$

$$+ \frac{-\rho_{-m}^0}{\left[\hbar(\omega - \omega_0) - (gm + g'm')\beta H_0\right]^2}$$

$$\left. + \frac{\rho_{-m}^0}{\left[\hbar(\omega + \omega_0) + (gm + g'm')\beta H_0\right]^2} \right\} \left(|E_+|^2 - |E_-|^2\right)$$

$$\quad (5.22)$$

$$+ \left\{ \frac{-\rho_m^0}{\left[\hbar(\omega - \omega_0) + (gm + g'm')\beta H_0\right]^2} \right.$$

$$+ \frac{-\rho_m^0}{\left[\hbar(\omega + \omega_0) - (gm + g'm')\beta H_0\right]^2}$$

$$+ \frac{\rho_{-m}^0}{\left[\hbar(\omega - \omega_0) - (gm + g'm')\beta H_0\right]^2}$$

$$\left.\left. + \frac{\rho_{-m}^0}{\left[\hbar(\omega + \omega_0) + (gm + g'm')\beta H_0\right]^2} \right\} \left(|E_+|^2 + |E_-|^2\right) \right)$$

Since ΔM^P arises from the induced population change in the ground states due to optical Stark shifts, it vanishes in a diamagnetic system which has only a singlet ground state that is populated at ordinary temperatures. It is therefore designated as the paramagnetic part of ΔM. Because of the finite relaxation time for the population distribution to reach new equilibrium, ΔM^P cannot respond instantaneously to a short incoming light pulse. In fact, from the time variation of ΔM^P, one should be able to deduce the T_1 relaxation time of the ground states. The ΔM^D term arising from the wavefunction mixing by the optical field exists even in a diamagnetic system and is designated as the diamagnetic part of ΔM. It responds almost instantaneously to the incoming light. The paramagnetic part is proportional to $1/kT$ for $|\Delta E_{\pm m}| \ll kT$, and the diamagnetic part is essentially independent of temperature. This is similar to the temperature behavior of ordinary paramagnetism and diamagnetism.[6] Both ΔM^P in (5.21) and ΔM^D in (5.22) have been explicitly decomposed into a part proportional to $(|E_+|^2 - |E_-|^2)$ and a part proportional to $(|E_+|^2 + |E_-|^2)$. The former corresponds to the inverse Faraday effect, and the latter to the inverse Cotton–Mouton effect. For light frequency far away from resonance, the inverse Cotton–Mouton effect is much smaller than the inverse Faraday effect. As seen from (5.21) and (5.22), the ratio of ΔM_{CM} to ΔM_F is about $|(g'm' + gm)\beta H_0/\hbar(\omega - \omega_0)|$ for the paramagnetic part, and is about $|(g'm' + gm)\beta H_0/\hbar(\omega - \omega_0)|$ or $(\rho_{-m}^0 - \rho_{-m}^0)/(\rho_m^0 + \rho_{-m}^0)$, whichever is larger, for the diamagnetic part. It may become comparable to 1 when $\hbar(\omega - \omega_0)$ approaches the Zeeman splitting energy. However, close to resonance, the induced dc magnetization due to optical pumping often becomes dominant.[7] In actual experiments, the inverse Cotton–Mouton effect is distinguishable from the inverse Faraday effect by the fact that with a reversal of the magnetic field H_0, ΔM_{CM} changes sign but ΔM_F does not. Finally, we realize that (5.21) and (5.22) can be derived from (5.14) if the microscopic expressions of χ_+ and χ_- for the system in Fig. 5.2, following (2.17) for χ_{ij}, are used. This is left as an exercise to the readers. The above calculation for ΔM can of course be generalized to a paramagnetic system with N ground states. In dense media, a local field correction factor should also be included.

The experimental scheme for observing the light-induced dc magnetization is seen in Fig. 5.3. The light pulse induces a pulsed $\Delta M(t)$ in the sample. The time-derivative $d(\Delta M)/dt$ then induces a voltage across the terminals of a pick-up coil around the sample. As an example, consider the case of $CaF_2 : 3\%Eu^{2+}$. The Faraday rotation at $\lambda = 7000$ Å is 2×10^{-4} rad/cm-Oe at 4.2 K. From (4.9) and (4.10), we obtain $[\partial(\chi_+ - \chi_-)/\partial H_0]_{H_0=0} = 1.8 \times 10^{-10}$ esu/Oe. Then (5.14) predicts that with a circularly polarized ruby laser beam of 10 MW/cm^2 in the sample, the induced dc magnetization is $\Delta M = \Delta M_f = 7 \times 10^{-6}$ erg/Oe-cm^3. This is equivalent to that induced by a dc field of about 0.01 Oe. If a Q-switched laser pulse with a 10-MW peak intensity and

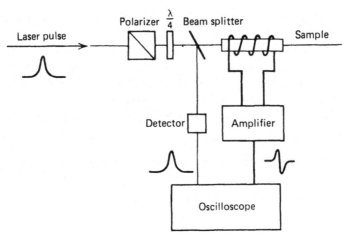

Fig. 5.3 Experimental arrangement for measurement of inverse magnetooptical effects.

a rise time of 2×10^{-8} sec is used, and $\Delta M(t)$ is assumed to follow instantaneously the intensity variation of the laser pulse, then the voltage induced across the terminals of a 30-turn pick-up coil will be 1.3 mv. This agrees with the experimental observation of van der Ziel et al.,[5] who demonstrated that the inverse relationship between the Faraday effect and the inverse Faraday effect indeed holds for many paramagnetic and diamagnetic substances.

5.4 INDUCED MAGNETIZATION BY RESONANT EXCITATION

The dc magnetization can of course also be induced by light through direct optical pumping, and is generally much stronger than the inverse magnetooptical effect discussed in Section 5.3. Optical pumping by circularly polarized light, for example, alters the population distribution in the magnetic sublevels of both the ground and the excited states. A net angular momentum and hence a magnetization result. Theoretically, rate equations can be used to calculate the population redistribution and hence the magnetization induced by the resonant optical excitation if transition probabilities and relaxation rates between levels are known. Optical pumping in gases and solids has long been a subject of extensive investigation. Polarized fluorescence is often a means for detection of the induced orientation of the angular momentum in the medium. With the setup in Fig. 5.3, however, it can also be studied by measuring the dc magnetization generated in the medium by the laser pulse.[7] This may be useful in some cases for studying relaxation between magnetic sublevels in condensed matter.

REFERENCES

1 M. Bass, P. A. Franken, J. F. Ward, and G. Weinreich, *Phys. Rev. Lett.* **9**, 446 (1962).

2 J. F. Ward, *Phys. Rev.* **143**, 569 (1966).

3 Y. R. Shen and N. Bloembergen, *Bull. Am. Phys. Soc.* **9**, 292 (1964).

4 P. S. Pershan, J. P. van der Ziel, and L. D. Malmstrom, *Phys. Rev.* **143**, 574 (1966).

5 J. P. van der Ziel, P. S. Pershan, and L. D. Malmstrom, *Phys. Rev. Lett.* **15**, 190 (1965)

6 J. H. van Vleck and M. H. Hebb, *Phys. Rev.* **46**, 17 (1934).

7 J. F. Holzrichter, R. M. MacFarlane, and A. L. Schawlow, *Phys. Rev. Lett.* **26**, 652 (1971)

6

Sum-Frequency Generation

Wave interaction in a nonlinear medium leads to wave mixing. The result is the generation of waves at sum and difference frequencies. Sum-frequency generation is one of the first three nonlinear optical effects discovered in the early days.[1] With the recent advances in tunable lasers, it has become one of the most useful nonlinear optical effects in extending the tunable range to shorter wavelengths. This chapter deals mainly with the basic principle of sum-frequency generation.

6.1 PHYSICAL DESCRIPTION

Bass et al.[1] in 1962 first observed optical sum-frequency generation in a crystal of triglycine sulfate. In their experiment, two ruby lasers, with their operating wavelengths 10 Å apart, were used to provide the input beams. The output, analyzed by a spectrograph, exhibited three lines around 3470 Å, two side lines arising from second harmonic generation and the middle one from sum-frequency generation by the two laser beams.

The physical interpretation of sum-frequency generation is straightforward. The laser beams at ω_1 and ω_2 interact in a nonlinear crystal and generate a nonlinear polarization $\mathbf{P}^{(2)}(\omega_3 = \omega_1 + \omega_2)$. The latter being a collection of oscillating dipoles acts as a source of radiation at $\omega_3 = \omega_1 + \omega_2$. In general, the radiation could appear in all directions; the radiation pattern depends on the phase-correlated spatial distribution of $\mathbf{P}^{(2)}(\omega_3)$. With appropriate arrangement, however, the radiation pattern can be strongly peaked in a certain direction. This can be determined by phase matching conditions. As discussed in Section 3.1, for effective energy transfer from the pump waves at ω_1 and ω_2 to the generated waves at ω_3, in the sum-frequency generation (Fig. 6.1), both energy and momentum conservation must be satisfied. The energy conservation requires $\omega_3 = \omega_1 + \omega_2$, while the momentum conservation requires $\mathbf{k}_3 = \mathbf{k}_1 + \mathbf{k}_2$. The latter indicates that the sum-frequency radiation is most effec-

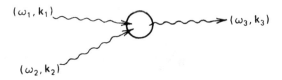

Fig. 6.1 Schematic description of sum-frequency generation.

tively generated in the so-called phase-matching direction defined by $\mathbf{k}_3 = \mathbf{k}_1 + \mathbf{k}_2$.[2] If the wave interaction length l is finite, momentum conservation needs to be satisfied only to within the uncertainty range of $1/l$. The radiation pattern should therefore be a finite phase-matching peak with an angular width corresponding to $\Delta k \sim 1/l$. Absorption in the medium, for instance, can limit the interaction length and broaden the phase-matching peak. In general, sum-frequency generation from the bulk, if allowed and phase-matched, dominates over that from the surface. In reflection, however, because of lack of phase matching, a surface layer $\sim \lambda/2\pi$ thick could contribute significantly to the output. This description gives a qualitative picture of sum-frequency generation, which needs to be substantiated with a more formal treatment.

6.2 FORMULATION

The coupled wave approach discussed in Section 3.1 finds a direct application here.[3] The three coupled waves are $\mathbf{E}(\omega_1)$, $\mathbf{E}(\omega_2)$, and the sum-frequency output $\mathbf{E}(\omega_3)$. Each field can be decomposed into a longitudinal and a transverse part $\mathbf{E}(\omega_i) = \mathbf{E}_{\parallel}(\omega_i) + \mathbf{E}_{\perp}(\omega_i)$. They obey the wave equations

$$\nabla^2 \mathbf{E}_{\perp}(\omega_i) + \frac{\omega_i^2}{c^2}\left[\boldsymbol{\varepsilon}(\omega_i)\cdot\mathbf{E}(\omega_i)\right]_{\perp} = -\frac{4\pi\omega_i^2}{c^2}\mathbf{P}_{\perp}^{(2)}(\omega_i)$$

and

$$\nabla \cdot \left[\mathbf{E}_{\parallel}(\omega_i) + 4\pi\mathbf{P}_{\parallel}^{(1)}(\omega_i) + 4\pi\mathbf{P}_{\parallel}^{(2)}(\omega_i)\right] = 0 \tag{6.1}$$

where $\mathbf{P}(\omega_i) = \mathbf{P}_{\parallel}(\omega_i) + \mathbf{P}_{\perp}(\omega_i)$, $\mathbf{P}^{(2)}(\omega_1) = \chi^{(2)}(\omega_1 = -\omega_2 + \omega_3):\mathbf{E}^*(\omega_2)$ $\mathbf{E}(\omega_3)$, $\mathbf{P}^{(2)}(\omega_2) = \chi^{(2)}(\omega_2 = \omega_3 - \omega_1):\mathbf{E}(\omega_3)\mathbf{E}^*(\omega_1)$, and $\mathbf{P}^{(2)}(\omega_3) = \chi^{(2)}(\omega_3 = \omega_1 + \omega_2):\mathbf{E}(\omega_1)\mathbf{E}(\omega_2)$. The general solution of (6.1) with boundary conditions is extremely complicated. Fortunately, in real situations, reasonable approximations often can be made to simplify the solution. To illustrate, consider a simple case with the following assumptions: (1) all waves are infinite plane waves; (2) depletion of energy from the pump waves can be neglected; (3) the nonlinear medium is semi-infinite with a plane boundary surface; (4) the nonlinear medium is cubic, or the beams are propagating along a symmetry axis. These assumptions are of course not essential, and in the appropriate circumstances can be relaxed, as we shall discuss later.

These assumptions drastically simplify the solution. Negligible depletion of pump field energy means that the nonlinear polarization terms responsible for wave coupling and energy transfer in the equations for $E(\omega_1)$ and $E(\omega_2)$ can be neglected. Thus the pump waves propagate essentially linearly in the nonlinear medium with $E(\omega_1)$ and $E(\omega_2)$ governed by the linear wave equations. In the infinite plane wave approximation we have in the nonlinear medium $E_T(\omega_1) = \mathscr{E}_{1T}\exp[i(\mathbf{k}_{1T} \cdot \mathbf{r} - \omega_1 t)]$ and $E_T(\omega_2) = \mathscr{E}_{2T}\exp[i(\mathbf{k}_{2T} \cdot \mathbf{r} - \omega_2 t)]$. The only equations left to be solved are those for $E(\omega_3)$ in (6.1) with $P^{(2)}(\omega_3) = \mathscr{P}_3^{(2)}\exp[i(\mathbf{k}_{3s} \cdot \mathbf{r} - \omega_3 t)]$ and $\mathbf{k}_{3s} \equiv \mathbf{k}_{1T} + \mathbf{k}_{2T}$. The solution for $E(\omega_3)$ in the medium is straightforward. It comprises a homogeneous solution (a free wave with wavevector $\mathbf{k}_{3T}$) and a particular solution (a driven wave with wavevector $\mathbf{k}_{3s}$),

$$E_T(\omega_3) = \left\{ A e^{i\mathbf{k}_{3T} \cdot \mathbf{r}} + \left[\frac{4\pi\omega_3^2}{c^2\left(k_{3s}^2 - k_{3T}^2\right)}\mathscr{P}_{3\perp}^{(2)} - \frac{4\pi\mathscr{P}_{3\|}^{(2)}}{\varepsilon_\|(\omega_3)} \right] e^{i\mathbf{k}_{3s} \cdot \mathbf{r}} \right\} e^{-i\omega_3 t}$$

$$(6.2)$$

where the amplitude A of the homogeneous solution is a coefficient to be determined from the boundary conditions, and we assume $E_{T,\|}(\omega_3) + 4\pi P_\|^{(1)}(\omega_3) = \varepsilon_\|(\omega_3)E_{T,\|}(\omega_3)$.

We now give a more complete description of the problem including the boundary conditions.[4] Let $z = 0$ be the boundary plane separating the semi-infinite nonlinear medium on the right and a linear medium on the left. All waves are propagating in the x–z plane with wavevectors described in Fig. 6.2. For each ω_i, there exists in the linear medium an incoming field $E_I(\omega_i)$ from the left and a reflected field $E_R(\omega_i)$ to the left, and in the nonlinear medium a transmitted field $E_T(\omega_i)$ to the right. They are related to one another by the boundary conditions. An immediate consequence of matching of the field components at the boundary is that at each ω_i, all the wavevector components parallel to the boundary surface must be equal. This leads to Snell's law of reflection and refraction for the pump waves. For the sum-frequency wave, we have

$$k_{3I,x} = k_{3R,x} = k_{3T,x} = k_{3s,x}.$$

$$(6.3)$$

In terms of the propagation angles described in Fig. 6.2, this relation becomes

$$\begin{aligned} k_{3I}\sin\theta_{3I} = k_{3R}\sin\theta_{3R} &= k_{3T}\sin\theta_{3T} \\ &= k_{3s}\sin\theta_{3s} \\ &= k_{1T}\sin\theta_{1T} + k_{2T}\sin\theta_{2T} \\ &= k_{1I}\sin\theta_{1I} + k_{2I}\sin\theta_{2I}. \end{aligned}$$

$$(6.4)$$

Equation (6.4) can be regarded as the nonlinear Snell law. When the wavevec-

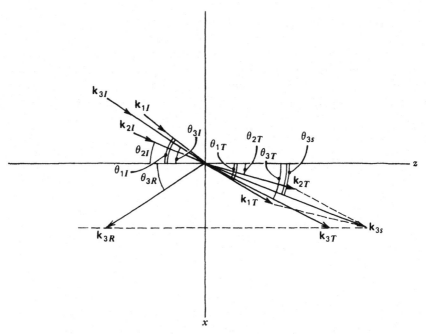

Fig. 6.2 Description of wavevectors of various waves involved in sum-frequency generation in a semi-infinite nonlinear medium with a boundary surface at $z = 0$.

tors of the incoming pump waves are known, it determines the propagation directions of the nonlinearly generated output waves.[4] To complete the solution, one must also find the amplitudes of the output waves.

In (6.2), $\mathscr{P}_3^{(2)}(\omega_3) = \chi^{(2)} : \mathscr{E}_{1T}\mathscr{E}_{2T}$. For a given nonlinear medium, $\chi^{(2)}$ is prescribed, and $\mathscr{E}_{1T}$ and $\mathscr{E}_{2T}$ are related to the incoming pump field amplitudes by the Fresnel coefficients. The only unknown in the expression of $E_T(\omega_3)$ in (6.2) is the coefficient A. Then we should also consider the incoming and reflected waves at ω_3, described by $E_I(\omega_3)$ and $E_R(\omega_3)$, respectively, in the linear medium

$$E_I(\omega_3) = \mathscr{E}_{3I}e^{i(\mathbf{k}_{3I}\cdot\mathbf{r} - \omega_3 t)}$$

and

$$E_R(\omega_3) = \mathscr{E}_{3R}e^{i(\mathbf{k}_{3R}\cdot\mathbf{r} - \omega_3 t)}. \tag{6.5}$$

The incoming field amplitude $\mathscr{E}_{3I}$ is given, but the reflected field amplitude $\mathscr{E}_{3R}$ is to be determined. Thus there are two unknown coefficients, A and $\mathscr{E}_{3R}$, to be fixed by the boundary conditions. Clearly, the requirement that both electric and magnetic field components parallel to the surface must be continuous provides enough relations to solve for A and $\mathscr{E}_{3R}$. We postpone the solution to a later section, considering first the case of sum-frequency generation in the bulk.

6.3   SIMPLE SOLUTION OF BULK SUM-FREQUENCY GENERATION

We are interested here only in sum-frequency generation in the bulk of a nonlinear medium, as described by the growth of $\mathbf{E}_T(\omega_3)$ along $\hat{z}$ in Fig. 6.2. Since the growth of the sum-frequency field is generally insignificant over a distance of a wavelength, the slowly varying amplitude approximation discussed in Section 3.3 is applicable here. With $\mathbf{E}_T(\omega_3) = \mathscr{E}_{3T}(z)\exp[i(\mathbf{k}_{3T} \cdot \mathbf{r} - \omega_3 t)]$, (6.1) then takes the form

$$\frac{\partial}{\partial z}\mathscr{E}_{3T,\perp}(z) = \frac{i2\pi\omega_3^2}{k_{3T,z}c^2}\mathscr{P}_{3\perp}^{(2)}e^{i\,\Delta k\,z}$$

$$\frac{\partial}{\partial z}\left[\varepsilon_\parallel(\omega_3)\mathscr{E}_{3T,\parallel} + 4\pi\mathscr{P}_{3\parallel}^{(2)}e^{i\,\Delta k\,z}\right] = 0$$

(6.6)

where

$$\Delta\mathbf{k} = \hat{z}\,\Delta k = \mathbf{k}_{1T} + \mathbf{k}_{2T} - \mathbf{k}_{3T}$$

(6.7)

is the phase mismatch. Solution of (6.6) yields

$$\mathscr{E}_{3T,\perp}(z) = \mathscr{E}_{3T,\perp}(0) + \frac{2\pi\omega_3^2}{c^2 k_{3T,z}\Delta k}\mathscr{P}_{3\perp}^{(2)}(e^{i\,\Delta k\,z} - 1)$$

$$\mathscr{E}_{3T,\parallel}(z) = \mathscr{E}_{3T,\parallel}(0) - \frac{4\pi\mathscr{P}_{3\parallel}^{(2)}}{\varepsilon_\parallel(\omega_3)}(e^{i\,\Delta k\,z} - 1).$$

(6.8)

As a further approximation, we may neglect the effect of nonlinear polarization on reflection and refraction at the boundary. Then, $\mathscr{E}_{3T}(0)$ is directly related to the incoming field $\mathscr{E}_{3I}(0)$ through the Fresnel coefficients.

The intensity of the generated sum-frequency wave at z is given by

$$I_3(z) = \frac{c\sqrt{\varepsilon(\omega_3)}}{2\pi}|\mathscr{E}_{3T}(z)|^2.$$

(6.9)

The corresponding output power is obtained from the integration of I_3 over the beam cross section. Here, the finite beam cross section seems to be in conflict with the infinite plane wave assumption, but as is well known, if the beam cross section is sufficiently large, then the ray approximation is valid, and each ray can be treated as an infinite plane wave. Thus with I_3 depending on the transverse coordinate ρ, the total output power at ω_3 is

$$\mathbb{P}(\omega_3, z) = \int I_3(z, \rho)\,d\rho$$

$$= \frac{c\sqrt{\varepsilon(\omega_3)}}{2\pi}\int|\mathscr{E}_{3T}(z, \rho)|^2\,d\rho.$$

(6.10)

A case of practical interest occurs in the absence of an input at ω_3, that is, $\mathscr{E}_{3I} = 0$. In the present approximation, $\mathscr{E}_{3T}(0)$ also vanishes. Then for $|k_3/\Delta k| \gg 1$, we have $|\mathscr{E}_{3T,\|}| \ll |\mathscr{E}_{3T,\perp}|$, and the intensity I_3 following (6.8) and (6.9) takes the form

$$I_3(z) = \frac{2\pi\omega_3^2}{c\sqrt{\varepsilon(\omega_3)}\cos^2\theta_{3T}}|\mathscr{P}_{3\perp}^{(2)}|^2\left[\frac{\sin(\Delta k\,z/2)}{\Delta k\,z/2}\right]^2 z^2. \tag{6.11}$$

As shown in Fig. 6.3, I_3 versus $\Delta k_z z$ given here peaks strongly at phase-matching $\Delta k_z = 0$. The peak value is

$$[I_3(z)]_{max} = \frac{2\pi\omega_3^2}{c\sqrt{\varepsilon(\omega_3)}\cos^2\theta_{3T}}|\mathscr{P}_{3\perp}^{(2)}|^2 z^2 \tag{6.12}$$

and the half-width between the first zeroes is

$$(\Delta k)_{HW}z = \frac{(\Delta k)_{HW}}{k_{3T}}k_{3T}z = 2\pi. \tag{6.13}$$

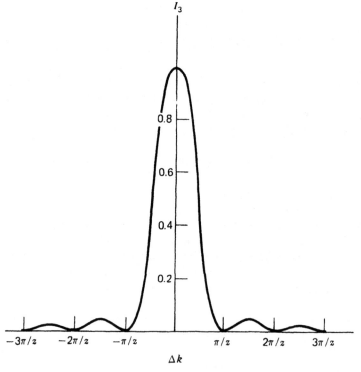

Fig. 6.3   Sum-frequency output as a function of the phase mismatch Δk near $\Delta k \sim 0$.

For $z = 1$ cm, $k_{3T} \sim 10^5$ cm^{-1} in a typical example, we find $(\Delta k)_{HW}/k_{3T} \sim 10^{-4}$, which indicates that in terms of Δk, the phase-matching peak is often extremely narrow.

The calculation of sum-frequency generation in anisotropic media requires slight modification. First, since $\mathbf{P}_\parallel^{(2)}$ in the second equation of (6.1) is usually negligible, $E_{T,\parallel}(\omega_i)/E_{T,\perp}(\omega_i) = \tan \alpha_i$ is a constant determined from linear wave propagation, where α_i is the angle between $\mathbf{E}_T(\omega_i)$ and $\mathbf{E}_{T,\perp}(\omega_i)$. With the infinte plane wave approximation, and the slowly varying amplitude approximation, (6.1) becomes

$$\frac{\partial}{\partial z} \mathscr{E}_{3T}(z) = \frac{i 2\pi \omega_3^2}{k_{3T,z} c^2 \cos^2 \alpha_3} \hat{e}_{3T} \cdot \mathscr{P}_3^{(2)} e^{i \Delta k z} \tag{6.14}$$

where $\hat{e}_{3T}$ is the unit vector along $\mathscr{E}_{3T}$. The solution of (6.14) is

$$\mathscr{E}_{3T}(z) = \mathscr{E}_{3T}(0) + \frac{2\pi \omega_3^2}{k_{3T,z} \Delta k c^2 \cos^2 \alpha_3} \hat{e}_{3T} \cdot \mathscr{P}_3^{(2)} (e^{i \Delta k z} - 1) \tag{6.15}$$

Within the range of our approximation, (6.15) is consistent with (6.8) for the isotropic medium.

6.4 SOLUTION WITH BOUNDARY REFLECTION

In the more general solution of (6.2) and (6.5), $\mathbf{E}_T(\omega_3)$ can be rewritten in the form

$$\mathbf{E}_T(\omega_3) = \mathscr{E}_{3T}(z) e^{i(\mathbf{k}_{3T} \cdot \mathbf{r} - \omega_3 t)}$$

$$\mathscr{E}_{3T,\perp} = \mathscr{E}_{3T,\perp}(0) + \frac{4\pi \omega_3^2}{c^2 (k_{3s}^2 - k_{3T}^2)} \mathscr{P}_{3\perp}^{(2)} (e^{i \Delta k z} - 1) \tag{6.16}$$

$$\mathscr{E}_{3T,\parallel} = \mathscr{E}_{3T,\parallel}(0) - \frac{4\pi \mathscr{P}_{3\parallel}^{(2)}}{\varepsilon_\parallel(\omega_3)} (e^{i \Delta k z} - 1)$$

with $\mathscr{E}_{3T,\perp}(0) = \mathbf{A} + 4\pi \omega_3^2 \mathscr{P}_{3\perp}^{(2)}/c^2 (k_{3s}^2 - k_{3T}^2)$ and $\mathscr{E}_{3T,\parallel}(0) = -4\pi \mathscr{P}_{3\parallel}^{(2)}/\varepsilon_\parallel(\omega_3)$. We then notice immediately that if the approximation

$$k_{3s}^2 - k_{3T}^2 = (k_{3s,z} + k_{3T,z}) \Delta k$$
$$\simeq 2 k_{3T,z} \Delta k \tag{6.17}$$

is used, (6.16) reduces to the simplified solution in (6.8). The above approximation is excellent when Δk is small, or equivalently, when the output in the backward direction with $\Delta k \sim k_{3T}$ can be neglected. As pointed out in Section 3.3, the latter is just what the slowly varying amplitude approximation means.

In finding $\mathscr{E}_{3T}(0)$, however, the more correct solution should include the effect of nonlinear polarization on boundary reflection of the sum-frequency wave. By requiring the tangential components of electric and magnetic fields be matched at the boundary $z = 0$ (see Fig. 6.4), we find[4]

$$\mathscr{E}_{3T,y}(0) = A_y + \frac{4\pi\omega_3^2\mathscr{P}_{3y}^{(2)}}{c^2\left(k_{3s}^2 - k_{3T}^2\right)}$$

$$= \mathscr{E}_{3I,y} + \mathscr{E}_{3R,y}$$

$$k_{3T}\cos\theta_{3T}A_y + k_{3s}\cos\theta_{3s}\frac{4\pi\omega_3^2\mathscr{P}_{3y}^{(2)}}{c^2\left(k_{3s}^2 - k_{3T}^2\right)} = k_{3R}\cos\theta_{3R}\left(\mathscr{E}_{3I,y} - \mathscr{E}_{3R,y}\right)$$

$$\mathscr{E}_{3T,x}(0) = A_h\cos\theta_{3T} + \frac{4\pi\omega_3^2\mathscr{P}_{3\perp,h}^{(2)}\cos\theta_{3s}}{c^2\left(k_{3s}^2 - k_{3T}^2\right)}$$

$$\qquad\qquad - \frac{4\pi\mathscr{P}_{3\parallel}^{(2)}\sin\theta_{3s}}{\varepsilon_\parallel(\omega_3)} \qquad\qquad (6.18)$$

$$= \left(\mathscr{E}_{3I,h} - \mathscr{E}_{3R,h}\right)\cos\theta_{3R}$$

$$k_{3T}A_h + k_{3s}\frac{4\pi\omega_3^2\mathscr{P}_{3\perp,h}^{(2)}}{c^2\left(k_{3s}^2 - k_{3T}^2\right)} = k_{3R}\left(\mathscr{E}_{3I,h} + \mathscr{E}_{3R,h}\right)$$

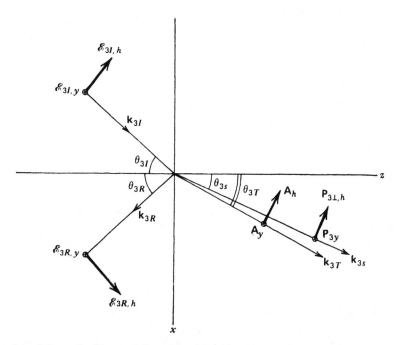

Fig. 6.4 Schematic diagram describing the incoming and reflected sum-frequency waves in the linear medium and the transmitted sum-frequency waves in the nonlinear medium. The boundary surface between the two media is at $z = 0$.

where the subscript h denotes components in the plane of incidence. This set of four equations can then be used to find the four unknown coefficients A_y, A_h, $\mathscr{E}_{3R,y}$, and $\mathscr{E}_{3R,h}$. The result is

$$
A_y = \frac{1}{k_{3T}\cos\theta_{3T} + k_{3R}\cos\theta_{3R}}
$$

$$
\times \left[2k_{3R}\cos\theta_{3R}\mathscr{E}_{3I,y} \right.
$$

$$
\left. - \frac{(k_{3s}\cos\theta_{3s} + k_{3R}\cos\theta_{3R})4\pi\omega_3^2\mathscr{P}_{3y}^{(2)}}{c^2(k_{3s}^2 - k_{3T}^2)} \right],
$$

$$
A_h = \frac{1}{k_{3R}\cos\theta_{3T} + k_{3T}\cos\theta_{3R}}
$$

$$
\times \left[2k_{3R}\cos\theta_{3R}\mathscr{E}_{3I,h} + \frac{4\pi k_{3R}\mathscr{P}_{3\parallel}^{(2)}\sin\theta_{3s}}{\varepsilon_\parallel(\omega_3)} \right. \tag{6.19}
$$

$$
\left. - \frac{(k_{3R}\cos\theta_{3s} + k_{3s}\cos\theta_{3R})4\pi\omega_3^2\mathscr{P}_{3\perp,h}^{(2)}}{c^2(k_{3s}^2 - k_{3T}^2)} \right],
$$

$$
\mathscr{E}_{3R,y} = \frac{1}{k_{3T}\cos\theta_{3T} + k_{3R}\cos\theta_{3R}}
$$

$$
\times \left[(k_{3R}\cos\theta_{3R} - k_{3T}\cos\theta_{3T})\mathscr{E}_{3I,y} \right.
$$

$$
\left. - \frac{(k_{3s}\cos\theta_{3s} - k_{3T}\cos\theta_{3T})4\pi\omega_3^2\mathscr{P}_{3y}^{(2)}}{c^2(k_{3s}^2 - k_{3T}^2)} \right],
$$

and

$$
\mathscr{E}_{3R,h} = \frac{1}{k_{3R}\cos\theta_{3T} + k_{3T}\cos\theta_{3R}}
$$

$$
\times \left[(k_{3T}\cos\theta_{3R} - k_{3R}\cos\theta_{3T})\mathscr{E}_{3I,h} \right.
$$

$$
\left. - \frac{(k_{3T}\cos\theta_{3s} - k_{3s}\cos\theta_{3T})4\pi\omega_3^2\mathscr{P}_{3\perp,h}^{(2)}}{c^2(k_{3s}^2 - k_{3T}^2)} + \frac{4\pi k_{3T}\mathscr{P}_{3\parallel}^{(2)}\sin\theta_{3s}}{\varepsilon_\parallel(\omega_3)} \right].
$$

With $\mathbf{A}$ and $\mathscr{E}_{3R}$ known, the solution in (6.2) and (6.5) is then complete. It shows that even in the absence of an input, $\mathscr{E}_{3I} = 0$, both $\mathscr{E}_{3R}$ and $\mathscr{E}_{3T}(0)$ are nonvanishing because of the nonlinear polarization effect on reflection and refraction. In fact, the reflected sum-frequency wave is easily detectable.[5] It can be shown that $\mathscr{E}_{3T}(0) = 0$ and $\mathscr{E}_{3R}$ is about kz times smaller than $\mathscr{E}_{3T}(z)$ at phase matching. Thus the reflected sum-frequency wave is essentially generated by the nonlinear polarization in a surface layer of the order of $\lambda/2\pi$ thick. With some modification, the solution here for a cubic nonlinear medium can be extended to anisotropic media.

6.5 PHASE-MATCHING CONSIDERATIONS

As shown in Fig. 6.3, the bulk sum-frequency generation is strong only when $|\Delta k z| \lesssim 1$. The phase mismatch, Δk, defines a coherent length, $l_{coh} = 1/\Delta k$. If the length of the medium l is below l_{coh}, the sum-frequency output increases more or less quadratically with l. If $l > l_{coh}$, the output tends to saturate and may even decrease as l increases. For efficient sum-frequency generation, we must therefore have l_{coh} sufficiently long, of the order of at least a few millimeters in practice.

In actual experiments, to avoid reduction of the effective beam interaction length due to finite cross sections, collinear phase matching is required:

$$\Delta k = k_{1T} + k_{2T} - k_{3T} = 0. \tag{6.20}$$

In terms of the refractive indices $n(\omega_i)$, (6.20) can be written as

$$\omega_1[n(\omega_3) - n(\omega_1)] + \omega_2[n(\omega_3) - n(\omega_2)] = 0. \tag{6.21}$$

Clearly, for isotropic or cubic materials with normal dispersion $n(\omega_3) > \{n(\omega_1), n(\omega_2)\}$, this relation can never be satisfied. Therefore, collinear phase matching can be achieved only with (1) anomalous dispersion or (2) birefringent crystals.[2] In the latter case, the medium should be a negative uniaxial crystal with $n_e(\omega_i) < n_0(\omega_i)$. By choosing the wave at ω_3 to be extraordinary, it is possible to find $[n_e(\omega_3) - n(\omega_1)]$ and $[n_e(\omega_3) - n(\omega_2)]$ with opposite signs so that (6.21) can be satisfied. Two types of collinear phase matching are commonly used. In Type I, both $n(\omega_1)$ and $n(\omega_2)$ are ordinary or extraordinary, while in Type II, either $n(\omega_1)$ or $n(\omega_2)$ is ordinary.

6.6 EFFECT OF ABSORPTION

Absorption is detrimental to sum-frequency generation since it limits the effective interaction length to roughly the attenuation length. It also broadens the phase-matching peak and lowers the peak value. This can be seen by including absorption in the derivation in Section 6.4. With absorption, the wavevectors become complex: $k = k' + i\beta$, where β is the attenuation coefficient. Equation (6.14) changes into

$$\left(\frac{\partial}{\partial z} + \beta_{3T}\right)\mathscr{E}_{3T}(z) = \frac{i2\pi\omega_3^2}{c^2 k'_{3T,z}\cos^2\alpha_3}\hat{e}_{3T} \cdot \mathscr{P}_3^{(2)} e^{i\,\Delta k'z - (\beta_{1T} + \beta_{2T})z} \tag{6.22}$$

with

$$\mathbf{E}_T(\omega_3) = \mathscr{E}_{3T}(z)e^{i(k'_{3T}z - \omega_3 t)}.$$

The resulting solution is

$$\mathscr{E}_{3T}(z) = \mathscr{E}_{3T}(0)e^{-\beta_{3T}z} + \frac{i2\pi\omega_3^2\hat{e}_{3T}\cdot\mathscr{P}_3^{(2)}}{c^2 k'_{3T,z}\cos^2\alpha_3\left[i\Delta k' - (\beta_{1T} + \beta_{2T} - \beta_{3T})\right]}$$

$$\times\left[e^{i\,\Delta k'z - (\beta_{1T} + \beta_{2T})z} - e^{-\beta_{3T}z}\right]. \tag{6.23}$$

If the absorption at either the pump frequencies ω_1 and ω_2 or the output frequency ω_3 is appreciable so that

$$|e^{i\Delta k'z - (\beta_{1T} + \beta_{2T})z} - e^{-\beta_{3T}z}|^2 \simeq 1,$$

then the output intensity can be approximated by

$$I_3(z) = \frac{2\pi\omega_3^2|\hat{e}_3\cdot\mathscr{P}_3^{(2)}|^2}{c\sqrt{\varepsilon(\omega_3)}\cos^2\theta_{3T}\cos^4\alpha_3\left[(\Delta k')^2 + \beta^2\right]} \tag{6.24}$$

where $\beta = \beta_{1T} + \beta_{2T}$ with $\beta_{3T} \sim 0$ or $\beta = \beta_{3T}$ with $\beta_{1T} + \beta_{2T} \sim 0$. The phase-matching curve I_3 versus Δk now takes on a Lorentzian shape with a half-width β. Compared with the zero absorption case, the peak value here is independent of z and is reduced by a factor of $1/\beta^2 z^2$. This shows that with absorption, the effective interaction length is reduced to $1/\beta$, which is just the attenuation length. When both $(\beta_{1T} + \beta_{2T})$ and β_{3T} are appreciable, the output intensity even decreases exponentially with z.

6.7 SUM-FREQUENCY GENERATION WITH HIGH CONVERSION EFFICIENCY

We saw in earlier sections that at perfect phase matching, the output power of sum-frequency generation in a nonabsorbing bulk medium is proportional to l^2, the square of the length of the medium. Then as $l \to \infty$, the output power would increase without limit, in violation of energy conservation. This is the consequence of the assumption of negligible pump power depletion, which is not valid when the output becomes significant compared to the pump. The set of three coupled equations in (3.4) or (6.1) must now be solved together to find a complete solution.

For sum-frequency generation with high conversion efficiency, the following conditions usually exist: (1) the coupled waves are collinearly phase matched; (2) the medium is nearly lossless; and (3) the slowly varying amplitude approximation is valid. The coupled equations can therefore be written as [similar to (6.14)]

$$\frac{\partial \mathscr{E}_{1T}}{\partial z} = \frac{i\omega_1^2}{k_{1T,z}\cos^2\alpha_1} K_1 \mathscr{E}_{2T}^* \mathscr{E}_{3T},$$

$$\frac{\partial \mathscr{E}_{2T}}{\partial z} = \frac{i\omega_2^2}{k_{2T,z}\cos^2\alpha_2} K_2 \mathscr{E}_{1T}^* \mathscr{E}_{3T},$$

(6.25)

and

$$\frac{\partial \mathscr{E}_{3T}^*}{\partial z} = \frac{-i\omega_3^2}{k_{3T,z}\cos^2\alpha_3} K_3 \mathscr{E}_{1T}^* \mathscr{E}_{2T}^*$$

with

$$K_1 = \frac{2\pi}{c^2} \hat{e}_{1T} \cdot \chi^{(2)}(\omega_1 = -\omega_2 + \omega_3) : \hat{e}_{2T}\hat{e}_{3T},$$

$$K_2 = \frac{2\pi}{c^2} \hat{e}_{2T} \cdot \chi^{(2)}(\omega_2 = \omega_3 - \omega_1) : \hat{e}_{3T}\hat{e}_{1T},$$

and

$$K_3 = \frac{2\pi}{c^2} \hat{e}_{3T} \cdot \chi^{(2)}(\omega_3 = \omega_1 + \omega_2)^* : \hat{e}_{1T}\hat{e}_{2T}.$$

From the permutation symmetry of $\chi^{(2)}$ in a lossless medium discussed in Section 2.5, we find $K_1 = K_2 = K_3 \equiv K$. Equation (6.25) can be solved exactly.[3] First, we can easily show from (6.25) that the total power flow W in the medium,

$$W = \frac{c^2}{2\pi} \left[\frac{k_{1T,z}\cos^2\alpha_1|\mathscr{E}_{1T}|^2}{\omega_1} + \frac{k_{2T,z}\cos^2\alpha_2|\mathscr{E}_{2T}|^2}{\omega_2} + \frac{k_{3T,z}\cos^2\alpha_3|\mathscr{E}_{3T}|^2}{\omega_3} \right],$$

(6.26)

is a constant independent of z. This is also known as the Manley–Rowe relation.[6] Then the number of photons created at ω_3 must be equal to the numbers of photons annihilated at ω_1 and ω_2:

$$\sqrt{\varepsilon_1} \frac{|\mathscr{E}_{1T}(0)|^2 - |\mathscr{E}_{1T}(z)|^2}{\hbar\omega_1} = \sqrt{\varepsilon_2} \frac{|\mathscr{E}_{2T}(0)|^2 - |\mathscr{E}_{2T}(z)|^2}{\hbar\omega_2}$$

$$= \sqrt{\varepsilon_3} \frac{|\mathscr{E}_{3T}(z)|^2 - |\mathscr{E}_{3T}(0)|^2}{\hbar\omega_3}.$$

(6.27)

In solving (6.25), we define

$$u_1 e^{i\phi_1} = \left(\frac{c^2 k_{1T,z} \cos^2 \alpha_1}{2\pi\omega_1^2 W} \right)^{1/2} \mathcal{E}_{1T}(z),$$

$$u_2 e^{i\phi_2} = \left(\frac{c^2 k_{2T,z} \cos^2 \alpha_2}{2\pi\omega_2^2 W} \right)^{1/2} \mathcal{E}_{2T}(z),$$

$$u_3 e^{i\phi_3} = \left(\frac{c^2 k_{3T,z} \cos^2 \alpha_3}{2\pi\omega_3^2 W} \right)^{1/2} \mathcal{E}_{3T}(z),$$

$$\theta(z) = \phi_3(z) - \phi_1(z) - \phi_2(z),$$

$$\zeta = K \left[\frac{2\pi W \omega_1^2 \omega_2^2 \omega_3^2}{c^2 k_{1T,z} k_{2T,z} k_{3T,z} \cos^2 \alpha_1 \cos^1 \alpha_2 \cos^2 \alpha_3} \right]^{1/2} z,$$

$$m_1 = u_2^2(0) + u_3^2(0) = u_2^2 + u_3^2,$$

$$m_2 = u_3^2(0) + u_1^2(0) = u_3^2 + u_1^2,$$

and

$$m_3 = u_1^2(0) - u_2^2(0) = u_1^2 - u_2^2.$$

Equation (6.25) becomes

$$\frac{du_1}{d\zeta} = -u_2 u_3 \sin\theta,$$

$$\frac{du_2}{d\zeta} = -u_3 u_1 \sin\theta, \tag{6.29}$$

$$\frac{du_3}{d\zeta} = u_1 u_2 \sin\theta,$$

and

$$\frac{d\theta}{d\zeta} = K \cot\theta \frac{d}{d\zeta} \ln(u_1 u_2 u_3).$$

The last equation in (6.29) can be integrated to yield

$$u_1 u_2 u_3 \cos\theta = \Gamma$$

where Γ is a constant independent of z. Then by eliminating $\sin\theta$ in (6.29), we find

$$\frac{du_1^2}{d\zeta} = \frac{du_2^2}{d\zeta} = -\frac{du_3^2}{d\zeta} = \pm 2\left[u_1^2 u_2^2 u_3^2 - \Gamma^2 \right]^{1/2}. \tag{6.30}$$

The choice of sign " + " or " – " in (6.30) depends on the initial value of θ. The general solution of (6.30) is given in Ref. 3.

We consider here the frequently encountered case where the boundary condition is $\mathcal{E}_{3T}(0) = 0$, or $u_3(0) = 0$, which leads to $\Gamma = 0$ and $\theta = \pi/2$. Equation (6.30) becomes

$$\frac{du_3^2}{d\zeta} = 2\left[u_3^2(m_1 - u_3^2)(m_2 - u_3^2)\right]^{1/2}. \tag{6.31}$$

The solution takes the form of a Jacobi elliptical integral

$$
\begin{aligned}
\zeta &= \frac{1}{u_1(0)}\left[\operatorname{sn}^{-1}\left(\frac{u_3}{u_2(0)}, \gamma\right) - \operatorname{sn}^{-1}\left(\frac{u_3(0)}{u_2(0)}, \gamma\right)\right] \\
&= \frac{1}{u_1(0)}\int_{u_3(0)/u_2(0)}^{u_3/u_2(0)} \frac{dy}{\left[(1 - y^2)(1 - \gamma^2 y^2)\right]^{1/2}}
\end{aligned}
\tag{6.32}
$$

assuming $m_2 = u_1^2(0) > m_1 = u_2^2(0)$ and $\gamma = u_2(0)/u_1(0)$. From (6.32) and (6.28), we find the intensities of the three waves as

$$
\begin{aligned}
u_3^2(\zeta) &= u_2^2(0)\operatorname{sn}^2\left[u_1(0)\zeta, \gamma\right], \\
u_2^2(\zeta) &= u_2^2(0) - u_2^2(0)\operatorname{sn}^2\left[u_1(0)\zeta, \gamma\right],
\end{aligned}
\tag{6.33}
$$

and

$$u_1^2(\zeta) = u_1^2(0) - u_2^2(0)\operatorname{sn}^2\left[u_1(0)\zeta, \gamma\right].$$

The elliptical function $\operatorname{sn}^2[u_1(0)\zeta, \gamma]$ is periodic in ζ with a period

$$L = \frac{2}{u_1(0)}\int_0^1 \frac{dy}{\left[(1 - y^2)(1 - \gamma^2 y^2)\right]^{1/2}}. \tag{6.34}$$

Physically, this indicates that as the interaction length increases, energy is transferred back and forth between the wave at ω_2 and the waves at ω_1 and ω_2 with a period L. While the process first pumps energy into the sum-frequency field, it reverses the energy flow after photons in one of the pump waves are depleted.

A simple case of physical interest is the up-conversion process used, for example, to convert an infrared image to the visible. It often occurs with $u_1(0) \gg u_2(0)$ and $u_3(0) = 0$. Since $\gamma \ll 1$, the elliptical integral of (6.32) reduces to a simple form and we find

$$
\begin{aligned}
u_3^2(\zeta) &= u_2^2(0)\sin^2\left[u_1(0)\zeta\right], \\
u_2^2(\zeta) &= u_2^2(0) - u_2^2(0)\sin^2\left[u_1(0)\zeta\right],
\end{aligned}
\tag{6.35}
$$

and

$$u_1^2(\zeta) = u_1^2(0) - u_2^2(0)\sin^2\left[u_1(0)\zeta\right] \simeq u_1^2(0).$$

They are plotted in Fig. 6.5, which shows explicitly the periodic variation of energy flow back and forth between the waves at ω_2 and ω_3. In this case, the depletion of the pump field at ω_1 is negligible. Therefore, the solution in (6.35) can also be obtained from (6.25) by letting $\mathscr{E}_{1T}$ be a constant.

Another case of interest is when $u_1(0) = u_2(0)$ so that $\gamma = 1$. The solution becomes

$$u_3^2(\zeta) = u_1^2(0)\tanh^2\left[u_1(0)\zeta\right],$$
$$u_1^2(\zeta) = u_2^2(\zeta) = u_1^2(0)\text{sech}^2\left[u_1(0)\zeta\right]. \tag{6.36}$$

It shows that the period of interaction L is infinite. As $\zeta \to \infty$, we have $u^3(\zeta) \to u_1(0)$ and $u_1(\zeta) = u_2(\zeta) \to 0$. This applies to the case of second harmonic generation, which we discuss in more detail in Chapter 7.

The foregoing discussion is based on the assumption of infinite plane waves. In reality, the beam cross sections are finite with intensity variation over the transverse profile. Accordingly, the results here have to be modified, using, for example, the ray approximation. As a result, complete depletion of photons in any beam is impossible. Focused beams are often used in actual experiments to increase the laser intensity, and the theoretical treatment of the problem

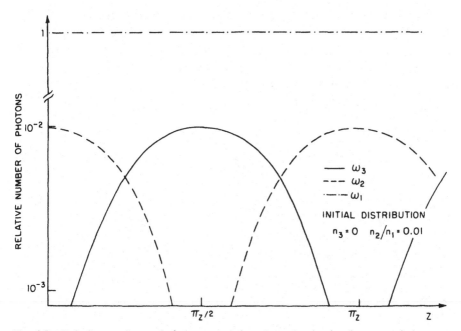

Fig. 6.5 Relative numbers of photons, as a function of z, in the three coupled waves with perfect phase matching ($\omega_3 = \omega_1 + \omega_2$, $k_3 = k_1 + k_2$) in an up-conversion process. The initial distribution of the photons in the three waves is $n_1 = 100\ n_2$ and $n_3 = 0$. (After Ref. 3.)

becomes more complicated. Boyd and Kleinman[7] have worked out the case with negligible depletion. Here we simply refer to their work and postpone our discussion on focused beams to the next chapter.

6.8   A PRACTICAL EXAMPLE

In most applications, efficient sum-frequency generation is desired. A number of rules should therefore be followed:

1   A nonlinear optical crystal with little absorption at ω_1, ω_2, and ω_3 is first chosen. It should have a sufficiently large nonlinear susceptibility $\chi^{(2)}$ and should allow collinear phase matching.

2   The phase matching directions in the crystal, generally in the form of a cone, are determined from the known refractive index tensor of the crystal.

3   The particular phase-matching direction with the appropriate set of polarizations for the three waves is selected to optimize the effective nonlinear susceptibility $\chi_{\text{eff}}^{(2)} = \hat{e}_3 \cdot \chi^{(2)} : \hat{e}_1 \hat{e}_2$.

4   The length of the crystal is finally chosen to give the desired conversion efficiency.

We consider here a practical example of sum-frequency generation in a KDP crystal with the pump beams at $\lambda_1 = 5320$ Å and $\lambda_2 = 6200$ Å. The sum-frequency generated is at $\lambda_3 = 2863$ Å. The ordinary refractive indices of KDP at room temperature are $n_0(\omega_1) = 1.5283$, $n_0(\omega_2) = 1.5231$, and $n_0(\omega_3) = 1.5757$. For a beam propagating in a direction at an angle $\textcircled{H}$ away from the optical axis, the extraordinary refractive index is given by

$$n_e\left(\omega_i, \textcircled{H}\right) = \frac{n_{em}(\omega_i) n_0(\omega_i)}{\left[n_0^2(\omega_i) \sin^2\textcircled{H} + n_{em}^2(\omega_i) \cos^2\textcircled{H} \right]^{1/2}}$$

with $n_{em}(\omega_1) = 1.4822$, $n_{em}(\omega_2) = 1.4783$, and $n_{em}(\omega_3) = 1.5231$. For type I phase matching, we have from (6.21)

$$n_e\left(\omega_3, \textcircled{H}\right) = \frac{\lambda_3}{\lambda_1} n_0(\omega_1) + \frac{\lambda_3}{\lambda_2} n_0(\omega_2)$$
$$= 1.5258$$

from which we can find

$$\textcircled{H} = \sin^{-1}\left[\frac{n_{em}(\omega_3)}{n_e(\omega_3)} \sqrt{\frac{n_0^2(\omega_3) - n_e^2(\omega_3)}{n_0^2(\omega_3) - n_{em}^2(\omega_3)}} \right]$$
$$= 76.6°.$$

Let the waves be propagating in a plane at an angle Φ from the X-axis of the crystal. In the X-Y-Z coordinates, the three polarization vectors are

$$\hat{e}_1 = \hat{e}_2 = (\sin\Phi, -\cos\Phi, 0)$$

and

$$\hat{e}_3 \cong \left(-\cos\left(H\right)\cos\Phi, -\cos\left(H\right)\sin\Phi, \sin\left(H\right)\right).$$

The KDP crystal has a $\bar{4}$ 2-m point group. Its nonvanishing $\chi^{(2)}$ elements are

$$\chi^{(2)}_{XYZ} = \chi^{(2)}_{YXZ} \cong \chi^{(2)}_{XZY} = \chi^{(2)}_{YZX} = 2.6 \times 10^{-9} \text{ esu}$$

and

$$\chi^{(2)}_{ZXY} = \chi^{(2)}_{ZYX} = 2.82 \times 10^{-9} \text{ esu}.$$

The effective nonlinear susceptibility* for type I phase matching is

$$\chi^{(2)}_{\text{eff}} = \hat{e}_3 \cdot \boldsymbol{\chi}^{(2)} : \hat{e}_1\hat{e}_2 = -\chi^{(2)}_{ZXY}\sin\left(H\right)\sin 2\Phi$$

$$= -2.74 \times 10^{-9}\sin 2\Phi \text{ esu}.$$

To optimize $|\chi^{(2)}_{\text{eff}}|$, we must choose $\Phi = 45°$. Finally, in the limit of negligible depletion of pump power, the output power is, following (6.12), given by

$$\mathbb{P}_3 = \frac{8\pi^3\omega_3^2}{c^3\sqrt{\varepsilon(\omega_1)\varepsilon(\omega_2)\varepsilon(\omega_3)}}|\chi^{(2)}_{\text{eff}}|^2 z^2\left(\frac{\mathbb{P}_1\mathbb{P}_2}{A}\right)$$

$$= 4 \times 10^{-4}z^2\left(\frac{\mathbb{P}_1\mathbb{P}_2}{A}\right) \text{ MW},$$

where A is the beam cross section in square centimeters, z in centimeters, and we have used the approximation $\mathbb{P}_i = I_iA$ in megawatts.

6.9 LIMITING FACTORS FOR HIGH CONVERSION EFFICIENCY

As a nonlinear effect, the output power of sum-frequency generation is expected to increase with the pump intensity if the pump power is kept the same. This seems to suggest that a tighter focusing of the pump beams should

---

*The expressions of $\chi^{(2)}_{\text{eff}}$ for type I and type II phase matching for the 13 uniaxial crystal classes can be found in Ref. 8.

be used to attain higher conversion efficiency, as long as the longitudinal focal dimension (the confocal parameter) is longer than the effective interaction length. There is, however, a limit to the focusing one can use. First, too high a laser intensity leads to optical damage in the crystal. Then, the reduced beam cross section due to focusing may decrease the effective interaction length even for collinearly propagating beams. This occurs in an anisotropic crystal. For an extraordinary wave, the directions of wave propagation and ray (energy) propagation are generally different. Therefore, although the waves are propagating collinearly, the rays are not. "Walk-off" of the rays effectively decreases the interaction length.

The walk-off effect can of course be minimized if the beams are propagating in a direction along which the wavevector and ray vector are parallel. This may be achieved for sum-frequency generation in a uniaxial crystal in a plane perpendicular to the optical axis, and is known as 90° phase matching. Such phase matching has been found possible in many crystals over a certain frequency range by temperature tuning.

The poor beam quality also reduces the conversion efficiency. A multimode laser beam can be considered crudely as a beam with hot spots. The small dimension of these hot spots increases the walk-off effect and decreases the interaction length. Therefore, for high conversion efficiency, beams with TEM_{00} mode should be used.

Good crystal quality is also important for efficient sum-frequency generation. Inhomogeneity prevents perfect phase matching throughout the crystal. Since $|\Delta kz| < \pi/2$ is needed for efficient energy conversion, the tolerable fluctuation of the refractive index due to inhomogeneity is $\Delta n < \lambda/4z = 2.5 \times 10^{-5}$ for $\lambda = 1$ μm and $z = 1$ cm. This means that the requirement on the crystal quality is stringent. For the same reason, temperature uniformity throughout the crystal length is also important. For a typical case with $dn/dT \simeq 5 \times 10^{-5}$, a temperature stability of $\Delta T \leq 0.5$ K throughout the crystal is necessary to achieve $\Delta n < 2.5 \times 10^{-5}$. This discussion generally applies to all mixing processes.

REFERENCES

1 M Bass, P. A Franken, A. E Hill, C. W. Peters, and G. Weinreich. *Phys Rev Lett* 8, 18 (1962)

2 P D Maker, R W. Terhune, M. Nisenhoff, and C. M. Savage, *Phys Rev. Lett* 8, 21 (1962), J A Giordmaine, *Phys Rev Lett* 8, 19 (1962)

3 J A Armstrong, N Bloembergen, J Ducuing, and P S Pershan, *Phys Rev* 127, 1918 (1962)

4 N Bloembergen and P. S Pershan, *Phys Rev* 128, 606 (1962)

5 J Ducuing and N Bloembergen, *Phys Rev Lett* 10, 474 (1963), R K Chang and N Bloembergen, *Phys Rev.* 144, 775 (1966)

6 J M Manley and H E Rowe, *Proc IRE* 47, 2115 (1959)

7 G. D. Boyd and D. A. Kleinman, *J. Appl. Phys.* 39, 3597 (1968).

8 F. Zernike and J. E. Midwinter, *Applied Nonlinear Optics* (Wiley, New York, 1973), pp. 64–65.

BIBLIOGRAPHY

Akhmanov, S. A., and R. V. Khokhlov, *Problems of Nonlinear Optics* (Gordon and Breach, New York, 1972).

Bloembergen, N., *Nonlinear Optics* (Benjamin, New York, 1965)

Zernike, F., and J. E. Midwinter, *Applied Nonlinear Optics* (Wiley, New York, 1973).

7

Harmonic Generation

In the history of nonlinear optics, the discovery of optical harmonic generation marked the birth of the field.[1] The effect has since found wide application as a means to extend coherent light sources to shorter wavelengths. This chapter summarizes the important aspects of harmonic generation. As it is a special case of optical mixing, most of the discussion in Chapter 6 can be applied here without much modification. The application of harmonic generation to the measurements of nonlinear optical susceptibilities and to the characterization of ultrashort pulses is also discussed.

7.1 SECOND HARMONIC GENERATION

The theory of second harmonic generation follows exactly that of sum-frequency generation discussed in Chapter 6. With $\omega_1 = \omega_2 = \omega$ and $\omega_3 = 2\omega$, the derivation and results in Sections 6.2 to 6.6 can be applied directly to the present case. In particular, the plane wave approximation with negligible pump depletion yields a second harmonic output power

$$\mathbb{P}_{2\omega}(z) = \frac{32\pi^3\omega^2}{c^3\epsilon(\omega)\sqrt{\epsilon(2\omega)}}|\hat{e}_{2\omega}\cdot\boldsymbol{\chi}^{(2)}:\hat{e}_\omega\hat{e}_\omega|^2 z^2 \frac{\sin^2(\Delta k\,z/2)}{(\Delta k\,z/2)^2}\frac{\mathbb{P}_\omega^2(0)}{A}. \quad (7.1)$$

For collinear phase matching in a crystal with normal dispersion, we must have, following (6.20), either

$$n_e(2\omega) = n_0(\omega) \qquad \text{(type I)} \qquad (7.2)$$

or

$$n_e(2\omega) = \tfrac{1}{2}[n_0(\omega) + n_e(\omega)] \qquad \text{(type II)}. \qquad (7.3)$$

86

The calculation in the limit of high conversion efficiency requires slight modification. Specifically, for type I collinear phase matching, the permutation symmetry of $\chi^{(2)}$, following (2.30), gives

$$K = \frac{2\pi}{c^2}\hat{e}_{2\omega}\cdot\chi^{(2)}(2\omega = \omega + \omega):\hat{e}_\omega\hat{e}_\omega$$

$$= \frac{\pi}{c^2}\hat{e}_\omega\cdot\chi^{(2)}(\omega = -\omega + 2\omega):\hat{e}_\omega\hat{e}_{2\omega}. \tag{7.4}$$

The coupled equations of (6.24) reduce to the form[1]

$$\frac{\partial\mathscr{E}_\omega}{\partial z} = \frac{i\omega^2}{k_{\omega,z}\cos^2\alpha_\omega}2K\mathscr{E}_\omega^*\mathscr{E}_{2\omega},$$

$$\frac{\partial\mathscr{E}_{2\omega}}{\partial z} = \frac{-i(2\omega)^2}{k_{2\omega,z}\cos^2\alpha_{2\omega}}K\mathscr{E}_\omega^2. \tag{7.5}$$

The conservation of power flow and the conservation of the number of photons in (6.26) and (6.27), respectively, become

$$W = \frac{c^2}{2\pi}\left[\frac{k_{\omega,z}\cos^2\alpha_\omega|\mathscr{E}_\omega|^2}{\omega} + \frac{k_{2\omega,z}\cos^2\alpha_{2\omega}|\mathscr{E}_{2\omega}|^2}{2\omega}\right]$$

$$\frac{|\mathscr{E}_\omega(0)|^2 - |\mathscr{E}_\omega(z)|^2}{\hbar\omega} = 2\frac{|\mathscr{E}_{2\omega}(z)|^2 - |\mathscr{E}_{2\omega}(0)|^2}{2\hbar\omega}. \tag{7.6}$$

Using the definition of u_ω and $u_{2\omega}$ [u_1 and u_3 in (6.28) with $\omega_1 = \omega_2 = \omega$ and $\omega_3 = 2\omega$], we obtain

$$\frac{du_\omega}{d\zeta} = -2u_\omega u_{2\omega}\sin\theta$$

and

$$\frac{du_{2\omega}}{d\zeta} = u_\omega^2\sin\theta \tag{7.7}$$

If we assume $u_{2\omega}(0) = 0$, then $\theta = \pi/2$, and the solution takes the form

$$u_\omega(\zeta) = u_\omega(0)\text{sech}\left[\sqrt{2}\,u_\omega(0)\zeta\right],$$

$$u_{2\omega}(\zeta) = \frac{1}{\sqrt{2}}u_\omega(0)\tanh\left[\sqrt{2}\,u_\omega(0)\zeta\right]. \tag{7.8}$$

The second harmonic output power is then given by

$$P_{2\omega}(z) = P_\omega(0)\tanh^2\left[C(P_\omega(0)/A)^{1/2}z\right] \tag{7.9}$$

where $C = K(2\omega/\sqrt{\varepsilon})(2\pi c/\sqrt{\varepsilon})^{1/2}$, assuming, for simplicity, $k_{\omega, z} = k_{\omega}$ $= \frac{1}{2}k_{2\omega} = \frac{1}{2}k_{2\omega, z}$ and $\alpha_{\omega} = \alpha_{2\omega}$. Following (7.9), Fig. 7.1 shows how $\mathbb{P}_{2\omega}(z)$ increases with z at the expense of $\mathbb{P}_{\omega}(z)$.

As a practical example, we consider the use of KDP as a second-harmonic generator for a Nd : YAG laser beam at 1.06 μm. Using the same calculation as in Section 6.8 with $n_0(\omega) = 1.4939$ and $n_{em}(2\omega) = 1.4706$ at room temperature, we find that for type I collinear phase matching, the beams should be propagating in a direction at an angle $\left(H\right) = 40.5°$ away from the Z-axis of the crystal. The pump field should be linearly polarized in a plane bisecting the X-Z and Y-Z planes in order to yield an optimum $\chi_{\text{eff}}^{(2)} = \chi_{ZXY}^{(2)}(2\omega)\sin\left(H\right)$ $= 1.5 \times 10^{-9}$ esu. With the plane wave approximation, the efficiency of second harmonic generation, following (7.9), is

$$\eta = \frac{\mathbb{P}_{2\omega}(z)}{\mathbb{P}_{\omega}(0)} = \tanh^2\left\{C\left[\frac{\mathbb{P}_{\omega}(0)}{A}\right]^{1/2} z\right\}$$

$$= \tanh^2\left\{4.7 \times 10^{-2}\left[\frac{\mathbb{P}_{\omega}(0)}{A}\right]^{1/2} z\right\} \quad (\mathbb{P}_{\omega} \text{ in MW}).$$

$$(7.10)$$

As seen from (7.10), the efficiency η reaches 58% when $[\mathbb{P}_{\omega}(0)/A]^{1/2}z = 21\sqrt{MW}$, or $\mathbb{P}_{\omega}(0)/A = 18$ MW/cm^2 for $z = 5$ cm. In an actual experiment, η is often less because of the finite dimension and transverse intensity variation of the pump beam. An efficiency as high as 40% with giant pulses or 85% with ultrashort pulses has, however, been achieved.[2] The foregoing calculation does indicate that in order to have an appreciable conversion efficiency ($\eta > 10\%$) in a crystal like KDP, a pump intensity of the order of 10 MW/cm^2 is needed

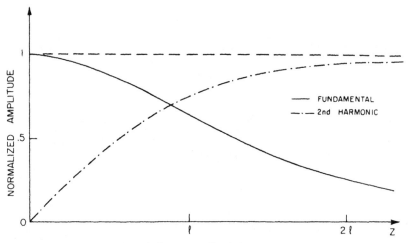

Fig. 7.1 Decay and growth of the normalized fundamental and second harmonic amplitudes, respectively, for the perfect phase-matching case. (After Ref. 1.)

with a crystal length of a few centimeters. (A much longer crystal is seldom practical.) In general, higher pump intensity leads to a larger η, except in the limit of very high conversion efficiency.[2] For a low-power pump beam, therefore, focusing is generally used to increase η and hence the second harmonic output.

Focusing, however, increases the walk-off effect, but as mentioned in Section 6.9, it also increases the spread of the beam propagation. The spread hurts the conversion efficiency as a portion of the beam now deviates from the exact phase matching direction. For type I collinear phase matching at the angle $\textcircled{H}$, for example, it can be shown readily that a small deviation $\Delta\theta$ of the beam propagation direction from $\textcircled{H}$ leads to a phase mismatch[3]

$$\Delta k \cong \tfrac{1}{2}k_\omega n_0^2(\omega)\left[\frac{1}{n_e^2(2\omega)} - \frac{1}{n_0^{2'}(2\omega)}\right]\sin 2\textcircled{H}\,\Delta\theta. \tag{7.11}$$

With $\Delta kl = \pi$ being the half-width of the phase-matching peak, the acceptance angle about $\textcircled{H}$ one can allow for beam convergence is, from (7.11),

$$\Delta\theta_A = \frac{2\pi}{k_\omega l}\left[\frac{n_0^2(2\omega)}{n_0^2(2\omega) - n_e^2(2\omega)}\right]\frac{1}{\sin 2\textcircled{H}}. \tag{7.12}$$

For a KDP crystal with $l = 1$ cm and $\textcircled{H} = 45°$ at $\lambda_\omega = 1.06$ μm, the acceptance angle $\Delta\theta_A$ is only 2.5 mrad. Equation (7.12) shows that $\Delta\theta$ diverges as $\textcircled{H}$ approaches 90°. This occurs because the higher order terms of $\Delta\theta$ were neglected in (7.11). The correct result for $\textcircled{H} = 90°$ is, assuming $n_0(2\omega) \sim n_e(2\omega)$,

$$\Delta\theta_A = \left\{\frac{n_e(2\omega)}{k_\omega l[n_0(2\omega) - n_e(2\omega)]}\right\}^{1/2}. \tag{7.13}$$

For a 1-cm KDP crystal at $\lambda_\omega = 1.06$ μm, the acceptance angle is 36 mrad, which is an order of magnitude larger than in the previous case. The large acceptance angle for 90° phase matching is clearly advantageous if beam focusing is used in second harmonic generation.

7.2 SECOND HARMONIC GENERATION BY FOCUSED GAUSSIAN BEAMS

Single-mode laser beams usually are required for efficient second harmonic generation. The conversion efficiency may be greatly enhanced by focusing of the pump beam into the nonlinear crystal. At the focal waist a single-mode beam has a Gaussian intensity profile described by $\exp(-\rho^2/W_0^2)$ with W_0 being the beam radius. The longitudinal dimension of the focal region is defined by the confocal parameter $b = kW_0^2$ as the distance between two points on the focusing axis where the beam radii are $\sqrt{2}$ times larger than that at the waist. Within the focal region, the beam has approximately a plane wave front, so that the plane wave approximation can be used.

We consider first the case of negligible double refraction or walk-off effect, e.g., the $90°$ phase-matching case. Obviously, if the crystal length l is smaller than the confocal parameter b, the conversion efficiency for second harmonic generation can be described by the result of plane wave approximation in (7.10) with $A = \pi W_0^2$,

$$\eta = \tanh^2\left\{ C\left[\frac{\mathbf{P}_\omega(0)}{\pi W_0^2} \right]^{1/2} l \right\}. \tag{7.14}$$

Here, as long as $b > l$, tighter focusing should increase $\mathbf{P}_\omega(0)/\pi W_0^2$ and improve the efficiency. If $b < l$, however, the approximation breaks down, and tighter focusing tends to reduce the efficiency. Thus optimum focusing occurs when the confocal parameter is about equal to the crystal length, $b \sim l$. Boyd and Kleinman[4] studied the focusing problem in detail with numerical calculation. They introduced an efficiency reduction factor $h_0(\xi)$ with $\xi = l/b$ to take into account the effect of focusing on the efficiency[4,5] They find

$$\eta = \tanh^2\left[\frac{C^2\mathbf{P}_\omega(0)k_\omega l h_0(\xi)}{\pi} \right]^{1/2}. \tag{7.15}$$

The function $h_0(\xi)$ is plotted in Fig. 7.2. For $\xi = l/b < 0.4$, we have $h_0(\xi) \simeq \xi$, and η in (7.15) reduces to the form in (7.14) as we expected. In the tight focusing limit, $\xi > 80$, the function $h_0(\xi)$ takes the asymptotic form $h_0(\xi) \simeq 1.187\pi^2/\xi$, and the efficiency drops rapidly with increasing ξ. The maximum value of $h_0(\xi) = 1.068$ appears at $\xi = 2.84$, with $h_0(\xi) \simeq 1$ in the range $1 \leq \xi \leq 6$. This shows that although optimum focusing occurs at $b = l/2.84$ for given l, the efficiency will not decrease appreciably even if $b \simeq l$.

With double refraction, the situation is more complicated. Boyd and Kleinman showed that in the limit of low conversion efficiency, (7.15) is still valid if $h_0(\xi)$ is replaced by $h(B, \xi)$ $[h(0, \xi) = h_0(\xi)]$, or

$$\eta = \frac{C^2\mathbf{P}_\omega(0)k_\omega l h(B, \xi)}{\pi} \tag{7.16}$$

where $B = \frac{1}{2}\rho(k_\omega l)^{1/2}$ is a double refraction parameter, and

$$\rho = \tan^{-1}\left\{ \frac{n_0^2(\omega)}{2}\left[\frac{1}{n_e^2(2\omega)} - \frac{1}{n_0^2(2\omega)} \right]\sin 2\bigcirc\!\!\!\!\!H \right\}$$

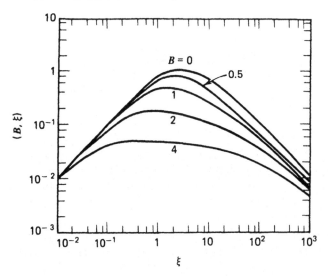

Fig. 7.2 Efficiency reduction factor $h(B, \xi)$ versus $\xi = l/b$ at various values of the double refraction parameter B. (After Ref. 4.)

is the walk-off angle between the Poynting vectors of the collinearly propagating fundamental and second harmonic waves along a direction at an angle (H) away from the optical axis. The function $h(B, \xi)$ for several values of B is seen in Fig. 7.2. Note that $h(B, \xi)$ depends only weakly on ξ near its maximum $h_M(B)$. The latter can be approximated to within 10% by the expression[3]

$$h_M(B) = \frac{h_M(0)}{1 + (4B^2/\pi)h_M(0)} \tag{7.17}$$

with $h_M(0) = 1.068$. This equation together with (7.16) indicate that the efficiency reduction due to double refraction becomes appreciable when $(4B^2/\pi)h_M(0) \sim 1$. We can define an effective length

$$l_{\text{eff}} = \frac{\pi}{k_\omega \rho^2 h_M(0)} \simeq \frac{\pi}{k_\omega \rho^2}. \tag{7.18}$$

Equation (7.17) becomes

$$h_M(B) = \frac{h_M(0)}{1 + l/l_{\text{eff}}}. \tag{7.19}$$

This shows explicitly that when $l_{\text{eff}} = l$ in the presence of double refraction, the efficiency for second harmonic generation with optimum focusing reduces by a factor of 2 as compared to the case without double refraction. When $l_{\text{eff}} \ll l$, this efficiency becomes

$$\eta_{\text{opt}} = \frac{\mathbf{C}^2 \mathbf{P}_\omega(0) k_\omega l_{\text{eff}} h_M(0)}{\pi}. \qquad (7.20)$$

The effect of double refraction on η_{opt} is insignificant only when $l \ll l_{\text{eff}}$. These results do not depend critically on focusing as long as $h(B, \xi) \simeq h_M(B)$.

One can use a more physical argument to understand these results. Because of double refraction, the phase-matched fundamental and second harmonic beams can overlap only over roughly a distance, $l_a = W_0\sqrt{\pi}/\rho$, often known as the aperture length. For optimum focusing, we like to have $l = b = k_\omega W_0^2$, but to avoid reduction of efficiency by double refraction, we must have $l < l_a = \sqrt{\pi l / k_\omega \rho^2}$, which leads to $l < l_{\text{eff}} = \pi/k_\omega \rho^2$, which is the same result described in the preceding paragraph.

This discussion assumes that the laser intensity in the crystal is not limited by optical damage. This is of course not always the case. As an example, let us

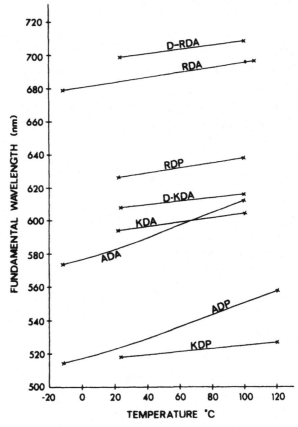

Fig. 7.3 Fundamental wavelength versus crystal temperature at $90°$ phase matching for some of the KDP isomorphs. (After Ref. 6.)

assume a crystal with $n \simeq 1.5$, $l = 1$ cm and $\rho = 30$ mrad at $\lambda_\omega = 1.06$ μm. Then, $B = 3.65$, and from Fig. 7.2, $h(B = 3.65, \xi) \simeq h_M(B)$ for $0.2 < \xi < 10$. Since $l_{\text{eff}} = 0.04$ cm is much less than l, we have $h_M(B) \simeq l_{\text{eff}}/l$, and according to (7.20) $\eta \propto l_{\text{eff}}$. Comparing with the case of no walk-off and optimum focusing $b \sim l$, we have $\eta_{\rho=0} \propto l$, and hence, $\eta_{\rho=0}/\eta_\rho = l/l_{\text{eff}} = 2n\rho^2 l/\lambda = 25$. This shows that it is of great advantage to use $90°$ phase matching to avoid the walk-off effect. The $90°$ phase matching for second harmonic generation can be achieved by temperature tuning in many crystals. In Fig. 7.3, the $90°$ phase-matched wavelength as a function of temperature is shown for a number of KDP isomorphs.[6]

7.3   THIRD HARMONIC GENERATION IN CRYSTALS

In a crystal with inversion symmetry, second harmonic generation is forbidden under the electric dipole approximation, although it can be induced by an applied dc electric field.[7] Third harmonic generation, on the other hand, is always allowed. The theory for third harmonic generation in the limit of negligible pump depletion is the same as that for second harmonic generation with $\mathbf{P}^{(2)}(2\omega)$ replaced by $\mathbf{P}^{(3)}(3\omega) = \chi^{(3)} (3\omega = \omega + \omega + \omega) : \mathbf{E}(\omega)\mathbf{E}(\omega)\mathbf{E}(\omega)$. Since $|\chi^{(3)}|$ is usually small [$\sim 10^{-12}$ to 10^{-15} esu as compared to $|\chi^{(2)}| \sim 10^{-7}$ to 10^{-9} esu typically], and the laser intensity is often limited by optical damage in crystals, the conversion efficiency for this third-order nonlinear process is small. In addition, phase matching is more difficult to achieve. It has therefore found little practical application.

An efficient third harmonic generator can, however, be constructed by having two nonlinear crystals in series.[8] The first generates a second harmonic beam. The transmitted fundamental beam and the second harmonic output beam are then combined in the second crystal to yield a third harmonic output by sum-frequency generation. Both processes are phase-matched (either type I or type II). With a sufficiently intense fundamental beam, the overall efficiency of third harmonic generation can be fairly high. Commercial units with efficiency as high as 20% are available.

In principle, this type of two-step third harmonic generation can occur in a single crystal. However, except in very special cases, it is not possible to have both second harmonic generation and sum-frequency generation simultaneously phase matched. Consequently, the overall conversion efficiency cannot be very significant.

7.4   HARMONIC GENERATION IN GASES

Third harmonic generation can also occur in gases. One would think that because of the much lower atomic or molecular density in gases than in liquids or solids, the third-order nonlinear susceptibility $|\chi^{(3)}|$ for a gas medium should be much smaller than that for a liquid or solid, and the efficiency for third harmonic generation in gases would be so low that it could never be significant. This conjecture turns out to be incorrect, as pointed out by Miles and Harris.[9] First, $|\chi^{(3)}|$ can be resonantly enhanced. The much sharper transitions in gases allow much stronger enhancement near resonances, especially those with higher transition matrix elements. Then, the limiting laser intensity in gases is orders of magnitude higher than in condensed matter ($>$ few GW/cm^2 in gases as compared to few hundred MW/cm^2 in solids). As a result, although $|\chi^{(3)}|$ is small, the induced nonlinear polarization $|\mathbf{P}^{(3)}|$ by a high-intensity laser field can be comparable to $|\mathbf{P}^{(2)}|$ induced in a solid with a moderately intense beam.

Consider sodium vapor. The third-order nonlinear susceptibility for Na can be fairly accurately estimated from the general expression of $\chi^{(3)}(3\omega)$ derived by the technique in Sections 2.2 and 2.3:

$$\chi^{(3)}_{ijkl}(3\omega = \omega + \omega + \omega) = \frac{Ne^4}{\hbar^3} \sum_{g,a,b,c} (r_i)_{ga}(r_j)_{ab}(r_k)_{bc}(r_l)_{cg}\rho^{(0)}_{gg}A_{abc}$$

where

$$
\begin{aligned}
A_{abc} = &\left[(\omega_{cg} - \omega)(\omega_{bg} - 2\omega)(\omega_{ag} - 3\omega)\right]^{-1} \\
&+ \left[(\omega_{cg} - \omega)(\omega_{bg} - 2\omega)(\omega_{ag} + \omega)\right]^{-1} \\
&+ \left[(\omega_{cg} - \omega)(\omega_{bg} + 2\omega)(\omega_{ag} + \omega)\right]^{-1} \\
&+ \left[(\omega_{cg} + 3\omega)(\omega_{bg} + 2\omega)(\omega_{ag} + \omega)\right]^{-1}.
\end{aligned}
\tag{7.21}
$$

Here N is the number of atoms per unit volume, and we assume that the frequencies are sufficiently far way from resonances so that the damping constants in the denominators can be neglected. For alkali atoms, the transition frequencies and the major matrix elements are often known. Therefore, $|\chi^{(3)}(3\omega)|$ can be calculated from (7.21). This has been done by Miles and Harris.[9] The result for Na is seen in Fig. 7.4 along with the energy level diagram for Na. It shows that even when 3ω is a few hundred cm^{-1} away from an $s \rightarrow p$ resonance, the near-resonance enhancement can make the value of $|\chi^{(3)}|/N$ larger than 10^{-33} esu. Then, with $N = 10^{17}$ atoms/cm^3, and $|E(\omega)| \sim 2 \times 10^3$ esu corresponding to a beam intensity of 1 GW/cm^2, the induced nonlinear polarization $|P^{(3)}| = |\chi^{(3)}EEE|$ can be larger than 10^{-6} esu. This is compared with $|P^{(2)}(2\omega)| = |\chi^{(2)}EE| \sim 10^{-5}$ esu induced in KDP with $|\chi^{(2)}| \sim 10^{-9}$ esu and $|E| \sim 10^2$ esu (2.5 MW/cm^2). Therefore, third harmonic generation in sodium vapor should be easily observable with 3ω near resonance, e.g., with a Nd laser at 1.06 μm.

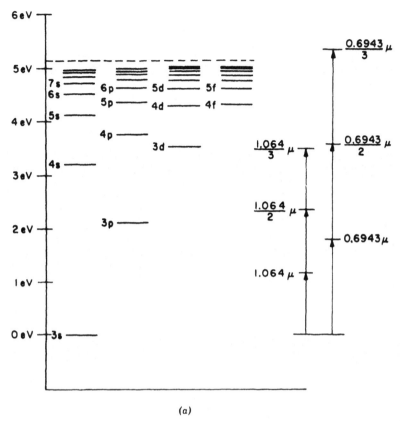

(a)

Fig. 7.4 (a) Energy levels of sodium. (b) Third-order nonlinear polarizability, $\chi^{(3)}(3\omega)/N$, versus fundamental wavelength for sodium. (After Ref. 9.)

To have high conversion efficiency, aside from resonant enhancement and sufficiently high pump intensity, the third harmonic generation process must be collinearly phase-matched with $n(\omega) = n(3\omega)$. Since a gas medium is isotropic, the usual method utilizing the birefringent property of a medium for phase matching is not applicable here. Then, phase matching for third harmonic generation (or optical mixing in general) is not always achievable in a gas medium. When anomalous dispersion exists between ω and 3ω, however, it can be achieved by using a buffer gas to compensate the difference of refractive indices at ω and 3ω. This is demonstrated in Fig. 7.5. With ω below and 3ω above a strong $s \rightarrow p$ transition of the alkali atom, the anomalous dispersion causes $n_A(\omega) > n_A(3\omega)$ in a pure alkali vapor. If a buffer gas (e.g., Xe) with normal dispersion $n_B(\omega) < n_B(3\omega)$ is mixed into the medium, then by adjusting the buffer gas density, it is possible to achieve phase matching with $n_A(\omega) + n_B(\omega) = n_A(3\omega) + n_B(3\omega)$.

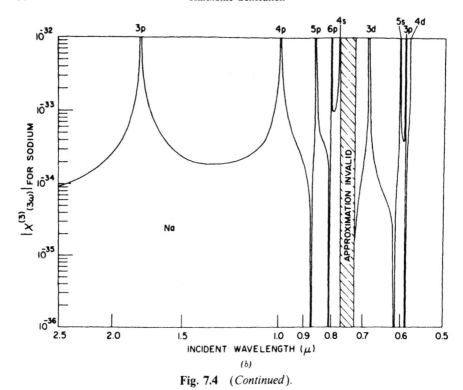

Fig. 7.4 (*Continued*).

There are several important advantages of using a gas medium for nonlinear optical mixing.

1   A homogeneous medium longer than 10 cm is easily available.
2   Since the medium is isotropic, the walk-off problem does not exist. Optimum focusing can then be used to increase conversion efficiency.
3   Aside from the high optical damage threshold, a gas medium also has self-healing capability. Except in special cases, no permanent change can be afflicted in the medium by laser-induced ionization or dissociation.
4   Atomic vapor is transparent to radiation at almost all frequencies below the ionization level except for a number of discrete absorption lines, and is the only nonlinear medium one can use in the extreme uv or soft X-ray region.

A gas medium may then appear to be ideal for third harmonic generation, especially for conversion to the uv range. High conversion efficiency could presumably be obtained by using a high laser intensity with a reasonably long gas cell. Unfortunately, there are also many factors that often limit the

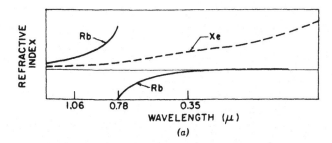

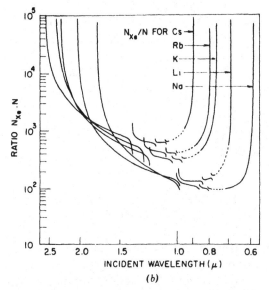

Fig. 7.5 (*a*) Refractive indices of Rb and Xe versus wavelength. (*b*) Required ratio of Xe to alkali atoms versus fundamental wavelength for phase-matched third harmonic generation. (After Ref. 9.)

efficiency through limitation of the laser intensity:

1 Linear absorption at ω and 3ω suppresses the efficiency (Section 6.6). Resonant enhancement of $|\chi^{(3)}|$ also enhances the linear absorption, although not proportionally.

2 Two-photon and multiphoton absorption may become important in limiting the efficiency when a high-intensity pump beam is used.

3 Population redistribution due to absorption can induce a phase mismatch to the optical mixing process.

4 A refractive index change caused by another laser-induced mechanism can also give rise to a phase mismatch.

5 Laser-induced breakdown of the medium may cut off the mixing process.

All these factors become rapidly more important when ω or 3ω gets closer to resonance. Usually, (3) turns out to be the limiting process and (5) may easily occur with long laser pulses.

Third harmonic generation in gases has been experimentally demonstrated in many cases.[10] With 30-psec, 300-MW Nd:YAG laser pulses optimally focused to a 10^{-3}-cm^2 spot in a 50-cm Rb(3 torr):Xe(2000 torr) cell, Bloom et al.[11] observed a phase-matched third harmonic output at 3547 Å with a 10% conversion efficiency. The same group also obtained phase-matched third harmonic generation at 3547 Å in Na:Mg with a 3.8% efficiency. Then, the uv third harmonic generation from 5320 to 1773 Å and from 3547 to 1182 Å was also observed in Cd:Ar and Xe:Ar gas mixtures by Kung et al.[12] with a maximum efficiency of 0.3%.

This discussion could be extended to higher-order harmonic generation in gases, although the conversion efficiency is expected to be very low because of the relatively small nonlinear optical susceptibilities. It was suggested by Harris[13] that coherent vacuum uv and soft X-ray radiation could be obtained from fifth and seventh harmonic generation in atomic vapor. This was later demonstrated by Reintjes et al.[14]

7.5 MEASUREMENT OF NONLINEAR OPTICAL SUSCEPTIBILITIES

The well-established theory of sum-frequency and harmonic generation allows us to deduce nonlinear optical susceptibilities $\chi^{(2)}$ ($\omega = \omega_1 + \omega_2$) and $\chi^{(n)}(n\omega)$ from sum-frequency and harmonic generation. We discuss here the measurement of $\chi^{(2)}(2\omega)$ as an example.

As shown in (7.1), the absolute magnitude of $|\hat{e}_{2\omega} \cdot \chi^{(2)} : \hat{e}_\omega \hat{e}_\omega|$ can be deduced from the measured second harmonic output power if $\mathbb{P}(\omega)$, A, Δk, z, etc. in (7.1) are known. Then, by choosing the set of polarizations $\hat{e}_\omega$ and $\hat{e}_{2\omega}$ properly aligned with the appropriate crystal orientation, the particular tensor element of $\chi^{(2)}(2\omega)$ can be found. For more accurate determination of $\chi^{(2)}(2\omega)$, the second harmonic output $\mathbb{P}(2\omega)$ as a function of Δkz is measured, and the effect of beam profile is taken into account in the calculation. An absolute measurement is always difficult, however, as the laser beam characteristics must be known to great accuracy. It has only been attempted in a few cases, mainly on ammonium dihydrogen phosphate (ADP).[15] The nonlinear susceptibilities of other crystals can then be measured in comparison with that of ADP. In particular, careful comparison between KDP, quartz, and ADP has been made,[16] and these three crystals are now often used as reference materials in the measurements of $\chi^{(2)}$ of given materials.

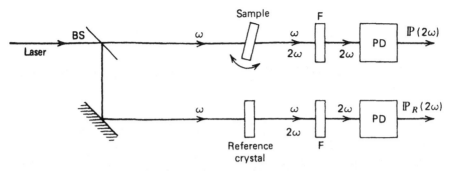

Fig. 7.6 Experimental arrangement for measurement of the relative second harmonic susceptibility of a sample.

In the relative experiment, the laser beam is split into two; one is used for generating $\mathbb{P}(2\omega)$ in the sample and the other for generating $\mathbb{P}_R(2\omega)$ in the reference crystal (Fig. 7.6). The ratio of the two is

$$\frac{\mathbb{P}(2\omega)}{\mathbb{P}_R(2\omega)} = \frac{|\hat{e}_{2\omega} \cdot \mathbf{\chi}^{(2)} : \hat{e}_\omega \hat{e}_\omega|^2 \varepsilon_R(\omega) \sqrt{\varepsilon_R(2\omega)} \sin^2(\Delta k\, l/2)}{|\hat{e}_{2\omega} \cdot \mathbf{\chi}_R^{(2)} : \hat{e}_\omega \hat{e}_\omega|^2 \varepsilon(\omega) \sqrt{\varepsilon(2\omega)} (\Delta k/2)^2} \frac{(\Delta k_R/2)^2}{\sin^2(\Delta k_R l_R/2)}$$

(7.22)

assuming that two arms have equal laser intensities. Here, the subindex R refers to the reference crystal. With other quantities known, the ratio $|\hat{e}_{2\omega} \cdot \mathbf{\chi}^{(2)} : \hat{e}_\omega \hat{e}_\omega|/|\hat{e}_{2\omega} \cdot \mathbf{\chi}_R^{(2)} : \hat{e}_\omega \hat{e}_\omega|$ can be determined from the measurement of $\mathbb{P}(2\omega)/\mathbb{P}_R(2\omega)$ versus Δkl. As seen in (7.22), the result is now independent of the laser characteristics. This makes the measurement much easier since the very difficult absolute measurement of the laser characteristics is no longer necessary.

The result of $\mathbb{P}(2\omega)/\mathbb{P}_R(2\omega) \propto \sin^2(\Delta kl/2)$ as a function of Δkl appears as a set of fringes, known as Maker fringes.[17] It is usually obtained by rotation of a plane-parallel slab of sample about an axis. The effective sample thickness is then given by $d \cos\theta$ with d being the slab thickness and θ the angle between the slab normal and the beam propagation direction. With $\mathbb{P}(2\omega) \propto \sin^2[(\Delta k) d \cos\theta/2]$, the Maker fringes arise through variation of θ. In practice, the crystal orientation is also chosen, if possible, to make Δk independent of θ. An example is seen in Fig. 7.7, where a slab of quartz is used with its $\hat{c}$-axis parallel to the face being the axis of rotation for variation of θ.

The nonlinear susceptibility is in general a complex quantity. The phase factor of $\chi_{ijk}^{(2)}$ can be measured from the interference of second harmonic generation in two slabs of crystals in series.[18] Let the two crystals of thickness d_1 and d_2, respectively, be separated by a distance s. Assume phase matching in the first crystal. The second harmonic field generated from the first crystal in

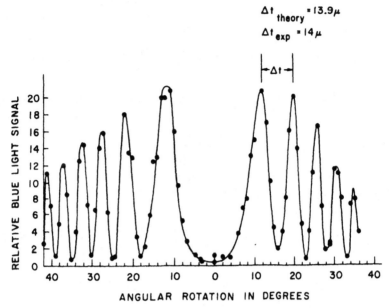

$$\Delta t_{\text{theory}} = 13.9\mu$$
$$\Delta t_{\text{exp}} = 14\mu$$

Fig. 7.7 Relative second harmonic intensity as a function of the optical thickness of the crystal, displaying the Maker fringes. Change of the optical thickness is achieved by the angular rotation of the crystal. (After Ref. 17.)

the normal direction is

$$E_{2\omega}(d_1) = \frac{i4\pi\omega}{cn_1(2\omega)}\chi^{(2)}_{1,\text{eff}}d_1\mathcal{E}^2_\omega e^{i2(\omega/c)n_1(\omega)d_1 - i2\omega t}. \tag{7.23}$$

The input fields at the entrance of the second crystal are

$$E_\omega(d_1 + s) = \mathcal{E}_\omega e^{i(\omega/c)[n_1(\omega)d_1 + n_0(\omega)s] - i\omega t}$$

and

$$E_{2\omega}(d_1 + s) = E_{2\omega}(d_1)e^{i(2\omega/c)n_0(2\omega)s} \tag{7.24}$$

where n_0 is the refractive index of the medium between the two crystals. Then the second harmonic output from the second crystal is

$$\mathbb{P}(2\omega) \propto |E_{2\omega}(d_1 + s) + \frac{4\pi\omega}{cn_2(2\omega)\Delta k}\chi^{(2)}_{2,\text{eff}}d_2 E^2_\omega(d_1 + s)(e^{i\Delta k d_2} - 1)|^2$$

$$= \left|\frac{4\pi\omega}{c}\mathcal{E}^2_\omega\right|^2 \left|\frac{id_1}{n_1(2\omega)}\chi^{(2)}_{1,\text{eff}}e^{i(2\omega/c)[n_0(2\omega) - n_0(\omega)]s}\right. \tag{7.25}$$

$$\left.+ \frac{d_2}{n_2(2\omega)\Delta k}(e^{i\Delta k d_2} - 1)\chi^{(2)}_{2,\text{eff}}\right|^2.$$

Expression (7.25) shows that $\mathbb{P}(2\omega)$ depends on the relative phase of the effective nonlinear susceptibilities of the two crystals $\chi^{(2)}_{1,\text{eff}}$ and $\chi^{(2)}_{2,\text{eff}}$. If the crystals are mounted in an enclosed chamber filled with a known gas and the gas pressure is varied, then because of the dispersion of the gas, $n_0(2\omega) \neq n_0(\omega)$, the relative phase of the two terms in (7.25) will vary, resulting in a set of interference peaks. This observed interference in $\mathbb{P}(2\omega)$ versus $[n_0(2\omega) - n_0(\omega)]s$ allows us to determine the relative phase of $\chi^{(2)}_{1,\text{eff}}$ and $\chi^{(2)}_{2,\text{eff}}$. Usually, $\chi^{(2)}_{ZXY}$ of KDP is used as the reference. For a nonabsorbing crystal, $\chi^{(2)}_{ijk}$ is real, either positive or negative. In Table 7.1, we list the values of $\chi^{(2)}_{ijk}$ for a number of commonly encountered nonlinear optical crystals.

Table 7.1
Selected Second Harmonic Nonlinear Susceptibilities of a Number of Crystals

Material	Symmetry Class	$\chi^{(2)} \left(\dfrac{3}{4\pi} \times 10^{-8} \text{ esu} \right)^a$	Fundamental Wavelength (μm)
α-SiO$_2$ (quartz)	$32-D_3$	$\chi^{(2)}_{xxx} = 0.8 \pm 0.04$ $\chi^{(2)}_{xyz} = 0.018$	1.0582
Te	$32-D_3$	$\chi^{(2)}_{xxx} \simeq 10^4$	10.6
Ba$_2$NaNb$_5$O$_{15}$	$mm2-C_{2v}$	$\chi^{(2)}_{zxx} = -29.1 \pm 1.5$ $\chi^{(2)}_{zyy} = -29.1 \pm 2.9$ $\chi^{(2)}_{zzz} = -40 \pm 2.9$	1.0642
LiNbO$_3$	$3m-C_{3v}$	$\chi^{(2)}_{yyy} = 6.14 \pm 0.56$ $\chi^{(2)}_{zxx} = -11.6 \pm 1.7$ $\chi^{(2)}_{zzz} = 81.4 \pm 21$	1.0582
BaTiO$_3$	$4mm-C_{4v}$	$\chi^{(2)}_{xzx} = -34.4 \pm 2.8$ $\chi^{(2)}_{zxx} = -36 \pm 2.8$ $\chi^{(2)}_{zzz} = -13.2 \pm 1$	1.0582
NH$_4$H$_2$PO$_4$ (ADP)	$42m-D_{2d}$	$\chi^{(2)}_{xyz} = 0.96 \pm 0.05$ $\chi^{(2)}_{zxy} = 0.97 \pm 0.06$	0.6943
KH$_2$PO$_4$ (KDP)	$\bar{4}2m-D_{2d}$	$\chi^{(2)}_{xyz} = 0.98 \pm 0.04$ $\chi^{(2)}_{zxy} = 0.94$	1.0582
ZnO	$6mm-C_{6v}$	$\chi^{(2)}_{zxx} = 4.2 \pm 0.4$ $\chi^{(2)}_{xzx} = 4.6 \pm 0.4$ $\chi^{(2)}_{zzz} = -14.0 \pm 0.4$	1.0582
LiIO$_3$	$6-C_6$	$\chi^{(2)}_{zzx} = -11.2 \pm 0.6$ $\chi^{(2)}_{zzz} = -8.4 \pm 2.8$	1.0642
CdSe	$6mm-C_{6v}$	$\chi^{(2)}_{xzx} = 62 \pm 15$ $\chi^{(2)}_{zxx} = 57 \pm 13$ $\chi^{(2)}_{zzz} = 109 \pm 25$	10.6
GaAs	$\bar{4}3m-T_d$	$\chi^{(2)}_{xyz} = 377 \pm 38$	10.6
GaP	$\bar{4}3m-T_d$	$\chi^{(2)}_{xyz} = 70$	3.39

^{a}The values of $\chi^{(2)}$ are obtained from R. J. Pressley, ed., *Handbook of Lasers* (Chemical Rubber Co., Cleveland, Ohio, 1971), p. 497. In the convention we have adopted, our $\chi^{(2)}$ here are two times larger than the d coefficients given in the literature. Note that $\chi^{(2)}$ (esu) here is related to $\chi^{(2)}$ (m/v) by $\chi^{(2)}$ (esu) $= 3/4\pi \times 10^4 \, \chi^{(2)}$ (m/v).

For absorbing crystals, $\chi_{ijk}^{(2)}$ is complex, and the measurement of second harmonic reflection from the surface, with the help of the theory developed in Section 6.4 is often used to find $\chi_{ijk}^{(2)}$.[19] Again, the interference technique can be adopted to measure the phase of $\chi_{ijk}^{(2)}$.

The foregoing methods allow accurate relative or absolute measurement of $\chi_{ijk}^{(2)}$, but the crystal to be studied must be of fair size and good quality. In practice, however, special effort often is needed to grow a crystal of large dimensions. It is therefore important that the nonlinear optical constants of the crystal can somehow be estimated beforehand. The powder method developed by Kurtz[20] is most helpful in this respect.

Figure 7.8 shows the experimental arrangement. Powder sample is packed into a thin cell of a definite thickness, and the second harmonic output from the sample over the entire 4π solid angle is collected. The output is measured relative to the second harmonic generation in a reference crystal. The desired information can then be obtained by the measurement of the second harmonic output as a function of the particle size of the powder. For an average particle size $\bar{r}$ much smaller than the average coherent length, defined by $l_{\text{coh}} = \pi/\overline{\Delta k}$ $= \pi c/\omega[\overline{n(2\omega)} - \overline{n(\omega)}]$, the second harmonic output $\mathbb{P}(2\omega)$ increases almost linearly with $\bar{r}$ since essentially all the particles in the beam are effectively phase matched while the number of particles in the light path decreases inversely with $\bar{r}$. As $\bar{r}$ becomes larger than l_{coh}, the output $\mathbb{P}(2\omega)$ can increase further if the material is phase-matchable. This is because some particles in the light path should have the correct orientation for phase matching. The output, however, shows saturation as the corresponding decrease in the number of

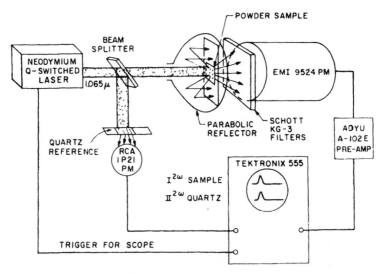

Fig. 7.8 Schematic layout of the apparatus used in the powder measurement of the second-order nonlinearity. [After S. K. Kurtz, *IEEE J. Quant. Electron.* **QE-4**, 578 (1968).]

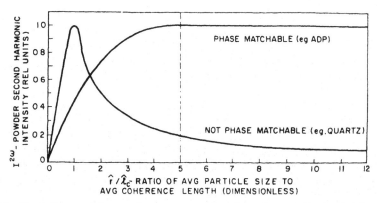

Fig. 7.9 Typical second harmonic output as a function of normalized particle size for powders of phase-matchable and non-phase-matchable crystals. [After S. K. Kurtz, *IEEE J. Quant. Electron.* **QE-4**, 578 (1968).]

particles in the light path suppresses the gain of $\mathbb{P}(2\omega)$ (Fig. 7.9). For non–phase-matchable materials, the output from each particle saturates when $\bar{r} > \bar{l}_{coh}$, and hence $\mathbb{P}(2\omega)$ should decrease inversely with $\bar{r}$ as a result of decrease in the number of particles in the light path as shown in Fig. 7.9. With this technique, numerous materials have been surveyed. They can be divided into five groups:[20] centrosymmetric, phase-matchable, non–phase-matchable, large nonlinear coefficient, and small nonlinear coefficient.

7.6 SECOND HARMONIC GENERATION WITH ULTRASHORT PULSES

Second harmonic generation with ultrashort pulses requires some special consideration. With the pulse length smaller than the length of the medium, the nonlinear polarization varies drastically along the length at a given time. The only simple case occurs when the group velocities of the forward propagating fundamental and second harmonic waves are the same. Then the two pulses will propagate together and interact with each other as if the problem were stationary. This is the quasi-stationary case. The solution is identical to that of the stationary case if $z - v_g t$ replaces z where v_g is the group velocity. If the group velocity dispersion is nonnegligible, then the solution becomes much more complicated. Physically, the velocity mismatch causes the fundamental pulse to displace against the second harmonic pulse as they propagate along. This reduces the effective interaction length and decreases the conversion efficiency.[21,22]

A rigorous mathematical treatment of the problem has been worked out by Akhmanov et al.[21] Infinite plane waves propagating along z with slowly varying amplitudes are assumed. As shown in Section 3.5, the pulse propagation of a wave, $\mathscr{E}(z, t)\exp[ikz - i\omega t]$, in a nonlinear medium can be described by

$$\frac{\partial \mathscr{E}}{\partial z} - \frac{1}{v_g}\frac{\partial \mathscr{E}}{\partial t} = \frac{i2\pi\omega^2}{kc^2}\mathbf{P}^{NL}(z, t)e^{-i(kz-\omega t)}. \qquad (7.26)$$

In the present case, the group velocities of the fundamental and second harmonic waves are v_{1g} and v_{2g}, respectively. If we use, as independent variables, z and $\eta \equiv t - z/v_{1g}$ instead of z and t, then the wave equations governing the fundamental and second harmonic wave amplitudes $\mathscr{E}_\omega(z, \eta)$ and $\mathscr{E}_{2\omega}(z, \eta)$ under the phase-matching condition become

$$\frac{\partial \mathscr{E}_\omega}{\partial z} = \sigma\mathscr{E}_\omega\mathscr{E}_{2\omega},$$

$$\frac{\partial \mathscr{E}_{2\omega}}{\partial z} + v\frac{\partial \mathscr{E}_{2\omega}}{\partial \eta} = -\sigma\mathscr{E}_\omega^2 \qquad (7.27)$$

where $v = v_{2g}^{-1} - v_{1g}^{-1}$ and $\sigma = (2\pi\omega^2/k_1c^2)\hat{e}_{2\omega}\cdot\mathbf{\chi}^{(2)}:\hat{e}_\omega\hat{e}_\omega$ assuming $\mathscr{E}_{2\omega}(z = 0) = 0$. The solution of (7.27) is nontrivial. Akhmanov et al.[21] showed that the coupled nonlinear equations can be combined into a single second-order differential equation

$$\frac{\partial^2}{\partial z^2}\left(\frac{1}{\mathscr{E}_\omega}\right) = F(\eta - vz)\frac{1}{\mathscr{E}_\omega} \qquad (7.28)$$

where $F(\eta - vz) = \sigma^2(\mathscr{E}_\omega^2 + \mathscr{E}_{2\omega}^2) + \sigma v\partial\mathscr{E}_{2\omega}/\partial\eta$ is a function of $(\eta - vz)$ only, as can be shown by the vanishing Jacobian

$$\begin{vmatrix} \partial F/\partial\eta, & \partial F/\partial z \\ \dfrac{\partial}{\partial\eta}(\eta - vz), & \dfrac{\partial}{\partial z}(\eta - vz) \end{vmatrix} = 0.$$

Equation (7.28) is now a linear equation with a varying coefficient F. It can be solved analytically for arbitrarily large conversion efficiency.

Let the fundamental pulse at $z = 0$ take the form

$$\mathscr{E}_\omega(t) = A/\left(1 + \frac{t^2}{\tau^2}\right). \qquad (7.29)$$

Here, τ is the pulsewidth. In addition, we define a number of characteristic lengths for the problem: L is the length of the medium; $L_{NL} = 1/\sigma A$ is the interaction length at which 75% of the fundamental power is converted into second harmonic power in the stationary case; $L_\nu = \tau/\nu$ is the propagating distance over which the overlapping fundamental and second harmonic pulses of width τ are clearly separated. With a new set of variables $\tilde\eta \equiv \eta/\tau$, $\tilde z \equiv z/L_\nu$, $\tau_{cr} \equiv \nu L_{NL}$, $f \equiv (\tau^2/\tau_{cr}^2 - 1)^{1/2}$ and $\xi \equiv [\tau^2/\tau_{cr}^2 - 1]^{1/2} \times [\tan^{-1}\tilde\eta - \tan^{-1}(\tilde\eta - \tilde z)]$, the solution of (7.27) has the form

$$\mathscr{E}_\omega = \frac{A}{(1 + \tilde\eta^2)^{1/2}[1 + (\tilde\eta - \tilde z)^2]^{1/2}}\left\{\cosh\xi + \frac{\tilde\eta}{f}\sinh\xi\right\}^{-1} \tag{7.30}$$

$$\mathscr{E}_{2\omega} = A\frac{\tau_{cr}}{\tau}\left\{\frac{\tilde z\cosh\xi + [f - \tilde\eta(\tilde\eta - \tilde z)/f]\sinh\xi}{\cosh\xi + (\tilde\eta/f)\sinh\xi}\right\}\bigg/\left[1 + (\tilde\eta - \tilde z)^2\right].$$

When $L_\nu \gg L_{NL}$ ($\tau \gg \tau_{cr}$), the group velocity dispersion is clearly negligible so far as second harmonic generation is concerned. In this quasi-stationary case, the solution in (7.30) reduces to

$$\mathscr{E}_\omega(z, \eta) = \frac{A}{1 + \tilde\eta^2}\operatorname{sech}\left[\frac{z}{(1 + \tilde\eta^2)L_{NL}}\right],$$

$$\mathscr{E}_{2\omega}(z, \eta) = \frac{A}{1 + \tilde\eta^2}\tanh\left[\frac{z}{(1 + \tilde\eta^2)L_{NL}}\right]. \tag{7.31}$$

They are exactly the same as the stationary solution for second harmonic generation given in (7.8) if we replace $\mathscr{E}_\omega(z = 0)$ there by $A/(1 + \tilde\eta^2)$. In the limit of negligible pump depletion, $z \ll (1 + \tilde\eta^2)L_{NL}$, the second harmonic field from (7.31) is proportional to the square of the fundamental field

$$\mathscr{E}_{2\omega}(z, \eta) = \mathscr{E}_\omega^2(\eta)z/AL_{NL}.$$

Then the second harmonic pulsewidth is approximately half of that of the fundamental.

When $L_\nu \lesssim L_{NL}$, the group velocity mismatch becomes important, and the general solution of (7.30) must be used. There is a relative displacement between the fundamental and the second harmonic pulses. Consequently, the conversion efficiency drops, and the second harmonic pulse broadens. This has been experimentally confirmed.[23] On the other hand, one can also use the group velocity mismatch to sharpen an input second harmonic pulse through amplification. If the fundamental pulse is appreciably longer than the second harmonic pulse, and if $v_{g2} > v_{g1}$, then the two pulses can be arranged in such a way that the leading edge of the second harmonic pulse always sees the undepleted part of the fundamental pulse and gets amplified more than the lagging edge, resulting in a sharper output pulse.

The group velocity mismatch is generally more appreciable at higher frequencies because of anomalous dispersion due to absorption bands in the uv region. For a 1-psec pulse propagation in KDP, for example, $L_\nu \cong 3$ cm at $\lambda = 1.06$ μm and $L_\nu = 0.3$ cm at $\lambda = 0.53$ μm. Thus the effect of group velocity mismatch is much more important for frequency doubling of picosecond pulses into the uv.

REFERENCES

1   J. A. Armstrong, N. Bloembergen, J. Ducuing, and P. S. Pershan, *Phys. Rev.* **127**, 1918 (1962); N. Bloembergen, *Nonlinear Optics* (Benjamin, New York, 1965), p. 85.

2   J. Reintjes and R. C. Eckardt, *Appl. Phys. Lett.* **30**, 91 (1977).

3   R. L. Byer and R. L. Herbst, in Y. R. Shen, ed., *Nonlinear Infrared Generation* (Springer-Verlag, Berlin, 1977), p. 81.

4   G. D. Boyd and D. A. Kleinman, *J. Appl. Phys.* **39**, 3597 (1968).

5   D. R. White, E. L. Dawes, and J. H. Marburger, *IEEE J. Quant. Electron.* **QE-6**, 793 (1970).

6   R S. Adhav and R. W. Wallace, *IEEE J. Quant. Electron.* **QE-9**, 855 (1973).

7   R. W. Terhune, *Solid State Design* **4**, 38 (1963).

8   See, for example, the review article by R. Piston, *Laser Focus* **14**(7), 66 (1978).

9   R. B. Miles and S. E. Harris, *Appl. Phys. Lett.* **19**, 385 (1971); *IEEE J. Quant. Electron.* **QE-9**, 470 (1973).

10 J. F. Young, G. C. Bjorklund, A. H. Kung, R. B. Miles, and S. E. Harris, *Phys. Rev. Lett.* **27**, 1551 (1971).

11 D M Bloom, G. W. Bekkers, J. F. Young, and S. E. Harris, *Appl. Phys. Lett.* **26**, 687 (1975); D. M. Bloom, J. F. Young, and S. E. Harris, *Appl. Phys. Lett.* **27**, 390 (1975).

12 A. H. Kung, J. F. Young, G. C. Bjorklund, and S. E. Harris, *Phys. Rev. Lett.* **29**, 985 (1972); A. H. Kung, J. F. Young, and S. E. Harris, *Appl. Phys. Lett.* **22**, 301 (1973) [Erratum: **28**, 239 (1976)].

13 S. E. Harris, *Phys. Rev. Lett.* **31**, 341 (1973).

14 J Reintjes, C. Y. She, R. C. Eckardt, N. E. Karangelen, R. C. Elton, and R. A. Andrews, *Phys. Rev. Lett.* **37**, 1540 (1976); *Appl. Phys. Lett.* **30**, 480 (1977).

15 G. E. Francois, *Phys. Rev.* **143**, 597 (1966); J. E. Bjorkholm and A. E. Siegman, *Phys. Rev.* **154**, 851 (1967).

16 J. Jerphagnon and S. K. Kurtz, *Phys. Rev.* **B1**, 1739 (1970).

17 P. D. Maker, R. W. Terhune, M. Nisenoff, and C. M. Savage, *Phys. Rev. Lett.* **8**, 21 (1962).

18 J. J. Wynne and N. Bloembergen, *Phys. Rev.* **188**, 1211 (1969); R. C. Miller and W. A. Nordland, *Phys. Rev.* **B2**, 4896 (1970).

19 J. Ducuing and N. Bloembergen, *Phys. Rev. Lett.* **10**, 474 (1963); R. K. Chang and N. Bloembergen, *Phys. Rev.* **144**, 775 (1966).

20 S. K. Kurtz and T. T. Perry, *J. Appl. Phys.* **39**, 3798 (1968); S. K. Kurtz, *IEEE J. Quant. Electron.* **QE-4**, 578 (1968).

21 S. A. Akhmanov, A. S. Chirkin, K. N. Drabovich, A. I. Kovrigin, R. V. Khokhlov, and A. P. Sukhorukov, *IEEE J. Quant. Electron.* **QE-4**, 598 (1968).

22 J. Comly and E. Garmire, *Appl. Phys. Lett.* **12**, 7 (1968).

23 S. Shapiro, *Appl. Phys. Lett.* **13**, 19 (1968).

BIBLIOGRAPHY

Akhmanov, S. A., A. I. Kovrygin, and A. P. Sukhorukov, in H. Rabin and C. L. Tang, eds., *Quantum Electronics* (Academic Press, New York, 1972), Vol. 1, p. 476.

Bloembergen, N., *Nonlinear Optics* (Benjamin, New York, 1965).

Kleinman, D. A., in F. T. Arecchi and E. O. Schulz-Dubois, eds., *Laser Handbook* (North-Holland Publishing Co., Amsterdam, 1972), p. 1229.

Pressley, R. J., ed., *Handbook of Lasers*, (Chemical Rubber Co., Cleveland, Ohio, 1971), p. 489.

Zernike, F. and J. E. Midwinter, "Applied Nonlinear Optics" (Wiley, New York, 1973)

8

Difference-Frequency Generation

Difference-frequency generation is theoretically not very different from sum-frequency generation, but the problem is of great technical importance in its own right, as it provides a means for generating intense coherent tunable radiation in the infrared. Traditionally, blackbody radiation has been the only practical infrared source. Yet, governed by the Planck distribution, it has weak radiative power in the medium- and far-infrared range. A 1-cm^2, 5000 K blackbody radiates 3500 W over the 4π solid range, but its far-infrared content in the spectral range of 50 ± 1 cm^{-1} is only 3×10^{-6} W/cm$^2 \cdot$ sterad. Infrared lasers may seem to have all the desired properties as infrared sources but their output frequencies generally are discrete with essentially zero tunability. Tunable infrared radiation can, however, be generated by difference-frequency generation. Being coherent with high average or peak intensity, it may find many applications in the field of infrared sciences. This chapter deals mainly with infrared generation by difference-frequency mixing. The diffraction effect is considered in the long wavelength limit. Far-infrared generation by ultrashort pulses is also discussed.

8.1 PLANE-WAVE SOLUTION

In the infinite plane-wave approximation, the theory for difference-frequency generation follows almost exactly that of sum-frequency generation if the pump intensities can be approximated as constant. Then the output power at $\omega_2 = \omega_3 - \omega_1$ generated from the bulk is

$$
\begin{aligned}
\mathbb{P}(\omega_2, z) \\
= \frac{8\pi^3 \omega_2^2}{c^3\sqrt{\varepsilon(\omega_1)\varepsilon(\omega_2)\varepsilon(\omega_3)}} |\hat{e}_2 \cdot \boldsymbol{\chi}^{(2)}(\omega_2 = \omega_3 - \omega_1) : \hat{e}_3\hat{e}_1|^2 \frac{\sin^2(\Delta kz/2)}{(\Delta kz/2)^2} z^2 \\
\times \frac{\mathbb{P}(\omega_1)\mathbb{P}(\omega_3)}{A}.
\end{aligned}
\tag{8.1}
$$

With phase matching and in the presence of appreciable pump depletion, the solution of Section 6.7 must be used. However, the usual initial boundary condition is $u_2(z = 0) = 0$ [i.e., $\mathscr{E}_2(0) = 0$; the notations here follow those in Section 6.7.] in the present case. The equation to be solved becomes ($\theta = -\pi/2$)

$$\frac{du_2^2}{d\zeta} = +2\left[u_2^2(m_1 - u_2^2)(m_3 + u_2^2)\right]^{1/2} \tag{8.2}$$

with $m_1 = u_3^2(0)$ and $m_3 = u_1^2(0)$. The solution takes the form

$$u_2^2(\zeta) = -u_1^2(0)\mathrm{sn}^2\left[\imath u_3(0)\zeta, \gamma\right],$$
$$u_1^2(\zeta) = u_1^2(0) - u_1^2(0)\mathrm{sn}^2\left[iu_3(0)\zeta, \gamma\right], \tag{8.3}$$
$$u_3^2(\zeta) = u_3^2(0) + u_1^2(0)\mathrm{sn}^2\left[\imath u_3(0)\zeta, \gamma\right]$$

where

$$\mathrm{sn}^{-1}\left(\frac{iu_2}{u_1(0)}, \gamma\right) = \int_0^{\imath u_2/u_1(0)} \frac{dy}{\left[(1 - y^2)(1 - \gamma^2 y^2)\right]^{1/2}}$$

$$\gamma = \frac{iu_1(0)}{u_3(0)}.$$

In the particular simply case where $u_2^2(\zeta) \ll u_3^2(0)$ or $|\gamma^2 y^2| \ll 1$, we have $\mathrm{sn}[iu_3(0)\zeta] \cong i\sinh[u_3(0)\zeta]$ and hence

$$u_2^2(\zeta) = u_1^2(0)\sinh^2\left[u_3(0)\zeta\right],$$
$$u_1^2(\zeta) = u_1^2(0)\cosh^2\left[u_3(0)\zeta\right], \tag{8.4}$$
$$u_3^2(\zeta) = u_3^2(0) - u_1^2(0)\sinh^2\left[u_3(0)\zeta\right] \cong u_3^2(0).$$

For $|u_3(0)\zeta| \ll 1$, this solution leads to (8.1) with $\Delta k = 0$. The above results can of course be obtained directly from the coupled wave equations of (6.25) by letting $\mathscr{E}_3$ be a constant. The more general solution with $\mathscr{E}_3 \cong$ constant, $\mathscr{E}_1(0) \neq 0$, $\mathscr{E}_2(0) \neq 0$, and $\Delta k \neq 0$ can also be obtained, but this is postponed to the next chapter in connection with a discussion of parametric amplification.

The plane-wave approximation adopted here is good as long as the output wavelength is much smaller than the beam cross section. The foregoing results should describe fairly well near-IR and mid-IR generation by difference-frequency mixing. Experimental reports on the subject are numerous, and have been summarized in recent review articles.[1,2] An important fact to realize is that the efficiency of infrared generation is expected to be low because of its dependence on the square of the output frequency as seen in (8.1).

8.2 FAR-INFRARED GENERATION BY DIFFERENCE-FREQUENCY MIXING

The infinite plane-wave approximation ceases to be valid for far-infrared generation in the long-wavelength limit as diffraction becomes important when the pump beam diameter appears comparable to the far-IR wavelength. We must look for a better solution of the wave equation

$$\left[\nabla \times (\nabla \times) - \frac{\omega_2^2}{c^2}\varepsilon(\omega_2)\cdot\right]\mathbf{E}(\omega_2) = \frac{4\pi\omega_2^2}{c^2}\mathbf{P}^{(2)}(\omega_2) \tag{8.5}$$

with $\mathbf{P}^{(2)}(\omega_2) = \chi^{(2)}(\omega_2 = \omega_3 - \omega_1) : \mathbf{E}(\omega_3)\mathbf{E}^*(\omega_1)$. Since the conversion efficiency is expected to be small because of the small ω_2, the depletion of the pump fields can be neglected and the amplitude of $\mathbf{P}^{(2)}$ can be regarded as independent of propagation.

If we neglect the boundary reflections by assuming that the nonlinear crystal is immersed in an infinite index-matched linear medium, the far-field solution of (8.5) has the familiar expression[3]

$$\mathbf{E}(\mathbf{r}, \omega_2) = \left(\frac{\omega_2}{c}\right)^2 \int_V d^3r'(1 - \hat{r}\hat{r})\cdot\frac{\mathbf{P}^{(2)}(\mathbf{r}', \omega_2)e^{ik_2|\mathbf{r} - \mathbf{r}'|}}{|\mathbf{r} - \mathbf{r}'|} \tag{8.6}$$

where V is the volume of interaction of the pump fields in the nonlinear medium. With $\mathbf{P}^{(2)}(\mathbf{r}', \omega_2)$ known, $\mathbf{E}(\mathbf{r}, \omega_2)$ can be calculated. Consider, for example, the case where the pump fields $\mathbf{E}(\omega_1) = \mathscr{E}_1(\mathbf{r})\exp(ik_1z - i\omega_1t)$ and $\mathbf{E}(\omega_3) = \mathscr{E}_3(\mathbf{r})\exp(ik_3z - i\omega_3t)$ can be approximated by $\mathscr{E}_1(\mathbf{r})$ and $\mathscr{E}_3(\mathbf{r})$ being constant in a cylinder defined by $(x^2 + y^2) \le a^2$ and vanishing elsewhere. The nonlinear polarization is assumed to have the form

$$(1 - \hat{r}\hat{r})\cdot\mathbf{P}^{(2)}(\mathbf{r}', \omega_2) \cong \vec{\mathscr{P}}^{(2)}e^{i(k_{2s}z' - \omega_2t)} \quad \text{for } (x^2 + y^2) \le a^2$$
$$= 0 \quad \text{for } (x^2 + y^2) > a^2 \tag{8.7}$$
$$k_{2s} \equiv k_3 - k_1.$$

With this expression of $\mathbf{P}^{(2)}$, the integral in (8.6) can be readily evaluated. Let

$$\mathbf{r} = \hat{z}r\cos\phi + \hat{x}r\sin\phi \quad \text{(see Fig. 8.1)},$$
$$\mathbf{r}' = \hat{x}\rho'\cos\theta + \hat{y}\rho'\sin\theta + \hat{z}z',$$

and

$$\exp\frac{(ik_2|\mathbf{r} - \mathbf{r}'|)}{|\mathbf{r} - \mathbf{r}'|} \cong \exp\frac{[ik_2r - ik_2(\mathbf{r}\cdot\mathbf{r}'/r)]}{r}$$

for the far field. Equation (8.6) then becomes[4,5]

$$
\begin{aligned}
\mathbf{E}(\mathbf{r}, \omega_2) &= \frac{\omega_2^2}{c^2} \vec{\mathscr{P}}^{(2)} \frac{e^{i(k_2 r - \omega_2 t)}}{r} \int_{-l/2}^{l/2} dz' e^{-ik_{2z}'(1 - \cos\phi + \Delta k/k_2)} \\
&\quad \times \int_0^a d\rho' \int_0^{2\pi} \rho' d\theta e^{-ik_2 \rho' \cos\theta \sin\phi} \\
&= \frac{2\pi\omega_2^2}{c^2} \vec{\mathscr{P}}^{(2)} \frac{e^{i(k_2 r - \omega_2 t)}}{r} \left(\frac{\sin\alpha}{\alpha}\right) l \int_0^a d\rho' \rho' J_0\left(\frac{\beta\rho'}{a}\right)
\end{aligned} \tag{8.8}
$$

with

$$
\alpha = \frac{k_2 l}{2}\left(1 + \frac{\Delta k}{k_2} - \cos\phi\right), \qquad \beta = k_2 a \sin\phi, \qquad \Delta k = k_{2s} - k_2,
$$

and J_n being Bessel's function. We now have

$$
|E(\mathbf{r}, \omega_2)|^2 = \left(\frac{\omega_2^2}{c^2}\right)|\chi_{\text{eff}}^{(2)}\mathscr{E}_1\mathscr{E}_3|^2 \frac{l^2(\pi a^2)^2}{r^2}\left(\frac{\sin\alpha}{\alpha}\right)^2\left[\frac{2J_1(\beta)}{\beta}\right]^2. \tag{8.9}
$$

Integration of $c\sqrt{\varepsilon(\omega_2)}\,|E(\mathbf{r}, \omega_2)|^2/2\pi$ over the detector surface (Fig. 8.1) yields the total far-infrared power $\mathbb{P}(\omega_2)$ collected by the detector as

$$
\mathbb{P}(\omega_2) = \int_0^{\phi_{\text{max}}} |E(\mathbf{r}, \omega_2)|^2 2\pi r^2 \sin\phi \, d\phi. \tag{8.10}
$$

This result is physically understandable. The $[2J_1(\beta)/\beta]^2$ term arises from diffraction from a circular aperture as usual, and the $(\sin\alpha/\alpha)^2$ term describes

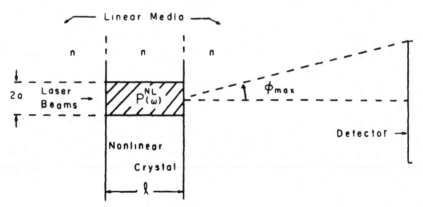

Fig. 8.1 Schematic for calculations of power output. The laser beams produce a nonlinear polarization in the crystal at the difference frequency; the polarization is then treated as a source for the difference-frequency generation.

the phase-matching condition. In the limit of $k_2 a \gg 1$ so that the diffraction effect is expected to be negligible, $2J_1(\beta)/\beta$ is appreciable only for $\phi \lesssim 1/k_2 a$, and then $(\sin\alpha/\alpha)^2$ reduces approximately to the usual phase-matching factor $[\sin(\Delta kl/2)/(\Delta kl/2)]^2$. Also, for $k_2 a \gg 1$, if the detector is large enough so that $\phi_{max} \gg 1/k_2 a$, we have $\int_0^{\phi max}[2J_1(\beta)/\beta]^2 \sin\phi \, d\phi \cong 2/k_2^2 a^2$. The output power $\mathbb{P}(\omega_2)$ calculated from (8.10) can then be shown to have an expression exactly the same as that in (8.1) derived from the infinite plane-wave approximation.

The theory here properly takes into account the diffraction effect. Equations (8.9) and (8.10) can in fact be used for an order-of-magnitude estimate of the far-infrared output. In the long-wavelength limit, the output approaches the ω_2^4 dependence on the frequency as one would expect from the dipole radiation theory. This suggests that the efficiency of difference-frequency generation should decrease drastically toward longer wavelengths in the far-IR region. Even so, with the commonly available lasers, the far-IR output from difference-frequency generation still can be much more intense than a blackbody radiation source.

A number of simplifying approximations have been employed in the derivation leading to (8.9). It is possible to use a more realistic expression for $\mathbb{P}^{(2)}(\mathbf{r}', \omega_2)$ in (8.6) and evaluate the integral numerically to yield a better result. However, the assumption that the nonlinear medium is immersed in an index-matched linear medium is fairly ideal and is usually a poor approximation. In practice, a nonlinear crystal in air has a very different refractive index at far-IR wavelengths than that of the air. Consequently, reflections of far-IR waves at the boundary surfaces are very important. In treating waves at the boundaries of the nonlinear crystal, one cannot use the far-field approximation. This makes the foregoing theoretical approach inappropriate for dealing with the boundary effects. In order to properly take into account the boundary effects, one should decompose the spatially dependent far-IR field into spatial Fourier components and impose the boundary conditions on each component separately. The calculation naturally becomes much more complicated, and numerical solution is often necessary for elucidation. We discuss here only some of the physical results and refer the readers to the literature for the details of the calculation.[5,6]

Since the far infrared refractive index of a solid is usually large (~ 5), reflection at a solid–air boundary can be high. Even multiple reflections can be significant, and in a crystal slab they give rise to a Fabry–Perot factor to each Fourier component in the output. The long wavelength of the far-IR field makes the phase-matching angle less critical, so that phase matching can be satisfied approximately by far-IR output over a fairly broad cone. This cone is substantially broadened outside the crystal through refraction at the boundary. Part of the far-IR radiation may not even be able to get out of the crystal because of total reflection. Focusing of the pump beams generally helps, but absorption hurts the far-IR output as expected. The output field in (8.6) incorporated with an average transmission coefficient can in fact be a very good approximation if the realistic $P^{(2)}(r', \omega_2)$ is used in the calculation. Equation (8.1) obtained from the infinite plane-wave approximation, however, gives a poor description of the far-IR generation.

Experimentally, far-IR generation by difference-frequency mixing has been observed in numerous cases,[2, 5] with output frequencies ranging from 1 to several hundred inverse centimeters. For example, in $LiNbO_3$, $\chi^{(2)}_{YYY}(\omega = \omega_1 - \omega_2) = 4.5 \times 10^{-8}$ esu for $\omega_1 \sim \omega_2$ around the ruby laser frequency. If the pump laser beams are of 1 MW each over an area of 0.2 cm^2, a far-IR power of ~ 3 mW at 10 cm^{-1} is expected from (8.10) to be generated from a $LiNbO_3$ crystal 0.05 cm thick under the phase-matching condition. In a real experiment, 1 mW at 8.1 cm^{-1} was detected from a crystal of 0.047 cm.[4] Discretely tunable CW far-IR output of 10^{-7} W has also been observed from mixing of two CO_2 lasers (25 W) in GaAs.[7] Tunable far-IR radiation can also be generated by stimulated polariton scattering and by spin-flip Raman transitions. We postpone their discussion to Chapter 10.

8.3 FAR-INFRARED GENERATION BY ULTRASHORT PULSES

The discussion in previous sections on infrared generation by optical mixing applies to cases where the pump beams are quasi-monochromatic. The two pump pulses are assumed to be sufficiently long, and the spectral purity of the infrared output, generally correlated with the laser spectral widths, is limited by the pulsewidth. Here, however, we consider the case of far-IR generation by a single short laser pulse.[8, 9] If the laser pulsewidth is as short as 1 psec, the corresponding spectral width should be at least 15 cm^{-1}. Then, in a nonlinear crystal, the various spectral components of the pulse can beat with one another and generate far-IR radiation up to the submillimeter range. One might consider this an optical rectification process in which a dc picosecond pulse is generated. However, unlike the case discussed in Section 5.1, we are here interested only in the radiative component of the rectified field. This generation of the radiative output is subject to the influences of phase matching, diffraction, boundary reflection, radiation efficiency, and so on.[10]

Far-IR generation by ultrashort pulses is, as usual, governed by the wave equation

$$\left[\nabla \times (\nabla \times E) + \frac{1}{c^2} \frac{\partial^2}{\partial t^2} \varepsilon \cdot E \right] = -\frac{4\pi}{c^2} \frac{\partial^2}{\partial t^2} P^{(2)}(r, t). \qquad (8.11)$$

Given $P^{(2)}(r, t)$, (8.11) with appropriate boundary conditions can, at least in principle, be solved. The far-IR output and its power spectrum can then be calculated. The solution of far-IR generation by a single short pulse in a thin slab of nonlinear crystal has actually been obtained through the Fourier

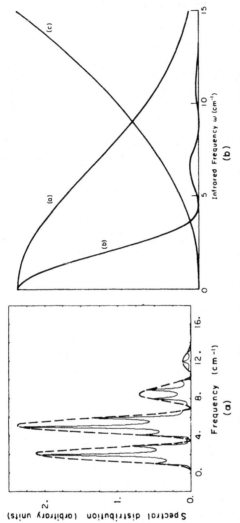

Fig. 8.2 (*a*) The calculated far-infrared output spectrum for a 2-psec Nd laser pulse normally incident on a 1-mm LiNbO₃ slab. The crystal is oriented with the *c*-axis parallel to the plane surfaces of the slab, and the laser pulse is polarized along the *c*-axis. The solid curve under the dashed envelope arises from the Fabry-Perot geometry of the slab. (*b*) Curves showing the three major contributions to the far-infrared output spectrum in (*a*). Curve *a* gives the power spectrum of the rectified input pulse. Curve *b* gives the phase-matching effect for phase matching at $\omega = 0$. Curve *c* gives the dipole radiation efficiency. (After Ref. 5.)

transform of $\mathbf{E}(\mathbf{r}, t)$ and $\mathbf{P}^{(2)}(\mathbf{r}, t)$, neglecting the dispersion of ε and $\chi^{(2)}$ in $\mathbf{P}^{(2)}(\mathbf{r}, t) = \chi^{(2)} : \mathbf{E}(\mathbf{r}, t)\mathbf{E}^*(\mathbf{r}, t)$.[10] We present here a physical description of the solution.

Figure 8.2a shows the calculated power spectrum of far-IR radiation generated by a 2-psec Nd laser pulse from a 1-mm LiNbO$_3$ slab. First, the Fabry–Perot geometry of the slab gives rise to the interference pattern under the dotted envelope. Then the dotted envelope of the spectrum is basically the product of three contributions seen in Fig. 8.2b: curve a represents the power spectrum of the rectified input pulse, curve c describes the ω^2 dependence of the radiation efficiency with a much sharper low-frequency cutoff as $\omega \to 0$ due to diffraction; curve b is the phase-matching curve with its phase-matching peak at $\omega = 0$ for the particular crystal orientation with the $\hat{c}$-axis along the face of the slab. Thus, the calculated power spectrum in Fig. 8.2a can be physically understood.

Such theoretical calculation actually gives a very good description of the experimental observation. Figure 8.3a shows a comparison between theory and experiment for far-IR generation from a 0.775-mm LiNbO$_3$ with the $\hat{c}$-axis in the slab face by a train of normally incident mode-locked Nd/glass laser pulses.[8] The Fabry–Perot pattern is absent here because the spectrum has been averaged over the actual instrument resolution. By orienting the crystal to

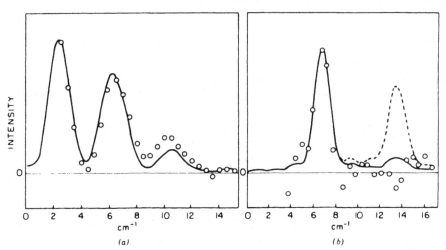

Fig. 8.3 (a) Far-infrared spectrum generated by mode-locked pulses in LiNbO$_3$ phase matched at zero frequency. The experimental points were obtained from the Michelson interferogram and the solid curve from the theoretical calculation assuming Gaussian laser pulses with a 1.8 psec pulsewidth. (b) Far-infrared spectrum generated by mode-locked pulses in LiNbO$_3$ oriented to have forward and backward phase matching at 13.5 and 6.7 cm^{-1}, respectively. The experimental points were obtained from the Michelson interferogram. The solid and dashed curves were calculated by assuming Gaussian laser pulses with a pulse-width of 2.3 and 1.8 psec, respectively. (After Ref. 8.)

achieve phase matching at finite ω, one expects, from the above discussion, that a single phase-matching peak at $\omega \neq 0$ may dominate the output spectrum. An example is seen in Fig. 8.3b. Again, theory and experiment agree well. The two peaks in the theoretical curves correspond to phase-matched generation of the far-IR radiation in the forward and backward directions, respectively. This figure suggests that we can have tunable far-IR output by simply rotating the nonlinear crystal. As shown, the pulse still has a fairly broad linewidth, indicating that it is also a pulse of picosecond duration. Nevertheless, since the output linewidth is appreciably narrower than the laser linewidth, the output pulse must be appreciably longer than the input pulse. That the output is still significant after the input has more or less decayed away is an interesting fact considering that the medium response to the input pulse is essentially instantaneous in this case. With an input peak power of 0.2 GW over a cross section of 1 cm^2, a far-IR output of 200 W peak power has been detected from a 0.78-mm LiNbO$_3$ crystal.[8]

REFERENCES

1   R. L. Byer, in Y. R. Shen, ed. *Nonlinear Infrared Generation* (Springer-Verlag, Berlin, 1977).

2   F. Zernike, in C. L. Tang, ed. *Methods of Experimental Physics*, Vol. XV: *Quantum Electronics*, Part B (Academic Press, New York, 1979), p. 143.

3   See, for example, J. D. Jackson, *Classical Electrodynamics*, 2nd ed. (Wiley, New York, 1975), Chapter 9.

4   D. W. Faries, K. A. Gehring, P. L. Richards, and Y. R. Shen, *Phys. Rev.* **180**, 363 (1969).

5   Y. R. Shen, *Prog. Quant. Electron.* **4**, 207 (1976).

6   J. R. Morris and Y. R. Shen, *Phys. Rev.* **A15**, 1143 (1977).

7   B. Lax, R. L. Aggarwal, and G. Favrot, *Appl. Phys. Lett.* **23**, 679 (1973); B. Lax and R. L. Aggarwal, *J. Opt. Soc. Am.* **64**, 533 (1974).

8   K. H. Yang, P. L. Richards, and Y. R. Shen, *Appl. Phys. Lett.* **19**, 285 (1971).

9   T. Yajima and N. Takeuchi, *Jap. J. Appl. Phys.* **10**, 907 (1971).

10   J. R. Morris and Y. R. Shen, *Optics Comm.* **3**, 81 (1971).

BIBLIOGRAPHY

Shen, Y. R., *Prog. Quant. Electron.* **4**, 207 (1976).

Shen, Y. R., ed., *Nonlinear Infrared Generation* (Springer-Verlag, Berlin, 1977).

Warner, J., in H. Rabin and C. L. Tang, eds. *Quantum Electronics* (Academic Press, New York, 1973), vol. 1, p. 703.

9

Parametric Amplification and Oscillation

The three-wave interaction discussed in previous chapters is manifested by energy flow from the two lower-frequency fields to the sum-frequency field or vice-versa. The latter happens in difference-frequency generation, which, in general, can be initiated with a single pump beam at the sum frequency. Difference-frequency generation can then be considered as the inverse process of sum-frequency generation, and is generally known as a parametric conversion process. Parametric amplification and oscillation in the radio frequency and microwave range were developed before the laser was invented.[1] The same process was expected in the optical region, and was actually demonstrated in 1965.[2] It has since become an important effect because it allows the construction of widely tunable coherent infrared sources through the controllable decomposition of the pump frequency. In this chapter we explore the theory of parametric fluorescence, amplification, and oscillation together with some practical considerations.

9.1 PARAMETRIC AMPLIFICATION

As an inverse process of sum-frequency generation, the general theory of parametric amplification is the same as that for difference-frequency generation. In fact, the only difference of the two processes is in the input conditions. Even there, the difference is not clear-cut, but we normally consider parametric amplification as a process initiated by a single pump beam while difference-frequency generation is initiated by two pump beams of more or less comparable intensities. The difference disappears after a significant fraction of the pump energy has been transferred to the two lower frequency fields. Thus the theoretical description of parametric amplification with infinite plane waves again starts from the set of three coupled wave equations (3.4). In the slowly varying amplitude approximation with $E(\omega_i) = \mathscr{E}_i(z)\exp[i(\mathbf{k}_i \cdot \mathbf{r} - \omega_i t + \phi_i)]$ and a plane boundary at $z = 0$, they become (see Sections 3.3 and 6.7)

$$\frac{\partial}{\partial z}\mathscr{E}_1 = \frac{i\omega_1^2}{k_{1z}} K^*\mathscr{E}_2^*\mathscr{E}_3 e^{i\,\Delta k\,z + i\theta_0},$$

$$\frac{\partial}{\partial z}\mathscr{E}_2^* = \frac{-i\omega_2^2}{k_{2z}} K\mathscr{E}_1\mathscr{E}_3^* e^{-i\,\Delta k\,z - i\theta_0}, \qquad (9.1)$$

$$\frac{\partial}{\partial z}\mathscr{E}_3 = \frac{i\omega_3^2}{k_{3z}} K\mathscr{E}_1\mathscr{E}_2 e^{-i\,\Delta k\,z - i\theta_0}$$

where

$$K = \frac{2\pi}{c^2}\hat{e}_3 \cdot \chi^{(2)}(\omega_3 = \omega_1 + \omega_2) : \hat{e}_1\hat{e}_2 \quad \text{and} \quad \Delta k = k_{3z} - k_{1z} - k_{2z},$$

and $\theta_0 = \phi_3 - \phi_1 - \phi_2$ is the initial phase difference of the fields at $z = 0$.[3] We assume here $\theta_0 = -\pi/2$. The solution of (9.1) with $\Delta k = 0$ has been discussed in previous chapters in connection with sum- and difference-frequency generation. In parametric amplification, $E(\omega_3)$ is known as the pump wave, $E(\omega_1)$ [or $E(\omega_2)$] the signal wave, and $E(\omega_2)$ [or $E(\omega_1)$] the idler wave. We consider here first the case of negligible pump depletion with $\Delta k \neq 0$.

The assumption of negligible pump depletion means that $\mathscr{E}_3$ can be regarded as a constant. Equation (9.1) then reduces to a set of two linearly coupled equations between $\mathscr{E}_1$ and $\mathscr{E}_2^*$. Writing $\mathscr{E}_1 = C_1 e^{i\gamma_1 z}$ and $\mathscr{E}_2 = C_2 e^{i\gamma_2 z}$, we find immediately $\gamma_1 = -\gamma_2^* + \Delta k$, and

$$\begin{vmatrix} i\gamma_1, & -\dfrac{\omega_1^2}{k_{1z}}K\mathscr{E}_3 \\[2mm] -\dfrac{\omega_2^2}{k_{2z}}K\mathscr{E}_3^*, & (i\gamma_1 - i\,\Delta k) \end{vmatrix} = 0. \qquad (9.2)$$

This leads to the solution

$$\gamma_{1\pm} = \tfrac{1}{2}[\Delta k \pm ig],$$

$$g = \left[g_0^2 - (\Delta k)^2\right]^{1/2}, \qquad g_0^2 = \left(\frac{4\omega_1^2\omega_2^2}{k_{1z}k_{2z}}\right)K^2\mathscr{E}_3^2,$$

$$\mathscr{E}_1 = \left[C_{1+}e^{-(1/2)gz} + C_{1-}e^{(1/2)gz}\right]e^{+i(1/2)\Delta kz},$$

$$\mathscr{E}_2^* = \left[C_{2+}e^{-(1/2)gz} + C_{2-}e^{(1/2)gz}\right]e^{-i(1/2)\Delta kz}, \qquad (9.3)$$

$$C_{1\pm} = \pm\frac{1}{2g}\left[(i\,\Delta k \pm g)\mathscr{E}_1(0) - \frac{\omega_1}{\omega_2}\left(\frac{k_{2z}}{k_{1z}}\right)^{1/2}g_0\mathscr{E}_2^*(0)\right],$$

$$C_{2\pm} = \mp\frac{1}{2g}\left[(i\,\Delta k \mp g)\mathscr{E}_2^*(0) + \frac{\omega_2}{\omega_1}\left(\frac{k_{1z}}{k_{2z}}\right)^{1/2}g_0\mathscr{E}_1(0)\right].$$

This solution shows the following physical properties. If $K\mathscr{E}_3$ is small so that $g_0^2 < (\Delta k)^2$, then g is purely imaginary. If $K\mathscr{E}_3$ is sufficiently large so that $g_0^2 > (\Delta k)^2$, then g is real and positive, and at large gz, both $\mathscr{E}_1$ and $\mathscr{E}_2$ grow exponentially with z. Thus $g_0 = (\Delta k)$ is the threshold for parametric amplification. The parametric gain is clearly a maximum with $g = g_0$ at phase matching, $\Delta k = 0$. Introduction of attenuation coefficients at ω_1 and ω_2 in the above formalism is straightforward. As expected, they increase the threshold and decrease the gain.

As an example, consider parametric amplification in $LiNbO_3$ with $\chi_{eff}^{(2)} = 2.7 \times 10^{-8}$ esu at $\lambda_1 \sim \lambda_2 \simeq 1.06$ μm with $n_1 = n_2 = 2.23$. The maximum gain is found to be $g_0 = 0.9 \times 10^{-2}\mathscr{E}_3$ cm^{-1}. For a pump field of $\mathscr{E}_3 = 100$ esu corresponding to 2.5 MW/cm^2, the gain is 0.9 cm^{-1}. Thus the overall exponential gain gl even in a crystal of length $l = 5$ cm is not very large. To achieve an overall gain of $g_0 l \sim 40$, we must either use a pump beam of much higher intensity (which is attainable only with picosecond pulses if optical damage to the crystal is to be avoided) or use an optical cavity to increase the effective length. In the latter case, the system may become an oscillator, as will be discussed later.

As noted in (9.3), the phase mismatch Δk suppresses the gain very effectively. Therefore, in the limit of high conversion efficiency, we need only consider the phase-matching case although the general solution of (9.1) with $\Delta k \neq 0$ has been worked out by Armstrong et al.[4] Following the notations and derivation in Section 6.7, we find, assuming $\theta_0 = -\pi/2$ in (9.1),

$$\frac{d}{d\zeta}u_2^2 = 2\left[u_2^2(m_1 - u_2^2)(m_3 - u_2^2)\right]^{1/2}, \qquad (9.4)$$

which has the solution [assuming $u_1^2(0) < u_2^2(0)$]

$$u_2^2(\zeta) = \left(u_2^2(0) - u_1^2(0)\right)\mathrm{sn}^2$$

$$\times \left[i\sqrt{u_3^2(0) + u_2^2(0)}\ \zeta + \mathrm{sn}^{-1}\left(i\frac{u_2(0)}{\sqrt{u_1^2(0) - u_2^2(0)}},\gamma\right),\gamma\right]$$

$$u_3^2(\zeta) = u_2^2(0) + u_3^2(0) - u_2^2(\zeta)$$

$$\qquad\qquad\qquad (9.5)$$

$$u_1^2(\zeta) = u_1^2(0) - u_2^2(0) + u_2^2(\zeta)$$

$$\gamma^2 = \frac{u_2^2(0) - u_1^2(0)}{u_2^2(0) + u_3^2(0)}.$$

In the case of $u_2(0) = 0$, this result can be shown to reduce to (8.3) derived for difference-frequency generation.

9.2 DOUBLY RESONANT PARAMETRIC OSCILLATOR

As mentioned earlier, an optical cavity can be used to increase the overall gain in parametric amplification. Then, parametric oscillation can also occur. That tunable output is possible in the absence of input makes the parametric oscillator a practically more useful device than the parametric amplifier.

As shown in Fig. 9.1, a parametric oscillator is composed of a nonlinear optical crystal sitting in an optical cavity. For simplicity, we assume that the cavity is formed by two plane parallel mirrors. Two types of cavity commonly are used. The doubly resonant cavity has the mirrors strongly reflecting at both ω_1 and ω_2, and the singly resonant cavity has the mirrors strongly reflecting at either ω_1 or ω_2. Usually, both mirrors are transparent to the pump wave. If the single-pass parametric gain is small, the pump intensity can be regarded as independent of the distance in the cavity.

Let us consider the doubly resonant case first.[3-5] The fields in the cavity can be written as

$$\mathbf{E}(\omega_1) = 2\mathscr{E}_1(t)\sin k_1 z e^{-i\omega_{10}t},$$
$$\mathbf{E}(\omega_2) = 2\mathscr{E}_2(t)\sin k_2 z e^{-i\omega_{20}t}, \qquad (9.6)$$
$$\mathbf{E}(\omega_3) = \mathscr{E}_3 e^{ik_3 z - i\omega_3 t}$$

with $\omega_1 \cong \omega_{10}$, $\omega_2 \cong \omega_{20}$, and $\omega_3 = \omega_1 + \omega_2 = \omega_{10} + \omega_{20} + \Delta\omega$. The cavity condition requires

$$k_1 = \frac{\omega_{10}n_1}{c} = \frac{m_1\pi + \Phi_1}{l} \quad \text{and} \quad k_2 = \frac{\omega_{20}n_2}{c} = \frac{m_2\pi + \Phi_2}{l} \qquad (9.7)$$

where m_1 and m_2 are positive integers, l is the cavity length (we assume here, for simplicity, that the crystal length is equal to the cavity length), and $2\Phi_1$ and $2\Phi_2$ are round-trip phase shifts at ω_1 and ω_2 due to boundary reflections and refractions. The coupled wave equations in this case become

$$\left[k_1^2 + \frac{n_1^2}{c^2}\left(\Gamma_1\frac{\partial}{\partial t} + \frac{\partial^2}{\partial t^2}\right)\right]\mathbf{E}(\omega_1) = +\frac{4\pi\omega_1^2}{c^2}\chi^{(2)} : \mathbf{E}^*(\omega_2)\mathbf{E}(\omega_3),$$

$$\left[k_2^2 + \frac{n_2^2}{c^2}\left(\Gamma_2\frac{\partial}{\partial t} + \frac{\partial^2}{\partial t^2}\right)\right]\mathbf{E}^*(\omega_2) = +\frac{4\pi\omega_2^2}{c^2}\chi^{(2)} : \mathbf{E}^*(\omega_3)\mathbf{E}(\omega_1). \quad (9.8)$$

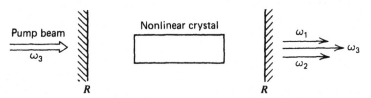

Fig. 9.1 Schematic diagram of a parametric oscillator.

Here, Γ_1 and Γ_2 are the damping constants at ω_1 and ω_2. They constitute the attenuation loss due to absorption and scattering in the cavity plus the mirror transmission loss. With an intensity attenuation coefficient of $\alpha_i(\omega_i)$ per unit length and two mirrors of equal reflectivity $R_i(\omega_i)$, Γ_i is obtained from the definition

$$e^{-2\Gamma_i n_i l/c} = R_i^2 e^{-2\alpha_i l}$$

or

$$\Gamma_i = \frac{c}{n_i}\left[\alpha_i - \frac{1}{l}\ln R_i\right]$$

$$\cong \frac{c}{n_i}\left[\alpha_i + \frac{1}{l}(1 - R_i)\right] \quad \text{if } R_i \cong 1. \tag{9.9}$$

We assume $\Gamma_1 = \Gamma_2$. With the slowly varying amplitude approximation, $|\partial^2 \mathcal{E}_i/\partial t^2| \ll |\omega_i \partial \mathcal{E}_i/\partial t|$, (9.8) reduces to

$$\left[\frac{\partial}{\partial t} + \frac{1}{2}\Gamma\right]\mathcal{E}_1(t)\sin k_1 z = i\omega_{10}\left(\frac{c}{n_1}\right)^2 K\mathcal{E}_2^*(t)\mathcal{E}_3 \sin k_2 z e^{ik_3 z}, \tag{9.10a}$$

$$\left[\frac{\partial}{\partial t} + \frac{1}{2}\Gamma\right]\mathcal{E}_2^*(t)\sin k_2 z = -i\omega_{20}\left(\frac{c}{n_2}\right)^2 K\mathcal{E}_1(t)\mathcal{E}_3 \sin k_1 z e^{ik_3 z} \tag{9.10b}$$

where

$$K = \frac{2\pi}{c^2}\hat{e}_1 \cdot \chi^{(2)}(\omega_1 = -\omega_2 + \omega_3) : \hat{e}_2 \hat{e}_3$$

and we treat $\mathcal{E}_3$ as a constant. Multiplying (9.10a) and (9.10b) by $\sin k_1 z$ and $\sin k_2 z$, respectively, and integrating the equations from $z = 0$ to $z = l$ we find

$$\left[\frac{\partial}{\partial t} + \frac{1}{2}\Gamma\right]\mathcal{E}_1(t) = i\tfrac{1}{2}\omega_{10}\left(\frac{c}{n_1}\right)^2 K\mathcal{E}_3\mathcal{E}_2^*(t)\frac{\sin(\Delta k \, l/2)}{(\Delta k \, l/2)}e^{i\Delta k \, l/2},$$

$$\left[\frac{\partial}{\partial t} + \frac{1}{2}\Gamma\right]\mathcal{E}_2^*(t) = -i\tfrac{1}{2}\omega_{20}\left(\frac{c}{n_2}\right)^2 K\mathcal{E}_3\mathcal{E}_1(t)\frac{\sin(\Delta k \, l/2)}{(\Delta k \, l/2)}e^{-i\Delta k \, l/2}. \tag{9.11}$$

The solution of this set of linearly coupled equations can be written in the form

$$\mathcal{E}_1(t) = \left(D_{1+}e^{s_+ t} + D_{1-}e^{s_- t}\right)e^{-i(1/2)\Delta\omega t},$$

$$\mathcal{E}_2^*(t) = \left(D_{2+}e^{s_+ t} + D_{2-}e^{s_- t}\right)e^{i(1/2)\Delta\omega t},$$

$$s_\pm = \tfrac{1}{2}[-\Gamma \pm G],$$

$$G = \left[G_0^2 - (\Delta\omega)^2\right]^{1/2}, \tag{9.12}$$

$$G_0^2 = \frac{\omega_{10}\omega_{20}c^4}{n_1^2 n_2^2}K^2\mathcal{E}_3^2\frac{\sin^2(\Delta k \, l/2)}{(\Delta k \, l/2)^2}$$

where $D_{1\pm}$ and $D_{2\pm}$ are coefficients to be determined from (9.11) together with the initial values $\mathscr{E}_1(0)$ and $\mathscr{E}_2^*(0)$. In the discussion of parametric oscillation, however, it is not possible to find $D_{1\pm}$ and $D_{2\pm}$. We are more interested in the threshold condition for oscillation. Equation (9.12) shows that oscillation starts when $G = G_{th} = \Gamma$ or the threshold pump intensity for parametric oscillation is

$$I_{th} = \frac{n_3 c}{2\pi}\left(\mathscr{E}_3^2\right)_{th}$$

$$= \frac{(\Delta\omega)^2 + \Gamma^2}{\dfrac{2\pi\omega_{10}\omega_{20}c^3}{n_1^2 n_2^2 n_3}K^2\dfrac{\sin^2(\Delta k\,l/2)}{(\Delta k\,l/2)^2}}. \tag{9.13}$$

Clearly, the threshold is a minimum for $\Delta\omega = 0$ and $\Delta k = 0$. Thus if we use the same LiNbO$_3$ crystal described in the example of parametric amplification and insert it between two plane mirrors of $R_1 = R_2 = 0.98$ separated by the crystal length $l = 5$ cm, we find a minimum pump threshold of

$$(I_{th})_{min} = 2.5 \text{ kW/cm}^2$$

assuming that the attenuation loss α_i is negligible. This is a fairly low intensity and is achievable even with a CW laser. The doubly resonant CW parametric oscillator was first demonstrated by Smith et al. in 1968.[6]

The solution in (9.12) also shows that the output frequencies of the signal and idler waves are

$$\omega_1 = \omega_{10} + \tfrac{1}{2}\Delta\omega \quad \text{and} \quad \omega_2 = \omega_{20} + \tfrac{1}{2}\Delta\omega \tag{9.14}$$

with ω_{10} and ω_{20} being the cavity resonant frequencies given by (9.7) and $\Delta\omega = \omega_3 - \omega_{10} - \omega_{20}$ is automatically minimized to achieve the lowest pump threshold possible. Tuning of the output frequencies is discussed in Section 9.4. One serious disadvantage of the doubly resonant parametric oscillator is that it has low stability.[7,8] Let us assume that originally the oscillator with a cavity length l operates at $\omega_1 = \omega_{10} = m_1\pi c/n_1 l$, $\omega_2 = \omega_{20} = m_2\pi c/n_2 l$, and $\omega_1 + \omega_2 = \omega_3$. A small change of the cavity length Δl due to external perturbation shifts the output frequencies to $\omega_1' = \omega_{10}' = (m_1 + \Delta m)\pi c/n_1(l + \Delta l)$, $\omega_2' = \omega_{20}' = (m_2 - \Delta m)\pi c/n_2(l + \Delta l)$, and $\omega_1' + \omega_2' = \omega_3$. We then find

$$\Delta m \cong \omega_3\left(\frac{\Delta l}{l}\right)\Big/\left(\frac{\pi c}{n_1 l}\right)\left(\frac{n_2 - n_1}{n_2}\right)$$

The shift in the output frequencies due to Δl is therefore given by

$$\delta\omega = \omega_1' - \omega_1 = \omega_2 - \omega_2'$$

$$= \left[\frac{n_2}{n_2 - n_1}\omega_3 - \omega_1\right]\frac{\Delta l}{l}$$

$$\cong \frac{n_2}{n_2 - n_1}\omega_3\frac{\Delta l}{l}. \tag{9.15}$$

Thus a change of $\Delta l/l = 10^{-7}$, which shifts the frequency of a cavity mode by only 10^{-7}, will cause the output frequencies of the doubly resonant parametric oscillator to change by more than $10^{-5}\omega_3$ if $|n_2 - n_1|/n_2 \leq 10^{-2}$. This shows that the output of the oscillator will be very unstable, subject to external vibration and thermal fluctuations.

For steady-state operation, an oscillator must have its gain clamped to the threshold value, since otherwise the output would continuously grow in time or decay to zero. This makes the calculation of conversion efficiency fairly simple.[5] The pump field in the cavity is self-adjusted through pump depletion by parametric conversion to yield an oscillator gain clamped to the threshold. The signal and idler fields increase with the increasing pump energy, but being standing waves, their amplitudes are constant across the cavity. The part of the pump power coupled into the signal and idler fields should show up as power loss in the cavity and in the signal and idler output through the cavity mirrors. Let us assume the phase-matching case, $\Delta k = 0$. In the slowly varying amplitude approximation, the equation for $\mathscr{E}_3$ reduces to that in (9.1). With $\theta_0 = -\pi/2$, the forward propagating pump field at $z = l$ is

$$\mathscr{E}_3^+(l) = \mathscr{E}_3^+(0) - \left(\frac{\omega_3 c}{n_3}\right) K \mathscr{E}_1 \mathscr{E}_2 l. \tag{9.16}$$

Then the generated fields at ω_1 and ω_2 in the cavity can also react back and generate a backward propagating wave at ω_3 with

$$\mathscr{E}_3^-(0) = \frac{\omega_3 c}{n_3} K \mathscr{E}_1 \mathscr{E}_2 l. \tag{9.17}$$

Energy conservation requires

$$n_3\{|\mathscr{E}_3^+(0)|^2 - |\mathscr{E}_3^+(l)|^2 - |\mathscr{E}_3^-(0)|^2\} = n_1|\mathscr{E}_1|^2(1 - e^{-2\Gamma_1 n_1 l/c})$$
$$+ n_2|\mathscr{E}_2|^2(1 - e^{-2\Gamma_2 n_2 l/c}) \tag{9.18}$$
$$\simeq \left(n_1^2|\mathscr{E}_1|^2\Gamma_1 + n_2^2|\mathscr{E}_2|^2\Gamma_2\right)2l/c$$

That the number of photons generated at ω_1 and ω_2 must be the same leads to

$$\frac{n_1^2|\mathscr{E}_1|^2\Gamma_1}{\omega_1} = \frac{n_2^2|\mathscr{E}_2|^2\Gamma_2}{\omega_2}. \tag{9.19}$$

Combining (9.16) to (9.19) gives

$$\frac{n_2^2\Gamma_2|\mathscr{E}_2|^2}{\omega_2 n_3|\mathscr{E}_3^+(0)|^2} = \frac{n_1^2\Gamma_1|\mathscr{E}_1|^2}{\omega_1 n_3|\mathscr{E}_3^+(0)|^2}$$
$$= \frac{c}{\omega_3 l}\left[\frac{1}{\sqrt{N}} - \frac{1}{N}\right] \tag{9.20}$$

where

$$N = \left[\frac{\omega_1\omega_2 c^4}{n_1^2 n_2^2} K^2 |\mathscr{E}_3^+(0)|^2 \right] \Big/ \Gamma_1\Gamma_2.$$

If we take $\omega_1 = \omega_{10}$, $\omega_2 = \omega_{20}$, and $\Gamma_1 = \Gamma_2$, and use the expression of the gain G_0 in (9.12) for $\Delta k = 0$, we have $N = G_0^2/\Gamma^2 = G_0^2/G_{th}^2$. Since G_0^2 is proportional to the pump intensity, N has the physical meaning of how many times the pump intensity is above the threshold value, with $N = 1$ corresponding to the threshold. The total power output (signal and idler, including attenuation loss in the cavity) is

$$\mathbb{P}_1 + \mathbb{P}_2 = \left(\frac{c}{2\pi} \right) \frac{\left[n_1^2 |\mathscr{E}_1|^2 2\Gamma_1 + n_2^2 |\mathscr{E}_2|^2 2\Gamma_2 \right] l}{c}.$$

The overall conversion efficiency is then given by

$$\eta = \frac{\mathbb{P}_1 + \mathbb{P}_2}{\mathbb{P}_3} = \frac{2}{N}(\sqrt{N} - 1), \tag{9.21}$$

which can reach 50% for $N = 4$.

9.3   SINGLY RESONANT PARAMETRIC OSCILLATOR

Because of its intrinsic instability, doubly resonant parametric oscillators are seldom used in practice even though they have lower oscillation thresholds. The instability can be eliminated by using a singly resonant cavity.[9,10] Let the mirrors be transparent at ω_1 and ω_3, and highly reflective only at ω_2. Then we can describe the three fields as

$$\mathbf{E}(\omega_1) = \mathscr{E}_1(z, t)e^{ik_1 z - i\omega_1 t},$$
$$\mathbf{E}(\omega_2) = 2\mathscr{E}_2(t)\sin k_2 z \, e^{-i\omega_{20} t}, \tag{9.22}$$
$$\mathbf{E}(\omega_3) = \mathscr{E}_3(z)e^{ik_3 z - i\omega_3 t}$$

with $\omega_{20} = m_2\pi c/n_2 l$ and $\omega_1 + \omega_{20} = \omega_3$. Clearly, in this case, a small fractional change of $\Delta l/l$ induces only a shift of $\delta\omega = \omega_{20}(\Delta l/l)$ in the output frequencies.

To find the threshold of oscillation, we can again solve the coupled wave equations for $\mathbf{E}(\omega_1)$ and $\mathbf{E}(\omega_2)$. We shall, however, take a somewhat different approach here. We can start with the solution of parametric amplification in (9.3) and impose the condition that for parametric oscillation, a single-pass gain must be equal to the round-trip loss. Assume the phase-matching case

$\Delta k = 0$. The round-trip attenuations of the two fields are $e^{-\Gamma_1 n_1 l/c}$ and $e^{-\Gamma_2 n_2 l/x}$, respectively. Then from (9.3) we have

$$\mathscr{E}_1(0) = e^{-\Gamma_1 n_1 l/c}\left[\mathscr{E}_1(0)\cosh\tfrac{1}{2}g_{th}l + i\left(\frac{n_2}{n_1}\right)^{1/2}\mathscr{E}_2^*(0)\sinh\tfrac{1}{2}g_{th}l\right],$$

$$\mathscr{E}_2^*(0) = e^{-\Gamma_2 n_2 l/c}\left[\mathscr{E}_2^*(0)\cosh\tfrac{1}{2}g_{th}l - i\left(\frac{n_1}{n_2}\right)^{1/2}\mathscr{E}_1(0)\sinh\tfrac{1}{2}g_{th}l\right]$$

(9.23)

and hence

$$\begin{vmatrix} \cosh\tfrac{1}{2}g_{th}l - e^{\Gamma_1 n_1 l/c}, & i\left(\dfrac{n_2}{n_1}\right)^{1/2}\sinh\tfrac{1}{2}g_{th}l \\[2mm] -i\left(\dfrac{n_1}{n_2}\right)^{1/2}\sinh\tfrac{1}{2}g_{th}l, & \cosh\tfrac{1}{2}g_{th}l - e^{\Gamma_2 n_2 l/c} \end{vmatrix} = 0.$$

(9.24)

This leads to the threshold condition for oscillation

$$\begin{aligned} \cosh\tfrac{1}{2}g_{th}l &= \frac{1 + e^{(\Gamma_1 n_1 + \Gamma_2 n_2)l/c}}{e^{\Gamma_1 n_1 l/c} + e^{\Gamma_2 n_2 l/c}} \\[2mm] &= 1 + \frac{(e^{\Gamma_1 n_1 l/c} - 1)(e^{\Gamma_2 n_2 l/c} - 1)}{e^{\Gamma_1 n_1 l/c} + e^{\Gamma_2 n_2 l/c}}. \end{aligned}$$

(9.25)

The result here is quite general and is applicable also to the doubly resonant case, where $\tfrac{1}{2}g_{th}l$, $\Gamma_1 n_1 l/c$, and $\Gamma_2 n_2 l/c$ are all much smaller than 1. Equation (9.25) reduces to $g_{th}^2 = 4n_1 n_2 \Gamma_1 \Gamma_2/c^2$, which can be shown to be the same as the threshold condition ($G_{th} = \Gamma$) derived in the previous section. Now, for the singly resonant oscillator, only $\tfrac{1}{2}g_{th}l$ and $\Gamma_2 n_2 l/c$ are much smaller than 1, but $\exp(\Gamma_1 n_1 l/c) \gg 1$ because of the large transmission loss of the mirrors. We then find the threshold condition to be

$$\begin{aligned} g_{th}^2 l^2 &= 8\frac{(e^{\Gamma_1 n_1 l/c} - 1)\Gamma_2 n_2 l/c}{1 + e^{\Gamma_1 n_1 l/c}} \\[2mm] &\simeq \frac{8\Gamma_2 n_2 l}{c}. \end{aligned}$$

(9.26)

In comparison with the doubly resonant case, the square gain threshold ratio, which is equal to the pump threshold ratio, is

$$\begin{aligned} \frac{(g_{th}^2)_{\text{Singly}}}{(g_{th}^2)_{\text{Doubly}}} &= \frac{(I_{th})_{\text{Singly}}}{(I_{th})_{\text{Doubly}}} \\[2mm] &\simeq \frac{2}{1 - R_1}. \end{aligned}$$

(9.27)

For $R_1 = 0.98$, this ratio indicates that the pump threshold for the singly resonant oscillator is 100 times higher than that for the doubly resonant oscillator. Thus singly resonant parametric oscillators usually require pulsed lasers as pump sources.

The conversion efficiency can again be calculated from energy or photon number consideration, knowing that the oscillator gain must be clamped to the threshold in the steady-state operation. In this case, $\mathscr{E}_2$ is a constant, and $\mathscr{E}_1(l)$ and $\mathscr{E}_3(l)$ are obtained from the coupled wave equations in (9.1)

$$\frac{\partial \mathscr{E}_1}{\partial z} = \frac{\omega_1 c}{n_1} K \mathscr{E}_2^* \mathscr{E}_3$$

and   (9.28)

$$\frac{\partial \mathscr{E}_3}{\partial z} = -\frac{\omega_3 c}{n_3} K \mathscr{E}_1 \mathscr{E}_2.$$

The backward waves at ω_1 and ω_2 are negligible. With $\mathscr{E}_1(0) = 0$, we find

$$|\mathscr{E}_1(l)|^2 = \left(\frac{\omega_1 n_3}{\omega_3 n_1}\right)|\mathscr{E}_3(0)|^2 \sin^2 \beta l$$

and   (9.29)

$$|\mathscr{E}_3(l)|^2 = |\mathscr{E}_3(0)|^2 \cos^2 \beta l$$

where

$$\beta^2 = \frac{\omega_1 \omega_3 c^2}{n_1 n_3} K^2 |\mathscr{E}_2|^2.$$

The photon number generated at ω_1 in a single-pass gain must be equal to the photon number at ω_2 generated and then lost in a round trip around the cavity, so that

$$\frac{(2\Gamma_2 n_2 l/c) n_2 |\mathscr{E}_2|^2}{\omega_2} = \frac{n_1 |\mathscr{E}_1(l)|^2}{\omega_1}$$

$$= \left(\frac{n_3}{\omega_3}\right)|\mathscr{E}_3(0)|^2 \sin^2 \beta l.$$ (9.30)

The last equality leads to the relation

$$\frac{\sin^2 \beta l}{\beta^2 l^2} = \frac{1}{N'}$$ (9.31)

where

$$N' = \frac{(\omega_1 \omega_2 c^2/n_1 n_2) K^2 |\mathscr{E}_3(0)|^2}{8(\Gamma_2 n_2 l/c)} = \frac{g_0}{g_{th}}$$

again has the physical meaning of how many times the pump intensity is above the threshold value. The conversion efficiency is then given by

$$
\begin{aligned}
\eta &= \frac{\mathbb{P}_1 + \mathbb{P}_2}{\mathbb{P}_3} \\[2mm]
&= \frac{n_1|\mathscr{E}_1(l)|^2 + (2\Gamma_2 n_2 l/c) n_2|\mathscr{E}_2|^2}{n_3|\mathscr{E}_3(0)|^2} \\[2mm]
&= \sin^2\beta l
\end{aligned} \tag{9.32}
$$

with $\sin^2\beta l$ determined from (9.31). For $N' = 4$, we find $\eta = 90\%$. This is of course somewhat ideal since we have used the steady-state plane wave approximation in the derivation. A more realistic calculation with Gaussian beam profiles has been worked out by Bjorkholm.[11] The output conversion efficiencies at ω_1 and ω_2, defined as output versus input, are

$$
\begin{aligned}
\eta_{\text{out}}(\omega_1) &= \left(\frac{\omega_1}{\omega_3}\right)\eta, \\[2mm]
\eta_{\text{out}}(\omega_2) &= \frac{1 - R_2}{\Gamma_2 n_2 l/c}\left(\frac{\omega_2}{\omega_3}\right)\eta.
\end{aligned} \tag{9.33}
$$

An overall output conversion efficiency, $\eta_{\text{out}}(\omega_1) + \eta_{\text{out}}(\omega_2)$, as high as 70% has been experimentally demonstrated.[12]

9.4 FREQUENCY TUNING OF PARAMETRIC OSCILLATORS

The output frequencies of a parametric oscillator are determined by energy and momentum conservation

$$
\omega_3 = \omega_1 + \omega_2 \quad \text{and} \quad k_3 = k_1 + k_2.
$$

Together they yield the relation

$$
\omega_3[n_3(\omega_3) - n_2(\omega_3 - \omega_1)] = \omega_1[n_1(\omega_1) + n_2(\omega_3 - \omega_1)]. \tag{9.34}
$$

Equation (9.34) dictates the signal frequency ω_1 if the dispersions of the refractive indices $n_i(\omega_i)$ are known. As discussed earlier in sum-frequency generation, (9.34) can be satisfied only in anisotropic crystals. With negative uniaxial crystals in the normal dispersion region, $n_3(\omega_3)$ must be extraordinary, while $n_1(\omega_1)$ and $n_2(\omega_2)$ can be either both ordinary (type I phase matching) or one ordinary and one extraordinary (type II phase matching).

Equation (9.34) shows that the output frequencies of the oscillator (or the frequencies for maximum parametric gain) can be tuned if $n(\omega)$ can be varied by external parameters. We consider here frequency tuning by angular rotation of the crystal and by temperature. We assume type I phase matching in the following discussion.

In angle tuning, let the output frequencies be ω_1 and ω_2 when the crystal is oriented with an angle θ between its $\hat{c}$ axis and the axis of the cavity. We have

$$\omega_3 n_3^e(\omega_3, \theta) = \omega_1 n_1^0(\omega_1) + \omega_2 n_2^0(\omega_2). \tag{9.35}$$

If the crystal is now rotated to $\theta + \Delta\theta$, the output frequencies should correspondingly change to $\omega_1 + \Delta\omega$ and $\omega_2 - \Delta\omega$. Equation (9.35) becomes

$$\omega_3 \left[n_3^e(\omega_3, \theta) + \frac{\partial n_3^e}{\partial \theta} \Delta\theta \right] = (\omega_1 + \Delta\omega) \left[n_1^0(\omega_1) + \frac{\partial n_1^0}{\partial \omega_1} \Delta\omega + O(\Delta\omega^2) \right]$$

$$+ (\omega_2 - \Delta\omega) \left[n_2^0(\omega_2) - \frac{\partial n_2^0}{\partial \omega_2} \Delta\omega + O(\Delta\omega^2) \right] \tag{9.36}$$

or

$$\Delta\omega \cong \omega_3 \frac{\partial n_3^e}{\partial \theta} \Delta\theta \bigg/ \left\{ \left[\omega_1 \frac{\partial n_1^0}{\partial \omega_1} - \omega_2 \frac{\partial n_2^0}{\partial \omega_2} \right] + n_1^0(\omega_1) - n_2^0(\omega_2) \right\} \tag{9.37}$$

where $O(\Delta\omega^2)$ are terms of orders higher than or equal to $(\Delta\omega)^2$. For uniaxial crystals, one finds

$$\frac{dn_3^e}{d\theta} = -\frac{(n_3^e)^3}{2} \sin 2\theta \left[\left(\frac{1}{n_3^{em}} \right)^2 - \left(\frac{1}{n_3^0} \right)^2 \right]. \tag{9.38}$$

Then, given the dispersions of $n_i(\omega_i)$, the angular tuning curve ω_1 (or ω_2) versus θ can be calculated from (9.37) and (9.38). As ω_1 approaches $\omega_3/2$, however, the quadratic terms of $\Delta\omega$ in (9.36) become nonnegligible. Keeping the $(\Delta\omega)^2$ terms in (9.36) near the degenerate operating point $\theta = \theta_0$ and $\omega_1 = \frac{1}{2}\omega_3$, we find, instead of (9.37),

$$\Delta\omega = \left\{ \omega_3 \left(\frac{\partial n_3^e}{\partial \theta} \right)_{\theta_0} \bigg/ \left[2 \frac{\partial n}{\partial \omega} + \frac{\omega_3}{2} \frac{\partial^2 n}{\partial \omega^2} \right]_{(1/2)\omega_3} \right\}^{1/2} (\Delta\theta)^{1/2}. \tag{9.39}$$

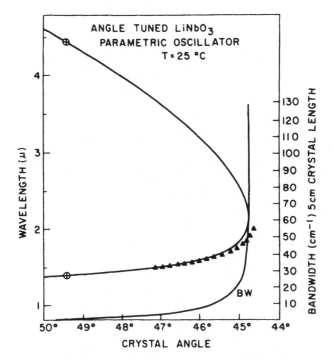

ANGLE TUNED LiNbO₃
PARAMETRIC OSCILLATOR
T = 25 °C

Fig. 9.2 Tuning range and gain bandwidth for the angle-tuned, singly resonant, LiNbO₃ parametric oscillator. The pump wavelength is 1.06 μm. (After Ref. 14.)

As examples, we show in Figs. 9.2 and 9.3 the angular tuning curves of LiNbO₃ pumped by a 1.06-μm laser beam and ADP pumped by a 0.347-μm beam. The tuning is of the order of 1000 cm⁻¹ per degree of rotation away from the degenerate point.

Temperature tuning follows a similar treatment. Assume at temperature T

$$\omega_3 n_3^e(\omega_3, T) = \omega_1 n_1^0(\omega_1, T) + \omega_2 n_2^0(\omega_2, T). \qquad (9.40)$$

At $T + \Delta T$, the output frequencies change to $\omega_1 + \Delta\omega$ and $\omega_2 - \Delta\omega$, and we have, away from the degenerate operating point,

$$\omega_3 \left[n_3^e(\omega_3, T) + \frac{\partial n_3^e}{\partial T} \Delta T \right] = (\omega_1 + \Delta\omega) \left[n_1^0(\omega_1, T) + \frac{\partial n_1^0}{\partial T} \Delta T + \frac{\partial n_1^0}{\partial \omega_1} \Delta\omega \right]$$

$$+ (\omega_2 - \Delta\omega) \left[n_2^0(\omega_2, T) + \frac{\partial n_2^0}{\partial T} \Delta T - \frac{\partial n_2^0}{\partial \omega_2} \Delta\omega \right]$$

$$\Delta\omega = \frac{\omega_3 (\partial n_3^e/\partial T) - \omega_1 (\partial n_1^0/\partial T) - \omega_2 (\partial n_2^0/\partial T)}{n_1^0 - n_2^0 + \omega_1 (\partial n_1^0/\partial \omega) - \omega_2 (\partial n_2^0/\partial \omega)}.$$

$$(9.41)$$

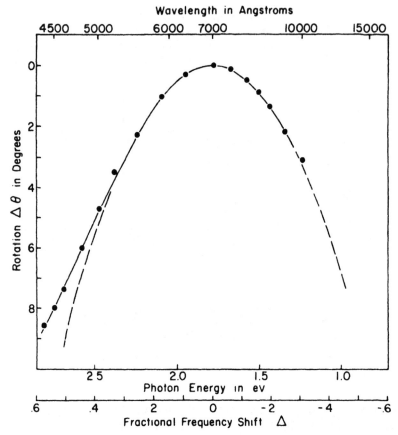

Fig. 9.3 Tuning curve for the angle-tuned ADP parametric oscillator. The pump wavelength is 0.347 μm. The angle $\Delta\theta$ is measured with respect to the angle at which the signal and idler frequencies are degenerate. The dotted curve is a theoretical curve. [After D. Magde and H. Mahr, *Phys. Rev. Lett.* **18**, 905 (1967).]

Around the degenerate operating point, $T = T_0$ and $\omega_1 = \omega_3/2$, we find

$$\Delta\omega = \left\{ \omega_3 \left[\frac{\partial n_3^e}{\partial T} - \frac{\partial n_1^e}{\partial T} \right]_{T_0} \bigg/ \left[2\frac{\partial n_1}{\partial \omega} + \frac{\omega_3}{2}\left(\frac{\partial^2 n}{\partial \omega^2} \right) \right]_{(1/2)\omega_3} \right\}^{1/2} (\Delta T)^{1/2}.$$

(9.42)

The temperature tuning curves of $LiNbO_3$ for a number of pump laser frequencies are shown in Fig. 9.4 as an example. With a pump wavelength at 0.53 μm, the degenerate point is at $T_0 \cong 49.3°C$ and tuning is 300 cm^{-1}/(°C).

While $\Delta k = k_3 - k_1 - k_2 = 0$ determines the output frequencies, $\Delta kl \cong 2\pi$ determines the gain bandwidth of a parametric oscillator. Since

$$\Delta kl = \left| n_1 - n_2 + \omega_1 \left(\frac{\partial n_1}{\partial \omega_1} \right) - \omega_2 \left(\frac{\partial n_2}{\partial \omega_2} \right) \right| l\delta\omega \bigg/ c$$

(9.43)

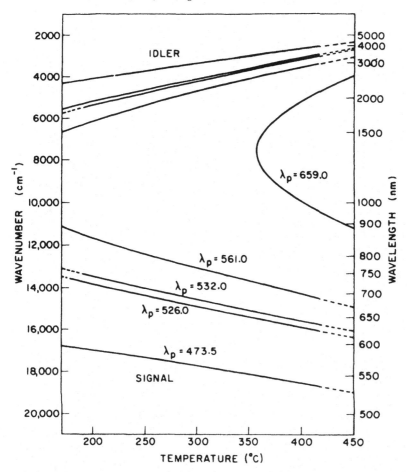

Fig. 9.4 Tuning curves for the temperature-tuned $LiNbO_3$ parametric oscillator at various pump frequencies. [After R. L. Byer, in H. Rabin and C. L. Tang, eds., *Quantum Electronics* (Academic Press, NY, 1975), vol. I, p. 631.]

the bandwidth is given by

$$\delta\omega = 2\pi c \left/ l \left| n_1 - n_2 + \omega_1 \left(\frac{\partial n_1}{\partial \omega_1} \right) - \omega_2 \left(\frac{\partial n_2}{\partial \omega_2} \right) \right| \right. . \qquad (9.44)$$

Around the degenerate operating point, a better approximation gives

$$\delta\omega = \left\{ \frac{2\pi c}{l} \left/ \left[2\frac{\partial n}{\partial \omega} + \frac{\omega_3}{2} \left(\frac{\partial^2 n}{\partial \omega^2} \right) \right]_{\omega_3/2} \right. \right\}^{1/2} \qquad (9.45)$$

The result here shows that near the degenerate point, the bandwidth can be fairly large, ~ 100 cm^{-1} for $l = 1$ cm. Away from the degenerate point, $\delta\omega$ is

Table 9.1
Representative Parametric Oscillators

Pump Laser	Nonlinear Material	Output Power and Pulse Width	Conversion Efficiency	Tuning Range	Reference
Nd: Glass and second and third harmonics at 0.532 and 0.35 μm	KDP	100 kW (20 ns)	3%	0.957–1.17 μm (0.532 μm pump) 0.48–0.58 m and 0.96–1.16 μm (0.35 μm pump)	A. A. Akmanov, A. I. Kovrigin, V. A. Kolosov, A. S. Piskarkas, V. V. Fadeav, and R. V. Khoklov, *JETP Lett.* **3**, 241 (1966) S. A. Akhmanov, O. N. Chunsev, V. V. Fadeav, R. V. Khoklov, D. H. Klyshki, A. I. Kovrigin, and A. S. Piskarskas, Presented at Symp. Modern Optics, Brooklyn, 1967
Nd: YAG operating at fourth harmonic 0.266 μm	ADP	100 kW (2 ns)	25%	0.42–0.73 μm	J. M. Yarborough and G. A. Massey, *Appl. Phys. Lett.* **18**, 438 (1971)
Nd: YAG operating at second harmonic wavelengths 0.472, 0.532, 0.579, 0.635 μm	LiNbP$_3$	0.1–10 kW (~200 ns)	~45%	0.55–3.65 μm	R. W. Wallace, *Appl. Phys. Lett.* **17**, 497 (1970)
Nd: YAG 1.06 μm	LiNbO$_3$	0.1–1 MW (~15 ns)	~40%	1.4–4.4 μm	R. L. Herbst, R. N. Fleming, and R. L. Byer, *Appl. Phys. Lett.* **25**, 520 (1974)
Ruby 0.6943 μm	LiNbO$_3$	3 W	10^{-6}	66–200 μm	B. C. Johnson, and H. E. Puthoff, J. Soo Hoo, and S. S. Sussman, *Appl. Phys. Lett.* **18**, 181 (1971)

Source	Crystal	Power (pulse)	Efficiency	Tuning range	Reference
Nd : Glass operating at second harmonic 0.53 μm	α–HIO$_3$, LiIO$_3$	–10 MW (20 ns)	10%	0.68–2.4 μm	A. I. Izrailenko, A. I. Kovrigin, and P. V. Nikles, *JETP Lett.* **12**, 331 (1970); and A. I. Kovrigin P. V. Nikles, *JETP Lett.* **13**, 313 (1971)
Ruby laser 0.6943 μm	LiIO$_3$	2 kW (20 ns)	–1%	0.77–4 μm	L. S. Goldberg, *Appl. Phys. Lett.* **17**, 489 (1970)
Ruby laser 0.6943 μm	LiIO$_3$	100 kW (15 ns)	10%	1.1–1.9 μm	A. J. Campillo and C. L. Tang, *Appl. Phys. Lett.* **19**, 36 (1971); A. J. Campillo, *IEEE J. Quant. Electron.* **8**, 809 (1972)
Ruby laser operating at second harmonic 0.347 μm	LiIO$_3$	10 kW (5 ns)	8%	0.415–2.1 μm	G. Nath and G. Pauli, *Appl. Phys. Lett.* **22**, 75 (1973)
Nd : YAG 1.833 μm	CdSe	–1 kW (100 ns)	40%	2.2–2.3 μm 10.5–9.7 μm	R. L. Herbst and R. L. Byer, *Appl. Phys. Lett.* **21**, 189 (1972)
CaF$_2$: Dy 2.36 μm	CdSe	5 kW (30 ns)	0.5%	3.3 μm 7.86 μm	A. A. Davydov, L. A. Kulevskii, A. M. Prokhorov, A. D. Savel'ev, and V. V. Smirnov, *JETP Lett.* **15**, 513 (1972)
HF laser 2.87 μm	CdSe	800 W (–300 ns)	10%	4.3–4.5 μm 8.1–8.3 μm	J. A. Weiss and L. S. Goldberg, *Appl. Phys. Lett.* **24**, 389 (1974)
Nd : CaWO$_4$ 1.065 μm	Ag$_3$AsS$_3$	100 W (25 ns)	–0.1%	1.22–8.5 μm	D. C. Hanna, B. Luther Davies, H. N. Rutt, and R. C. Smith, *Appl. Phys. Lett.* **20**, 34 (1972); D. C. Hanna, B. Luther Davies, and B. C. Smith, *Appl. Phys. Lett.* **22**, 440 (1973)

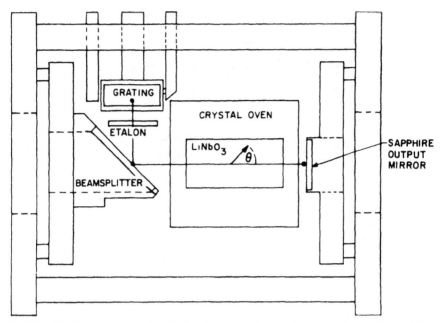

Fig. 9.5 Singly resonant cavity design for an angle-tuned parametric oscillator. (After Ref. 14.)

typically around 5–10 cm^{-1}. A more dispersive crystal yields a narrower bandwidth.

The output linewidth can be narrowed appreciably by using frequency selection elements in the cavity. Figure 9.5 shows an angle-tuned singly resonant LiNbO$_3$ parametric oscillator designed by Byer.[13,14] The beam splitter transmits 90% of the pump beam and reflects 99% of the signal beam. The angular tuning curve of this oscillator is shown in Fig. 9.2 together with the gain bandwidth. If a 600-line/mm grating blazed at 1.8 μm is used as the back mirror of the cavity, the signal output has a linewidth around 1 cm^{-1} for a beam spot size of 1.6 mm. With a prism beam expander in the cavity to increase the beam spot size on the grating, the linewidth can be narrowed by another order of magnitude. Alternatively, a 1-mm tilted etalon can be inserted in the cavity to reduce the linewidth to less than 0.1 cm^{-1}. Other frequency-selective elements, such as multielement birefringent filters and etalon assembly, can also be used in the cavity for reduction of the output linewidth.

In Table 9.1 we reproduce the list prepared by Byer and Herbst,[14] describing the operating characteristics of a number of representative parametric oscillators.

9.5 PARAMETRIC FLUORESCENCE

Parametric oscillation occurs through amplification of noise photons initiated by parametric scattering or fluorescence. In general, in the parametric process, a photon at ω_3 is scattered into a photon at ω_1 and a photon at ω_2 with $\omega_1 + \omega_2 = \omega_3$ and $\mathbf{k}_1 + \mathbf{k}_2 \cong \mathbf{k}_3$. Parametric scattering or fluorescence refers to the parametric scattering process where the initial numbers of photons at ω_1 and ω_2 are zero. This nonlinear optical emission can be properly described only by quantizing the fields.[15]

In the parametric process with negligible pump depletion, the intense pump field can be treated as a constant classical field. The Hamiltonian governing the problem can be written as

$$\mathscr{H} = \sum_{i=1,2} \hbar\omega_i \left(a_i^+ a_i + \tfrac{1}{2}\right) + \tfrac{1}{2}\hbar G_0 \left[a_1^+ a_2^+ e^{-i\omega_3 t} + a_1 a_2 e^{i\omega_3 t}\right] \quad (9.46)$$

where a_i^+ and a_i are creation and annihilation operators for photons at ω_i, respectively, and G_0 is given in (9.12). The Heisenberg equation of motion for an operator X is

$$\frac{dX}{dt} = \frac{1}{i\hbar}[X, \mathscr{H}] \quad (9.47)$$

from which we obtain

$$\frac{da_1^+}{dt} = i\omega_1 a_1^+ + i\tfrac{1}{2}G_0 a_2 e^{i\omega_3 t}$$

$$\frac{da_2}{dt} = -i\omega_2 a_2 - i\tfrac{1}{2}G_0 a_1^+ e^{-i\omega_3 t}. \quad (9.48)$$

The above set of equations can be solved to yield

$$a_1^+(t) = \left[a_1^+(0)\cosh\tfrac{1}{2}G_0 t + ia_2(0)\sinh\tfrac{1}{2}G_0 t\right]e^{i\omega_1 t}$$

$$a_2(t) = \left[a_2(0)\cosh\tfrac{1}{2}G_0 t - ia_1(0)\sinh\tfrac{1}{2}G_0 t\right]e^{-i\omega_2 t}. \quad (9.49)$$

Then, [assuming $\langle a_1(0)\rangle = \langle a_2(0)\rangle = 0$] the average numbers of photons at ω_1 and ω_2 are, respectively,

$$\langle n_1(t)\rangle = \langle a_1^+(t)a_1(t)\rangle$$
$$= \langle n_1(0)\rangle\cosh^2\tfrac{1}{2}G_0 t + (1 + \langle n_2(0)\rangle)\sinh^2\tfrac{1}{2}G_0 t$$
$$\langle n_2(t)\rangle = \langle a_2^+(t)a_2(t)\rangle$$
$$= \langle n_2(0)\rangle\cosh^2\tfrac{1}{2}G_0 t + (1 + \langle n_1(0)\rangle)\sinh^2\tfrac{1}{2}G_0 t. \quad (9.50)$$

The result here shows clearly that the number of photons at ω_1 and ω_2 can grow out of zero in the parametric process. For $\langle n_1(0)\rangle = 0$ and $\langle n_2(0)\rangle = 0$, we have

$$\langle n_1(t)\rangle = \langle n_2(t)\rangle = \sinh^2\tfrac{1}{2}G_0 t. \quad (9.51)$$

While parametric fluorescence leads to parametric oscillation, it can also provide initial photons as input of a parametric amplifier. If the single-pass gain of the amplifier is large, the output can be appreciable. This single-pass parametric amplification of noise photons in a nonlinear crystal often is known as parametric superfluorescence.[15, 16] To find the output power, we note in (9.50) that the output at ω_1 (or ω_2) actually grows out of the initial noise photon at ω_2 (or ω_1). We can assume one photon per mode is created at ω_2 (or ω_1) by parametric scattering at the input end, $z = 0$, and use (9.3) to calculate the output of the parametric amplifier at ω_1 (or ω_2). For one photon in each mode, the corresponding input intensity at ω_2 is $\hbar\omega_2 c/n_2 V$, and the output at ω_1 at $z = l$ is

$$I_s(\omega_1) = \left(\frac{g_0^2 \hbar \omega_1 c}{g^2 V}\right) \sinh^2 \tfrac{1}{2}gl. \tag{9.52}$$

The number of modes in the frequency range ω_2 to $\omega_2 + d\omega$ and in a solid angle extended by $\mathbf{k}_2$ from ψ_2 to $\psi_2 + d\psi_2$ (see Fig. 9.6) is, for small ψ_2,

$$
\begin{aligned}
dN &= \frac{2\pi k_2^2 \sin\psi_2\, dk_2\, d\psi_2}{8\pi^3/V} \\
&= \frac{V n_2}{4\pi^2 c} k_1^2 \psi_1\, d\psi_1\, d\omega.
\end{aligned}
\tag{9.53}
$$

The total output power in a beam of cross section A between ω_1 and $\omega_1 - d\omega$ collected by a faraway detector with a small collection angle θ is then given by

$$
\begin{aligned}
\mathbb{P}(\omega_1)\, d\omega &= \int I_s(\omega_1) A\, dN \\
&= (d\omega)\frac{n_1^2 n_2 \hbar \omega_1^3 g_0^2 l^2 A}{16\pi^2 c^2} \int_0^\theta d\psi_1 \frac{\sinh^2 \tfrac{1}{2}gl}{\left(\tfrac{1}{2}gl\right)^2} \psi_1.
\end{aligned}
\tag{9.54}
$$

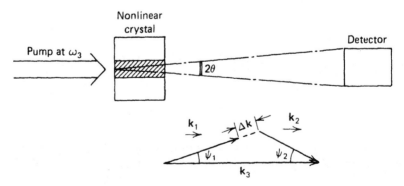

Fig. 9.6 Geometry of parametric fluorescence collected by a photodetector.

Here $g^2 = g_0^2 - (\Delta k)^2$ is a function of ψ_1 through Δk. In the forward direction, $\psi_1 = 0$, the phase-matching condition $\Delta k = 0$ is satisfied at $\omega_1 = \omega_1^0$ and $\omega_2 = \omega_2^0$ such that $\omega_1^0 n_1 + \omega_2^0 n_2 = \omega_3 n_3$. From Fig. 9.6, we find for small ψ_1 and ψ_2,

$$k_1^2 - (k_2 + \Delta k - k_3)^2 = 2k_2 k_3 (1 - \cos\psi_1)$$

$$\Delta k \cong (k_3 - k_1 - k_2) + \left(\frac{k_1 k_3}{2k_2}\right)\psi_1^2 \qquad (9.55)$$

$$= -a\,\Delta\omega + b\psi_1^2$$

where

$$a = \left[\left(\frac{dk_1}{d\omega_1}\right)_{\omega_1^0} - \left(\frac{dk_2}{d\omega_2}\right)_{\omega_2^0}\right],$$

$$\Delta\omega = \omega_1 - \omega_1^0 = \omega_2^0 - \omega_2,$$

$$b = \frac{k_1 k_3}{2k_2}.$$

Equation (9.54) becomes

$$\mathbb{P}(\omega_1) = \frac{n_1^2 n_2 \hbar \omega_1^3 g_0^2 l^2 A}{16\pi^2 c^2} \int_0^\theta \frac{\sinh^2\left\{\left[g_0^2 - \left(-a\Delta\omega + b\psi_1^2\right)^2\right]^{1/2} l/2\right\}}{\left\{\left[g_0^2 - \left(-a\Delta\omega + b\psi_1^2\right)^2\right]^{1/2} l/2\right\}^2} \psi_1\,d\psi_1.$$

$$(9.56)$$

This result shows that $\mathbb{P}(\omega_1)$ is a maximum at $\omega_1 = \omega_1^0$, that is, when the waves at ω_1, ω_2, and ω_3 are phase matched in the forward direction. The half-power points appear at $\omega_1 = \omega_1^0 + |\Delta\omega_+|$ and $\omega_1 = \omega_1^0 - |\Delta\omega_-|$, where $|\Delta\omega_+|$ and $|\Delta\omega_-|$ are given approximately by

$$\frac{\sinh^2\left\{\left[g_0^2 - \left(b\theta^2 \mp a|\Delta\omega_\pm|\right)^2\right]^{1/2} l/2\right\}}{\left\{\left[g_0^2 - \left(b\theta^2 \mp a|\Delta\omega_\pm|\right)^2\right]^{1/2} l/2\right\}^2} \bigg/ \frac{\sinh^2(g_0 l/2)}{(g_0 l/2)^2}$$

$$\simeq e^{[g_0^2 - (b\theta^2 \mp a|\Delta\omega_\pm|^2)]^{1/2} l}/e^{g_0 l} = \tfrac{1}{2},$$

which yields, for $g_0 \gg b\theta^2$,

$$|\Delta\omega_+| = \frac{1}{a}\left[\left(\frac{2g_0}{l}\ln 2\right)^{1/2} + b\theta^2\right]$$

$$|\Delta\omega_-| = \frac{1}{a}\left[\left(\frac{2g_0}{l}\ln 2\right)^{1/2} - b\theta^2\right].$$

$$(9.57)$$

Therefore, the bandwidth of parametric superfluorescence observed by the detector is

$$\Delta\omega_{BW} = |\Delta\omega_+| + |\Delta\omega_-|$$

$$= \frac{2}{a}\left(\frac{2g_0}{l}\ln 2\right)^{1/2}. \tag{9.58}$$

A crude estimate using (9.56) yields, for $g_0 l = 50$, and ω_1 in the near infrared, a $\mathbb{P}(\omega_1)/A$ as large as 10^7 W/cm^2 per cm^{-1} in a 10-mrad forward cone. Even in a 5-cm LiNbO$_3$ crystal, $g_0 l = 50$ requires a pump beam intensity of about 150 MW/cm^2. This intensity, however, is easily obtained with mode-locked laser pulses without damaging the crystal. Indeed, parametric superfluorescence has become the most realistic scheme for producing tunable picosecond pulses in the near-infrared region.[17] The general time-dependent description of parametric amplification or superfluorescence by ultrashort pump pulses has been given by Akhmanov et al.,[18] but the qualitative features are the same as those discussed in Section 7.6. In particular, if the group velocity mismatch can be neglected, then the quasi-stationary solution is still applicable in the moving frame of the propagating pulses.

Parametric fluorescence or superfluorescence can also be used to determine experimentally the frequency tuning curves of parametric oscillators.[15, 16] This is most useful when the refractive index data of the crystals are not readily available.

9.6   BACKWARD PARAMETRIC OSCILLATOR

The case of parametric amplification with counterpropagating signal and idler waves shows interesting characteristics and deserves special consideration. The two waves grow in opposite directions and, through parametric interaction, impose a positive feedback on each other. Then, with sufficient gain, the system may become a mirrorless oscillator, that is, it may yield a finite output with a zero input.[19] The principle is similar to the backward-traveling wave tube.[20]

The equations governing backward parametric amplification are a simple modification of (9.1). With negligible pump depletion and linear attenuation, they are

$$\frac{\partial}{\partial z}\mathscr{E}_1 = \frac{i\omega_1^2}{k_1}K^*\mathscr{E}_2^*\mathscr{E}_3 e^{i\,\Delta k\,z + i\theta_0}$$

and   (9.59)

$$\frac{\partial}{\partial z}\mathscr{E}_2^* = \frac{i\omega_2^2}{k_2}K\mathscr{E}_1\mathscr{E}_3^* e^{-i\,\Delta k\,z - i\theta_0}$$

assuming $E_1 = \mathscr{E}_1\exp[ik_1 z - i\omega_1 t + i\phi_1]$, $E_2 = \mathscr{E}_2\exp[-ik_2 z - i\omega_2 t + i\phi_2]$, $\mathscr{E}_3$

= constant, and $\Delta k = k_3 - k_1 + k_2$. The input boundary conditions are $\mathscr{E}_1 = \mathscr{E}_1(0)$ at $z = 0$ and $\mathscr{E}_2^* = \mathscr{E}_2^*(l)$ at $z = l$. The solution of (9.59) for $\Delta k = 0$ is ($\theta_0 = \pi/2$ is assumed for simplicity)

$$
\begin{aligned}
\mathscr{E}_1(z) &= \mathscr{E}_1(0)\left[\cos\frac{g_0(z-l)}{2}\bigg/\cos\frac{g_0 l}{2}\right] \\
&\quad + i\frac{\omega_1}{\omega_2}\left(\frac{k_2}{k_1}\right)^{1/2}\mathscr{E}_2^*(l)\left[\sin\frac{g_0 z}{2}\bigg/\cos\frac{g_0 l}{2}\right], \\
\mathscr{E}_2^*(z) &= i\frac{\omega_2}{\omega_1}\left(\frac{k_1}{k_2}\right)^{1/2}\mathscr{E}_1(0)\left[\sin\frac{g_0(z-l)}{2}\bigg/\cos\frac{g_0 l}{2}\right] \\
&\quad + \mathscr{E}_2^*(l)\left[\cos\frac{g_0 z}{2}\bigg/\cos\frac{g_0 l}{2}\right]
\end{aligned}
\tag{9.60}
$$

with g_0 given in (9.3), and hence the output is

$$
\begin{aligned}
\mathscr{E}_1(l) &= \mathscr{E}_1(0)\bigg/\cos\frac{g_0 l}{2} + i\frac{\omega_1}{\omega_2}\left(\frac{k_2}{k_1}\right)^{1/2}\mathscr{E}_2^*(l)\tan\frac{g_0 l}{2}, \\
\mathscr{E}_2^*(0) &= -i\frac{\omega_2}{\omega_1}\left(\frac{k_1}{k_2}\right)^{1/2}\mathscr{E}_1(0)\tan\frac{g_0 l}{2} + \mathscr{E}_2^*(l)\bigg/\cos\frac{g_0 l}{2}.
\end{aligned}
\tag{9.61}
$$

If $g_0 l/2 \to \pi/2$, both $\mathscr{E}_1(l)$ and $\mathscr{E}_2^*(0)$ become infinite unless the input $\mathscr{E}_1(0)$ and $\mathscr{E}_2^*(l)$ are zero. This is therefore the oscillation threshold. The actual output will grow drastically as $g_0 l \to \pi$ and will be determined only by taking into account the pump depletion through the nonlinear coupling between the waves.

Experimentally, backward parametric oscillation has not yet been observed. The reason is that the phase-matching condition $\Delta k = 0$ cannot be satisfied in the usual cases. It is possible to obtain $\Delta k = 0$ with ω_1 (or ω_2) in the far infrared range,[21] but then the gain coefficient $g_0 l$ is too small to reach the oscillation threshold.

REFERENCES

1 W. H. Louisell, *Coupled Mode and Parametric Electronics* (Wiley, New York, 1960).

2 J. A. Giordmaine and R. C. Miller, *Phys. Rev. Lett.* **14**, 973 (1965).

3 R. G. Smith, *J. Appl. Phys.* **41**, 4121 (1970).

4 J. A. Armstrong, N. Bloembergen, J. Ducuing, and P. S. Pershan, *Phys. Rev.* **127**, 1918 (1962).

5 A. E. Siegman, *Appl. Opt.* **1**, 739 (1962).

6 R. G. Smith, J. E. Geusic, H. J. Levinstein, S. Singh, and L. G. van Uitert, *J. Appl. Phys.* **39**, 4030 (1968).

7   J. A. Giordmaine and R. C. Miller, in P. L. Kelley, B. Lax, and P. E. Tannenwald eds., *Physics of Quantum Electronics* (McGraw-Hill, New York, 1966); *Appl. Phys. Lett* **9**, 298 (1966).

8   J. E. Bjorkholm, *Appl. Phys. Lett.* **13**, 399 (1968).

9   J. E. Bjorkholm, *Appl. Phys. Lett.* **13**, 53 (1968).

10 L. B. Kreuzer, *Appl. Phys. Lett.* **15**, 263 (1969).

11 J. E. Bjorkholm, *IEEE J. Quant. Electron.* **QE-7**, 109 (1971).

12 J. Falk and J. E. Murray, *Appl. Phys. Lett.* **14**, 245 (1969).

13 S. J. Brosnan and R. L. Byer, *IEEE J. Quant. Electron.* **QE-15**, 415 (1979).

14 R. L. Byer and R. L. Herbst, in Y. R. Shen, ed., *Nonlinear Infrared Generation* (Springer-Verlag, Berlin, 1977), p. 81.

15 W. H. Louisell, A. Yariv, and A. E. Siegman, *Phys. Rev.* **124**, 1646 (1961); T. G. Giallorenzi and C. L. Tang, *Phys. Rev.* **166**, 225 (1968).

16 S. E. Harris, M. K. Oshman, and R. L. Byer, *Phys. Rev. Lett.* **18**, 732 (1967); R. L. Byer and S. E. Harris, *Phys. Rev.* 168, 1064 (1968).

17 T. A. Rabson, H. J. Ruiz, P. L. Shah, and F. K. Tittel, *Appl. Phys. Lett.* **21**, 129 (1972); A. H. Kung, *Appl. Phys. Lett.* **25**, 653 (1974); A. Seilmeier, K. Spanner, A. Laubereau, and W. Kaiser, *Opt. Comm.* **24**, 237 (1978).

18 S. A. Akhmanov, A. I. Kovrigin, A. P. Sukhorukov, R. V. Khokhlov, and A. S. Chirkin, *JETP Lett.* **7**, 182 (1968); S. A. Akhmanov, A. S. Chirkin, K. N. Drabovich, A. I. Kovrigin, R. V. Khokhlov, and A. P. Sukhorukov, *IEEE J. Quant. Electron.* **QE-4**, 598 (1968); S. A. Akhmanov, A. P. Sukhorukov, and A. S. Chirkin, *Sov. Phys.–JETP* **28**, 748 (1969).

19 S. E. Harris, *Appl. Phys. Lett.* 9, 114 (1966).

20 J. R. Pierce, *Travelling Wave Tubes* (Van Nostrand, Princeton, NJ, 1950).

21 K. H. Yang, P. L. Richards, and Y. R. Shen, *Appl. Phys. Lett.* **19**, 320 (1971).

BIBLIOGRAPHY

Akhmanov, S. A., and R. V. Khokhlov, *Uspekhi* **9**, 210 (1966).

Brosnan, S. J., and R. L. Byer, *IEEE J. Quant. Electron.* **QE-15**, 415 (1979); R. A. Baumgartner and R. L. Byer, *J. Quant. Electron.* **QE-15**, 432 (1979).

Byer, R. L., in H. Rabin and C. L. Tang, eds., *Treatise in Quantum Electronics*, Vol. I, Part B (Academic Press, New York, 1975), 587.

Byer, R. L., and R. L. Herbst, in Y. R. Shen, ed., *Nonlinear Infrared Generation* (Springer-Verlag, Berlin, 1977), p. 81.

Giordmaine, J. A., in R. J. Glauber, ed., *Quantum Optics* (Academic Press, New York, 1969), p. 493.

Harris, S. E. *Proc. IEEE* **57**, 2096 (1969).

Smith, R. G., in F. T. Arrechi and E. O. Schulz-Dubois, eds., *Laser Handbook* (North-Holland Publishing Co., Amsterdam, 1972), p. 837.

Tang, C. L., in H. Rabin and C. L. Tang, eds., *Quantum Electronics*, Vol. I, Part A (Academic Press, New York, 1972), p. 419.

Yariv, A., *Quantum Electronics* (Wiley, New York, 1975), 2nd ed., Chapter 17.

10

Stimulated
Raman Scattering

The wave coupling problems discussed in previous chapters are certainly not limited to electromagnetic waves. They can be generalized to involve both electromagnetic waves and other types of waves, allowing us to imagine a host of new nonlinear optical effects. In this chapter, we show that stimulated Raman scattering can be described classically as a parametric generation process with one of the electromagnetic waves replaced by a material excitational wave. More correctly, one would, of course, treat the material excitation quantum mechanically. Stimulated Raman scattering is then considered a two-photon stimulated process grown out of spontaneous Raman emission.

Stimulated Raman scattering is one of the few nonlinear optical effects discovered in the early 1960s. Over the years, many useful applications have been developed out of this process. Some of them are discussed in this chapter.

10.1 HISTORICAL REMARKS

In 1962 Woodbury and Ng,[1] while studying Q-switching of a ruby laser with a nitrobenzene Kerr cell, detected an infrared component in the laser output. Its frequency was 1345 cm^{-1} downshifted from the laser frequency. This frequency shift coincided with the vibrational frequency of the strongest Raman mode of nitrobenzene. It was then recognized by Woodbury and Eckhardt[2] that the infrared output must result from stimulated Raman emission in nitrobenzene. This was soon verified in a large number of liquids by several research workers. Similar effects were also found in gases and solids. Table 10.1 shows a list of Raman modes of some materials in which stimulated Raman scattering was observed.

An early theoretical description of stimulated Raman scattering was given by Hellwarth,[3] who treated it as a two-photon process with a full quantum

Table 10.1
Frequency Shift, Linewidth, and Scattering Cross Section of Spontaneous Raman Scattering for a Number of Substances and the Corresponding Stimulated Raman Gain[a]

Substance	Raman Shift (cm^{-1})	Linewidth 2Γ (cm^{-1})	Cross Section $d\sigma/d\Omega \times 10^8$ $(cm^{-1} - ster^{-1})$	Raman Gain $G_R \times 10^3$ (cm/MW)
Gas H_2^b	4155	0.2		1.5 (300 K, 10 atm)
Liquid O_2	1522	0.177	0.48 ± 0.14	14.5 ± 4
Liquid N_2	2326.5	0.067	0.29 ± 0.09	17 ± 5
Benzene	992	2.15	3.06	2.8
CS_2	655.6	0.50	7.55	24
Nitrobenzene	1345	6.6	6.4	2.1
$LiNbO_3$	258	7	262	28.7
$InSb^c$	$0 - 300$	0.3	10	1.7×10^4

[a] After Y. R. Shen, in M. Cardona, ed., *Light Scattering in Solids* (Springer-Verlag, Berlin, 1975), p. 275.
[b] E. E. Hagenlocker, R. W. Minck, W. G. Rado, *Phys. Rev.* **154**, 226 (1967).
[c] For a carrier concentration $n_e \simeq 10^{16}$ cm^{-3}.

mechanical calculation. The simple theory, however, cannot account for the observed anti-Stokes radiation which often appears with intensity almost as high as the Stokes radiation. Garmire et al.[4] and Bloembergen and Shen[5] later used the coupled wave approach to describe stimulated Raman scattering and were able to explain the anti-Stokes generation as well as the higher-order Stokes and anti-Stokes output.

Yet the theories still could not explain many other important observations. These included an observed stimulated Raman gain much larger than the value predicted by the theories, extremely sharp threshold for stimulated Raman emission, asymmetry of forward–backward Raman intensity, and appreciable spectral broadening of the Raman radiation. It was later realized that these anomalies were actually induced by self-focusing of the laser beam in the medium, which is discussed in Chapter 17. Without self-focusing, the theories predicted experimental results satisfactorily.

Early interest in stimulated Raman scattering arose because it could provide intense coherent radiation at new frequencies and because it was a possible loss mechanism in propagating high-power laser beams in a medium, for example, in the atmosphere or in a fusion plasma. More recently, stimulated Raman scattering was used to generate tunable coherent infrared radiation by either tuning the material excitation as in stimulated polariton scattering[6] and stimulated spin-flip Raman scattering,[7] or by tuning the exciting laser frequency with, for example, a tunable dye laser. Spectroscopic applications of stimulated Raman scattering have also been developed, with emphasis on high-resolution studies.[8] Transient stimulated Raman scattering has been applied to the study of relaxation of material excitations in the picosecond region with mode-locked laser pulses.[9]

10.2   QUANTUM THEORY OF STIMULATED RAMAN SCATTERING

Raman scattering is a two-photon process in which one photon at $\omega_1(\mathbf{k}_1)$ is absorbed and one photon at $\omega_2(\mathbf{k}_2)$ is emitted, while the material makes a transition from the initial state $|i\rangle$ to the final state $|f\rangle$, as shown in Fig. 10.1. Energy conservation requires $\hbar(\omega_1 - \omega_2) = \mathbb{E}_f - \mathbb{E}_i \equiv \hbar\omega_{fi}$, which is the energy difference between the final and initial states. Stokes and anti-Stokes scattering correspond to $\omega_{fi} > 0$ and $\omega_{fi} < 0$, respectively.

To find the Raman transition probability, we use the standard second-order perturbation calculation.[10] The interaction Hamiltonian in the electric-dipole approximation is

$$\mathcal{H}_{int} = -e\mathbf{r} \cdot \mathbf{E} + \text{adjoint} \tag{10.1}$$

where $\mathbf{E} = \mathscr{E}_1 e^{i(\mathbf{k}_1 \cdot \mathbf{r} - \omega_1 t)} + \mathscr{E}_2 e^{i(\mathbf{k}_2 \cdot \mathbf{r} - \omega_2 t)}$. The Raman transition probability per unit time per unit volume per unit energy interval is found to be

$$\frac{dW_{fi}}{d(\hbar\omega_1)} = \frac{dW_{fi}}{d(\hbar\omega_2)} = \frac{8\pi^3 N\omega_1\omega_2}{\varepsilon_1\varepsilon_2}|\langle f|M|i\rangle|^2|\langle \alpha_f|a_2^\dagger a_1|\alpha_i\rangle|^2 g(\hbar\Delta\omega)$$

$$M = \sum_s \left[\frac{e\mathbf{r} \cdot \hat{e}_2|\, s\rangle\langle s\,|e\mathbf{r} \cdot \hat{e}_1}{\hbar(\omega_1 - \omega_{si})} - \frac{e\mathbf{r} \cdot \hat{e}_1|\, s\rangle\langle s\,|e\mathbf{r} \cdot \hat{e}_2}{\hbar(\omega_2 + \omega_{si})} \right]. \tag{10.2}$$

Here N is the density of molecules or unit cells in the medium, ε is the

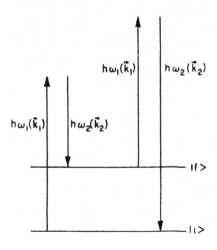

Fig. 10.1   Schematic drawing showing the Stokes ($\omega_2 < \omega_1$) Raman transition from the initial state $|i\rangle$ to a final excited state $|f\rangle$, and the anti-Stokes ($\omega_2 > \omega_1$) Raman transition from $|f\rangle$ to $|i\rangle$.

dielectric constant, $\hat{e}$ denotes the field polarization, $|s\rangle$ is the intermediate state of the material system, $|\alpha\rangle$ denotes the state of the radiation field, a^+ and a are the creation and annihilation operators of the field, and finally $g(\hbar\Delta\omega)$ describing the lineshape is the joint density of states of the Raman transition. For a Lorentzian lineshape, we have

$$g(\hbar\Delta\omega) = \frac{\hbar\Gamma/\pi}{\hbar^2(\Delta\omega)^2 + \hbar^2\Gamma^2} \tag{10.3}$$

where $\hbar\Gamma$ is the halfwidth in energy units.

The transition probability in (10.2) is proportional to $|\langle\alpha_f|a_2^+ a_1|\alpha_i\rangle|^2$. If $\langle\alpha_i| = \langle n_1, n_2|$ and $\langle\alpha_f| = \langle n_1 - 1, n_2 + 1|$ with n_1 and n_2 being integers, then $dW_{fi}/d(\hbar\omega_2) \propto n_1(n_2 + 1)$; spontaneous and stimulated Raman scattering correspond to $n_2 = 0$ and $n_2 \neq 0$, respectively. The states of the radiation fields are often more complex in general. Then there exists no simple expression of $\langle\alpha_f|a_2^+ a_1|\alpha_i\rangle$. However, if the average numbers of photons $\bar{n}_1$ and $\bar{n}_2$ at ω_1 and ω_2 are much larger than 1, the approximation $|\langle\alpha_f|a_2^+ a_1|\alpha_i\rangle|^2 \cong \bar{n}_1\bar{n}_2$ is excellent.[11] In a certain sense, this is equivalent to saying that the field can be treated classically.

Consider now the propagation of the ω_1 and ω_2 beams in the Raman medium. The Raman transition leads to the absorption of the ω_1 beam and amplification of the ω_2 beam. From a simple physical argument, the change of the average number of Raman photons in a single mode at ω_2 per unit length of propagation is given by[6]

$$\frac{dn_2}{dz} = \left(\frac{dW_{fi}}{d\omega_2}\rho_i - \frac{dW_{if}}{d\omega_2}\rho_f\right)\frac{\epsilon_2^{1/2}}{c} - \alpha_2\bar{n}_2$$
$$= (G_R - \alpha_2)\bar{n}_2 \tag{10.4}$$

where ρ_i and ρ_f are the populations in $|i\rangle$ and $|f\rangle$, and α_2 is the attenuation coefficient at ω_2. Since $W_{fi} = W_{if}$ from detail balancing, the quantity G_R takes the form

$$G_R = \frac{dW_{fi}}{d\omega_2}(\rho_i - \rho_f)\frac{\epsilon_2^{1/2}}{c\bar{n}_2}$$
$$\cong \frac{8\pi^3 N\omega_1\omega_2\hbar}{\epsilon_1\epsilon_2^{1/2}c}|M_{fi}|^2\bar{n}_1 g(\hbar\Delta\omega)(\rho_i - \rho_f), \qquad \text{for } \bar{n}_1, \bar{n}_2 \gg 1 \tag{10.5}$$

which is a constant if $\bar{n}_1$ can be treated as a constant. This is the case when depletion of the ω_1 pump beam is negligible. The solution of (10.4) is then simply an exponentially growing function of z:

$$\bar{n}_2(z) = \bar{n}_2(0)e^{(G_R - \alpha)z} \tag{10.6}$$

with G_R playing the role of a stimulated gain coefficient proportional to the incoming laser intensity at ω_1.

Since the gain is proportional to the Raman transition probability, one expects that it should also be proportional to the spontaneous Raman cross section. By definition, the differential Raman cross section $d^2\sigma/d(\hbar\omega_2)\,d\Omega$ is the scattering probability of an incoming photon at ω_1 per unit area into a Raman photon of a certain polarization at ω_2 per unit solid angle around Ω and unit energy interval around $\hbar\omega_2$:

$$
\begin{aligned}
\frac{d^2\sigma}{d(\hbar\omega_2)\,d\Omega} &= g_E \rho_i \frac{dW_{fi}}{d(\hbar\omega_2)} \Big/ N |\langle \alpha_f | a_2^+ a_1 | \alpha_i \rangle|^2 c \\
&= \left(\frac{\omega_1 \omega_2^3 \epsilon_2^{1/2}}{c^4 \epsilon_1} \right) |M_{fi}|^2 g(\hbar\Delta\omega)\rho_i
\end{aligned}
\tag{10.7}
$$

where

$$
g_E = k_2^2 \left(\frac{dk_2}{d\omega_2} \right) \Big/ (2\pi)^3
$$

is the density of radiation modes per unit solid angle at ω_2. From (10.5) and (10.7), we immediately find the relation

$$
\begin{aligned}
G_R &= N \frac{4\pi^2 c^3 \epsilon_1}{\omega_1 \omega_2^2 \epsilon_2 \rho_i} (\rho_i - \rho_f) \left(\frac{d^2\sigma}{d(\hbar\omega_2)\,d\Omega} \right) |\mathscr{E}_1|^2 \\
&= N \frac{4\pi^2 c^3 \epsilon_1}{\omega_1 \omega_2^2 \epsilon_2 \rho_i} (\rho_i - \rho_f) \left(\frac{d\sigma}{d\Omega} \right) |\mathscr{E}_1|^2 g(\hbar\Delta\omega).
\end{aligned}
\tag{10.8}
$$

Thus, given the spontaneous Raman cross section $d\sigma/d\Omega$ and the halfwidth Γ, the stimulated Raman gain in a medium can be estimated easily. Table 10.1 shows the cross sections, the halfwidths, and the estimated gains for the Raman lines of some materials in which stimulated Raman scattering has been observed.

As an example, let us consider stimulated Raman scattering in benzene. From Table 10.1, we find that the Raman gain for the 992 cm^{-1} mode of benzene is 2.8×10^{-3} cm/MW. Thus, in order to generate e^{30} Raman photons from one noise photon (corresponding to an output of ~ 100 kW/cm^2) in a 10-cm benzene cell, a laser beam of 1 GW/cm^2 is needed. This shows that a high-power pulsed laser is necessary for the study of stimulated Raman scattering. In actual experiments, stimulated Raman scattering with a power gain of $\sim e^{30}$ in benzene and in many other liquids has been observed with a laser beam of ~ 100 MW/cm^2 or less. The order-of-magnitude discrepancy between theory and experiment is the result of self-focusing of the laser beam in the medium (see Section 17.3).

Note that we use the approximation $|\langle\alpha_f|a_2^+ a_1|\alpha_i\rangle|^2 \cong \bar{n}_1\bar{n}_2$ in describing stimulated Raman amplification. This approximation is certainly not valid when $\bar{n}_1$ or $\bar{n}_2$ is small. Therefore, strictly speaking, this theory is rather crude for describing stimulated Raman amplification starting from noise or spontaneous scattering. The complete quantum theory of stimulated Raman scattering by spontaneous scattering is a difficult but challenging problem in quantum optics and has not yet been fully developed. In some respects, it is a two-photon analog of the superfluorescence problem (Chapter 21).

10.3   COUPLED WAVE DESCRIPTION OF STIMULATED RAMAN SCATTERING

Coupling of Pump and Stokes Waves

From the wave interaction point of view, a two-photon transition is a third-order process. It can be seen from the relation that the net transition rate is equal to the rate of generation or loss of photons at ω_1 or ω_2:

$$\frac{dW_{fi}}{d\omega}(\rho_i - \rho_f) = 2\left|\text{Re}\left[\frac{\partial}{\partial t}\mathbf{P}^{(3)*}(\omega_1)\cdot\mathbf{E}_1\right]\right|/\hbar\omega_1$$

$$= 2\left|\text{Re}\left[\frac{\partial}{\partial t}\mathbf{P}^{(3)*}(\omega_2)\cdot\mathbf{E}_2\right]\right|/\hbar\omega_2. \tag{10.9}$$

Since $dW_{fi}/d\omega \propto |E_1|^2|E_2|^2$, we have $\mathbf{P}^{(3)}(\omega_1) \propto |E_2|^2E_1$ and $\mathbf{P}^{(3)}(\omega_2) \propto |E_1|^2E_2$, both of which are therefore third-order nonlinear polarizations.

Stimulated Raman scattering can then be described as a third-order wave coupling process between the pump and Stokes waves. The wave equations are

$$\nabla \times(\nabla \times \mathbf{E}_1) - \frac{\omega_1^2}{c^2}\varepsilon_1\mathbf{E}_1 = \frac{4\pi\omega_1^2}{c^2}\mathbf{P}^{(3)}(\omega_1)$$

and   (10.10)

$$\nabla \times(\nabla \times \mathbf{E}_2) - \frac{\omega_2^2}{c^2}\varepsilon_2\mathbf{E}_2 = \frac{4\pi\omega_2^2}{c^2}\mathbf{P}^{(3)}(\omega_2).$$

For simplicity, we consider the special case of an isotropic medium with $\mathbf{E}_1$ and $\mathbf{E}_2$ polarized in the same direction. The nonlinear polarizations take the form

$$P^{(3)}(\omega_1) = \left[\chi_1^{(3)}|E_1|^2 + \chi_{R1}^{(3)}|E_2|^2\right]E_1$$

and   (10.11)

$$P^{(3)}(\omega_2) = \left[\chi_{R2}^{(3)}|E_1|^2 + \chi_2^{(3)}|E_2|^2\right]E_2.$$

As seen in (10.11), the $\chi_1^{(3)}$ and $\chi_2^{(3)}$ terms in $P^{(3)}$ only act to modify the dielectric constants ε_1 and ε_2 in (10.10). They are responsible for the field-induced birefringence, self-focusing, etc., but have no direct effect on stimulated Raman scattering. We therefore neglect them in the following discussion. The $\chi_R^{(3)}$ terms in $P^{(3)}$, on the other hand, effectively couple E_1 and E_2 in (10.10) and cause energy transfer between the two fields. They are then responsible for the stimulated Raman process, and they are known as the Raman susceptibilities.

The microscopic expressions for $\chi_R^{(3)}$ can of course be obtained from the usual procedure outlined in Chapter 2. Each $\chi_R^{(3)}$ should have a resonant term and a nonresonant term. Only the resonant term is connected to the Raman process. In fact, the microscopic expression of the resonant term $[\chi_R^{(3)}]_R$ can be obtained directly from the relation in (10.9)

$$
\begin{aligned}
\frac{dW_{fi}}{d\omega_2}(\rho_i - \rho_f) &= \frac{2\,\mathrm{Im}\,\chi_{R1}^{(3)}|E_1|^2|E_2|^2}{\hbar} \\
&= -\frac{2\,\mathrm{Im}\,\chi_{R2}^{(3)}|E_1|^2|E_2|^2}{\hbar}.
\end{aligned}
\tag{10.12}
$$

From (10.5), (10.8), and the relation $|E|^2\varepsilon/2\pi = \hbar\omega\bar{n}$, we find

$$
G_R = -\frac{4\pi\omega_2^2}{c^2 k_2}\left(\mathrm{Im}\,\chi_{R2}^{(3)}\right)|E_1|^2,
$$

$$
\begin{aligned}
\mathrm{Im}\,\chi_{R2}^{(3)} &= -\mathrm{Im}\,\chi_{R1}^{(3)} \\
&= -N\frac{c^4\varepsilon_1(\rho_i - \rho_f)}{\omega_1\omega_2^3\varepsilon_2^{1/2}\rho_i}\frac{d\sigma}{d\Omega}\pi g(\hbar\Delta\omega) \\
&= -N|M_{fi}|^2(\rho_i - \rho_f)\pi g(\hbar\Delta\omega).
\end{aligned}
\tag{10.13}
$$

If $g(\hbar\Delta\omega)$ is a Lorentzian, then from the Kramers–Kronig relation, the real part of $[\chi_R^{(3)}]_R$ can be explicitly derived. We then have the microscopic expression

$$
\chi_{R2}^{(3)} = \left[\chi_{R2}^{(3)}\right]_{NR} + \left[\chi_{R2}^{(3)}\right]_R,
$$

$$
\left[\chi_{R2}^{(3)}\right]_R = -\frac{N|M_{fi}|^2(\rho_i - \rho_f)}{\hbar\left[(\omega_1 - \omega_2 - \omega_{fi}) - i\Gamma\right]},
\tag{10.14}
$$

$$
\chi_{R1}^{(3)} = \chi_{R2}^{(3)*}.
$$

Knowing $\chi_R^{(3)}$, we can now solve (10.10) with (10.11). For plane-wave propagation along $\hat{z}$, and assuming slowly varying amplitude approximation, the wave equations reduce to

$$\left(\frac{\partial}{\partial z} + \frac{\alpha_1}{2}\right)\mathscr{E}_1 = i\left(\frac{2\pi\omega_1^2}{c^2 k_1}\right)\chi_{R1}^{(3)}|E_2|^2\mathscr{E}_1$$

and   (10.15)

$$\left(\frac{\partial}{\partial z} + \frac{\alpha_2}{2}\right)\mathscr{E}_2 = i\left(\frac{2\pi\omega_2^2}{c^2 k_2}\right)\chi_{R2}^{(3)}|E_1|^2\mathscr{E}_2$$

They can be transformed into

$$\left(\frac{\partial}{\partial z} + \alpha_1\right)|E_1|^2 = -\frac{4\pi\omega_1^2}{c^2 k_1}\,\mathrm{Im}\,\chi_{R1}^{(3)}|E_1|^2|E_2|^2$$

and   (10.16)

$$\left(\frac{\partial}{\partial z} + \alpha_2\right)|E_2|^2 = -\frac{4\pi\omega_2^2}{c^2 k_2}\,\mathrm{Im}\,\chi_{R2}^{(3)}|E_1|^2|E_2|^2.$$

These equations can now be identified with the equation for $\bar{n}_2$ in (10.4) and a similar equation for $\bar{n}_1$ if $|E_1|^2$ and $|E_2|^2$ are replaced by $2\pi\hbar\omega_1\bar{n}_1/\varepsilon_1$ and $2\pi\hbar\omega_2\bar{n}_2/\varepsilon_2$, respectively, with G_R given in (10.13). Thus, when the depletion of the pump field $|E_1|^2$ is negligible, we again have the exponentially growing solution of $|E_2(z)|^2 = |E_2(0)|^2\exp(G_R z - \alpha z)$.

If the attenuation constants α_1 and α_2 are vanishingly small, then even the exact solution of (10.16) can be obtained easily with the help of the conservation of the total number of photons $[\varepsilon_1^{1/2}|\mathscr{E}_1|^2/\omega_1 + \varepsilon_2^{1/2}|\mathscr{E}_2|^2/\omega_2] = K$. The solution takes the form

$$\frac{|\mathscr{E}_1(z)|^2}{|\mathscr{E}_1(z)|^2 - \omega_1 K/\varepsilon_1^{1/2}} = \frac{|\mathscr{E}_2(0)|^2}{|\mathscr{E}_1(0)|^2 - \omega_1 K/\varepsilon_1^{1/2}}\exp\left[\frac{-\omega_1 K G_R z}{|\mathscr{E}_1(0)|^2\varepsilon_1^{1/2}}\right],$$

$$\frac{|\mathscr{E}_2(z)|^2}{|\mathscr{E}_2(z)|^2 - \omega_2 K/\varepsilon_2^{1/2}} = \frac{|\mathscr{E}_2(0)|^2}{|\mathscr{E}_2(0)|^2 - \omega_2 K/\varepsilon_2^{1/2}}\exp\left[\frac{+\omega_2 K G_R z}{|\mathscr{E}_1(0)|^2\varepsilon_1^{1/2}}\right].$$

(10.17)

Parametric Coupling of Optical and Material Excitational Waves

Stimulated Raman scattering can also be considered as a parametric generation process in which the optical pump wave generates a Stokes wave and a material excitation wave.[4,5] This can be seen as follows:

We recall that the expressions for the nonlinear polarizations in (10.10) can also be derived using the density matrix formalism of Chapter 2. Let us consider here only the Raman resonant term in $P^{(3)}$. We can write, in the notation of Fig. 10.1,

$$P_{res}^{(3)}(\omega_2) = N\sum_s \left[\langle s|e\mathbf{r} \cdot \hat{e}_2|i\rangle \rho_{is}^{(3)}(\omega_2) + \langle f|e\mathbf{r} \cdot \hat{e}_2|s\rangle \rho_{sf}^{(3)}(\omega_2) \right]$$

$$= N\sum_s \left[\frac{\langle f|e\mathbf{r} \cdot \hat{e}_1|s\rangle\langle s|e\mathbf{r} \cdot \hat{e}_2|i\rangle}{\hbar(\omega_2 + \omega_{si})} E(\omega_1)\rho_{if}^{(2)}(\omega_2 - \omega_1) \qquad (10.18) \right.$$

$$\left. - \frac{\langle f|e\mathbf{r} \cdot \hat{e}_2|s\rangle\langle s|e\mathbf{r} \cdot \hat{e}_1|i\rangle}{\hbar(\omega_2 - \omega_{sf})} E(\omega_1)\rho_{if}^{(2)}(\omega_2 - \omega_1) \right]$$

which, with $\omega_2 - \omega_{sf} = \omega_1 - \omega_{si}$ and (10.2), reduces to

$$P_{res}^{(3)}(\omega_2) = -NM_{fi}E(\omega_1)\rho_{if}^{(2)}(\omega_2 - \omega_1). \qquad (10.19)$$

The resonant second-order density matrix element $\rho_{if}^{(2)}(\omega_2 - \omega_1)$, on the other hand, can be obtained from the equation

$$\left(\frac{\partial}{\partial t} + i\omega_{if} + \Gamma_{if} \right)\rho_{if}^{(2)}(\omega_2 - \omega_1) = \frac{1}{i\hbar}\left\{ [-e\mathbf{r} \cdot \hat{e}_1^\dagger E_1^*, \rho^{(1)}(\omega_2)] \right.$$

$$\left. + [-e\mathbf{r} \cdot \hat{e}_2 E_2, \rho^{(1)}(-\omega_1)] \right\}_{if} \qquad (10.20)$$

$$= \frac{1}{i\hbar}M_{fi}^*E_1^*E_2(\rho_i - \rho_f).$$

Then, in the steady state, we have

$$\rho_{if}^{(2)}(\omega_2 - \omega_1) = \frac{M_{fi}^*(\rho_i - \rho_f)}{\hbar(\omega_2 - \omega_1 - \omega_{if} + i\Gamma_{if})}E_1^*E_2,$$

$$\qquad (10.21)$$

$$P_{res}^{(3)}(\omega_2) = [\chi_{R2}^{(3)}]_R|E_1|^2E_2$$

with $[\chi_{R2}^{(3)}]_R$ having exactly the same expression as in (10.14). One can find $P_{res}^{(3)}(\omega_1)$ similarly.

The formalism here shows explicitly $\rho_{if}^{(2)}(\omega_2 - \omega_1)$ as a material excitation resonantly driven by optical mixing $E_1^*E_2$. Stimulated Raman scattering can therefore be considered a result of coupling of three waves $E_1(\omega_1)$, $E_2(\omega_2)$, and

$\rho_{fi}(\omega_1 - \omega_2)$, as governed by the wave equations

$$\left[\nabla \times (\nabla \times) + \frac{\varepsilon_1}{c^2} \frac{\partial^2}{\partial t^2}\right]\mathbf{E}_1 = \frac{4\pi\omega_1^2}{c^2} \mathbf{P}^{(3)}(\omega_1)$$

$$= \frac{4\pi\omega_1^2}{c^2} \hat{e}_1 \left\{ \chi_1^{(3)}|E_1|^2 E_1 + \left[\chi_{R1}^{(3)}\right]_{NR}|E_2|^2 E_1 \right.$$

$$\left. + NM_{fi}^* E_2(\omega_2)\rho_{fi}(\omega_1 - \omega_2) \right\},$$

$$\hspace{6cm} (10.22)$$

$$\left[\nabla \times (\nabla \times) + \frac{\varepsilon_2}{c^2} \frac{\partial^2}{\partial t^2}\right]\mathbf{E}_2 = \frac{4\pi\omega_2^2}{c^2} \mathbf{P}^{(3)}(\omega_2) = \frac{4\pi\omega_2^2}{c^2} \hat{e}_2 \left\{ \left[\chi_{R2}^{(3)}\right]_{NR}|E_1|^2 E_2 \right.$$

$$\left. + \chi_2^{(3)}|E_2|^2 E_2 + NM_{fi} E_1(\omega_1)\rho_{fi}^*(\omega_1 - \omega_2) \right\},$$

$$\left(\frac{\partial}{\partial t} - \imath\omega_{fi} + \Gamma_{fi}\right)\rho_{fi}^*(\omega_1 - \omega_2) = \frac{-i}{\hbar} M_{fi}^*(\rho_i - \rho_f) E_1^* E_2$$

where E_1 is the pump wave, and E_2 and ρ_{fi} are the generated waves. Aside from the $\chi^{(3)}$ terms, which are not essential for stimulated Raman scattering, (10.22) is very similar to (3.4), which was used in Chapter 9 to describe the parametric generation process, except that here the dynamic equation for $\rho_{fi}^{(2)}(\omega_1 - \omega_2)$ has replaced the idler wave equation.

We assumed in the above discussion a localized Raman excitation between two energy levels. This is a good approximation for most Raman excitations including molecular vibration, optical phonon, electronic excitation, spin-flip transition, and optical magnon and plasmon, even though the dynamic equations for different excitations are generally different. Since $\rho_{fi}^{(2)}$ has no dispersion, phase matching for the wave coupling is automatically satisfied here. This makes the solution look different from that of the optical parametric process. The general formalism, however, is valid for any material excitation if the dynamic equation for $\rho_{fi}^{(2)}$ is replaced by the appropriate dynamic equation for the excitation, with or without dispersion.

We note that as long as the response of $\rho_{fi}^{(2)}$ can be considered as stationary, as given in (10.21), elimination of $\rho_{fi}^{(2)}$ in (10.22) with $(\partial^2/\partial t^2)E(\omega) = -\omega^2 E(\omega)$ immediately leads to (10.10), which was used earlier to describe stimulated Raman scattering. However, the set of equations in (10.22) is clearly more general as it also describes the transient case where $\rho_{fi}^{(2)}$ does not respond instantaneously to the time variation of the driving force $E_1 E_2^*$. The population difference $\rho_i - \rho_f$ in (10.22) can be approximated by its thermal equilibrium value under weak excitation, but in general, from physical argument, it should obey the relaxation equation

$$\frac{1}{2}\left(\frac{\partial}{\partial t} + \frac{1}{T_1}\right)\left[(\rho_i - \rho_f) - \left(\rho_i^{(0)} - \rho_f^{(0)}\right)\right] = \frac{dW_{fi}}{Nd\omega_2}(\rho_i - \rho_f) \quad (10.23)$$

where the right-hand side is the transition rate, which, from (10.9) and (10.19), is found to be

$$\frac{dW_{f_i}}{Nd\omega_2}(\rho_i - \rho_f) = \frac{i}{\hbar}\left[M_{f_i}E_1E_2^*\rho_{f_i}^* - M_{f_i}^*E_1^*E_2\rho_{f_i}\right]. \qquad (10.24)$$

In the general description of stimulated Raman scattering, (10.23) should be solved together with (10.22). Strong excitation of the population leads to saturation in the Raman gain. This description, however, applies only to the localized two-level excitation. For the boson-type excitation, (10.23) should be replaced by the equation

$$\left(\frac{\partial}{\partial t} + \frac{1}{T_1}\right)(n_B - \bar{n}_B) = \frac{dW_{f_i}}{d\omega_2}(\rho_i - \rho_f) \qquad (10.25)$$

where n_B and $\bar{n}_B$ are the average numbers of bosons present with and without the driving fields, respectively.

Stimulated Raman Scattering by Molecular Vibrations or Optical Phonons

The most common case of stimulated Raman scattering is by a molecular vibration or optical phonon, which is often described as oscillation of the normal coordinate Q. In agreement with the conventional definition of Q, we can identify $\rho_{f_i}^{(2)}(\omega)$ in (10.22) with $(\hbar/2\omega)^{1/2}Q$ and replace its dynamic equation by the driven harmonic oscillator equation for Q

$$\left(\frac{\partial^2}{\partial t^2} + 2\Gamma\frac{\partial}{\partial t} + \omega_{f_i}^2\right)Q(\omega_1 - \omega_2) = \left[\frac{2(\omega_1 - \omega_2)}{\hbar}\right]^{1/2}M_{f_i}E_1E_2^*(\rho_i - \rho_f),$$
$$(10.26)$$

which, with $\omega_1 - \omega_2 \simeq \omega_{f_i}$, reduces to

$$\left(\frac{\partial}{\partial t} + i\omega_{f_i} + \Gamma\right)Q = i[2\hbar(\omega_1 - \omega_2)]^{-1/2}M_{f_i}E_1E_2^*(\rho_i - \rho_f).$$
$$(10.26a)$$

Then the equations for E_1 and E_2 in (10.22) together with (10.26) properly describe stimulated Raman scattering by phonons. In particular, in the steady-state case, we again obtain the expression for the resonant Raman susceptibility $[\chi_{R2}^{(3)}]_R$ in (10.14).

10.4   STOKES–ANTI-STOKES COUPLING

We have thus far discussed only the stimulated Stokes emission. Experimentally, however, both Stokes and anti-Stokes waves are simultaneously generated with comparable intensity in the stimulated Raman process even at very low temperature. This is very different from the case in spontaneous Raman scattering and cannot be understood as a stimulated two-photon process described in Section 10.2 since the anti-Stokes wave then would be absorbed instead of generated. Yet it can be explained readily by the coupled wave description.[4,5] A third-order nonlinear polarization $P^{(3)}(\omega_a)$ at the anti-Stokes frequency $\omega_a = 2\omega_l - \omega_s$ can be induced in the medium through mixing of the pump wave at ω_l and the Stokes wave at ω_s via the Raman resonance at $\omega_l - \omega_s = \omega_a - \omega_l$, as we see below.

With the simultaneous presence of $E_l(\omega_l)$, $E_s(\omega_s)$, and $E_a(\omega_a)$, the material excitation ρ_{fl}, now driven by both $E_l E_s^*$ and $E_a E_l^*$, obeys the equation

$$\left(\frac{\partial}{\partial t} - i\omega_{fl} + \Gamma_{fl}\right)\rho_{fl}^* = \frac{-i}{\hbar}\left[M_{fl,s}^* E_l^* E_s + M_{fl,a}^* E_a^* E_l\right](\rho_i - \rho_f) \qquad (10.27)$$

where $M_{fl,s}$ and $M_{fl,a}$ are the Raman matrix elements for the transitions incurred by E_l, E_s and E_a, E_l respectively. Mixing of the material excitation with the em waves then gives rise to the resonant part of the nonlinear polarizations

$$P_{res}^{(3)}(\omega_l) = N\left(M_{fl,s}^* E_s \rho_{fl} + M_{fl,a} E_a \rho_{fl}^*\right),$$

$$P_{res}^{(3)}(\omega_s) = N M_{fl,s} E_l \rho_{fl}^*, \qquad (10.28)$$

$$P_{res}^{(3)}(\omega_a) = N M_{fl,a}^* E_l \rho_{fl}.$$

If the response of ρ_{fl} to the fields is stationary, then ρ_{fl} can be solved from (10.27) and substituted into (10.28). The resultant $P_{res}^{(3)}$ combined with the nonresonant part of $P^{(3)}$ can be cast into the form $P^{(3)}(\omega = \omega_\alpha + \omega_\beta + \omega_\gamma) = \chi^{(3)} E_\alpha(\omega_\alpha) E_\beta(\omega_\beta) E_\gamma(\omega_\gamma)$ with $\chi^{(3)}$ containing both a resonant and a nonresonant term. The set of wave equations describing the steady-state anti-Stokes generation then is given by

$$\left[\nabla \times (\nabla \times) - \frac{\omega_l^2}{c^2}\varepsilon_l\right]\mathbf{E}_l = \frac{4\pi\omega_l^2}{c^2}\hat{e}_l\left[\chi_{ss}^{(3)*}|E_s|^2 E_l\right.$$

$$\left. + \left(\chi_{sa}^{(3)} + \chi_{sa}^{(3)*}\right)E_s E_a E_l^* + \chi_{aa}^{(3)}|E_a|^2 E_l\right],$$

$$\left[\nabla \times (\nabla \times) - \frac{\omega_s^2}{c^2}\varepsilon_s\right]\mathbf{E}_s = \frac{4\pi\omega_s^2}{c^2}\hat{e}_s\left[\chi_{ss}^{(3)}|E_l|^2 E_s + \chi_{sa}^{(3)} E_l^2 E_a^*\right], \qquad (10.29)$$

$$\left[\nabla \times (\nabla \times) - \frac{\omega_a^2}{c^2}\varepsilon_a\right]\mathbf{E}_a^* = \frac{4\pi\omega_a^2}{c^2}\hat{e}_a\left[\chi_{sa}^{(3)} E_l^2 E_s^* + \chi_{aa}^{(3)}|E_l|^2 E_a^*\right]$$

where

$$\chi_{ss}^{(3)} = \left(\chi_{ss}^{(3)}\right)_{NR} - \frac{N|M_{fi,s}|^2(\rho_i - \rho_f)}{\hbar(\omega_l - \omega_s - \omega_{fi} - i\Gamma_{fi})},$$

$$\chi_{sa}^{(3)} = \left(\chi_{sa}^{(3)}\right)_{NR} - \frac{NM_{fi,s}M_{fi,a}^*(\rho_i - \rho_f)}{\hbar(\omega_l - \omega_s - \omega_{fi} - i\Gamma_{fi})}, \qquad (10.30)$$

$$\chi_{aa}^{(3)} = \left(\chi_{aa}^{(3)}\right)_{NR} - \frac{N|M_{fi,a}|^2(\rho_i - \rho_f)}{\hbar(\omega_l - \omega_s - \omega_{fi} - i\Gamma_{fi})}$$

and we have neglected terms of $\chi^{(3)}|E_l|^2 E_l$ in (10.29).

The set of equations in (10.29) actually describes a four-wave parametric generation process with E_l being the pump wave, E_s and E_a being the signal and idler waves, and χ_{sa} acting as the coupling constant between the signal and idler waves. The solution of (10.29) in the limit of no pump depletion therefore follows closely that of parametric generation described in Chapter 9. We skip the derivation and present only the results here. Assuming an isotropic medium with a plane boundary at $z = 0$ and slowly varying amplitudes for E_s and E_a, we find[5]

$$E_s = \left[\mathscr{E}_{s+}\exp(i\Delta K_+ z) + \mathscr{E}_{s-}\exp(i\Delta K_- z)\right]\exp(i\mathbf{k}_s \cdot \mathbf{r} - \alpha_z z)$$

and
$$(10.31)$$

$$E_a^* = \left[\mathscr{E}_{a+}^*\exp(i\Delta K_+ z) + \mathscr{E}_{a-}^*\exp(i\Delta K_- z)\right]\exp\left[-i\mathbf{k}_a \cdot \mathbf{r} - (i\Delta k + \alpha_z)z\right]$$

where

$$k^2 = \frac{\omega^2 \varepsilon'}{c^2},$$

$$0 = 2k_{l\lambda,y} - k_{s\lambda,y} - k_{ax,y},$$

$$\Delta k = 2k_{lz} - k_{sz} - k_{az}, \qquad k_z \equiv \mathbf{k} \cdot \hat{z},$$

$$\Delta K_\pm = \frac{\Delta k}{2} \pm \left[\left(\frac{\Delta k}{2}\right)^2 - (\Delta k)\Lambda\right]^{1/2},$$

$$\Lambda = \left(\frac{2\pi\omega_s^2}{c^2 k_{sz}}\right)\chi_{ss}^{(3)}|E_l|^2,$$

$$\alpha_z = \frac{\omega^2 \varepsilon''}{c^2 k_z} = \alpha\left(\frac{k}{k_z}\right),$$

$$G_R = -2\,\mathrm{Im}(\Lambda).$$

For the sake of simplicity, we neglect the dispersions of the absorption

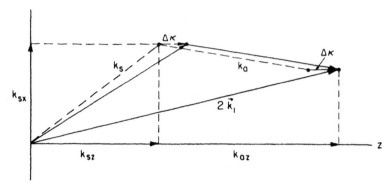

Fig. 10.2 General relationship between the wavevectors of Stokes, anti-Stokes, and laser waves.

coefficients α_z and the coupling coefficients $(2\pi\omega^2/c^2 k_z)\chi^{(3)}$. The relationship between the various wavevectors is shown in Fig. 10.2. With $\mathscr{E}_s = \mathscr{E}_{s0}$ and $\mathscr{E}_a = \mathscr{E}_{a0}$ at $z = 0$, we have

$$\frac{\mathscr{E}_{a\pm}^*}{\mathscr{E}_{s\pm}} = \frac{\Delta K_\pm - \Lambda}{\Lambda},$$

$$\mathscr{E}_{s\pm} = \frac{(-\Delta K_\mp + \Lambda)\mathscr{E}_{s0} + \Lambda \mathscr{E}_{a0}^*}{\Delta K_\pm - \Delta K_\mp}.$$

(10.32)

A number of important physical results come out of the above solution. First, through the Stokes–anti-Stokes coupling, two composite waves $(\mathscr{E}_{s+}, \mathscr{E}_{a+})$ and $(\mathscr{E}_{s-}, \mathscr{E}_{a-})$ are formed with the eigenvectors $\Delta K_\pm$ and the Stokes–anti-Stokes amplitude ratio $(\mathscr{E}_s/\mathscr{E}_a)_\pm$. One of them may experience an exponential gain and the other a loss if $\mathrm{Im}(\Delta K_-) < 0$. The coupling clearly depends on phase matching. If the phase mismatch is sufficiently large that $|\Delta k| \gg |\Lambda|$, then the Stokes and anti-Stokes waves are effectively decoupled. This is explicitly seen by the fact that $\Delta K_\pm$ and $\mathscr{E}_{a\pm}/\mathscr{E}_{s\pm}$ reduce to

$$\Delta K_- = \Lambda - 2(\Lambda^2/\Delta K), \qquad \Delta K_+ = \Delta k - \Lambda$$

and

$$\left|\frac{\mathscr{E}_{a-}^*}{\mathscr{E}_{s-}}\right| = \left|\frac{2\Lambda}{\Delta k}\right| \ll 1, \qquad \left|\frac{\mathscr{E}_{a+}^*}{\mathscr{E}_{s+}}\right| = \left|\frac{\Delta k}{\Lambda}\right| \gg 1.$$

The first denotes a nearly pure Stokes wave with an exponential power gain of $2\,\mathrm{Im}(\Delta K_-) = G_R$, while the second denotes a nearly pure anti-Stokes wave with an attenuation $-G_R$. These are the results expected when the Stokes and anti-Stokes waves are decoupled from one another. The Stokes–anti-Stokes coupling is maximum when $\Delta k = 0$. The solution yields $\Delta K_\pm = 0$ and

$|\mathscr{E}_{a\pm}^{*}/\mathscr{E}_{s\pm}| = 1$. Then neither Stokes nor anti-Stokes waves should experience an exponential growth. This is because through coupling, the Stokes gain is completely canceled by the anti-Stokes attenuation, a result well known in the parametric amplifier theory where no gain can be expected at $\omega_s = \omega_l - \omega_t$ if the other sideband at $\omega_s = \omega_l + \omega_t$ is not suppressed. As $|\Delta k|$ gradually deviates from zero, the gain $|2\,\mathrm{Im}(\Delta K_-)|$ increases rapidly from 0 toward G_R, as shown in Fig. 10.3, while the anti-Stokes–Stokes ratio $|\mathscr{E}_{a-}^{*}/\mathscr{E}_{s-}|$ decreases from 1 to 0. Appreciable anti-Stokes generation is therefore expected in the region $|\Delta k| \sim |\Lambda|$ where both $|\mathrm{Im}(\Delta K_-)|$ and $|\mathscr{E}_{a-}^{*}/\mathscr{E}_{s-}|$ are sufficiently finite. The anti-Stokes radiation should appear in the form of a double cone around the phase-matching direction, as shown in Fig. 10.4. We should, however, remember that infinite plane waves are assumed in this theory. In reality, a pump beam of finite cross section has a spread of wavevectors. Consequently, the sharp dip in the gain curve in Fig. 10.3 may get smeared out, and instead of the double cone in Fig. 10.4, one may observe only a single cone of anti-Stokes radiation.[12]

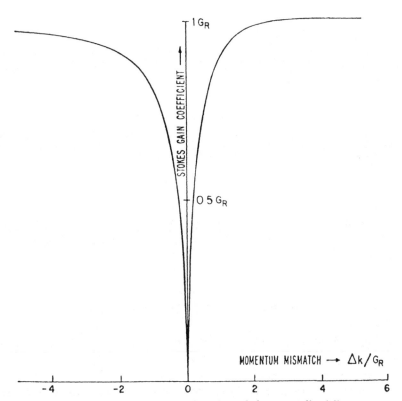

Fig. 10.3 The Stokes power gain as a function of the normalized linear momentum mismatch $\Delta k/G_R$ in the z direction. The asymmetry is due to the nonresonant part $\chi_{NR}^{(3)} = 0.1|\mathrm{Im}\,\chi_R^{(3)}|_{\max}$. (After Ref. 5.)

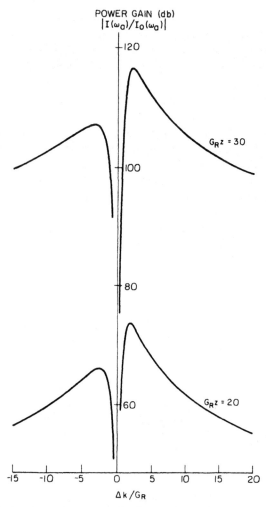

Fig. 10.4 Anti-Stokes intensity versus the linear phase mismatch Δk (normalized by the Stokes power gain G_R). The asymmetry is due to $\chi_{NR} = 0.1|\mathrm{Im}\,\chi_R^{(3)}|_{max}$. (After Ref. 5.)

10.5　HIGHER-ORDER RAMAN EFFECTS

Intense higher-order Stokes and anti-Stokes radiation often shows in experiments accompanying first-order Stokes and anti-Stokes generation.[12] Unlike the overtones in spontaneous scattering, they appear with frequencies at $\omega \pm n\omega_{f_i}$, where n is an integer. This characteristic suggests that they are generated more or less successively. For example, when the first Stokes $E_s(\omega_s)$ becomes sufficiently intense, it may also act as a pump wave to generate the second Stokes E_{s2} ($\omega_{s2} = \omega_s - \omega_{f_i}$). In general, however, we must find the answer from the coupled wave approach, because an nth-order Stokes or anti-Stokes wave may couple to several Stokes and anti-Stokes waves of various orders through third-order nonlinear polarizations. Again, coupling of a particular set of waves is strongest if phase matching is satisfied. Unfortunately, with many complex waves nonlinearly coupled together, the problem becomes extremely complex.

We consider here, as an example, a special case where only the Stokes generation in the $+\hat{z}$ direction needs to be taken into account.[5] This can be achieved in a real situation with short pulsed laser excitation such that the backward Stokes generation is suppressed (see Section 10.9), while the anti-Stokes generation can be neglected. The set of coupled wave equations is then given by

$$\left(\frac{\partial^2}{\partial z^2} + \frac{\omega_l^2 \epsilon_l}{c^2}\right) E_l = -\left(\frac{4\pi\omega_l^2}{c^2}\right)\chi_{s1}^{(3)*}|E_s|^2 E_l,$$

$$\left(\frac{\partial^2}{\partial z^2} + \frac{\omega_s^2 \epsilon_s}{c^2}\right) E_s = -\left(\frac{4\pi\omega_s^2}{c^2}\right)\left[\chi_{s1}^{(3)}|E_l|^2 E_s + \chi_{s2}^{(3)*}|E_{s2}|^2 E_s\right], \quad (10.33)$$

$$\left(\frac{\partial^2}{\partial z^2} + \frac{\omega_{s2}^2 \epsilon_{s2}}{c^2}\right) E_{s2} = -\left(\frac{4\pi\omega_{s2}^2}{c^2}\right)\left[\chi_{s2}^{(3)}|E_s|^2 E_{s2} + \chi_{s3}^{(3)*}|E_{s3}|^2 E_{s2}\right],$$

etc. The solution of (10.33) can be obtained by numerical calculation, as seen in Fig. 10.5. As the length of the medium or the pump power increases, the first Stokes power increases gradually at the beginning, and then suddenly builds up to a maximum while the pump power plunges to nearly zero by depletion. Then the first Stokes power remains roughly constant for a while and gets depleted into the second Stokes, and so on. This is in fact what one would

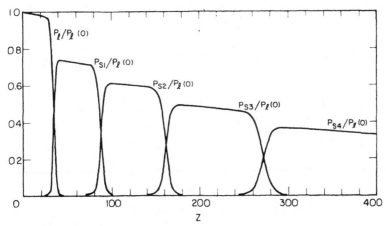

Fig. 10.5 Generation of higher-order Stokes waves as a function of normalized cell length $Z = (16\pi^3\omega_l^2 \operatorname{Im} \chi_{s1}^{(3)}/k_l c^3)P_l(0)z$. (After Ref. 5.)

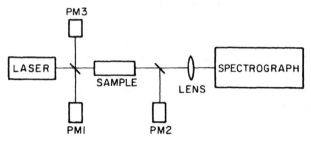

Fig. 10.6 Schematic of a set-up for investigation of stimulated Raman scattering. PM1, PM2, and PM3 are photodetectors measuring the laser, the forward Raman, and backward Raman radiation, respectively.

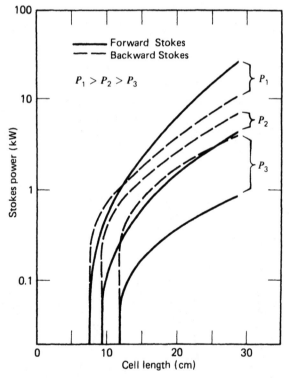

Fig. 10.7 First-order forward and backward Stokes power versus the toluene cell length at three laser powers $P_1 = 80$, $P_2 = 67$, and $P_3 = 53$ MW/cm^2. [After Y. R. Shen and Y. J. Shaham, *Phys. Rev.* **163**, 224 (1967).]

expect from the simple physical picture. The calculation here, however, assumes infinite plane waves. With a pump beam of finite cross section, the rise and fall of an nth-order Stokes wave should be more gradual, as demonstrated both theoretically and experimentally by von der Linde et al.[13]

10.6 EXPERIMENTAL OBSERVATIONS AND APPLICATIONS

Stimulated Raman Scattering in Self-Focusing Media

A typical setup for studying stimulated Raman scattering is seen in Fig. 10.6. As mentioned in Section 10.1, earlier results on the Stokes intensity measurements disagreed strongly with theory. Most of those experiments were on liquids with large Kerr constants and the results showed a sharp threshold for stimulated Raman scattering and an effective Raman gain an order of magnitude larger than the predicted gain. An example is given in Fig. 10.7, which also shows a forward–backward asymmetry in the Stokes output that was not predicted by the theory. Other anomalies such as the Raman spectral broadening and the anomalous anti-Stokes ring pattern were also observed. It was later realized that these anomalies were initiated by self-focusing, which readily occurred in Kerr media (see Chapter 17). Self-focusing has a threshold; it increases the laser intensity dramatically at the focal region, and breaks the forward–backward symmetry of the Raman amplification. This then explains qualitatively the results of Fig. 10.7. Raman anomalies constituted a subject of great confusion in the past. We do not go into any detailed discussion on the subject here. Interested readers should consult Section 17.3 and the relevant literature (see Bibliography).

Stimulated Raman Scattering in Non-Self–Focusing Media

Even without self-focusing, stimulated Raman scattering often shows a certain gain anomaly. An example is seen in Fig. 10.8, where the Stokes output is plotted against the laser input in liquid nitrogen. Self-focusing was not observed in this case. As the laser power increases, the Stokes output first increases linearly because it is generated from spontaneous scattering, and then grows quasi-exponentially when stimulated scattering sets in. At a certain input power, the output rises suddenly with an effective gain much larger than the theoretical prediction. It finally levels off as a result of depletion of the laser power.

The sharp rise of the Stokes output is presumably due to feedback in the Stokes amplification from Rayleigh scattering or diffuse reflection from walls and cell windows. Experiments of the kind in Fig. 10.8 actually describe the build-up of Raman oscillation, that is, amplification from noise. As is well known, the output of an oscillator without saturation depends critically on small perturbation or feedback. This makes the quantitative interpretation of

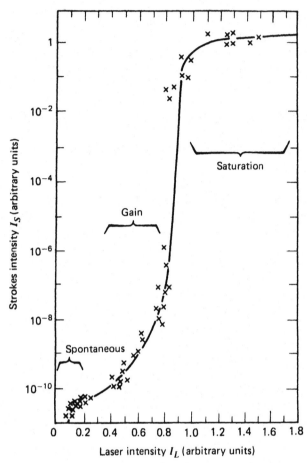

Fig. 10.8 Experimental result of the first-order Stokes output as a function of the laser intensity in liquid N_2. [After J. B. Grun, A. K. McGuillan, and B. P. Stoicheff, *Phys. Rev.* **180**, 61 (1969).]

the oscillator output extremely difficult, especially if the perturbation or feedback cannot be well characterized. With careful elimination of feedback, Haidacher and Maier have shown that the sharp rise of the Stokes output can be greatly suppressed.[14]

Raman Gain Measurements

The theory developed earlier is clearly a theory of Raman amplification rather than oscillation. To check the theory, one should carry out experiments on Raman amplifiers.[15] This can be done with the oscillator-amplifier setup in Fig. 10.9. The backward Stokes emission from the oscillator provides a Stokes input to the amplifier, and the amplification gain of the backward Raman scattering

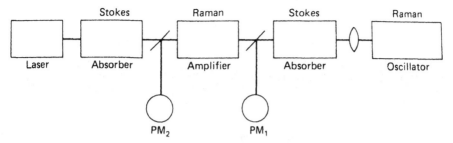

Fig. 10.9 Experimental set-up for measuring the backward Stokes gain. (After Ref. 15.)

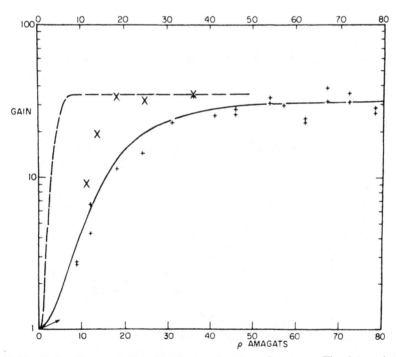

Fig. 10.10 Stokes Raman gain in H_2 gas as a function of pressure. The data points X for forward gain should be compared with the dashed theoretical curve (cell length: 80 cm; peak input intensity: 20 MW/cm^2). The data points $+$ for backward gain should be compared with the solid theoretical curve (cell length: 30 cm; peak input intensity: 60 MW/cm^2 at 0.6943 μm.) (After Ref. 15.)

in the amplifier is measured. Figure 10.10 shows the result on hydrogen gas exhibiting good agreement between theory and experiment. Such a Raman gain experiment was, however, not every successful in self-focusing media, where self-focusing in both the oscillator and the amplifier led to complications.

Anti-Stokes and Higher-Order Raman Radiation

Anti-Stokes radiation of many order often can be observed in stimulated Raman scattering.[16,17] In condensed matter they appear in the form of bright multicolored rings on a plane perpendicular to the laser beam. Rings of different color correspond to different orders of anti-Stokes. Chiao and Stoicheff[12] showed in calcite that the nth-order anti-Stokes radiation is emitted in the direction given by the phase-matching condition $\mathbf{k}_{a,n} = \mathbf{k}_{a,n-1} + \mathbf{k}_l - \mathbf{k}_{s,n}$. This is expected if we assume that the higher-order anti-Stokes is generated successively from the lower-order Stokes and anti-Stokes. The first-order anti-Stokes ring should be defined by $\mathbf{k}_a = 2\mathbf{k}_l - \mathbf{k}_s$ according to the theory.

In self-focusing liquids the situation is more complicated. The directions of the anti-Stokes rings now deviate from those defined by $\mathbf{k}_{a,n} = \mathbf{k}_{a,n-1} + \mathbf{k}_l - \mathbf{k}_{s,n}$ as they are now affected by self-focusing. Garmire[16] had some success in interpreting these rings by assuming the existence of thin filaments of pump light resulting from self-focusing in the medium. The problem, however, remains since the assumption of filaments is not quite valid.

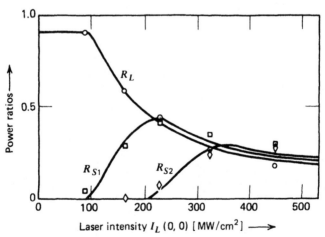

Fig. 10.11 Normalized transmitted laser (R_L), first (R_{S1}) and second Stokes (R_{S2}) power as a function of the incident laser intensity $I_L(0,0)$. The experimental data of R_L, R_{S1}, and R_{S2} are represented by circles, rectangles, and diamonds, respectively. The curves are calculated according to the theory in Section 10.5 with the finite beam cross section taken into account. (After Ref. 13.)

Higher-order Stokes radiation appears mostly along the axis in the forward and backward directions. Quantitative studies generally are difficult. Using subnanosecond laser pulses, von der Linde et al.[13] were able to carry out quantitative study in a special case. The short input pulse excited only the Stokes radiation in the forward direction, as already described in Section 10.5. Their results are shown in Fig. 10.11. Theoretical calculation following (10.33) but with a Gaussian beam profile shows good agreement with the experiment.

Stimulated Raman Scattering by Broadband Pump Source

Raman gain measurements in the absence of self-focusing may still show a forward–backward asymmetry. This results from the finite linewidth of the pump beam. If the pump linewidth is $2\Gamma_p$, and the Raman linewidth is 2Γ, theories predict that the maximum backward Raman gain is proportional to $(\Gamma + \Gamma_p)^{-1}$ and the maximum forward Raman gain is proportional to Γ^{-1}, assuming no relative dispersion between the Stokes and pump frequencies.[17] This surprising result can be understood qualitatively from the following picture. In the forward direction, as a short Δt section of the Stokes wave propagates in the medium, it always coherently interacts with the same Δt section of the pump wave. On the other hand, if a short section of the Stokes wave propagates in the backward direction, it constantly encounters a new wavefront of the pump wave. Consequently, the forward Raman gain follows the pump intensity variation and is proportional to Γ^{-1} as predicted by the stationary theory described earlier. The backward Raman gain is reduced because the amplification process averages over the amplitude and phase variation of the pump field. Explicit results can be obtained, for example, by considering the special case of stimulated Raman scattering by a short pump pulse (see Section 10.9) where the pump linewidth is given by the inverse of the pulsewidth. More generally, statistical theories should be used to describe the problem.[17]

Competition between Raman Modes

In stationary stimulated Raman scattering, only the mode with the maximum gain appears to participate in the process. It is usually the mode with both a large Raman cross section and a narrow Raman linewidth. Effective depletion of laser power into this Raman mode forbids the occurrence of stimulated scattering into other modes. Only in transient cases (see Section 10.10) will several competing modes show up simultaneously.

Inverse Raman Effect

A loss in pump radiation always accompanies the gain in the Stokes radiation in a stimulated Raman process. Thus with both the pump and the Stokes waves sent into a medium, one can observe simultaneously the gain of the

Stokes wave and the attenuation of the pump wave. The absorption of the pump radiation in a stimulated Raman process was first demonstrated by Jones and Stoicheff and is known as the inverse Raman effect.[18]

Tunable Infrared Sources Obtained from Stimulated Raman Scattering

Stimulated Raman scattering with its Raman-shifted output has long been considered a viable method for generating coherent radiation at new frequencies. Special interest is in tunable coherent sources. Since the Raman frequency of a medium is usually fixed, the tunability is often achieved by using a tunable pump laser.

Two systems have received much attention. One is the atomic vapor system, mostly alkali and alkali-earth metal vapor. Raman transitions involved are often of the $S \rightarrow P \rightarrow S$ or $S \rightarrow P \rightarrow D$ type.[19,20] The tunable pump source is near resonance with the $S \rightarrow P$ transition, so that the Raman cross section is greatly enhanced. As a result, even though the atomic vapor density is small ($\leq 10^{17}$ atoms/cm^3 at a pressure of ~ 10 torr), the Raman gain is significant. Stimulated Raman scattering with emission in the near infrared can be readily observed. With a dye laser input of a few tens of kilowatts in an alkali vapor, the infrared output can have a continuous tuning range of several hundred inverse centimeters and a peak photon conversion efficiency as high as 50%.[20] The tuning range can be further extended by using different Raman transitions. For example, in Cs, with a tunable pump dye laser in the blue and uv, the observed Stokes output appears in the range of 2.5–4.75 μm, 5.67–8.65 μm, and 11.65–15 μm resulting from the Raman transitions $6S \rightarrow 7S$, $6S \rightarrow 8S$, and $6S \rightarrow 9S$ respectively. It can be extended to ~ 20 μm using the $6S \rightarrow 10S$ transition. If a broadband dye laser is used as the input, the infrared output will have the same broad bandwidth. With a pulsed dye laser, Bethune et al.[21] generated the broadband infrared beam and used it to obtain single-shot absorption spectra of molecules. To increase the detection sensitivity, they up-converted the infrared signal transmitted through the sample into visible through nonlinear mixing in another alkali vapor cell. The technique allows the recording of infrared spectra of transient chemical species with nanosecond resolution using nanosecond laser pulses.

At high pump power, the output power from stimulated Raman scattering in atomic vapor is often limited by the occurrence of multiphoton absorption and ionization, population saturation, field-induced spectral broadening, and other nonlinear optical processes.[20] However, output as high as 30 mJ at ~ 2.9 μm with a photon conversion efficiency of 40% has been observed in Ba vapor. The Stokes output from atomic vapor tends to have a linewidth increasing with the pump intensity. This line broadening may result from the Stark effect caused by the photoionized atoms, or saturation in the Raman transition, or others; the dominant mechanism has not yet been identified.

Another system of immense practical interest is the molecular gas system, such as H_2, N_2. These simple molecules have very strong Raman modes.

Intense Raman output can be expected from a gas cell of a few tens of centimeters long at a few atmospheric pressure with several megawatts per square centimeter laser intensity in the visible.[22] Molecular hydrogen is probably most interesting because of its large Raman shift (4155 cm^{-1}). With a pump source at $\lambda \geq 7500$ Å, the third-order Stokes output has a wavelength larger than 10 μm. Figure 10.12 shows the accessible wavelength range the various orders of Stokes and anti-Stokes output from H_2 and D_2 can cover with a tunable pump source between 2000 and 8000 Å. The output power can be significant, as indicated in Table 10.2 for a commercial unit. It can be

Table 10.2
Raman Output from RS-1[a]

Wavelength (nm)[b]	Energy (mJ)	Pressure (psi)[c]
With 560-nm Pump Beam[d]		
195 (AS_8)	.0031	125
213 (AS_7)	.0091	125
234 (AS_6)	.024	110
259 (AS_5)	.054	115
290 (AS_4)	.10	145
330 (AS_3)	.26	160
382 (AS_2)	.78	190
454 (AS_1)	2.1	200
730 (S_1)	17	90
1048 (S_2)	6.2	300
1855 (S_3)	.60	275
With 280-nm Pump Beam[e]		
207 (AS_3)	.038	100
227 (AS_2)	.19	125
251 (AS_1)	.54	125
317 (S_1)	2.2	40
365 (S_2)	3.1	300
430 (S_3)	1.2	160
524 (S_4)	0.24	110
669 (S_5)	0.060	100

[a]After Quanta-Ray, Inc., advertising brochure on RS-1 Raman Shifter.
[b]S_i denotes ith Stokes wavelength; AS_i denotes ith Anti-Stokes wavelength.
[c]Gas: H_2 at 300°K
[d]Pump: 85 mJ at 560 nm from a Quanta-Ray PDL-1 dye laser pumped by a DCR-1A Nd : YAG laser.
[e]Pump: 17 mJ at 280 nm from a frequency doubled Quanta-Ray PDL-1 dye laser pumped by a DCR-1A Nd : YAG laser.

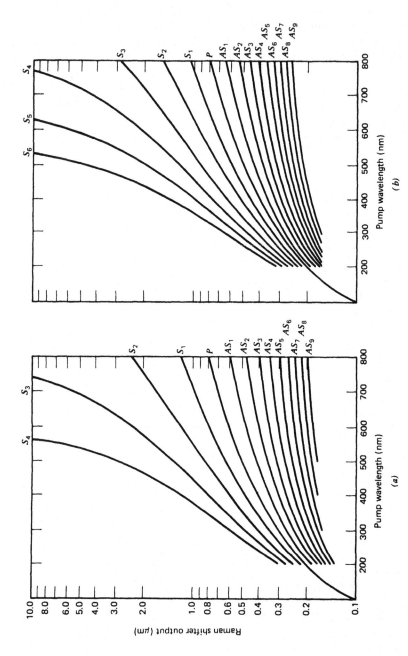

Fig. 10.12 The families of curves show the theoretically possible wavelengths accessible from use of (a) H_2 with a Raman shift of 4155 cm^{-1}, and (b) D_2 with a Raman shift of 2987 cm^{-1}, as pump gases. (After the advertising brochure on RS-1 Raman shifter by Quanta-Ray, Inc.)

improved by using the oscillator-amplifier scheme as in the laser-pumped dye laser case. Conversion efficiency as high as 80% in the Stokes output has been achieved.[23]

High-power infrared CO_2 lasers can also be used to obtain stimulated Raman scattering in molecular gases. It yields, for example, a Stokes output around 16 μm from CH_4. A conversion efficiency of ~ 10% can be achieved.[24] The 16-μm radiation is most important for laser isotope separation of uranium through vibrational excitation of UF_6. It can also be obtained by stimulated Raman scattering via rotational transition in para-hydrogen molecules using CO_2 lasers. An output energy in excess of 1 J and a peak power of ~ 20 MW with a photon conversion efficiency of 85% has been observed.[25] Tunable far-infrared output down to 257 μm has also been obtained from stimulated Raman scattering via rotational transitions $Q(J)$ in HF using the tunable infrared output from stimulated Raman scattering in H_2 by a flash-pumped dye laser as a pump source.[26]

Tunable UV Source Obtained from Anti-Stokes Scattering

Stimulated anti-Stokes Raman scattering is possible if a Raman transition has an inverted population [$\rho_i < \rho_f$ in (10.5) leading to a positive exponential gain

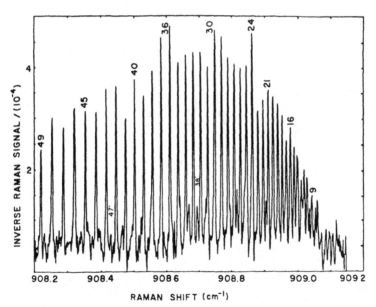

Fig. 10.13   Inverse Raman spectrum in the vicinity of ν_1 fundamental of CF_4 at 4 Torr. Fundamental and hot band transitions are labeled by J and J', respectively. Pump and probe laser powers of 2 MW and 100 mW were used, respectively. A 3-sec time constant was used to average the 10 pps signs. [After A. Owyoung, in W. O. N. Guimaraes, C. T. Lin, and A. Mooradian, eds., *Lasers and Applications* (Springer-Verlag, Berlin, 1981), p. 67.]

for the E_1 field]. This has been suggested as a means to obtain a potentially powerful uv source which is broadly tunable. To achieve the inverted population, it is generally necessary to choose a metastable state as the upper state in the Raman transition. The population can be pumped into the metastable state by various methods. In a recent demonstration,[27] photodissociation was used to obtain an inverted population between a metastable state and the ground state of a dissociation fragment, and stimulated anti-Stokes Raman scattering from the metastable state was then observed. For example, through ArF laser pumping, TlCl was dissociated into Tl and Cl. The thallium product was in the $6p^2P_{3/2}^0$ metastable state. It could reach a concentration of 4×10^{16} atoms/cm^3 out of the original TlCl concentration of 6.9×10^{16} molecules/cm^3. If the second- or third-harmonic output from a Q-switched Nd : YAG laser was then propagated into this photodissociated system, stimulated anti-Stokes Raman emission from $6p^2P_{3/2}^0$ to the ground state $6p^2P_{1/2}^0$ of Tl was readily seen. A 10% conversion efficiency was achieved in a cell 25 cm long.

Even spontaneous anti-Stokes Raman scattering can be useful as a radiation source.[28] Using a high-lying metastable state, the emission can be tunable over narrow regions in the XUV. Such a source can have the unique properties of

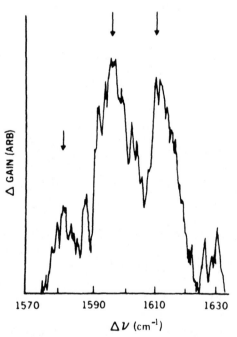

Fig. 10.14 The Raman gain spectrum of a monolayer of p-nitrobenzoic acid (PNBA) on a thin film of aluminum oxide supported by sodium fluoride. Three principal features are marked. [After J. P. Heritage and D. L. Allara, *Chem. Phys. Lett.* **74**, 507 (1980).]

narrow linewidth, ultrashort pulsewidth, and relatively high peak spectral brightness.

Stimulated Raman Scattering as a High-Resolution Spectroscopy Technique

The theory of stimulated Raman scattering in Section 10.2 or 10.3 shows a Stokes amplification gain $G_R(\omega_l - \omega_s)$ and a corresponding pump attenuation both proportional to the Raman lineshape. Therefore, measurements of the Stokes gain versus $\omega_l - \omega_s$ (known as stimulated Raman gain spectroscopy) or measurements of the pump attenuation versus $\omega_l - \omega_s$ (known as inverse Raman spectroscopy) should yield a Raman spectrum identical to that obtained from spontaneous Raman scattering. The coherent spectroscopic technique, however, has two important advantages. First, no spectrometer is needed, so that the spectral resolution is limited only by the laser linewidths which can be many orders of magnitude better than the resolution of a spectrometer. It can be used to obtain high-resolution Raman spectra of gases not realizable by spontaneous Raman scattering.[8] An example is seen in Fig. 10.13. Second, with CW mode-locked laser pulse and a locked-in detection scheme, the coherent technique can be extremely sensitive and can be used to study Raman spectra of thin films and adsorbed molecules.[29] Monolayer detection is possible, as has been demonstrated by Heritage shown in Fig. 10.14.

10.7   STIMULATED POLARITON SCATTERING

The material excitation ρ_{fi} discussed in Section 10.3 can in general be both infrared and Raman active, that is, it can be excited by both the two-photon Raman process and the one-photon infrared absorption process. This is the case, for example, with phonons in polar crystals. The direct coupling between infrared and phonon waves actually forms a mixed excitational wave which is usually known as polariton.[30] A typical polariton dispersion curve of a polar crystal is seen in Fig. 10.15. Because of the strong dispersion in the $k \sim \omega \varepsilon^{1/2}/c$ region, Raman scattering by polaritons shows a Raman frequency shift that depends on the scattering angle, as dictated by the frequency and wavevector matching conditions $\omega_l = \omega_s + \omega_3$ and $\mathbf{k}_l = \mathbf{k}_s + \mathbf{k}_3$, where ω_3 and $\mathbf{k}_3$ are the frequency and wavevector of the polariton.

Stimulated Raman scattering by polaritons (or stimulated polariton scattering) occurs when the pump excitation is sufficiently strong. It can again be described by the coupled wave approach. Four interacting waves are now involved in the problem: the pump $\mathbf{E}_l$, the Stokes $\mathbf{E}_s$, the infrared $\mathbf{E}_3$, and the material excitation ρ_{fi}. The process can be considered a combination of the parametric generation process discussed in Chapter 9 and the stimulated

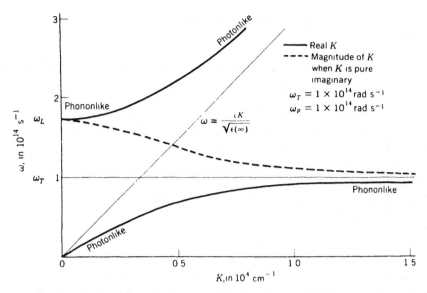

Fig. 10.15 Coupled modes of photons and transverse optical phonons in an ionic crystal. The fine horizontal line represents oscillators of frequency ω_T in the absence of coupling to the electromagnetic field, and the fine line labeled $\omega = cK/\sqrt{\epsilon(\infty)}$ corresponds to electromagnetic waves in the crystal, but uncoupled to the lattice oscillators at ω_T. The heavy lines are the dispersion relations in the presence of coupling between the lattice oscillators and the electromagnetic wave. One effect of the coupling is to create the frequency gap between ω_L and ω_T; within this gap the wavevector is purely imaginary of magnitude given by the broken line in the figure. [After C. Kittel, *Introduction to Solid State Physics*, 5th ed. (Wiley, New York, 1976), p. 304.]

Raman process discussed in previous sections. The wave equations for the four waves[31] are

$$\left[\nabla \times (\nabla \times) - \frac{\omega_l^2}{c^2}\epsilon_l\right]\mathbf{E}_l = \frac{4\pi\omega_l^2}{c^2}\hat{e}_l\left[\chi^{(2)*}E_sE_3 + NM_{fi}^*E_s\rho_{fi}\right],$$

$$\left[\nabla \times (\nabla \times) - \frac{\omega_s^2}{c^2}\epsilon_s\right]\mathbf{E}_s = \frac{4\pi\omega_s^2}{c^2}\hat{e}_s\left[\chi^{(2)}E_lE_3^* + NM_{fi}E_l\rho_{fi}^*\right],$$

$$\left[\nabla \times (\nabla \times) - \frac{\omega_3^2}{c^2}\epsilon_3\right]\mathbf{E}_3 = \frac{4\pi\omega_3^2}{c^2}\hat{e}_3\left[\chi^{(2)}E_lE_s^* + NA_{fi}^*\rho_{fi}\right], \qquad (10.34)$$

and

$$\left(\frac{\partial}{\partial t} + i\omega_{fi} + \Gamma_{fi}\right)\rho_{fi} = \frac{i}{\hbar}\left(A_{fi}E_3 + M_{fi}E_lE_s^*\right)(\rho_i - \rho_f)$$

where $\chi^{(2)}$ is the usual second-order nonlinear susceptibility and $A_{fi} = \langle f| - e\mathbf{r} \cdot \hat{e}_3|i\rangle$ is the transition matrix element for the infrared excitation of the material system from $|i\rangle$ to $|f\rangle$. The third-order nonresonant terms $\chi^{(3)}_{NR}|E_l|^2 E_j$ in (10.34) have been omitted for simplicity.

If the response of ρ_{f_l} is stationary, then by eliminating ρ_{f_l}, (10.34) reduces to

$$\left[\nabla \times (\nabla \times) - \frac{\omega_l^2}{c^2}\epsilon_l\right]\mathbf{E}_l = \frac{4\pi\omega_l^2}{c^2}\hat{e}_l\left[\chi^{(2)*}_{\text{eff}}E_\text{,}E_3 + \chi^{(3)*}_R|E_\text{,}|^2 E_l\right],$$

$$\left[\nabla \times (\nabla \times) - \frac{\omega_s^2}{c^2}\epsilon_s\right]\mathbf{E}_\text{,} = \frac{4\pi\omega_s^2}{c^2}\hat{e}_\text{,}\left[\chi^{(2)}_{\text{eff}}E_l E_3^* + \chi^{(3)}_R|E_l|^2 E_\text{,}\right], \quad (10.35)$$

$$\left[\nabla \times (\nabla \times) - \frac{\omega_3^2}{c^2}\epsilon_{3,\text{eff}}\right]\mathbf{E}_3 = \frac{4\pi\omega_3^2}{c^2}\hat{e}_3\chi^{(2)}_{\text{eff}}E_l E_\text{,}^*$$

where

$$\chi^{(2)}_{\text{eff}} = \chi^{(2)} - \frac{NA^*_{f_l}M_{f_l}(\rho_l - \rho_f)}{h(\omega_3 - \omega_{f_l} + i\Gamma_{f_l})}$$

$$\chi^{(3)}_R = -\frac{N|M_{f_l}|^2(\rho_l - \rho_f)}{h(\omega_3 - \omega_{f_l} + i\Gamma_{f_l})} \quad (10.36)$$

$$\epsilon_{3,\text{eff}} = \epsilon_3 - \frac{N|A_{f_l}|^2(\rho_l - \rho_f)}{h(\omega_3 - \omega_{f_l} + i\Gamma_{f_l})}$$

where $k_3 = (\omega_3/c)\epsilon^{1/2}_{3,\text{eff}}$ actually describes the polariton dispersion curve. The solution of (10.35) again resembles that of parametric generation. With a plane boundary at $z = 0$, and assuming no pump depletion, it takes the form

$$E_\text{,}^* = \left[\mathscr{E}^*_{s+}\exp(i\Delta K_+ z) + \mathscr{E}^*_{s-}\exp(i\Delta K_- z)\right]\exp(-i\mathbf{k}_s \cdot \mathbf{r}),$$

$$E_3 = \left[\mathscr{E}_{3+}\exp(i\Delta K_+ z) + \mathscr{E}_{3-}\exp(i\Delta K_- z)\right]\exp(i\mathbf{k}_3 \cdot \mathbf{r} + \Delta k z), \quad (10.37)$$

where

$$k = \frac{\omega}{c}(\epsilon_{3,\text{eff}})^{1/2},$$

$$\Delta k = k_{lz} - k_{sz} - k_{3z}, \qquad k_z = \mathbf{k} \cdot \hat{z},$$

$$\Delta K_\pm = \tfrac{1}{2}(\gamma_s - \gamma_3) \pm \tfrac{1}{2}\left[(\gamma_s + \gamma_3)^2 - 4\Lambda\right]^{1/2},$$

$$\gamma_s = \frac{k_s}{2k_{sz}}(i\alpha_s + 2k_R), \quad (10.38)$$

$$k_R = \frac{\omega_s^2}{2k_{sz}c^2}4\pi\chi^{(3)}_R|E_l|^2,$$

$$\gamma_3 = -\Delta k - i \left(\frac{k_3}{2k_{3z}} \right) \alpha_3,$$

$$\Lambda = \frac{4\pi^2 \omega_s^2 \omega_3^2}{c^2 k_{sz} k_{3z}} \left(\chi_{\text{eff}}^{(2)} \right)^2 |E_l|^2,$$

$$\left| \frac{\mathscr{E}_3}{\mathscr{E}_s^*} \right| = \left(\frac{\omega_3^2 k_{sz}}{\omega_s^2 k_{3z}} \right)^{1/2} |\Lambda^{1/2} (\Delta K_{\pm} + \gamma_3)|.$$

As one would expect, the solution here reduces to that of simple Raman–Stokes generation and parametric amplification, respectively, when the coupling of the infrared field E_3 to the other waves vanishes [$A_{f_l} = 0$ and $\chi^{(2)} = 0$] and when the nonlinear coupling of ρ_{f_l} with E_l and E_s vanishes ($M_{f_l} = 0$). Numerical examples on GaP and LiNbO$_3$ can be found in Henry and Garrett and in Sussman.[32]

By adjusting properly the relative angle between $\mathbf{k}_l$ and $\mathbf{k}_s$ to tune along the polariton dispersion curve, the output of stimulated polariton scattering is tunable. This has been demonstrated in LiNbO$_3$.[33] In a resonator, up to 70% of the laser power can be converted into Stokes. The infrared output is tunable from 50 to 238 cm^{-1}. Its peak power was found to be 5 W when a 1-MW pump beam from a Q-switched ruby laser with a beam diameter of 2 mm was focused into a 3.3-cm LiNbO$_3$ crystal by an $f = 50$ cm lens. This output is limited because LiNbO$_3$ has a low damage threshold. In practice, single-mode lasers should be used to avoid hot spots in the beam which increase the damage probability. Observation of far-infrared output from stimulated polariton scattering in quartz has also been reported.

10.8   STIMULATED SPIN-FLIP RAMAN SCATTERING

An alternative way to obtain tunable output from stimulated Raman scattering is to use a fixed pump frequency and tune the resonant frequency of the material excitation. Stimulated Raman scattering between Zeeman levels is an example: the resonant frequency is tuned by the applied magnetic field. Unfortunately, the tuning range is usually small. The Zeeman splitting $2\mu_B g B = gB/21.4$ cm^{-1}, with B in kOersted, is only ~ 1 cm^{-1} for $g = 2$ and $B = 10$ kOersted. In some solids, however, the effective g factor can be much larger, for example, $g \simeq 50$ in InSb. Then, when B is varied from 0 to 100 kOersted, the Zeeman splitting can be tuned over ~ 240 cm^{-1}. This is a reasonably broad tuning range.

It happens that InSb is also a good Raman scatterer in the infrared. Zeeman levels of band electrons in semiconductors are usually known as Landau levels, and Raman scattering between spin-up and spin-down states is called spin-flip Raman scattering, described schematically in Fig. 10.16. Wolff and Yafet[34]

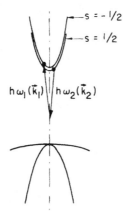

Fig. 10.16 Schematic of spin-flip Raman process in n-InSb.

showed that the spin-flip Raman cross section is given by

$$\left(\frac{d\sigma}{d\Omega}\right)_{SF} \cong \left(\frac{e^2}{m_s c^2}\right)^2 \left(\frac{\omega_s}{\omega_l}\right)\left[\frac{E_g \hbar\omega_l}{E_g^2 - \hbar^2\omega_l^2}\right]^2 \tag{10.39}$$

for pump and Stokes polarizations perpendicular to each other, where $m_s = 2m/|g|$ is the spin electron mass, m is the natural electron mass, and E_g is the energy gap. In InSb, we have $g \cong 50$ and $E_g = 1900$ cm^{-1}. Then, with $E_g \hbar\omega_l/(E_g^2 - \hbar^2\omega_l^2) \sim 1$, the spin-flip Raman cross section is already ~ 600 times larger than the Thomson scattering cross section for free electrons, which is $(e^2/mc^2)^2 \sim 10^{-26}$ cm^2/ster. This is what was actually observed in InSb with a CO_2 laser ($\omega_l = 940$ cm^{-1}). As a comparison, the Raman cross section for the strongest mode of benzene at 992 cm^{-1} is 3×10^{-30} cm^2/ster in the visible. For $\hbar\omega_l \simeq E_g$, $(d\sigma/d\Omega)_{SF}$ can be even much larger as a result of resonant enhancement, as seen in Fig. 10.17. With a CO laser at $\omega_l = 1800$ cm^{-1}, $(d\sigma/d\Omega)_{SF}$ was found to be $\sim 10^5$ times stronger than the Thomson scattering cross section.

According to (10.8), the stimulated Raman gain G_R is proportional to $N(d\sigma/d\Omega)\Gamma^{-1}$, and the density N here for spin-flip scattering refers to the carrier density. In the case of n-type semiconductors (Fig. 10.16), N is the electron density in the conduction band. As a result, G_R here is badly hurt by N, which is always much smaller than the atomic or molecular density in condensed matter. Even so, for $N > 10^{16}$/cm^3, it is still larger than those in other condensed media if the halfwidth Γ is assumed to be the same. Actually, Γ of n-InSb is quite narrow at low temperature and depends on the carrier concentration and $\mathbf{k}_s \cdot \mathbf{H}$. It can be as narrow as 0.15 cm^{-1} with $N = 1 \times 10^{16}$/cm^3.[35]

Assuming $\Gamma = 2$ cm^{-1}, $N = 3 \times 10^{16}$/cm^3, and $\rho_i - \rho_f = 1$ in (10.8), we find in n-InSb a spin-flip Raman gain $G_R = 1.7 \times 10^{-5}I$ cm^{-1}, where I is the

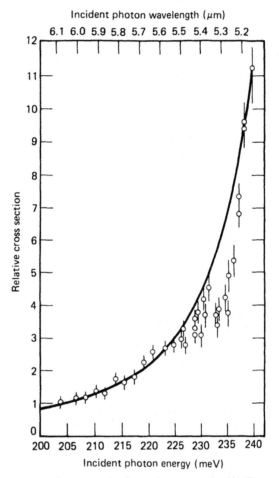

Fig. 10.17 Resonance enhancement of spontaneous spin-flip Raman scattering as a function of input photon energy ($n = 1 \times 10^{16}$ cm^{-3}, H = 40 kOe, and $T \sim 30$ K). [After S. R. J. Brueck and A. Mooradian, *Phys. Rev. Lett.* **28**, 161 (1972).]

CO_2 laser intensity in watts per square centimeter. This is the largest known Raman gain for all materials. The gain can be increased further by adjusting N properly to yield an optimum value for $N\Gamma^{-1}$ and by moving $\hbar\omega_1$ closer to E_g. At the CO laser frequency, the gain becomes $6 \times 10^{-4}I$ cm^{-1}. From these estimates, one expects that stimulated spin-flip Raman scattering should be observable in InSb of a few millimeters in length with a pump beam of $\sim 10^6$ W/cm^2 at 10.6 μm or $\sim 10^4$ W/cm^2 at 5.3 μm. In practice, optical feedback at the air–sample interfaces can result in Raman oscillation. Patel and Shaw,[7] using a Q-switched CO_2 laser of 1 kW at 10.6 μm focused to a spot of 10^{-3} cm^2 in a 5-mm n-InSb sample with $N \approx 10^{16}$/cm^3 at $T = 18$ K, observed a Stokes output of 10 W. The output was tunable from 10.9 to 13.0 μm with B varied from 15 to 100 kOrsted. The output linewidth was less than 0.03 cm^{-1}. Using a single-mode CW CO laser at 5.3 μm focused to $\sim 5 \times 10^{-5}$ cm^2 in a 4.8-mm n-InSb sample with $n \simeq 10^{16}$/cm^3 at $T \simeq 30$ K, Brueck and Mooradian[36] obtained a Raman oscillation threshold at a pump power less than 50 mW, a power conversion efficiency of 50%, and a maximum Stokes output in excess of 1 W. The output linewidth can be as narrow as 1 kHz. With samples in a low magnetic field, a conversion efficiency of 80% has been achieved.[37] Anti-Stokes radiation and Stokes radiation up to the fourth order have also been observed. Detailed operation characteristics of InSb spin-flip Raman lasers are listed in Ref. 38.

Stimulated spin-flip Raman scattering can also occur in other semiconductors. Among those reported in the literatures are CdS pumped by a dye laser, InAs pumped by an HF laser, and $Pb_xSn_{1-x}Te$, $Hg_xCd_{1-x}Te$, and $Hg_xMn_{1-x}Te$ pumped by a CO_2 TEA laser.

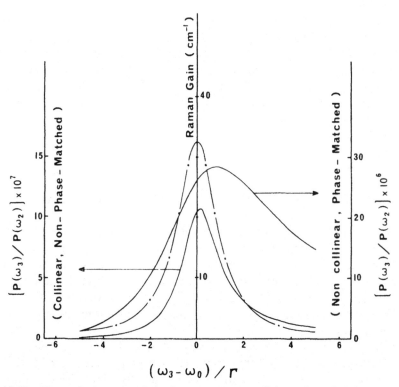

Fig. 10.18 Theoretical curves of the Raman gain g, and the ratios of the far-infrared output $P(\omega_3)$ to the Raman output $P(\omega_2)$ for the collinear phase-mismatched case and for the noncollinear phase-matched case. (After Ref. 39.)

Actually, the spin-flip transition can be excited by both the Raman process and the direct one-photon absorption process, although the latter is weak because it is a magnetic-dipole transition. Therefore, strictly speaking, stimulated spin-flip Raman scattering is a special case of stimulated polariton scattering,[39] and the theory developed in the previous section should apply here. In addition to the Stokes output, a far-infrared output at the spin-flip transition frequency is expected. In the present case, the calculation is, however, relatively simple, because the free carrier absorption at ω_3 is quite strong and we always have $(\gamma_s + \gamma_3)^2 \gg \Lambda$ in (10.38). As a result, the gain is almost exactly equal to the stimulated Raman gain with $\Lambda = 0$. We show in Fig. 10.18 a calculated example of both the gain and the ratio of the far-infrared output to the Stokes output. Collinear phase matching is not possible in this example, so that the far-infrared output in the forward direction is relatively low. It becomes stronger in the noncollinear phase-matching direction. In fact, because the direct excitation of the spin-flip transition is weak, the far-infrared output can be calculated through iteration by first finding the Stokes output from stimulated Raman scattering and then the difference-frequency output from the pump and Stokes mixing.

Far-infrared output from a spin-flip Raman oscillator has not yet been reported. Only the collinear phase-mismatched case has been tried. Far-infrared generation by optical mixing of pump and Stokes waves in InSb has, however, been observed with its maximum appearing at the spin-flip resonance.[40] The results agree very well with the theory. This far-infrared output is, of course, tunable over the same range as the Stokes output, and constitutes a potential tunable coherent source in the far-infrared which can be both intense and narrow in linewidth.

10.9   TRANSIENT STIMULATED RAMAN SCATTERING

Pulsed lasers often are used in stimulated Raman scattering experiments. We must therefore consider the time dependence of the output. If the pulsewidth is much longer than the relaxation time of the Raman excitation and the time required for light to traverse the medium, we can expect from physical argument that the output pulse will follow the temporal variation of the input pulse. This is the quasi-steady-state case. Otherwise, the output should exhibit a transient behavior.

To describe the transient effect, we should, in general, use the coupled wave approach of Section 10.3. In this approach, the dynamic equation for the material excitation explicitly takes into account the possible transient response of the medium. Even in the case of strong Raman gain, the slowly varying amplitude approximation of the fields is usually still valid. For Stokes generation in the forward direction along $\hat{z}$, the set of coupled equations in (10.22), following the derivation of Section 3.5, can be written as

$$\left(\frac{\partial}{\partial z} + \frac{1}{v_1}\frac{\partial}{\partial t}\right)\mathscr{E}_1(z,t) = i\left(\frac{2\pi\omega_1^2}{c^2 k_1}\right)NM_{f_i}^*\mathscr{E}_2 A$$

$$\left(\frac{\partial}{\partial z} + \frac{1}{v_2}\frac{\partial}{\partial t}\right)\mathscr{E}_2(z,t) = i\left(\frac{2\pi\omega_2^2}{c^2 k_2}\right)NM_{f_i}\mathscr{E}_1 A^* \qquad (10.40)$$

$$\left(\frac{\partial}{\partial t} + \Gamma_{f_i}\right)A^*(z,t) = \frac{-i}{\hbar}M_{f_i}^*(\rho_i - \rho_f)\mathscr{E}_1^*\mathscr{E}_2$$

with $E_i(\omega_i) = \mathscr{E}_i(z,t)\exp(ik_i z - i\omega_i t)$ and $\rho_{f_i}(\omega_1 - \omega_2) = A(z,t)\exp[i(k_1 - k_2)z - i(\omega_1 - \omega_2)t]$ in (10.22). Here, v_1 and v_2 are the group velocities at ω_1 and ω_2, respectively, and we have neglected for simplicity the nonresonant driving terms in the equations of $\mathscr{E}_1$ and $\mathscr{E}_2$.

Consider first the simpler case where the amplitude variations of $\mathscr{E}_1$ and $\mathscr{E}_2$ are sufficiently slow so that $|\partial A/\partial t|$ is negligible compared to $|\Gamma A|$. Then $A(z,t) = iM_{f_i}(\rho_i - \rho_f)\mathscr{E}_1\mathscr{E}_2^*/\hbar\Gamma_{f_i}$, and if we assume for simplicity $v_1 = v_2$, and use the transformation of variable $z' = z$ and $t' = t - z/v$, (10.40) reduces to

$$\frac{\partial\mathscr{E}_1}{\partial z'} = i\left(\frac{2\pi\omega_1^2}{c^2 k_1}\right)\chi_R^{(3)*}|\mathscr{E}_2|^2\mathscr{E}_1,$$

$$\frac{\partial\mathscr{E}_2}{\partial z'} = i\left(\frac{2\pi\omega_2^2}{c^2 k_2}\right)\chi_R^{(3)}|\mathscr{E}_1|^2\mathscr{E}_2 \qquad (10.41)$$

where $\chi_R^{(3)} = N|M_{f_i}|^2(\rho_i - \rho_f)/i\hbar\Gamma$. These equations are identical to (10.15) with $\alpha = 0$ except that $\mathscr{E}_1$ and $\mathscr{E}_2$ are now functions of z and $t - z/v$. In other words, $\mathscr{E}_1$ and $\mathscr{E}_2$ should obey the steady-state solution in the retarded time coordinate. Physically this result follows from the fact that a differential section of the laser pulse always interacts with one and the same differential section of the Stokes pulse throughout the medium. This is the quasi-steady-state solution.

For backward scattering, we must replace v_2 in (10.40) by $-v_2$, and (10.41) is no longer valid. The quasi-steady-state solution applies only when the amplitude variations of the input pulses are negligible in the time duration for light to traverse the entire length of the medium. However, the general solution for this case can still be found for $|v_1| = |v_2|$ as[41]

$$|\mathscr{E}_2|^2\left(t + \frac{z}{v}, t - \frac{z}{v}\right) = \frac{|\mathscr{E}_2|^2(t + z/v, 0)}{F_2(t + z/v) + \exp[-F_1(t - z/v)]},$$

$$F_1\left(t - \frac{z}{v}\right) = \int_0^{t-z/v} g|\mathscr{E}_1(0,y)|^2\,dy, \qquad (10.42)$$

$$F_2\left(t + \frac{z}{v}\right) = \int_0^{t+z/v} g|\mathscr{E}_2(y,0)|^2\,dy,$$

and

$$g = -\left(\frac{4\pi\omega_2^2}{c^2 k_2}\right) \mathrm{Im}(\chi_R^{(3)}).$$

It reduces to the quasi-steady-state solution when both the laser and the Stokes input pulses are so long that the integrands in F_1 and F_2 can be approximated by constants. On the other hand, if both input pulses are much shorter compared to the time for light to traverse the length of the medium, then the backward Raman amplification is greatly reduced in comparison with the forward Raman amplification due to the limited interaction range of the laser and the Stokes pulses. This is seen in (10.42) that $F_1(t - z/v) < g|\mathscr{E}_1(t - z/v)|^2 l$, where l is the length of the medium. Another case of interest is the amplification of a backward Stokes pulse in a long medium against a relatively long laser pulse. If the leading edge of the laser pulse is sufficiently steep, Stokes pulse sharpening may occur as the wavefront of the backward Stokes pulse continuously sees the fresh undepleted incoming laser beam and gets full amplification while the lagging part of the pulse does not.[41] An example is seen in Fig. 10.19. This pulse-sharpening phenomenon can be observed in some liquids when the initial Stokes pulse is generated by self-focusing near the end of the cell.

We now consider the more general case where $|\partial A/\partial t|$ is no longer negligible compared with $|\Gamma A|$. This happens when $\mathscr{E}_1\mathscr{E}_2^*$ varies rapidly so that the material excitation cannot respond instantaneously, or more quantitatively, when the laser pulsewidth T_p is smaller than or comparable with the dephasing

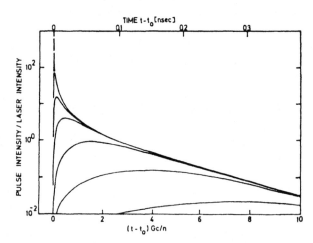

Fig. 10.19 Calculated normalized Raman pulse intensity as a function of time for an initial condition $|E_s| = |E_{s0}|(t - t_0)$ for $t > t_0$. The curves show the pulse development at length intervals of $\Delta l = 2.77/G$. G is the Raman gain and was determined to be 0.7 cm^{-1} in CS$_2$. Lower scale is in dimensionless units; upper scale describes the experimental conditions. (After Ref. 39.)

time $T_2 = 1/\Gamma_{f_t}$ of the Raman excitation (or more correctly, as we shall see later,[42] when $T_p < G_{Rm}/T_2$, where G_{Rm} is the steady-state Raman gain from (10.8) by assuming the laser intensity to be the peak intensity of the input pulse and l is the length of the medium). Although Q-switched pulses may be short enough for studying transient stimulated Raman scattering in gases, picosecond pulses are needed for liquids since T_2 is usually of the order of picoseconds. With a picosecond pump pulse, the backward Raman scattering is hardly detectable because of the very limited length of interaction between the backward Raman and the incoming pump pulse. We discuss here only the forward Raman scattering.

Consider the case where both the depletion of pump power and the induced population change are negligible. Then (assuming $v_1 = v_2$) (10.40) reduces to

$$\left(\frac{\partial}{\partial z} + \frac{1}{v}\frac{\partial}{\partial t}\right)\mathscr{E}_2 = i\eta_1\mathscr{E}_1\left(t - \frac{z}{v}\right)A^*,$$

$$\left(\frac{\partial}{\partial t} + \Gamma\right)A^* = -i\eta_2\mathscr{E}_1^*\left(t - \frac{z}{v}\right)\mathscr{E}_2 \tag{10.43}$$

where

$$\eta_1 = \left(\frac{2\pi\omega_2^2}{c^2 k_2}\right)NM_{f_t},$$

$$\eta_2 = \frac{M_{f_t}^*(\rho_t - \rho_f)}{\hbar},$$

and $\mathscr{E}_1(t - z/v)$ is given by the initial condition. The solution of (10.43) describes the transient stimulated Raman scattering. Its derivation is somewhat long and tedious. We shall therefore only sketch the result here and refer the readers to Ref. 42 for details.

With $t' = t - z/v$ and $z' = z$, (10.43) can be transformed into a second-order partial differential equation

$$\left[\frac{\partial^2}{\partial t'\partial z'} - \eta_1\eta_2|\mathscr{E}_1(t')|^2\right]U = 0 \tag{10.44}$$

where $U = F\exp(\Gamma t')$ and F stands for either $\mathscr{E}_2$ or A^*. The equation can be further simplified to

$$\left(\frac{\partial^2}{\partial\tau\partial z'} - \eta_1\eta_2\right)U = 0 \tag{10.45}$$

by defining $\tau = \int_{-\infty}^{t'}|\mathscr{E}_1(t'')|^2\,dt''$. Equation (10.45) is in the standard form of a hyperbolic equation which can be solved with arbitrary initial conditions. In the present case, the solution takes the form

$$\mathcal{E}_2(z',t') = \mathcal{E}_2(0,t') + (\eta_1\eta_2 z')^{1/2}\mathcal{E}_1(t')$$

$$\times \int_{-\infty}^{t'} e^{-\Gamma(t'-t'')}\left\{\mathcal{E}_1^*(t'')\mathcal{E}_2(0,t'')[\tau(t')-\tau(t'')]^{-1/2}\right.$$

$$\left.\times I_1\left(2[\eta_1\eta_2(\tau(t')-\tau(t''))z']^{1/2}\right)\right\}dt'', \quad (10.46)$$

$$A^*(z',t') = i\eta_1 \int_{-\infty}^{t'} e^{-\Gamma(t'-t'')}\left\{\mathcal{E}_1^*(t'')\mathcal{E}_2(0,t'')\right.$$

$$\left.\times I_0\left(2[\eta_1\eta_2(\tau(t')-\tau(t''))z']^{1/2}\right)\right\}dt'',$$

where the input conditions are $A^*(z') = 0$ at $t' \to -\infty$ and $\mathcal{E}_2(z',t') = \mathcal{E}_2(0,t')$ at $z = z' = 0$, and I_i is the ith order Bessel function of imaginary argument. The solution in (10.46) has the following characteristics:

1 Since $I_0(x) \simeq 1$ and $I_1(x) \simeq x$ for $x \ll 1$ and $I_i(x) \simeq (2\pi x)^{-\frac{1}{2}}\exp(x)$ for $x \gg 1$, the Stokes amplitude $\mathcal{E}_2$ first increases linearly with z and then, in the limit of large amplification, increases exponentially in the form

$$\mathcal{E}_2(z',t') \propto \mathcal{E}_1(t')\int_{-\infty}^{t'}\mathcal{E}_1^*(t'')\mathcal{E}_2(0,t'')[\tau(t')-\tau(t'')]^{-1}$$

$$(10.47)$$

$$\times \exp\left\{-\Gamma(t'-t'') + 2[\eta_1\eta_2(\tau(t')-\tau(t''))z']^{1/2}\right\}.$$

2 If the pump pulse is sufficiently long, then $\mathcal{E}_2$ takes on a quasi-steady-state exponential gain. This can be seen from (10.47) when $(t-t_0) > G_R z T_2$ for a rectangular pump pulse starting at t_0. For this reason, $T_p < G_{Rm} l T_2$ can be used as the condition for the observation of transient stimulated Raman scattering, as mentioned earlier.

3 If $T_p < T_2$, the factor $\exp[-\Gamma(t'-t'')]$ can be approximated by 1 in the integrals of (10.46) and (10.47). The Stokes amplitude grows rapidly only toward the middle part of the pump pulse. It then drops off following the pump pulse at the tail. The Stokes pulse is therefore always narrower than the pump pulse. The material excitation A behaves in the similar manner but has an exponential decay tail $\exp(-\Gamma t)$ after the pump pulse is switched off or dropped to nearly zero.

4 In the limit of large amplification, (10.47) gives

$$(\mathcal{E}_2)_{max} \propto \exp\left(\frac{G_T z}{2}\right) \quad (10.48)$$

where G_T is a transient gain given by

$$G_T = 4\left[\eta_1\eta_2\langle|\mathscr{E}_1|^2\rangle\frac{T_p}{z}\right]^{1/2},$$

$$\langle|\mathscr{E}_1|^2\rangle T_p = \int_{-\infty}^{\infty}|\mathscr{E}_1(t)|^2\,dt.$$

(10.49)

The transient gain here is independent of the laser pulseshape. For a pulse of the form $\mathscr{E}_1(t') = \mathscr{E}_{1m}\exp(-|t'/T|^n)$, the peak of the Stokes pulse is delayed from the peak of the pump pulse by a time $t_D = T(\frac{1}{2}\log G_{Rm}z)^{1/n}$.

Numerical calculations of transient stimulated Raman scattering have been carried out by Carmen et al.[42] for various pump shapes. The results confirm the characteristic features presented above.

Transient behavior of stimulated Raman scattering was first noticed in gases by Hagenlocker et al.[43] Later, with picosecond pulses, it was also observed in liquids. Quantitative measurements have shown good agreement with theoretical predictions.[44] A better experiment yet to be carried out is to measure the temporal variation of Stokes amplification in an amplifier cell (see Section 10.6C). The transient gain G_T is different from the steady-state gain G_R in the fact that the former depends only on the Raman cross section ($\propto \eta_1\eta_2$), while the latter is also inversely proportional to the halfwidth Γ. Therefore, it is possible to observe in transient stimulated Raman scattering some Raman modes which are suppressed in the steady-state case. More than one Raman mode can in fact simultaneously show up in transient Raman scattering.[45]

10.10 MEASUREMENTS OF EXCITATIONAL RELAXATION TIMES

Relaxation of a material excitation can be measured directly by probing the decay of the excitation. In analogy to the magnetic resonance cases, two relaxation times are often used to characterize the relaxation process: the longitudinal relaxation time T_1, which governs the decay of the induced population change in the excited state, and the transverse relaxation time T_2, which is the dephasing time of the excitational wave (see Section 2.1). In condensed matter, T_1 and T_2 are usually of the order of picoseconds, and therefore picosecond pulses are needed to excite and to probe the material excitation in the measurements of T_1 and T_2. We consider here only Raman-allowed excitations, with both excitation and probing achieved through Raman transitions. The general principle, however, is applicable to other types of transitions. We also note that only in the limit of homogeneous broadening is T_2 equal to the inverse halfwidth, but even then T_2 can be very different from T_1.

In the preceding section, it was seen that transient stimulated Raman scattering yields a material excitational wave A which decays exponentially as $\exp(-t/T_2)$ even after the pump pulse is switched off. This material excitational wave at $(\omega_1 - \omega_2) \cong \omega_{ex}$ can be probed by mixing A with a probe pulse E_3 at k_3 and ω_3 to generate a coherent anti-Stokes wave $E_a = \mathscr{E}_a \exp(i\mathbf{k}_a \cdot \mathbf{r} - i\omega_a t)$ with $\mathbf{k}_a = \mathbf{k}_1 - \mathbf{k}_2 + \mathbf{k}_3$ and $\omega_a = \omega_1 - \omega_2 + \omega_3$. The equation governing the forward anti-Stokes pulse amplitude is

$$\left(\frac{\partial}{\partial z} + \frac{1}{v_a}\frac{\partial}{\partial t}\right)\mathscr{E}_a(z,t) = i\left(\frac{2\pi\omega_a^2}{c^2 k_2}\right)NM^*_{fi,a}\mathscr{E}_3(z,t)A(z,t) \quad (10.50)$$

from which we find, with the help of transformation of variables $z' = z$ and $t' = t - z/v_a$, the solution

$$\mathscr{E}_a(l,t) \propto \int_0^l \mathscr{E}_3(z',t')A(z',t')\,dz'.$$

The time-integrated coherent anti-Stokes signal is therefore given by

$$\begin{aligned}S_{coh} &\propto \int_{-\infty}^{\infty} |\mathscr{E}_a(l,t)|^2\,dt \\ &\propto \int_{-\infty}^{\infty} \left|\int_0^l \mathscr{E}_3(z',t')A(z',t')\,dz'\right|^2 dt.\end{aligned} \qquad (10.51)$$

This signal is, of course, a function of the time delay t_D between the exciting and the probing pulses. If $t_D \gg T_2 \geq$ pulsewidth T_p, then it is clear from (10.51) that $S_{coh} \propto \exp(-2t_D/T_2)$. Therefore, T_2 can be easily deduced from the result of S_{coh} versus t_D. More rigorously, the effect of the finite pulsewidth T_p should be taken into account in the time convolution in deducing T_2.

The longitudinal relaxation time T_1 describes the decay of the induced population change $\Delta\rho$ as governed by (10.23) or, more explicitly,

$$\left(\frac{\partial}{\partial t} + \frac{1}{T_1}\right)\Delta\rho = \frac{iN}{2\hbar}\left[M_{fi}\mathscr{E}_1\mathscr{E}_2^*A^* - M_{fi}^*\mathscr{E}_1^*\mathscr{E}_2 A\right]\left(\rho_i^0 - \rho_f^0\right) \quad (10.52)$$

with $\Delta\rho = \frac{1}{2}(\rho_i - \rho_f - \rho_i^0 + \rho_f^0) \ll (\rho_i^0 - \rho_f^0)$. The equation shows that after the pump pulse is over $\Delta\rho$ should decay exponentially as $\exp(-t/T_1)$. Incoherent (spontaneous) anti-Stokes scattering is directly proportional to $\Delta\rho$ and therefore can be used to probe the decay of $\Delta\rho$. With a probe pulse E_2 at ω_2, the time-integrated signal at $\omega_3 = \omega_2 + \omega_{ex}$ is given by

$$S_{inc} \propto \int |\mathscr{E}_2(z,t)|^2\Delta\rho(z,t)\,dz\,dt, \qquad (10.53)$$

which is a function of the time delay t_D between the exciting and the probing pulses. When $t_D \gg T_1 \geq T_p$, this signal is proportional to $\exp(-t_D/T_1)$ so that T_1 can be deduced from S_{inc} versus t_D.

The method was first used by DeMartini and Ducuing[46] to measure T_1 of the 4155 cm^{-1} vibrational excitation of gaseous H_2. At 0.03 atmospheric pressure, $T_1 \simeq 30$ μsec, so that Q-switched laser pulses are short enough to

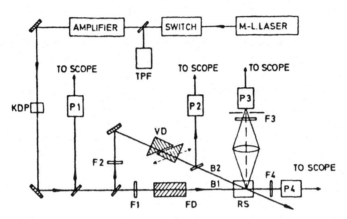

Fig. 10.20 Schematic of the experimental system for photon lifetime measurement. The pump beam B1 at $\lambda \simeq 1.06$ μm and the probe beam B2 at $\lambda \simeq 0.53$ μm interact in the Raman sample RS. Glass rod for fixed optical delay, FD; glass prisms for variable delay, VD; filter, F; photodetector, P; two-photon fluorescence system, TPF. [After A. Laubereau, D. von der Linde, and W. Kaiser, *Phys. Rev. Lett.* **28**, 1162 (1972).]

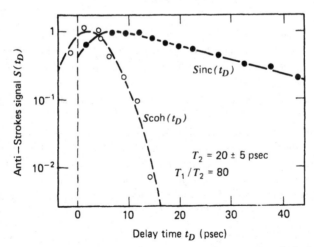

Fig. 10.21 Measured incoherent scattering $S_{inc}(t_D)/S_{inc_{max}}$, (closed circles) and coherent scattering $S_{coh}(t_D)/S_{coh_{max}}$, (open circles) versus delay time t_D for ethyl alcohol. The solid and dashed curves are calculated. [After A. Laubereau, D. von der Linde, and W. Kaiser, *Phys. Rev. Lett.* **28**, 1162 (1972).]

carry out the measurements. In condensed matter, however, T_1 and T_2 are in the picosecond range, and picosecond mode-locked laser pulses must be used, as pioneered by Alfano and Shapiro and by Kaiser and associates.[9] Figure 10.20 is a typical experimental arrangement. The mode-locked pulse from an Nd laser is used to excite the material excitation by stimulated Raman scattering, and the second harmonic of the mode-locked pulse, after an adjustable time delay, is used to probe the excitation. The results on ethyl alcohol in Figure 10.21 is an example. The exponential tails of the $S_{coh}(t_D)$ and $S_{inc}(t_D)$ curves in the figure yield T_1 and T_2 readily. The technique can be extended to the study of decay routes of an excitation, and the dephasing dynamics of an excitation in large molecules or condensed matter.[47]

Note, however, that this discussion of dephasing time measurements applies only to a homogeneously broadened Raman transition.[48] In the case of inhomogeneous broadening, the material excitation has a distribution of resonant frequencies ω_{ex}, and $A(z, t)$ in (10.50) and (10.51) should be replaced by an integration of the excitational waves over the distribution of ω_{ex}. The decay of S_{coh} with time is no longer in the form of $\exp(-2t_D/T_2)$. If the width of the inhomogeneous broadening is considerably larger than $1/T_p$, the decay time of S_{coh} will be dominated by T_p; then no information about T_2 can be obtained. It is, however, possible that when the pump pulse is so intense as to cause coherent saturation in the Raman transition, the T_2 value can still be deduced from the decay of S_{coh}.[49] More details on theory and experiments of vibrational relaxation studied by ultrashort pulses can be found in Refs. 47 and 48.

REFERENCES

1 E. J. Woodbury and W. K. Ng, *Proc. IRE* **50**, 2347 (1962).

2 E. J. Woodbury and G. M. Eckhardt, U.S. Patent No. 3,371,265 (27 February 1968).

3 R. W. Hellwarth, *Phys. Rev.* **130**, 1850 (1963); *Appl. Opt.* **2**, 847 (1963)

4 E. Garmire, E. Pandarese, and C. H. Townes, *Phys. Rev. Lett.* **11**, 160 (1963).

5 N. Bloembergen and Y. R. Shen, *Phys. Rev. Lett.* **12**, 504 (1964); Y R. Shen and N Bloembergen, *Phys. Rev.* **137**, A1786 (1965).

6 S. K Kurtz and J. A. Giordmaine, *Phys. Rev. Lett.* **22**, 192 (1969); J Gelbwachs, R. H. Pantell, H. E. Puthoff, and J. M. Yarborough, *Appl. Phys. Lett.* **14**, 258 (1969).

7 C. K. N. Patel and E. D. Shaw, *Phys. Rev. Lett.* **24**, 451 (1970); *Phys. Rev.* **B3**, 1279 (1971).

8 A Owyoung, *IEEE J. Quant. Electron.* **QE-14**, 192 (1978); in H. Walther and K. W. Rothe, eds., *Laser Spectroscopy*, vol. IV (Springer-Verlag, Berlin, 1979), p. 175; in W. O. N. Guimaraes, C. T. Lin, and A. Mooradian, eds., *Lasers and Applications* (Springer-Verlag, Berlin, 1981), p. 67.

9 D. von der Linde, A. Laubereau, and W. Kaiser, *Phys. Rev. Lett.* **26**, 954 (1971); R. R. Alfano and S. L. Shapiro, *Phys. Rev. Lett.* **26**, 1247 (1971).

10 See, for example, W. Heitler, *Quantum Theory of Radiation*, 3rd ed. (Cambridge University Press, Cambridge, 1954), p. 192.

11 R. Glauber, *Phys. Rev.* **130**, 2529 (1963); **131**, 2766 (1963).

12 R. W. Terhune, *Solid State Design* **4**, 38 (1963); R. Y. Chiao and B. P. Stoicheff, *Phys. Rev. Lett.* **12**, 290 (1964); H. J. Zeiger, P. E. Tannenwald, S. Kern, and R. Burendeen, *Phys. Rev. Lett.* **11**, 419 (1963).

13 D. von der Linde, M. Maier, and W. Kaiser, *Phys. Rev.* **178**, 11 (1969).

14 G. Haidacher and M. Maier, VII Int. Quant. Electron. Conf., San Francisco (1974), post-deadline paper Q.7.

15 N. Bloembergen, G. Bret, P. Lallemand, A. Pine, and P. Simova, *IEEE J. Quant. Electron.* **QE-3**, 197 (1967); P. Lallemand, P. Simova, and G. Bret, *Phys. Rev. Lett.* **17**, 1239 (1966).

16 E. Garmire, *Phys. Lett.* **17**, 251 (1965).

17 N. Bloembergen and Y. R. Shen, *Phys. Rev. Lett.* **13**, 720 (1964); R. L. Carman, F. Shimizu, C. S. Wang, and N. Bloembergen, *Phys. Rev.* **A2**, 60 (1970); S. A. Akhmanov, Yu. E. D'yakov, and L. I. Pavlov, *Sov. Phys. JETP* **39**, 249 (1974); M. G. Raymer, J. Mostowski, and J. L. Carlsten, *Phys. Rev.* **A19**, 2304 (1979); I. A. Walmsley and M. G. Raymer, *Phys Rev. Lett* **50**, 962 (1983).

18 W. J Jones and B. P. Stoicheff, *Phys. Rev. Lett.* **13**, 657 (1964)

19 P. P. Sorokin, N. S. Shiren, J. R. Lankard, E. C. Hammond, and T. G. Kazyaka, *Appl. Phys. Lett.* **22**, 342 (1973); M. Rokni and S. Yatsiv, *Phys. Lett.* **24A**, 277 (1967); J. J. Wynne and P P. Sorokin, in Y. R. Shen, ed. *Nonlinear Infrared Generation* (Springer-Verlag, Berlin, 1977), p. 159.

20 D C. Hanna, M A. Yuratich, and D. Cotter, *Nonlinear Optics of Free Atoms and Molecules* (Springer-Verlag, Berlin, 1979), Chapter 5, and the references therein.

21 D. S. Bethune, J. R. Lankard, and P. P. Sorokin, *Opt. Lett.* **4**, 103 (1980).

22 R. Frey and F. Pradere, *Opt. Comm.* **12**, 98 (1974); V. Wilkie and W. Schmidt. *Appl. Phys.* **16**, 151 (1978); T. R. Loree, R. C. Sze, D. L. Barker, and P. B. Scott, *IEEE J. Quant. Electron.* **QE-15**, 337 (1979).

23 R. Frey and F. Pradere, *Opt. Lett.* **5**, 374 (1980).

24 J. J. Tice and C. Wittig, *Appl. Phys. Lett.* **30**, 420 (1977).

25 R. L. Byer, *IEEE J. Quant. Electron.* **QE-12**, 732 (1976); R. L. Byer and W. R. Trunta, *Opt. Lett.* **3**, 144 (1978); P. Rabinowitz, A. Stein, R. Brickman, and A. Kaldor, *Appl. Phys. Lett.* **35**, 739 (1979).

26 R. Frey, F Pradere, J. Lukasik, and J Ducuing, *Opt. Comm.* **22**, 355 (1977); R. Frey, F. Pradere, and J. Ducuing, *Opt. Comm* **23**, 65 (1977); A. DeMartino, R Frey, and F. Pradere, *Opt. Comm.* **27**, 262 (1978).

27 J C. White and D. Henderson, *Phys. Rev. A* **25**, 1226 (1982); J. C. White, in H. Weber and W. Lüthy, eds., *Laser Spectroscopy*, VI (Springer-Verlag, Berlin, 1983).

28 S. E Harris, *Appl. Phys. Lett.* **31**, 498 (1977); L. J. Zych, L. Lugasik, J. F. Young, and S. E. Harris, *Phys. Rev. Lett.* **40**, 1493 (1978); S. E. Harris, R. W. Falcone, M. Gross, R. Normandin, K D. Pedrotti, J. E. Rothenberg, J. C. Wang, J. R Willison, and J. F. Young, in A R. M. McKellar, T. Oka, and B. P. Stoicheff, eds., *Laser Spectroscopy*, V, (Springer-Verlag, Berlin, 1981); S. E. Harris, R. G. Caro, R. W. Falcone, D. E. Holmgren, K. D. Pedrotti, J C. Wang, and J. F. Young, in H. Weber and W. Lüthy, eds., *Laser Spectroscopy*, VI, (Springer-Verlag, Berlin, 1983).

29 B. F. Levine, C. V. Shank, and J. P. Heritage, *IEEE J. Quant. Electron.* **QE-15**, 1418 (1979); J P. Heritage and D. L. Allara, *Chem. Phys. Lett.* **74**, 507 (1980); B. F. Levine, C. G. Bethea, A. R Tretola, and M. Korngor, *Appl. Phys. Lett.* **37**, 595 (1980).

30 K. Huang, *Proc Roy. Soc. (London)* **A208**, 353 (1951); M. Born and K. Huang, *Dynamic Theory of Crystal Lattices* (Oxford University Press, Oxford, 1954), Chapter II; J. J. Hopfield, *Phys. Rev.* **112**, 1555 (1958).

31 R. Loudon, *Proc. Phys. Soc.* **82**, 393 (1963); Y. R. Shen, *Phys. Rev.* **138**, A1741 (1965).

32 C H. Henry and C. G. B Garrett, *Phys. Rev* **171**, 1058 (1968); S S Sussman, Microwave Lab Report No. 1851, Stanford University (1970).

33 J M Yarborough, S. S Sussman, H. E. Puthoff, R. H. Pantell, and B C Johnson, *Appl Phys. Lett.* **15**, 102 (1969).

34 P A Wolff, *Phys. Rev. Lett.* **16**, 225 (1966); Y. Yafet, *Phys. Rev.* **152**, 858 (1966).

35 S R J. Brueck and A. Mooradian, *Phys. Rev. Lett.* **28**, 1458 (1972).

36 A. Mooradian, S. R. J. Brueck, and F. A Blum, *Appl. Phys. Lett.* **17**, 481 (1970); S R. J. Brueck and A. Mooradian, *Appl. Phys. Lett.* **18**, 229 (1971)

37 C. S DeSilets and C K. N Patel, *Appl. Phys. Lett.* **22**, 543 (1973); C. K. N. Patel, *Phys. Rev. Lett.* **28**, 649 (1972).

38 H. Pascher, G. Appold, R. Ebert, and H. G. Hafele, *Appl. Phys.* **15**, 53 (1978); B. Walker, G. W. Chantry, D. G. Moss, and C C Bradley, *J Phys.* **D9**, 1501 (1976); M. J Colles and C. R. Pidgeon, *Rep. Prog. Phys.* **38**, 329 (1975); S. D. Smith, R. B. Dennis, and R. G. Harrison, *Prog Quant. Electron* **5**, 205 (1977)

39 Y. R Shen, *Appl. Phys Lett.* **26**, 516 (1973).

40 V T. Nguyen and T. J. Bridges, *Phys. Rev. Lett* **29**, 359 (1972); *Appl Phys Lett* **23**, 107 (1973).

41 M Maier, W. Kaiser, and J. A. Giordmaine, *Phys. Rev.* **177**, 580 (1969).

42 R L Carmen, F. Shimizu, C. S. Wang, and N. Bloembergen, *Phys. Rev.* **A2**, 60 (1970); C S. Wang, in H. Rabin and C. L. Tang, eds., *Quantum Electronics*, vol. I (Academic Press, New York, 1975), p. 447.

43 E. E. Hagenlocker, R. W. Minck, and W. G. Rado, *Phys. Rev.* **154**, 226 (1967).

44 R. L. Carman and M. E. Mack, *Phys. Rev.* **A5**, 341 (1972).

45 R. L. Carman, M. E. Mack, F. Shimizu, and N. Bloembergen, *Phys. Rev. Lett.* **23**, 1327 (1969); M C Mack, R L. Carman, R. Reintjes, and N. Bloembergen, *Appl. Phys. Lett.* **16**, 209 (1970).

46 F. DeMartini and J Ducuing, *Phys. Rev. Lett.* **17**, 117 (1966).

47 A Laubereau and W. Kaiser, *Rev. Mod. Phys.* **50**, 607 (1978).

48 W. Smith, H. J. Polland, A. Laubereau, and W. Kaiser, *Appl. Phys. B* **26**, 77 (1981).

49 S. M George and C B. Harris, *Phys Rev.* **A28**, 863 (1983).

BIBLIOGRAPHY

Bloembergen, N , *Am J. Phys.* **35**, 989 (1967).

Shen, Y R , in M Cardona, ed , *Light Scattering in Solids* (Springer-Verlag, Berlin, 1975), p 275

Kaiser, W., and M. Maier, in F. T. Arecchi and E. O. Schulz-Dubois, eds , *Laser Handbook* (North-Holland Publishing Co., Amsterdam, 1972), p 1077

11

Stimulated
Light Scattering

In Chapter 10 stimulated Raman scattering was treated as a result of parametric interaction between light and material excitation. Examples were given in which the material excitation was either electronic or vibrational. The Raman shift involved could in principle range from zero to a frequency as large as the pump laser frequency. Some material excitations have very low frequencies, of the order of ≤ 1 cm^{-1}. They are usually related to atomic or molecular motion. Light scattering by such material excitations frequently is called something other than Raman scattering. For example, Brillouin scattering involves acoustic wave excitation, Rayleigh scattering deals with entropy wave excitation, and Rayleigh-wing scattering relates to molecular orientational excitation.[1] With sufficiently high pump laser intensity, all these spontaneous light scattering processes could become stimulated. Some of them are discussed briefly in this chapter.

11.1 STIMULATED BRILLOUIN SCATTERING

Stimulated Brillouin scattering results from parametric coupling between light and acoustic waves. The theory follows the general formalism given in Section 10.3 for stimulated Raman scattering, with the material excitational wave here referring to the acoustic wave. We consider only Brillouin scattering in liquid. The coupled wave equations, similar to (10.22), are

$$\left[\nabla \times (\nabla \times) + \frac{\varepsilon_1}{c^2} \frac{\partial^2}{\partial t^2} \right] \mathbf{E}_1 = \frac{4\pi\omega_1^2}{c^2} \mathbf{P}^{\mathrm{NL}}(\omega_1)$$

and

$$\left[\nabla \times (\nabla \times) + \frac{\varepsilon_2}{c^2} \frac{\partial^2}{\partial t^2} \right] \mathbf{E}_2 = \frac{4\pi\omega_2^2}{c^2} \mathbf{P}^{\mathrm{NL}}(\omega_2)$$

(11.1)

together with the driven acoustic wave equation

$$\left[\frac{\partial^2}{\partial t^2} - 2\Gamma_B\frac{\partial}{\partial t} - v^2\nabla^2\right]\Delta\rho = -\nabla\cdot\mathbf{f}. \tag{11.2}$$

We use the density variation $\Delta\rho$ to describe the acoustic wave; v is the acoustic velocity, and Γ_B is the acoustic damping coefficient or the halfwidth of the Brillouin line in spontaneous scattering. The driving force f and the nonlinear polarizations $\mathbf{P}^{NL}$ arise from the nonlinear coupling of the three waves and can be obtained as follows:

$$\mathbf{P}^{NL}(\omega_1) = \frac{\partial}{\partial\rho}\left[\frac{\varepsilon\mathbf{E}_2}{4\pi}\right]\Delta\rho = \frac{1}{4\pi}\frac{\partial\varepsilon}{\partial\rho}\mathbf{E}_2\Delta\rho;$$

$$\mathbf{P}^{NL}(\omega_2) = \frac{1}{4\pi}\frac{\partial\varepsilon}{\partial\rho}\mathbf{E}_1\Delta\rho^*, \tag{11.3}$$

$$\mathbf{f} = \nabla p = \nabla\left(\frac{1}{2\pi}\gamma\mathbf{E}_1\cdot\mathbf{E}_2^*\right)$$

where p is the electrostrictive pressure, $\gamma = \rho_0\,\partial\varepsilon/\partial\rho$ is the electrostrictive coefficient, and ρ_0 is the mass density of the liquid.

The problem is just another example of parametric wave interaction, and the solution of the coupled wave equations (11.1) and (11.2) follows those described repeatedly in Chapters 9 and 10. We consider here only backward stimulated Brillouin scattering in the steady-state case. The forward Brillouin scattering does not occur as it has a zero frequency shift. Let $\mathbf{E}_1 = \hat{e}_1\mathscr{E}_1\exp(ik_1z - i\omega_1t)$, $\mathbf{E}_2 = \hat{e}_2\mathscr{E}_2\exp(-ik_2z - i\omega_2t)$, and $\Delta\rho = A\exp(ik_az - i\omega_at)$, with $\omega_1 = \omega_2 + \omega_a$ and $k_a = \omega_a/v$. Following the slowly varying amplitude approximation, (11.1) and (11.2) reduce to

$$\left(\frac{\partial}{\partial z} + \frac{\alpha}{2}\right)\mathscr{E}_1 = \frac{i\omega_1^2}{2k_1c^2}\frac{\partial\varepsilon}{\partial\rho}(\hat{e}_1^\dagger\cdot\hat{e}_2)\mathscr{E}_2Ae^{-i\Delta kz}$$

$$\left(\frac{\partial}{\partial z} - \frac{\alpha}{2}\right)\mathscr{E}_2^* = \frac{i\omega_2^2}{2k_2c^2}\frac{\partial\varepsilon}{\partial\rho}(\hat{e}_1^\dagger\cdot\hat{e}_2)\mathscr{E}_1^*Ae^{-i\Delta kz}, \tag{11.4}$$

$$\left(\frac{\partial}{\partial z} + \frac{\Gamma_B}{v}\right)A = \frac{ik_a}{4\pi v^2}\rho_0\left(\frac{\partial\varepsilon}{\partial\rho}\right)(\hat{e}_1^\dagger\cdot\hat{e}_2)^*\mathscr{E}_1\mathscr{E}_2^*e^{i\Delta kz}$$

where $\Delta k = k_1 + k_2 - k_a$. This is analogous to the backward parametric amplification case described in Section 9.6. If the damping constants α and Γ_B are sufficiently small, the amplification can in principle go into oscillation. For the acoustic waves involved in backward Brillouin scattering in liquids, however, $\omega_a/2\pi$ is typically of the order of 5 GHz and the corresponding $\Gamma_B/2\pi$ is around 0.1 GHz (see Table 11.1). the attenuation coefficient Γ_B/v is $\sim 10^4$

Table 11.1

Frequency Shift ν_B, Linewidth Γ_B, and Maximum Steady-State Gain Factor of Stimulated Brillouin Scattering, $(g_B^e)_{max} \propto \beta_B^e$, for a Number of Liquids[a]

Substance	Frequency shift ν_B (MHz)	Linewidth Γ_B (spont.) (MHz)	Gain Factor	
			Calculated $g_B^e(max)$ (cm/MW)	Measured $g_B^e(max)$ (cm/MW)
CS_2	5850	52.3	0.197	0.13
Acetone	4600	224	0.017	0.020
n-Hexane		222	0.027	0.026
Toluene	5910	579	0.013	0.013
CCl_4	4390	520	0.0084	0.006
Methanol	4250	250	0.013	0.013
Benzene	6470	289	0.024	0.018
H_2O	5690	317	0.0066	0.0048
Cyclohexane	5550	774	0.007	0.0068

[a]After I. L. Fabellinskii, *Molecular Scattering of Light* (Plenum, New York, 1968).

cm^{-1} with $v \sim 10^5$ cm/sec. This is often much larger than the gain coefficient of stimulated Brillouin scattering estimated below. Consequently, the acoustic wave excitation here can be considered as highly damped, and (11.4) can be solved by first eliminating A using the approximation $\partial A / \partial z \cong \iota \Delta k A$. We then have

$$\left(\frac{\partial}{\partial z} + \frac{\alpha}{2}\right)\mathscr{E}_1 = \frac{i 2 \pi \omega_1^2}{k_1 c^2} \chi_B^{(3)} |\mathscr{E}_2|^2 \mathscr{E}_1,$$

$$\left(\frac{\partial}{\partial z} - \frac{\alpha}{2}\right)\mathscr{E}_2^* = \frac{i 2 \pi \omega_2^2}{k_2 c^2} \chi_B^{(3)} |\mathscr{E}_1|^2 \mathscr{E}_2^*$$

(11.5)

where

$$\chi_B^{(3)} = \left[\frac{(\partial\varepsilon/\partial\rho)(\hat{e}_1^\dagger \cdot \hat{e}_2)}{4\pi v}\right]^2 \frac{k_a \rho_0}{\Delta k - i\Gamma_B/v}.$$

Equation (11.5) looks exactly the same as (10.15) in the stimulated Raman scattering case except that the Brillouin susceptibility $\chi_B^{(3)}$ now replaces the Raman susceptibility $\chi_R^{(3)}$ in (10.15). Since $Im\chi_B^{(3)} > 0$, (11.5) shows that $\mathscr{E}_2$ would grow in the backward $(-z)$ direction if $(2\pi\omega_1/k_1c^2)Im\chi_B^{(3)}|\mathscr{E}_1|^2 > \alpha/2$, while $\mathscr{E}_1$ would decay in the forward direction. In the case of negligible pump depletion, the solution of (11.5) is

$$|\mathscr{E}_2(z)|^2 = |\mathscr{E}_2(l)|^2 e^{(G_B - \alpha)(l-z)}$$

(11.6)

where the Brillouin gain G_B is given by

$$G_B = \frac{4\pi\omega_2^2}{|k_2|c^2}\, \text{Im}\, \chi_B^{(3)}|\mathscr{E}_1|^2. \tag{11.7}$$

Calculated values of G_B for a number of liquids are listed in Table 11.1.[2] It is seen that with 100 MW/cm^2 pump intensity in a 10-cm cell, the exponential gain $G_B l$ can be of the order of 10 (much larger in CS$_2$ because of the narrower Brillouin width), and therefore stimulated Brillouin scattering should be easily observable.

The solution of (11.5) including the pump depletion, but with $\alpha = 0$, can also be obtained readily.[3] One finds the following algebraic relations between the inputs $|E_1(0)|^2$, $|E_2(l)|^2$ and the outputs $|E_1(l)|^2$, $|E_2(0)|^2$:

$$|E_1(0)|^2 - |E_1(l)|^2 \cong |E_2(0)|^2 - |E_2(l)|^2$$

and $\hspace{9cm}$ (11.8)

$$\frac{|E_2(l)|^2}{|E_2(0)|^2} = \frac{1 - |E_2(0)/E_1(0)|^2}{\exp\big[\big[1 - |E_2(0)/E_1(0)|^2\big]G_B|E_1(0)|^2 l\big] - |E_2(0)/E_1(0)|^2}.$$

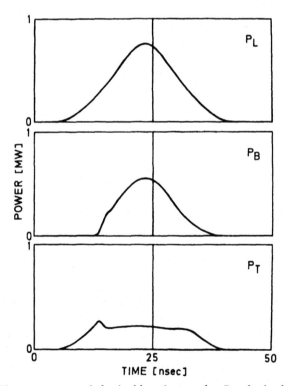

Fig. 11.1 Oscilloscope traces of the incident laser pulse P_L, the backward Brillouin pulse P_B, and the transmitted laser pulse P_T in ethyl ether. [After M. Maier, *Phys. Rev.* **166**, 113 (1968).]

Significant pump depletion in stimulated Brillouin scattering is in fact a common phenomenon. A typical example is seen in Fig. 11.1, where depletion of the pump pulse into a backscattered Brillouin pulse is clearly demonstrated.[4] Energy conversion as high as 90% has been reported.

The lifetime of acoustic waves is $\tau_B = 1/2\Gamma_B$, which is of the order of 1 nsec for $\omega_a/2\pi \sim 5$ GHz. Transient stimulated Brillouin scattering is expected if the pump pulse has a width comparable to 1 nsec or less. The transient solution again resembles that of the Raman case. However, we do not dwell on it here, but refer the readers to Kroll.[5] Experiments actually found strong dependence of the gain on the lifetime of the acoustic wave for pulsewidth less than $\sim 100 \tau_B$. This is especially so near the threshold for stimulated scattering. Careful measurements have shown quantitative agreement between theory and experiment.[6]

Stimulated Brillouin scattering was first observed by Chiao et al.[7] in quartz and sapphire using a Q-switched ruby laser. They analyzed the reflected light from the medium with a Fabry–Perot interferometer and found the Brillouin-shifted component. Because of the high conversion efficiency, the backscattered Brillouin pulse frequently is so intense that it can be detected by eye. Without proper isolation between the sample and the laser system, the backscattered Brillouin pulse will propagate into the laser medium and be further amplified. The result is that a Brillouin-shifted component will now appear in the laser output. Such a process can repeat a number of times, and the laser output will then have a spectrum containing several orders of Brillouin-shifted components.[8] This is what one would avoid in experiments requiring a single-mode laser beam.

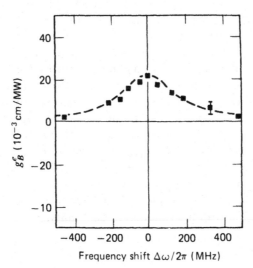

Fig. 11.2 Experimental Brillouin gain factor, $g_B^e \propto \beta_B^e$, versus frequency shift $\Delta\omega/2\pi$ for a nonabsorbing liquid (66% CS_2 and 34% CCl_4). The theoretical fit is a Lorentzian curve. (After Ref. 15.)

As in the Raman case, the best way to test the theory of stimulated Brillouin scattering is to conduct gain measurements with an oscillator–amplifier arrangement similar to that in Fig. 10.9. An example is shown in Fig. 11.2.[9] A theoretical fit to the data using (10.7) allows the deduction of Γ_B, which compares well with Γ_B obtained from spontaneous scattering.

Comparison of the Brillouin gains in Table 11.1 with the Raman gains in Table 10.1 shows that stimulated Brillouin scattering should usually dominate over stimulated Raman scattering in most liquids in the steady-state case. With a strong pump laser beam, especially one that is focused, stimulated Brillouin scattering can actually have enough gain to occur in all directions. Experimentally, acoustic waves of various frequencies are generated at the focal point in different directions. An audible sound can be heard when it happens. It is likely that a shock wave is initiated at the focal point. Cell windows often are shattered by the strong pressure waves generated, but the detailed mechanism involved is not yet understood.

11.2 STIMULATED THERMAL BRILLOUIN AND RAYLEIGH SCATTERING

We have assumed that the acoustic wave is described by the density variation $\Delta\rho$. This is, however, only an approximation first used by Einstein[10] and Brillouin[11] to describe spontaneous light scattering by low-frequency thermodynamic fluctuations in a single-component medium. Actually, ρ is a function of pressure p and entropy S, and one can write $\Delta\rho = (\partial\rho/\partial p)_s\Delta p + (\partial\rho/\partial S)_p\Delta S$. Then $\Delta p(t)$ describes the acoustic wave, while $\Delta S(t)$ describes the entropy wave at zero frequency with a diffusion-type equation of motion. In spontaneous scattering, Δp is responsible for the Brillouin doublet of the spectrum, and ΔS for the central Rayleigh component.[1] Therefore, for the stimulated Brillouin scattering discussed in the previous section, a more correct formalism will have to replace $\Delta\rho$ by Δp and $\partial\varepsilon/\partial\rho$ by $(\partial\varepsilon/\partial\rho)_S$. In some cases, however, it is more convenient to use the independent thermodynamic variables ρ and T instead of p and S. This is particularly true when the temperature T varies in direct response to the external heating of the medium. In the equations of motion, we expect that $\Delta\rho(t)$ and $\Delta T(t)$ are coupled since both, being linear combinations of $\Delta p(t)$ and $\Delta S(t)$, are mixtures of acoustic and entropy waves. Light scattering under the effect of heating due to optical absorption is known as thermal light scattering.[12]

The equations of motion for $\Delta\rho$ and ΔT are, respectively, the Navier–Stokes equation in conjunction with the equation of continuity,[12]

$$\rho_0\frac{\partial\mathbf{v}}{\partial t} + \frac{v^2}{\delta}\nabla(\Delta\rho) + \frac{v^2\beta_T\rho_0}{\delta}\nabla(\Delta T) - \eta\nabla^2\mathbf{v}$$
$$= \frac{\gamma}{2\pi}\nabla\left(\mathbf{E}_1\cdot\mathbf{E}_2^*\right) - \frac{1}{2\pi}\left(\frac{\partial\varepsilon}{\partial T}\right)_\rho(\mathbf{E}_1\cdot\mathbf{E}_2^*)\nabla(\Delta T), \qquad (11.9)$$

$$\frac{\partial}{\partial t}\Delta\rho + \rho_0\nabla\cdot\mathbf{v} = 0$$

and the energy transport equation

$$\left(\rho_0 C_v \frac{\partial}{\partial t} - \lambda_T \nabla^2\right)\Delta T - \frac{C_v(\delta - 1)}{\beta_T} \frac{\partial}{\partial t}\Delta\rho$$

$$= \frac{nc\alpha}{\pi}(\mathbf{E}_1 \cdot \mathbf{E}_2^*) + \frac{1}{\pi}\left(\frac{\partial\varepsilon}{\partial T}\right)_\rho T_0 \mathbf{E}_1 \cdot \frac{\partial}{\partial t}\mathbf{E}_2^* \tag{11.10}$$

where v is the acoustic wave velocity, $\delta = C_p/C_v$ is the ratio of heat capacities at constant pressure and at constant volume, β_T is the isothermal compressibility, η characterizes the acoustic damping, $\gamma = \rho_0(\partial\varepsilon/\partial\rho)_T$, λ_T is the thermal conductivity, and α is the linear absorption coefficient. The two equations in (11.9) can be combined to yield

$$\left(-\frac{\partial^2}{\partial t^2} + \frac{v^2}{\delta}\nabla^2 + \frac{\eta\partial}{\rho_0 \partial t}\nabla^2\right)\Delta\rho + \frac{v^2\beta_T\rho_0}{\delta}\nabla^2(\Delta T)$$

$$= \frac{\gamma}{2\pi}\nabla^2(\mathbf{E}_1 \cdot \mathbf{E}_2^*) - \frac{1}{2\pi}\left(\frac{\partial\varepsilon}{\partial T}\right)_\rho \nabla \cdot [\mathbf{E}_1 \cdot \mathbf{E}_2^* \nabla(\Delta T)]. \tag{11.11}$$

We notice that if the approximations $\delta \simeq 1$, $\alpha \simeq 0$, and $(\partial\varepsilon/\partial T)_\rho \simeq 0$ are used, (11.10) gives $\Delta T = 0$ and (11.11) reduces to the acoustic wave equation in (11.2). Stimulated thermal Brillouin and Rayleigh scattering is now described by the coupling of (11.10) and (11.11) with the wave equations for $\mathbf{E}_1$ and $\mathbf{E}_2$ in (11.1) having

$$\mathbf{P}^{NL}(\omega_1) = \frac{\gamma}{4\pi\rho_0}E_2\Delta\rho + \frac{1}{4\pi}\left(\frac{\partial\varepsilon}{\partial T}\right)_\rho E_2\Delta T$$

and $\tag{11.12}$

$$\mathbf{P}^{NL}(\omega_2) = \frac{\gamma}{4\pi\rho_0}E_1\Delta\rho^* + \frac{1}{4\pi}\left(\frac{\partial\varepsilon}{\partial T}\right)_\rho E_1\Delta T^*$$

The solution of the coupled wave equations is similar to those of stimulated Raman and Brillouin scattering. We consider only the steady-state solution for backward scattering here, and assume that both excitations $\Delta\rho$ and ΔT are highly damped. We can then replace $\partial/\partial t$ by $-i\omega_a = -i(\omega_1 - \omega_2)$ and $\nabla = \partial/\partial z$ by $i(k_1 + k_2)$ in (11.10) and (11.11) and solve for $\Delta\rho$ and ΔT. Substituting the expressions of $\Delta\rho$ and ΔT into (11.12) and then $\mathbf{P}^{NL}$ into (11.1), and using the slowly varying amplitude approximation for E_1 and E_2^*, we again find the amplitude equations for $\mathscr{E}_1$ and $\mathscr{E}_2^*$ in the form of (11.5), or

$$\left(\frac{\partial}{\partial z} + \alpha\right)|\mathscr{E}_1|^2 = -\beta|\mathscr{E}_1|^2|\mathscr{E}_2|^2$$

$$\left(\frac{\partial}{\partial z} - \alpha\right)|\mathscr{E}_2|^2 = -\beta|\mathscr{E}_1|^2|\mathscr{E}_2|^2. \tag{11.13}$$

With the approximation $(\partial\varepsilon/\partial\rho)_T\rho \gg (\partial\varepsilon/\partial T)_\rho T$, we have

$$\beta = \beta_B^e \frac{1}{1 + (\Delta\Omega/\Gamma_B)^2} + \beta_B^a \frac{2(\Delta\Omega/\Gamma_B)}{1 + (\Delta\Omega/\Gamma_B)^2} + (\beta_{RL}^e + \beta_{RL}^a)\frac{2\omega_a/\Gamma_{RL}}{1 + (\omega_a/\Gamma_{RL})^2}$$

where

$$\Delta\Omega = \frac{(k_1 + |k_2|)v}{\delta^{1/2}} - (\omega_1 - \omega_2),$$

$$\beta_B^e = \frac{\omega_2^2\gamma^2}{4\pi c^2\rho_0 v\Gamma_B},$$

$$\beta_B^a = \frac{\omega_2^2\gamma\gamma^a}{8\pi c^2\rho_0 v\Gamma_B},$$

$$\beta_{RL}^e = \frac{\omega_2^2\gamma\gamma^R}{4\pi c^2\rho_0 v\Gamma_{RL}}, \qquad\qquad (11.14)$$

$$\beta_{RL}^a = \frac{\omega_2^2\gamma\gamma^a}{4\pi c^2\rho_0 v\Gamma_{RL}}$$

$$\gamma^a = \frac{\alpha v c^2\beta_T}{c_p\omega_2},$$

$$\gamma^R = \frac{(\delta - 1)c\gamma\Gamma_{RL}}{4nv\omega_2},$$

$$\Gamma_B = \frac{\eta(k_1 + |k_2|)^2}{\rho_0},$$

and

$$\Gamma_{RL} = \frac{\lambda_T(k_1 + |k_2|)^2}{\rho_0 C_p}.$$

In the limit of negligible pump depletion, $|\mathscr{E}_2|^2$ grows exponentially in the backward direction with a gain $G - \alpha = \beta|\mathscr{E}_1|^2 - \alpha$.

The first two terms in (11.14) are responsible for stimulated Brillouin scattering, and the last term for stimulated Rayleigh scattering. The β_B^e term is for normal stimulated Brillouin scattering. It leads to the same Brillouin gain derived in the previous section. The gain spectrum is centered at $\Delta\Omega = 0$ or $\omega_1 - \omega_2 = (k_1 + |k_2|)v/\delta^{1/2}$, which is the frequency of the acoustic wave excitation involved in the backward Brillouin scattering. The β_B^a term corresponds to thermal Brillouin scattering, since it vanishes if the absorption

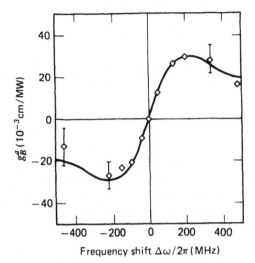

Fig. 11.3 Experimental thermal Brillouin gain factor, $g_B^a \propto \beta_B^a$, versus frequency shift $\Delta\omega/2\pi$ for an absorbing liquid (66% CS_2 and 34% CCl_4 with a small amount of I_2) with an absorption coefficient $\alpha = 0.83$ cm^{-1}. The theoretical curve is the dispersive counterpart of a Lorentzian. (After Ref. 15.)

coefficient α is zero. Its gain spectrum is antisymmetric about $\Delta\Omega = 0$. The β_{RL}^a term also vanishes when $\alpha = 0$, and corresponds to thermal Rayleigh scattering with a gain maximum at $\omega_a = \omega_1 - \omega_2 = \Gamma_{RL}$. Finally, the β_{RL}^e term corresponds to ordinary Rayleigh scattering with the same gain spectrum as β_{RL}^a.

Experimentally, stimulated Rayleigh scattering is most difficult to observe because of the small β_{RL}^e, but has actually been observed.[13] The maximum Rayleigh gain in liquids is estimated to be two orders of magnitude lower than the Brillouin gain.[2] With absorption, Rayleigh scattering can be greatly enhanced through the β_{RL}^a term. Indeed, stimulated thermal Rayleigh scattering can be readily observed in absorbing gases and liquids.[14] Stimulated thermal Brillouin scattering in absorbing media is also easily observable.[15] Its occurrence is evidenced by a small upshift of the Brillouin-shifted frequency, since the combined gain spectrum exhibited by the β_B^e and β_B^a terms in (11.14) has a maximum at $\Delta\Omega > 0$. The best way to study the effect of thermal Brillouin scattering is to measure the Brillouin gain as described in Section 11.1. The measured gain spectrum can then be compared directly with the theoretical spectrum in (11.14). An example is presented in Fig. 11.3, which shows a good agreement between theory and experiment.

11.3   STIMULATED RAYLEIGH-WING SCATTERING

Fluctuations of molecular orientation and distribution in a fluid medium result in fluctuations of the dielectric constant and lead to spontaneous light scattering.[1] This is known as Rayleigh-wing scattering, which has a spectrum similar to Rayleigh scattering but much broader in width. Stimulated Rayleigh-wing scattering can also be expected at high pump intensity. The physical picture is as follows: mixing of E_1 and E_2 reorient and redistributes the molecules; the molecular reorientation and redistribution, which vary in space and time, in turn beat with E_1 to enforce E_2. Stimulated Rayleigh-wing scattering is then simply the result of coupling between E_1, E_2, and the induced variation in molecular reorientation and redistribution. For quantitative description, we must find the equation of motion for molecular reorientation and redistribution. We consider here only the reorientation mechanism.

We assume anisotropic molecules with cylindrical symmetry. The optical polarizabilities parallel and perpendicular to the molecular axis are denoted by $\alpha_\parallel$ and $\alpha_\perp$, respectively. Let the molecular axis tilt at an angle θ from the $\hat{x}$-axis (Fig. 11.4). Then a linearly polarized E field along $\hat{x}$ induces a dipole p on the molecule with $p_x = \alpha_{xx}E$ and

$$\begin{aligned}\alpha_{xx} &= \alpha_\parallel\cos^2\theta + \alpha_\perp\sin^2\theta \\ &= \bar{\alpha} + \Delta\alpha(\cos^2\theta - \tfrac{1}{3})\end{aligned} \tag{11.15}$$

where

$$\bar{\alpha} = \tfrac{1}{3}(\alpha_\parallel + 2\alpha_\perp) \quad \text{and} \quad \Delta\alpha = \alpha_\parallel - \alpha_\perp .$$

The applied field E now interacts with the induced dipole and tends to align the molecule along $\hat{x}$ against the thermal randomization. Let us assume a collection of noninteracting molecules with a density N and a random orientational distribution in the absence of E. With the applied E, the orientational

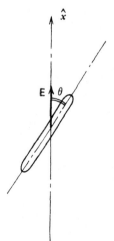

Fig. 11.4 Sketch showing a uniaxial molecule lying at an angle θ from the electric field E along $\hat{x}$.

distribution function at equilibrium becomes

$$f(\theta) = \frac{\exp(-2\alpha_{xx}|E|^2/k_BT)}{Z} \tag{11.16}$$

where

$$Z = \int_0^\pi \exp\left(-\frac{2\alpha_{xx}|E|^2}{k_BT}\right)\sin\theta\,d\theta,$$

k_B is the Boltzmann constant, and the factor 2 instead of $1/2$ appears in the exponential because of our convention on the amplitude of E (Section 2.9). Then the tensor component χ_{xx} of the optical susceptibility, following (11.15) and (11.16), is given by

$$\chi_{xx} = \bar{\chi} + \frac{2}{3}N\Delta\alpha Q$$

with

$$\begin{aligned}
\bar{\chi} &= N\bar{\alpha}, \\
Q &= \tfrac{3}{2}\langle\cos^2\theta - \tfrac{1}{3}\rangle \\
&= \frac{\displaystyle\int_0^\pi \tfrac{3}{2}(\cos^2\theta - \tfrac{1}{3})ecp[-2\Delta\alpha|E|^2(\cos^2\theta - \tfrac{1}{3})/k_BT]\sin\theta\,d\theta}{\displaystyle\int_0^\pi \exp[-2\Delta\alpha|E|^2(\cos^2\theta - \tfrac{1}{3})/k_BT]\sin\theta\,d\theta}
\end{aligned} \tag{11.17}$$

where $\langle\ \rangle$ denotes the orientational average. Similarly, we find

$$\chi_{yy} = \chi_{zz} = \bar{\chi} - \frac{1}{3}N\Delta\alpha Q. \tag{11.18}$$

Physically, the quantity Q, which describes the degree of molecular alignment, often is known as the orientational order parameter. For random distribution, $Q = 0$, and for perfect alignment $Q = 1$. The polarization induced by **E** takes the form

$$\begin{aligned}
\mathbf{P} &= \hat{x}\chi_{xx}E \\
&= \hat{x}\left(\bar{\chi}E + \frac{2}{3}N\Delta\alpha QE\right).
\end{aligned} \tag{11.19}$$

Since Q is also a function of E, the second term in (11.19) represents a nonlinear polarization $\mathbf{P}^{NL}$

$$\mathbf{P}^{NL} = \hat{x}\frac{2}{3}N\Delta\alpha QE \tag{11.20}$$

For relatively weak fields, we have $Q \propto |E|^2$, and then $\mathbf{P}^{NL}$ is a third-order nonlinear polarization.

In stimulated Rayleigh-wing scattering, the molecular orientational distribution changes in response to the beating of two optical fields $\mathbf{E}_1$ and $\mathbf{E}_2$. The equation of motion for molecular reorientation by this combined action of $\mathbf{E}_1$ and $\mathbf{E}_2$ is provided by the Debye rotational diffusion equation for the distribution function $f(\theta)$:[16]

$$\nu \frac{\partial f}{\partial t} = \frac{1}{\sin \theta} \frac{\partial}{\partial \theta} \left[\sin \theta \left(\frac{\partial f}{\partial \theta} + \frac{4 \Delta \alpha |E|^2 \sin \theta \cos \theta}{k_B T} f \right) \right] \qquad (11.21)$$

Equation (11.21) can be reduced to a simple equation of motion for Q by multiplying both sides by $\frac{3}{2}(\cos^2 \theta - \frac{1}{3})$ followed by an integration over θ, and neglecting the field-dependent terms of higher orders than $|E|^2/k_B T$:

$$\frac{\partial Q}{\partial t} = -\frac{Q}{\tau_D} + \frac{4}{3\nu} \Delta \alpha |E|^2 \qquad (11.22)$$

where $\tau_D = \nu/5 k_B T$ is the Debye relaxation time and ν is a viscosity coefficient for an individual molecule. Now that $\mathbf{E} = \mathbf{E}_1 + \mathbf{E}_2 = \hat{x}\mathcal{E}_1 \exp(i \mathbf{k}_1 \cdot \mathbf{r} - i\omega_1 t) + \hat{x}\mathcal{E}_2 \exp(i \mathbf{k}_2 \cdot \mathbf{r} - i\omega_2 t)$ and we are interested in the orientational redistribution excited by the beating of $\mathbf{E}_1$ and $\mathbf{E}_2$, the $|E|^2$ term in (11.22) should be replaced by $\mathbf{E}_1 \cdot \mathbf{E}_2^*$. Equation (11.22) is then coupled to the wave equations (11.1) for $\mathbf{E}_1$ and $\mathbf{E}_2$ via (11.20). In the steady-state case, one finds

$$Q = \frac{4\tau_D \Delta \alpha E_1 E_2^*}{3\nu(1 - i\omega\tau_D)}$$

$$\mathbf{P}^{NL}(\omega_2) = \chi_{RW}^{(3)} |\mathcal{E}_1|^2 \mathcal{E}_2 \qquad (11.23)$$

$$\chi_{RW}^{(3)} = \frac{8N\tau_D (\Delta \alpha)^2}{9\nu(1 + i\omega\tau_D)}$$

with $\omega = \omega_1 - \omega_2$. The Rayleigh-wing susceptibility $\chi_{RW}^{(3)}$ has a negative imaginary part. In analogy to the other stimulated light scattering cases, this indicates that E_2 can experience an exponential gain $\exp(G_{RW} - \alpha)z$ with

$$G_{RW} = \frac{2\pi\omega_2}{cn} \mathrm{Im}(\chi_{RW}^{(3)}) |\mathcal{E}_1|^2$$

$$= 16\pi \frac{\omega_2}{cn} \frac{N(\Delta\alpha)^2 \omega\tau_D |\mathcal{E}_1|^2}{45k_B T(1 + \omega^2\tau_D^2)}, \qquad (11.24)$$

which has its maximum at $\omega = 1/\tau_D$.

For a liquid with $\tau_D \sim 10^{-11}$ sec and $16\pi N(\Delta\alpha)^2/45k_B T \sim 10^{-11}$ cm^3/erg in a typical case, we have $(G_{RW})_{max} \sim 10^{-3}$ cm/MW, comparable to Raman gains in many liquids.[2] Stimulated Rayleigh-wing scattering is therefore expected to be easily observable, as is indeed the case.[17,18] In fact, it is very much analogous to stimulated Raman scattering by vibration: the material excitation $|Q|$ is independent of the wavevector so that the stimulated gain is isotropic. Consequently, Stokes–anti-Stokes coupling (see Section 10.4) that leads to the generation of anti-Stokes radiation in the near forward direction in stimulated Raman scattering also occurs in stimulated Rayleigh-wing scattering.[19] The results are somewhat different because of the difference in the resonant frequencies of the material excitations. Unlike the Raman case, the maximum gain with Stokes–anti-Stokes coupling in stimulated Rayleigh-wing scattering appears at $\omega_l - \omega_s = \omega_{as} - \omega_l = 0$ with $\mathbf{k}_s$ and $\mathbf{k}_{as}$ making an angle θ_{opt} with k_l.[19] In other words, the laser beam, generates through stimulated Rayleigh-wing scattering a cone of radiation of the same frequency at an angle θ_{opt} from the laser beam. In reality, a laser beam of finite cross section has a spread of k_l. The effect of stimulated Rayleigh-wing scattering is to broaden this spread of $\mathbf{k}_l$ or, equivalently, to reduce the laser beam cross section. The laser beam therefore appears to self-focus as it propagates in the medium. This is an unconventional way of describing self-focusing of light. The conventional way will be discussed in Chapter 17. We note that it is amplification of the existing off-axis $\mathbf{k}_l$ components that leads to self-focusing. If stimulated Raman scattering also occurs in the medium, it is initiated by amplification of noise. Since the Raman gain and Rayleigh-wing gain are comparable in many liquids, we can expect that the occurrence of self-focusing often precedes that of stimulated Raman scattering.

11.4 OTHER STIMULATED LIGHT SCATTERING

In a multicomponent system, local concentrations of the components can fluctuate, causing variation in the dielectric constant and scattering of light, known as concentration scattering.[1] In thermodynamics, concentrations together with ρ, T or p, S form a set of thermodynamic variables. Variation of the dielectric constant of a two-component system can be written for example, as,

$$\Delta\varepsilon = \left(\frac{\partial\varepsilon}{\partial\rho}\right)_{C,T}\Delta\rho + \left(\frac{\partial\varepsilon}{\partial T}\right)_{C,\rho}\Delta T + \left(\frac{\partial\varepsilon}{\partial C}\right)_{\rho,T}\Delta C \qquad (11.25)$$

where C is the relative concentration. In stimulated concentration scattering, ΔC is excited by the beating of $\mathbf{E}_1$ and $\mathbf{E}_2$. It obeys a driven diffusion equation[20]

$$\left(\frac{\partial}{\partial t} - D\nabla^2\right)\Delta C = -\frac{D(\partial\varepsilon/\partial C)_{\rho,T}\nabla^2\left(\mathbf{E}_1\cdot\mathbf{E}_2^*\right)}{8\pi\rho_0(\partial\mu/\partial C)_{\rho,T}} \qquad (11.26)$$

In (11.26) we neglected the coupling of ΔC to $\Delta \rho$ and ΔT; D is the diffusion coefficient and μ is the chemical potential. In a more rigorous treatment, the coupling between ΔC, $\Delta \rho$, and ΔT can be taken into account.[20] Analogous to the other stimulated light scattering cases, stimulated concentration scattering is then described by the solution of the coupled equations (11.26) and (11.1) with $\mathbf{P}^{NL}(\omega_1) = (\partial \varepsilon / \partial C)_{\rho, T} \mathbf{E}_2 \Delta C / 4\pi$ and $\mathbf{P}^{NL}(\omega_2) = (\partial \varepsilon / \partial C)_{\rho, T} \mathbf{E}_1 \Delta C^* / 4\pi$. The stimulated gain has a spectrum proportional to $\omega \tau_C (1 + \omega^2 \tau_C^2)$ with $\tau_C^{-1} = Dk^2$ and $\mathbf{k} = \mathbf{k}_1 - \mathbf{k}_2$. Details of theory and experiment of stimulated concentration scattering can be found in Ref. 20.

There are, of course, many other types of light scattering: light scattering by molecular libration (rotational oscillation), by sheer waves, by spin waves, by surface waves, etc.[1] In principle, with sufficient pump intensity, they can all become stimulated: knowing the dynamic equation for the material excitation, the theoretical treatment again follows the coupled-wave approach. However, the threshold for a certain stimulated scattering may be higher than the optical damage threshold; if so, such stimulated scattering will not be observable.

A very different type of stimulated light scattering is stimulated Compton scattering, first proposed by Pantell et al.[21] By backscattering microwaves from a relativistic electron beam, tunable far-infrared radiation could be generated. Tunability could be achieved by varying the electron energy. Sukhamte and Wolff[22] showed that stimulated Compton scattering could be greatly enhanced in a magnetic field if the microwave frequency is equal to the cyclotron resonance frequency. Experiments on stimulated Compton scattering have not yet been reported. However, in a related problem, tunable microwave and far-infrared radiation have been generated by relativistic electrons performing cyclotron motion in a magnetic field.[23] Also, intense microwave emission with peak power $\geq$ 500 MW and conversion efficiency $\sim$ 17% has been observed in coherent Cherenkov radiation from a relativistic electron beam interacting with a slow wave structure.[24]

REFERENCES

1 See, for example, I. L. Fabellinskii, *Molecular Scattering of Light* (Plenum, New York, 1968).

2 W. Kaiser and M. Maier, in F. T. Arrecchi and E. O. Schulz-DuBois, eds., *Laser Handbook* (North Holland Publishing Co., Amsterdam, 1972), p. 1077.

3 C. L. Tang, *J. Appl. Phys.* **37**, 2945 (1966).

4 M. Maier, *Phys. Rev.* **166**, 113 (1968).

5 N. M. Kroll, *J. Appl. Phys.* **36**, 34 (1965).

6 M. Maier and G. Renner, *Phys. Lett.* **34A**, 299 (1971); *Opt. Comm.* **3**, 301 (1971)

7 R. Y. Chiao, C. H. Townes, and B. P. Stoicheff, *Phys. Rev. Lett.* **12**, 592 (1964).

8 E. Garmire and C. H. Townes, *Appl. Phys. Lett.* **5**, 84 (1964).

9 D. Pohl and W. Kaiser, *Phys. Rev.* **B1** 31 (1970).

10 A. Einstein, *Ann. Phys.* **33**, 1275 (1910).

11 L. Brillouin, *Ann. Phys.* (Paris) **17**, 88 (1922).

12 R. M. Herman and M. A. Gray, *Phys. Rev. Lett.* **19**, 824 (1967).

13 G. I. Zaitsev, Yu. I. Kyzylasov, V. S. Starunov, and I. L. Fabellinskii, *JETP Lett.* **6**, 255 (1967); I. L. Fabellinskii, D. I. Mash, V. V. Morozov, and V. S. Starunov, *Phys. Lett.* **27A**, 253 (1968).

14 D. H. Rank, C. W. Cho, N. D. Foltz, and T. A. Wiggins, *Phys. Rev. Lett.* **19**, 828 (1969).

15 D. Pohl and W. Kaiser, *Phys. Rev.* **B1**, 31, (1970).

16 P. Debye, *Polar Molecules* (Dover, New York, 1929).

17 D. I. Mash, V. V. Morozov, V. S. Starunov, and I. L. Fabellinskii, *JETP Lett* **2**, 25 (1965).

18 M. Denariez and G. Bret, *Compt. Rend.* **265**, 144 (1967); *Phys. Rev.* **171**, 160 (1968).

19 R. Y. Chiao, P. L. Kelley, and E. Garmire, *Phys. Rev Lett.* **17**, 1158 (1966); R. L. Carman, R. Y. Chiao, and P. L. Kelley, *Phys. Rev. Lett.* **17**, 1281 (1966).

20 W. H. Lowdermilk and N. Bloembergen, *Phys. Rev.* **A5**, 1423 (1972).

21 R. H. Pantell, G. Soncini, and H. E. Puthoff, *IEEE J. Quant. Electron.* **QE-4**, 905 (1968).

22 V. P. Sukhamte and P. A. Wolff, *IEEE J. Quant. Electron.* **QE-10**, 870 (1974)

23 V. L. Granatstein, M. Herndon, R. K. Parker, and S. P. Schlesinger, *IEEE J. Quant. Electron.* **QE-10**, 651 (1974)

24 Y. Carmel, J. Ivers, R. E. Kribel, and J. Nation, *Phys. Rev. Lett.* **33**, 1278 (1974).

BIBLIOGRAPHY

Fabellinskii, I. L., *Molecular Scattering of Light* (Plenum, New York, 1968)

Kaiser, W., and M Maier, in F. T. Arrecchi and E. O. Schulz-Dubois, eds., *Laser Handbook* (North Holland Publishing Co., Amsterdam, 1972), p. 1077.

12

Two-Photon Absorption

One-photon and two-photon transitions follow different selection rules. They are therefore complementary to each other as spectroscopic tools. A well-known example is infrared absorption versus Raman scattering. In a two-photon absorption process, two photons are simultaneously absorbed to excite a material system. Being a higher-order process, its absorption cross section often is many orders of magnitude smaller than that of a one-photon absorption. Even so, two-photon absorption is readily observable with lasers and has become a valuable spectroscopic technique complementary to linear absorption spectroscopy. This chapter briefly describes the basic theory, measuring techniques, and various applications of two-photon absorption.

12.1 THEORY

The transition probability of a two-photon process was first derived by Göppert-Mayer using second-order perturbation theory.[1] The derivation was given in Section 10.2 for the case of Raman scattering. For two-photon absorption, the transition probability per unit time per unit volume per unit energy interval closely resembles that of (10.2) and is given by

$$\frac{dW_{fi}}{d(\hbar\omega_1)} = \frac{dW_{fi}}{d(\hbar\omega_2)} = \frac{8\pi^3 N\omega_1\omega_2}{\epsilon_1\epsilon_2}|\langle f|M|i\rangle|^2|\langle\alpha_f|a_2 a_1|\alpha_i\rangle|^2 g(\hbar\,\Delta\omega),$$

$$M = \sum_s \left[\frac{e\mathbf{r}\cdot\hat{e}_2|s\rangle\langle s|e\mathbf{r}\cdot\hat{e}_1}{\hbar(\omega_1 - \omega_{si})} + \frac{e\mathbf{r}\cdot\hat{e}_1|s\rangle\langle s|e\mathbf{r}\cdot\hat{e}_2}{\hbar(\omega_2 - \omega_{si})} \right] \tag{12.1}$$

(see Fig. 12.1). Notation here follows Section 10.2, with $\Delta\omega = \omega_1 + \omega_2 - \omega_{fi}$. In the semiclassical approximation, $|\langle\alpha_f|a_2 a_1|\alpha_i\rangle|^2 \simeq \bar{n}_1\bar{n}_2$ can be replaced by $(\epsilon_1\epsilon_2)|E_1|^2|E_2|^2/(2\pi)^2(\hbar\omega_1)(\hbar\omega_2) = I_1 I_2(\epsilon_1\epsilon_2)^{1/2}/c^2(\hbar\omega_1)(\hbar\omega_2)$, where I_1 and I_2 are the beam intensities at ω_1 and ω_2, respectively. The two beams

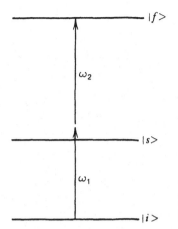

Fig. 12.1 Two-photon excitation of a system from $|i\rangle$ to $|f\rangle$ via the virtual intermediate state $|s\rangle$.

propagating along $\hat{z}$ in such a nonlinear absorbing medium have their attenuations governed by the equation

$$\frac{dI_1}{dz} = -\omega_1\gamma I_1 I_2, \qquad \frac{dI_2}{dz} = -\omega_2\gamma I_1 I_2 \tag{12.2}$$

with

$$\gamma = \frac{\left[dW_{fi}/d(\hbar\omega_2)\right](\rho_i - \rho_f)}{I_1 I_2}$$

$$= \frac{8\pi^3 N}{\varepsilon_1^{1/2}\varepsilon_2^{1/2}c^2}|M_{fi}|^2 g(\hbar\,\Delta\omega)(\rho_i - \rho_f).$$

As in the Raman case, the above equation for γ can also be derived from the coupled wave approach. It is easily shown, following a derivation similar to that of Section 10.3, that the two-photon absorption coefficient γ is linearly proportional to the imaginary part of the third-order nonlinear susceptibility $\chi^{(3)}$ for the two-photon absorption process:

$$\gamma = \frac{8\pi^2}{c^2\varepsilon_1^{1/2}\varepsilon_2^{1/2}}\,\mathrm{Im}\,\chi^{(3)}$$

$$\mathrm{Im}\,\chi^{(3)} = N\pi|M_{fi}|^2 g(\hbar\,\Delta\omega)(\rho_i - \rho_f). \tag{12.3}$$

The same result can of course be obtained by treating two-photon absorption as a wave-mixing process in which the two optical waves at ω_1 and ω_2 jointly excite the material excitational wave $\rho_{fi}^{(2)}(\omega_1 + \omega_2)$. The derivation follows that in Section 10.3.

The coupled equations in (12.2) can be solved analytically, noting that

$$\frac{1}{\omega_1}\frac{dI_1}{dz} = \frac{1}{\omega_2}\frac{dI_2}{dz},$$

which is the consequence of having equal numbers of photons absorbed at ω_1 and ω_2 in a two-photon absorption process. If I_{10} and I_{20} are the beam intensities at the entrance of the medium, we have

$$\frac{I_{10} - I_1}{\omega_1} = \frac{I_{20} - I_2}{\omega_2}. \tag{12.4}$$

The solution of (12.2) can then be obtained by first eliminating either I_1 or I_2. Assuming $I_{10} > I_{20}$, we find

$$I_1 = I_{10}\frac{(I_{10}/\omega_1) - (I_{20}/\omega_2)}{(I_{10}/\omega_1) - (I_{20}/\omega_2)\exp(-Kz)},$$

$$I_2 = I_{20}\frac{[(I_{10}/\omega_1) - (I_{20}/\omega_2)]\exp(-Kz)}{(I_{10}/\omega_1) - (I_{20}/\omega_2)\exp(-Kz)}, \tag{12.5}$$

$$K = \omega_1\omega_2\gamma\left(\frac{I_{10}}{\omega_1} - \frac{I_{20}}{\omega_2}\right).$$

If $I_{10} \gg I_{20}$, then the attenuation of I_1 is negligible, and the solution reduces to

$$I_1 \simeq I_{10},$$

$$I_2 \simeq I_{20}\exp(-Kz). \tag{12.6}$$

A special case of interest is when $\omega_1 = \omega_2$. The conventions of Section 2.9 should be used in dealing with the two-photon absorption coefficients in this case. Equation (12.2) becomes

$$\frac{dI_1}{dz} = -\omega_1\gamma I_1^2 \tag{12.7}$$

and the solution takes the form

$$I_1 = \frac{I_{10}}{1 + I_{10}\omega_1\gamma z}. \tag{12.8}$$

In the case of weak absorption, it reduces to

$$I_1 = I_{10}(1 - I_{10}\omega_1\gamma z). \tag{12.9}$$

The two-photon absorption coefficient γ and the corresponding third-order susceptibility $\chi^{(3)}$ are in general tensor quantities. Analogous to Raman scattering, the selection rules can be derived from group theory. They have been obtained by Inoue and Toyozawa[2] for the 32 crystal point groups and by McClain[3] for molecular fluids. Spin-orbit coupling has been included by Bader and Gold[4] in an extension of Inoue and Toyozawa's calculation.

12.2 EXPERIMENTAL TECHNIQUES

Two-photon absorption can be measured directly from beam attenuation if the absorption is sufficiently strong. Let us assume, as an example, a typical two-photon absorption coefficient with $\text{Im}\,\chi^{(3)} = 10^{-12}$ esu for a condensed medium. Then, from (12.3) and (12.5), the induced attenuation coefficient K is of the order of 10^{-2} for $I_1 \sim$ few MW/cm^2. This corresponds to a $\sim 1\%$ attenuation of the ω_2 beam in traversing through a medium 1 cm long, and it should be easily measurable. Direct attenuation measurement of two-photon absorption is therefore fairly straightforward with pulsed lasers unless $\chi^{(3)}$ is orders of magnitude smaller than 10^{-12} esu.

For two-photon spectroscopic work, one of the two input beams must be tunable. In the early days, only fixed-frequency lasers were available. The tunable beam was provided by an incandescent or arc lamp in conjunction with a monochromator. A two-photon absorption spectrum was obtained by measuring the laser-induced attenuation as a function of the frequency of the tunable beam in the medium. A typical experimental arrangement is shown in Fig. 12.2.[5] Over the years, several research groups have constructed more sophisticated, automated versions of the setup.[6] The incoherent lamp can now be replaced by a tunable laser with great improvement on the signal-to-noise ratio. Unfortunately, the tunability of a laser is still limited. Replacement of the lamp by a tunable laser is preferable only if a narrow spectral range is of interest.

Weak beam attenuation is generally difficult to measure. One would like to find other methods of higher sensitivity for two-photon absorption measurement. In many media, luminescence may appear following excitation. This is nearly always the case in gases, and is also fairly common in condensed matter although the quantum yield could be small. Since luminescence is easily detectable, it provides a means to monitor two-photon absorption with a sensitivity many orders of magnitude higher than the beam attenuation measurement. The first two-photon absorption experiment was actually done with this technique.[7] However, in two-photon absorption spectroscopy, one must be sure that the quantum yield of luminescence does not depend strongly on the excitation frequency; otherwise, the spectrum will appear distorted.

Two-photon excitation near or above the ionization level of an atom or molecule may lead to ionization, and the resulting electrons and ions are easily detectable.[8] Therefore, photoionization can also be a sensitive method for

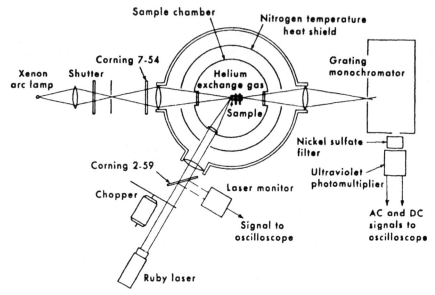

Fig. 12.2 Schematic diagram of an experimental set-up for two-photon absorption spectroscopy. (After Ref. 5.)

detecting two-photon absorption. Its application, however, is limited to the case where the final excited state is near or above the ionization level. The observed spectrum is the two-photon absorption spectrum weighted by the ionization rate, which generally depends on the energy of the final excited state.

By the same token, photoconductivity can also be used to detect two-photon absorption in a solid. If heat released through relaxation after the two-photon excitation can be monitored, it can also be used to measure two-photon absorption. An example is photoacoustic spectroscopy, in which heat released appears as an acoustic signal detectable by either a microphone or a transducer. Less conventional methods of detecting two-photon absorption include photoemission, photodissociation, photochemical reaction, and optogalvanic effect.

12.3   TWO-PHOTON ABSORPTION SPECTROSCOPY

Solids

The first spectroscopic measurements of two-photon absorption were carried out by Hopfield et al.[5] on alkali halides near the band edges, using the setup shown in Fig. 12.2. Since the crystals have inversion symmetry, the states near the band edges have more or less definite parities. Thus one-photon and

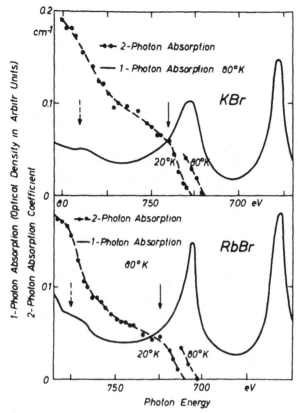

Fig. 12.3 One- and two-photon absorption spectra of KBr and RbBr. [After D. Fröhlich and B. Staginnus, *Phys. Rev. Lett.* **19**, 496 (1967).]

two-photon absorption spectra are expected to be different, as seen in Fig. 12.3. In particular, no exciton peak is present in the two-photon absorption spectrum. The result was used by Hopfield et al. to test the validity of the various exciton models for the alkali halides.

Two-photon absorption in semiconductors is the subject of numerous studies.[9] It was hoped that the technique would lead to new information about the band structures. Thus far, the results have been disappointing mainly because, first, the band structures of those semiconductors are already very well known; second, the two-photon absorption data are not very accurate due to laser fluctuations; and third, the spectral ranges covered by two-photon absorption are very limited. Two-photon absorption is, however, a useful tool to study excitions and exciton-polaritons in a semiconductor. With one-photon absorption, only the existence of the exciton-polaritons can be shown by the observation of the reststrahlung band. With two-photon absorption, the dispersion curve of the exciton-polaritons can be measured.[10] An example is given in Fig. 12.4. In this case the excitons can be excited by both one- and two-photon

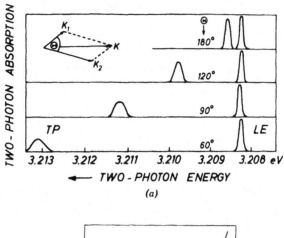

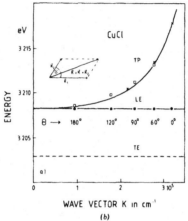

Fig. 12.4 (*a*) Two-photon absorption peaks near the first excitonic excitation of CuCl for different angles θ between the two incoming beams at 1.5 K. (*b*) Dispersion curves of the transverse exciton polariton (TP) and the longitudinal exciton (LE) of CuCl. The lines are theoretical curves calculated from reflectivity data. *The open squares and closed circles are measured results from two-photon absorption.* (and the crosses are from second harmonic generation.) [After D. Fröhlich, *Festkörperprobleme XXI*, 363 (1981).]

transitions. The observed two-photon absorption is due partly to excitation of excitons and partly to sum-frequency (or second harmonic) generation.[11] The correct theoretical treatment of the problem follows closely the derivation in Section 10.7.[12]

Two-photon absorption has also been used to probe the states of excitonic molecules[13] that cannot be reached by one-photon excitation. In other applications, two-photon absorption can be used to yield a uniform excitation of carriers in the bulk. This could be useful in both physics and device studies.

Molecular Fluids and Gases

Two-photon absorption can be used to probe excited states that cannot be reached by one-photon excitation. In molecules with centers of symmetry, the electronic states can be divided into gerade (g) and ungerade (u) states. One-photon transitions from g to g or from u to u are forbidden, but two-photon transitions are allowed. Thus with two-photon absorption, it is now possible to study a new set of electronic, vibrational, and rotational states which cannot be reached by one-photon absorption. Numerous examples are cited in Ref. 14. McClain[3] has pointed out that even though the molecules are randomly oriented in a gas or liquid, two-photon absorption with $\omega_1 \neq \omega_2$ still shows polarization properties that allow us to determine the symmetry of the excited states of the molecules. Therefore, two-photon absorption has become an important tool in the field of molecular spectroscopy, as evidenced by the large number of references cited in Ref. 14.

Atoms

Two-photon absorption also can be used to study excited electronic states of an atom that cannot be probed by one-photon absorption. Examples are the ns and nd states of an alkali atom. Because of the large transition matrix elements between the atomic states, two-photon absorption in atomic gases is generally much stronger than in molecular gases. Yet it is still too weak to be observed by the measurement of beam attenuation. Fortunately, other methods, such as photoluminescence and photoionization, can be used. They are sensitive enough to detect two-photon absorption in a vapor of less than 1 torr pressure. With counterpropagating beams of the same frequency, two-photon absorption in gases can yield Doppler-free spectral lines. This is described in Chapter 13. Applications of two-photon absorption to the atomic studies of high Rydberg states, quantum defect theory, and autoionizations are discussed in Chapter 18.

REFERENCES

1   M. Göppert-Mayer, *Ann. Physik* **9**, 273 (1931).

2   M. Inoue and Y. Toyozawa, *J. Phys. Soc. Japan* **20**, 363 (1965).

3   W. M. McClain, *J. Chem. Phys.* **55**, 2789 (1971).

4   T. R. Bader and A. Gold, *Phys. Rev.* **171**, 997 (1968).

5   J. J. Hopfield, J. M. Worlock, and K. J. Park, *Phys. Rev. Lett.* **11**, 414 (1963); J. J. Hopfield and J. M. Worlock, *Phys. Rev.* **137**, A1455 (1965).

6   See, for example, B. Staginnus, D. Fröhlich, and T. Caps, *Rev.Sci Inst.* **39**, 1129 (1968); M. W. Dowley and W. L. Peticolas, *IBM Res. Dev.* **12**, 188 (1968); R L. Swofford and W M McClain, *Rev. Sci. Inst.* **44**, 978 (1973).

7   W Kaiser and C. G. B. Garrett, *Phys. Rev. Lett.* **7**, 229 (1961).

8   See Chapter 18 on multiphoton ionization.

9   See, for example, C. C. Lee and H. Y. Fan, *Phys. Rev.* **B9**, 3502 (1974).

10   D. Fröhlich, E. Mohler, and P. Wiesner, *Phys. Rev. Lett.* **26**, 554 (1971).

11   D. C. Haueisen and H. Mahr, *Phys. Rev. Lett.* **26**, 838 (1971).

12   D. Boggett and R. Loudon, *Phys. Rev. Lett.* **28**, 1051 (1972).

13   G. M. Gale and A. Mysyrowicz, *Phys. Rev. Lett.* **32**, 727 (1974); L. L. Chase, N. Peyghambarian, G. Grynberg, and A. Mysyrowicz, *Phys. Rev. Lett.* **42**, 1231 (1979).

14   M. W. McClain, *Ann. Rev. Phys. Chem.* **31**, 559 (1980).

BIBLIOGRAPHY

Gold, A., in R. Glauber, ed., *Quantum Optics* (Academic Press, New York, 1969), p. 397.

McClain, W. M., *Acc. Chem. Res.* **7**, 129 (1974).

McClain, W. M., *Ann. Rev. Phys. Chem.* **31**, 559 (1980).

Mahr, H., in H. Rabin and C. L. Tang, eds., *Treatise in Quantum Electronics* (Academic Press, New York, 1975), p. 472.

Worlock, J. M., in F. T. Arrecchi and E. O. Schulz-Dubois, eds., *Laser Handbook* (North-Holland Publishing Co., Amsterdam, 1972), p. 1323.

13

High-Resolution Nonlinear Optical Spectroscopy

Lasers are known to have extremely narrow intrinsic linewidths. They are thus ideal tools for high-resolution spectroscopic studies. For a He–Ne laser at 3.39 μm, a linewidth as narrow as 3 Hz has been reported.[1] This represents a resolving power of 2×10^{13}, which is only two orders of magnitude less than the Mössbauer effect. In ordinary spectroscopy, however, the studies of spectroscopic details often are limited by inhomogeneous broadening rather than by instrument resolution. The Doppler width of the sodium D lines at room temperature is ~ 1.3 GHz, while the hyperfine splittings of the lines are only several hundred megahertz. In solids, the inhomogeneous width of a line can be even much higher. Thus for high-resolution spectroscopy it is of prime importance to find ways to reduce the effect of inhomogeneous broadening. This chapter introduces a number of nonlinear optical spectroscopic techniques which serve the purpose. These methods have, in recent years, revolutionized the field of atomic and molecular spectroscopy and stimulated a great deal of interest in the area of high-resolution solid-state spectroscopy.

13.1 GENERAL DESCRIPTION

Inhomogeneous broadening of a spectral transition arises because atoms, molecules, or ions in an ensemble do not all have the same local environment. Consider a transition between two states $|n\rangle$ and $|n'\rangle$ with a resonance frequency $\omega_{n'n}$. In general, $\omega_{n'n}$ is a function of a number of parameters, α, β, $\gamma, \ldots$ describing the local environment. These local parameters are random variables and should obey a certain statistical distribution function, say, $g(\alpha, \beta, \gamma, \cdots)$ with $\int g \, d\alpha \, d\beta \, d\gamma \cdots = 1$. A physical quantity X, which is a function of $\omega_{n'n}$, should then have its average value given by

$$X = \int X[\omega_{n'n}(\alpha, \beta, \gamma, \ldots)] g(\alpha, \beta, \gamma, \ldots) d\alpha \, d\beta \cdots. \qquad (13.1)$$

For example, a Lorentzian absorption line with inhomogeneous broadening has the expression

$$S(\omega) = \int \frac{S_0 \Gamma g(\alpha, \beta, \ldots)}{[\omega - \omega_{n'n}(\alpha, \beta, \ldots)]^2 + \Gamma^2} d\alpha \, d\beta \cdots. \qquad (13.2)$$

The number of local parameters necessary to characterize the local environment can be many. For impurity ions in a solid, each ion has its own local environment, and sees a local crystal field resulting from the Coulomb force of neighboring atoms or ions.[2] The crystal field can be described by a set of local parameters, and the distribution of ions over the local sites therefore, can be characterized by the distribution function of these local parameters. The total number of such local parameters depends on the local symmetry of the ion site. It can be very large for a low-symmetry site, for example, larger than 10 for a C_2 symmetry. The inhomogeneous broadening of the impurity ion spectrum is in principle determined by the statistical variation of the many local parameters characterizing the ion site. In practice, however, one or a few parameters describing the high-symmetry components of the crystal field may dominate the rest.

In gases, the situation is most fortunate, since the velocity of atoms or molecules is the only local parameter contributing to the inhomogeneous broadening, which is just the Doppler broadening. The thermal velocity obeys the Maxwellian distribution

$$g(v) = \frac{1}{\sqrt{\pi}\,u} e^{-v^2/u^2} \qquad (13.3)$$

where $u^2 = 2kT/m$, and m is the mass of a single atom or molecule. The Doppler width is then given by the well-known expression

$$(\Delta\omega_D)_{n'n} = 2\omega_{n'n}[(2kT/mc^2)\ln 2]^{1/2}$$
$$= 7.163 \times 10^{-7}(T/A)^{1/2}\omega_{n'n} \qquad (13.4)$$

where T is in degrees Kelvin and A is the atomic or molecular weight. For $A = 100$, we find $\Delta\omega_D \sim 0.02$ cm^{-1} (0.6 GHz) in the green, and $\Delta\omega_D \sim 10^{-3}$ cm^{-1} (30 MHz) in the infrared around 10 μm. These inhomogeneous broadenings may seem to be narrow by the standard of ordinary spectroscopy, but they often are much broader than homogeneous linewidths. The natural lifetime broadening is 10^5 to 10^7 Hz for atomic transitions, and 10 to 10^3 Hz for molecular vibrational transitions. The pressure broadening due to atomic or molecular collisions is around 10^4 Hz at 1 mtorr, and broadening due to collisions with walls of a container of a few centimeters in dimension is 10^3 to 10^4 Hz. The power broadening (or saturation broadening) of an atomic transition can be ~ 10 MHz/mW/cm^2. If a spectroscopic technique has sufficient resolution, then elimination of the Doppler broadening allows us to measure the homogeneous linewidth and lineshape and to probe the various physical mechanisms for the homogeneous broadening. Furthermore, level shifts and splittings smaller than the Doppler width but larger than or comparable to the homogeneous width can also be studied. These include many interesting problems such as Zeeman and Stark effects, collisional effects, hyperfine splittings, isotope shifts, Lamb shifts, quantum defects of Rydberg states, measurements of rotational splittings, and the like.

To eliminate Doppler broadening, the classical approach is to use a mono-energetic atomic or molecular beam. For a small beam divergence of 2ϕ, the residual Doppler width seen by a perpendicular probe beam is $(\delta\omega_D)_{n'n} = (2u/c)\omega_{n'n}\phi$ where u is the forward beam velocity. With $\phi = 10^{-2}$ rad, $(\delta\omega_D)_{n'n}$ can be more than one order of magnitude smaller than $(\Delta\omega_D)_{n'n}$.

High-resolution nonlinear optical spectroscopy, however, uses nonlinear optical methods to reduce the effect of inhomogeneous broadening. There exist a number of such techniques. They generally follow the basic idea of either using a resonant effect that is independent of inhomogeneous broadening or selectively studying only a group of molecules with the same resonant frequency. Most of the techniques discussed in the following sections are applicable, in principle, to both gases and condensed matter, although the spectral lines in condensed matter frequently are too broad to require high-resolution spectroscopic measurements.

13.2 QUANTUM BEATS[3]

Consider a system with two closely spaced excited states as seen in Fig. 13.1. If a laser pulse with an inverse pulsewidth larger than the spacing between the two excited states is used to resonantly excite the system, then after the pulse

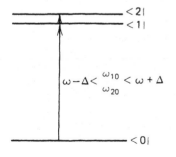

Fig. 13.1 A three-level system having two closely spaced upper levels coherently excited by a laser pulse with a spectral width 2Δ larger than the spacing ω_{21} between the upper levels.

excitation, the system is in a coherent superposition state

$$\langle\psi| = \langle\psi_0| + a_1\langle\psi_1|e^{-i\omega_{10}t-\Gamma_1 t} + a_2\langle\psi_2|e^{-i\omega_{20}t-\Gamma_2 t} \tag{13.5}$$

where $\langle\psi_0|$, and $\langle\psi_1|$, $\langle\psi_2|$ are the ground and excited eigenstates, respectively, and a_1 and a_2 are coefficients depending on the pulse excitation. The system in the coherent superposition state will radiate to go back to the ground state. The radiation power is given by

$$\mathscr{P}(t) \propto |\langle\psi(t)|e\mathbf{r}|\psi_0\rangle|^2$$
$$= |a_1\langle\psi_1|e\mathbf{r}|\psi_0\rangle e^{-(i\omega_{10}+\Gamma_1)t} + a_2\langle\psi_2|e\mathbf{r}|\psi_0\rangle e^{-(i\omega_{20}+\Gamma_2)t}|^2 \tag{13.6}$$
$$= A_1 e^{-2\Gamma_1 t} + A_2 e^{-2\Gamma_2 t} + B e^{-(\Gamma_1+\Gamma_2)t}\cos\left[(\omega_{20}-\omega_{10})t + \phi\right],$$

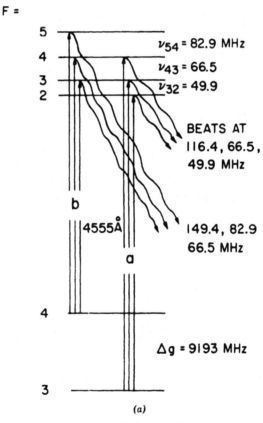

(a)

Fig. 13.2 (a) Hyperfine structure of the $6\,^2S_{1/2}$ and $7\,^2P_{3/2}$ levels of cesium. The sets of quantum beat frequencies expected for the a and b excitations are indicated. (b) Observed oscilloscope traces of the quantum beats in fluorescence resulting from the a and b excitations, respectively. The corresponding theoretical plots of the beats are shown under the experimental traces. (c) Time recording and frequency analysis of the $I_\pi - I_\sigma$ signal in (b). (After Ref. 4.)

EXCITATION FROM THE F = 3
GROUND STATE

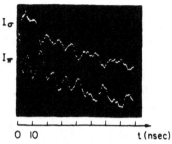

I_σ

I_π

0 10   t (nsec)

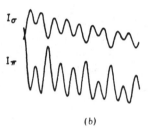

I_σ

I_π

(b)

EXCITATION FROM THE F = 3
GROUND STATE

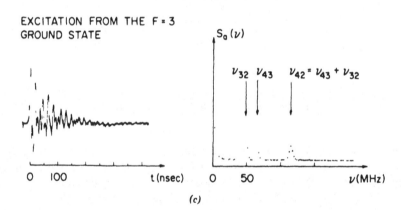

$S_0(\nu)$

ν_{32} ν_{43} $\nu_{42} = \nu_{43} + \nu_{32}$

0   100   t (nsec)   0   50 ν(MHz)

(c)

Fig. 13.2 (*Continued*).

which shows a damped oscillation with a beat frequency ($\omega_{20} - \omega_{10}$).

For an ensemble of such systems, each system may have slightly different ω_{10} and ω_{20} due to the Doppler effect or other inhomogeneous effects, but the oscillation frequency ($\omega_{20} - \omega_{10}$) should be the same for all systems. Consequently, the spontaneous radiation power from the ensemble after the pulse excitation is still given by (13.6). The oscillation is observable as long as $|\omega_{20} - \omega_{10}| \gtrsim (\Gamma_1 + \Gamma_2)$. The oscillation frequency yields directly the level spacing $\omega_{21} = \omega_{20} - \omega_{10}$.

This quantum beat technique can be extended to systems with several closely spaced transitions. The time-varying signal becomes more complicated, but if the level spacings are much larger than the damping coefficients Γ, the spectrum can be obtained readily from a Fourier transform of the time-varying signal. The technique is most useful for finding small level splittings, so that the signal can be accurately measured using the conventional transient detection system. Both the time-varying signal and its Fourier spectrum can be directly displayed on the oscilloscope. Figure 13.2 is an example of the results from a quantum beat experiment.[4]

13.3 SATURATION SPECTROSCOPY

The basic idea of saturation spectroscopy is as follows. A monochromatic light beam resonantly excites only a small group of atoms or molecules under the inhomogeneously broadened profile and induces in them a population change. This group of atoms or molecules, marked by the population change, can then be selectively studied by either absorption or luminescence. The effect of inhomogeneous broadening is thus suppressed.

Saturation in Excitation

Consider first the induced population change due to resonant excitation in a two-level system shown in Fig. 13.3. The rate equation for the population change is

$$\left(\frac{\partial}{\partial t} + \frac{1}{T_1} \right)(\Delta\rho - \Delta\rho^0) = -2W_{12}\,\Delta\rho \qquad (13.7)$$

where $\Delta\rho = \rho_{22} - \rho_{11}$ is the population difference between the two levels, $\Delta\rho^0$ is the corresponding thermal equilibrium value, $W_{12} = 2\pi\Omega^2 g(\omega)$ is the transition rate, $\Omega = (1/\hbar)|\langle 1|\,\mathbf{er}\cdot\mathbf{E}|2\rangle|$ is the Rabi frequency for the transition, and $g(\omega)$ is the lineshape function of the transition. We assume that $g(\omega) =$

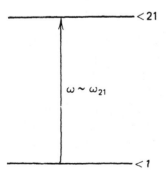

Fig. 13.3 A two-level system with a resonant excitation.

$\Gamma/\pi[(\omega - \omega_{21})^2 + \Gamma^2]$. The steady-state solution of (13.7) can be written as

$$\Delta\rho - \Delta\rho^0 = \frac{-\left(\Gamma^2 I/I_s\right)\Delta\rho^0}{\left(\omega - \omega_{21}\right)^2 + \Gamma^2(1 + I/I_s)} \tag{13.8}$$

with $I/I_s = 4\Omega^2 T_1/\Gamma$. Here $I = c|E|^2/2\pi n$ is the intensity and $I_s = c\Gamma|E|^2/8\pi n\Omega^2 T_1$ is defined as the saturation intensity. Physically, $\frac{1}{2}(\Delta\rho - \Delta\rho^0)$ of the population is resonantly pumped from the lower state into the excited state. When I/I_s approaches infinity, $\Delta\rho$ approaches 0. This is known as the pump saturation effect. The absorption of the pump beam is proportional to $W_{12}\Delta\rho$ and can be described by an absorption coefficient

$$\alpha = \frac{\alpha_0 \Gamma^2}{\left(\omega - \omega_{21}\right)^2 + \Gamma^2(1 + I/I_s)} \tag{13.9}$$

where α_0 is the peak value at $\omega = \omega_{21}$ and $I/I_s \to 0$. Equation (13.9) shows that for nonnegligible I/I_s, the absorption line is broadened by a factor of $(1 + I/I_s)^{1/2}$. This is the well-known power broadening effect.

In the weak saturation limit of $I/I_s \ll 1$, (13.8) and (13.9) reduce, respectively, to

$$\Delta\rho - \Delta\rho^0 \simeq \frac{-\left(\Gamma^2 I/I_s\right)\Delta\rho^0}{\left(\omega - \omega_{21}\right)^2 + \Gamma^2} \tag{13.10}$$

and

$$\alpha \simeq \frac{\alpha_0 \Gamma^2}{\left(\omega - \omega_{21}\right)^2 + \Gamma^2}\left[1 - \frac{\Gamma^2 I/I_s}{\left(\omega - \omega_{21}\right)^2 + \Gamma^2}\right]. \tag{13.11}$$

The change in the absorption coefficient α then is proportional to I or $|E|^2$ and can be regarded as a third-order nonlinear optical effect.

With inhomogeneous broadening, a monochromatic laser beam can resonantly excite only a small fraction of the atoms or molecules. Consider a gas medium in which the Doppler effect dominates the inhomogeneous broadening. For the group of atoms or molecules with a velocity component v_z along the laser beam propagation, the resonant frequency in the laboratory frame is $\omega_{21} - kv_z$, where ω_{21} is the resonant frequency in the rest frame. Then, following (13.8), the population excitation in this group of atoms or molecules is given by

$$\Delta\rho(v_z) - \Delta\rho^0(v_z) = \frac{\left(\Gamma^2 I/I_s\right)\Delta\rho^0(v_z)}{\left(\omega - \omega_{21} + kv_z\right)^2 + \Gamma^2(1 + I/I_s)}. \tag{13.12}$$

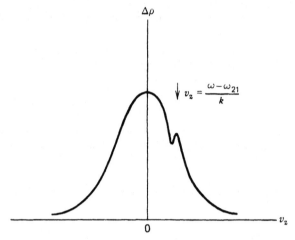

Fig. 13.4 Hole burning in the Maxwellian distribution resulting from resonant excitation.

Here $\Delta\rho^0(v_z) = (\Delta\rho^{00}/\sqrt{\pi}\,u)e^{-v_z^2/u^2}$ and $\Delta\rho^{00}$ is the thermal population difference between the two levels $|1\rangle$ and $|2\rangle$. Clearly, appreciable pumping of the population, $\Delta\rho(v_z) \neq \Delta\rho^0(v_z)$, can occur only in those atoms or molecules with $v_z \simeq (\omega - \omega_{21})/k$, as shown by the dip in the population distribution in Fig. 13.4. This is the hole-burning effect.[6] The pump excitation has modified $\Delta\rho^0(v_z)$ in such a way that it creates a hole with a halfwidth $\Gamma(1 + I/I_s)^{1/2}/k$ in the Maxwellian distribution.

Absorption of a Weak Probe in the Presence of a Strong Pump Beam

One would think that the hole-burning effect can be probed by a weak beam with a frequency ω' scanned across the hole. The absorption coefficient of the probe beam would be dominated by those atoms or molecules with $v_z \simeq (\omega' - \omega_{21})/k$ for co-propagating pump and probe beams, and therefore as $\omega' \simeq \omega$ the probe beam should feel the presence of the hole in $\Delta\rho(v_z)$. This simple description, however, neglects the coherent interaction between the pump and probe beams. As we shall see below, the coherent interaction can significantly modify the absorption spectrum, especially in the strong saturation limit when $I/I_s \gtrsim 1$.

The calculation of probe absorption can begin with the density matrix element $\rho_{21}(\omega')$, knowing that the absorption coefficient is given by

$$\alpha(\omega') = \left(\frac{4\pi\omega'}{c}\right)\operatorname{Im}\chi(\omega')$$

with   (13.13)

$$\chi(\omega') = \frac{Np_{12}\rho_{21}(\omega')}{E(\omega')}$$

and $p_{12} = \langle 1|er|2\rangle$. In the presence of the strong pumping field $E(\omega)$, we want to find $\rho_{21}(\omega')$ to all orders of $E(\omega)$, but linear in the probe field $E(\omega')$.[5] This is possible for a two-level system. Directly from the Liouville equation (2.6) for the density matrix, we can obtain the following set of couple equations (see Section 2.1):

$$\hbar\left(\omega' - \omega_{21} + i\Gamma\right)\rho_{21}(\omega')$$
$$= -p_{21}E(\omega')\left[\rho_{11}(0) - \rho_{22}(0)\right]$$
$$\quad - p_{21}E(\omega)\left[\rho_{11}(\omega' - \omega) - \rho_{22}(\omega' - \omega)\right],$$

$$\hbar\left(\omega' - \omega + i\frac{1}{T_1}\right)\left[\rho_{11}(\omega' - \omega) - \rho_{22}(\omega' - \omega)\right]$$
$$= -2p_{12}E^*(\omega)\rho_{21}(\omega') + 2p_{21}E(\omega')\rho_{12}(-\omega)$$
$$\quad + 2p_{21}E(\omega)\rho_{12}(\omega' - 2\omega), \tag{13.14}$$

$$\hbar\left(\omega' - 2\omega - \omega_{12} + i\Gamma\right)\rho_{12}(\omega' - 2\omega)$$
$$= -p_{12}E^*(\omega)\left[\rho_{22}(\omega' - \omega) - \rho_{11}(\omega' - \omega)\right],$$

$$\left[\rho_{11}(0) - \rho_{22}(0)\right] \equiv \Delta\rho = \Delta\rho^0 \Big/ \left[1 + \frac{\Gamma^2 I/I_s}{(\omega - \omega_{21})^2 + \Gamma^2}\right],$$

$$\rho_{12}(-\omega) = \frac{-\hbar^{-1}p_{12}E(-\omega)\Delta\rho}{\omega - \omega_{21} - i\Gamma}.$$

We ignore $\rho_{11}(2\omega)$ and $\rho_{22}(2\omega)$ in (13.14) because they are small in comparison. The solution of $\rho_{21}(\omega')$ from (13.14) is

$$\rho_{21}(\omega') = \left\{ -\frac{1}{\hbar}p_{21}E(\omega')\Delta\rho \right.$$

$$\times \left[\left(\omega' - \omega_{21} + i\Gamma\right) - \frac{2\Omega^2}{\left(\omega - \omega' + \dfrac{i}{T_1}\right) - \dfrac{2\Omega^2}{\omega' - 2\omega + \omega_{21} + i\Gamma}}\right]^{-1}$$

$$+ \left. \frac{\dfrac{2}{\hbar^2}p_{21}^2 E(\omega)E(\omega')\rho_{12}(-\omega)}{\left(\omega - \omega' + i\dfrac{1}{T_1}\right) - \dfrac{2\Omega^2}{\omega' - 2\omega + \omega_{21} + i\Gamma}} \right\}, \tag{13.15}$$

which can be written as

$$
\rho_{21}(\omega') = \frac{-(1/\hbar)p_{21}E(\omega')\Delta\rho}{\omega' - \omega_{21} + i\Gamma}
$$

$$
+ \frac{-(1/\hbar)p_{21}E(\omega')(\omega' - 2\omega + \omega_{21} + i\Gamma)(\omega + \omega' - 2\omega_{21})2\Omega^2\,\Delta\rho}{(\omega - \omega_{21} - i\Gamma)(\omega' - \omega_{21} + i\Gamma)D}
$$

$$
D = (\omega' - \omega_{21} + i\Gamma)\left[\left(\omega' - \omega + \frac{i}{T_1}\right)(\omega' - 2\omega + \omega_{21} + i\Gamma) - 2\Omega^2\right]
$$

$$
- 2\Omega^2(\omega' - 2\omega + \omega_{21} + i\Gamma),
$$

$$
\Omega^2 = \frac{|p_{21}E(\omega)|^2}{\hbar^2}. \tag{13.16}
$$

The first term in $\rho_{21}(\omega')$ of (13.16) arises from the hole-burning effect, while the second term comes from the coherent interference of the pump and probe beams. Physically, this interference sets up an oscillation in the populations $\rho_{11}(\omega' - \omega)$ and $\rho_{22}(\omega' - \omega)$, which in turn scatters the pump beam to yield a coherent output at ω'. The denominator of $\rho_{21}(\omega')$ now has three zeroes, corresponding to three distinct resonances when $E(\omega)$ is sufficiently strong. This result of strong interaction of light with a resonant two-level system can also be understood from the picture of dynamic (or ac) Stark splitting, or the dressed atom picture, which will be described in Chapter 22.

In the presence of Doppler broadening, there is a $\rho_{21}(\omega')$ for each velocity group of atoms or molecules. (We assume, for simplicity, that the Doppler broadening is much larger than the homogeneous broadening.) The expression of $\rho_{21}(\omega', v_z)$ can be obtained from (13.15) or (13.16) by replacing ω and ω' by $\omega - \mathbf{k}\cdot\mathbf{v}_z$ and $\omega' - \mathbf{k}'\cdot\mathbf{v}_z$, respectively. To find the absorption coefficient seen by the probe beam in this case, we must integrate $\rho_{21}(\omega', v_z)$ over the Doppler profile. In other words, $\rho_{21}(\omega')$ in (13.16) should be replaced by $\int_{-\infty}^{\infty} dv_z\,\rho_{21}(\omega', v_z)$. For counterpropagating pump and probe beams, the hole-burning term in (13.16) yields an absorption coefficient

$$
\alpha_{HB}(\omega') = \frac{4\pi N\omega'|p_{12}|^2}{\hbar c}
$$

$$
\times \int_{-\infty}^{\infty} \frac{\Gamma\Delta\rho^0(v_z)\,dv_z}{\left[(\omega' - \omega_{21} - k'v_z)^2 + \Gamma^2\right]\left[1 + \dfrac{\Gamma^2 I/I_s}{(\omega - \omega_{21} + kv_z)^2 + \Gamma^2}\right]}
$$

$$
= \alpha_0(\omega')\left\{1 - \left[1 - \frac{1}{(1 + I/I_s)^{1/2}}\right]\right.
$$

$$
\left. \times \frac{\tilde{\Gamma}^2}{\left[(\omega' - \omega) + 2(\omega - \omega_{21})\right]^2 + \tilde{\Gamma}^2} \right\} \tag{13.17}
$$

where

$$\alpha_0(\omega') = \left(\frac{4\pi\omega'}{c}\right)\left(\frac{N|p_{12}|^2}{\hbar\tilde{\Gamma}}\right)\Delta\rho^0\left(v_z = \frac{\omega' - \omega_{21}}{k'}\right),$$

$$\tilde{\Gamma} = \Gamma\left[1 + \left(1 + \frac{I}{I_s}\right)^{1/2}\right],$$

and

$$k' \cong k.$$

The above result shows that over the entire Doppler profile, $\alpha_{HB}(\omega')$ is approximately equal to $\alpha_0(\omega') \propto \Delta\rho^0[v_z = (\omega' - \omega_{21})/k']$ except around $\omega' = 2\omega_{21} - \omega$, where it exhibits a dip with a halfwidth $\tilde{\Gamma}$. Both the amplitude and the halfwidth of the dip increase as I/I_s increases. This resonant dip seen by the probe beam at $\omega' = 2\omega_{21} - \omega$ originates from the hole at $\omega' - kv_z = \omega_{21}$ created by the pump beam with frequency $\omega = \omega_{21} - kv_z$.

The weak saturation limit ($I/I_s \ll 1$) is of particular interest. Equation (13.17) reduces to

$$\alpha_{HB}(\omega') \cong \alpha_0(\omega')\left\{1 - \frac{2\Gamma^2 I/I_s}{[(\omega' - \omega) + 2(\omega - \omega_{21})]^2 + 4\Gamma^2}\right\}, \quad (13.18)$$

which shows that the absorption dip now has a halfwidth equal to the natural width 2Γ of the individual atoms. For co-propagating pump and probe beams, the formulation and results are essentially the same except that the probe beam sees a hole $\omega' = \omega$ as expected.

The term in $\rho_{21}(\omega')$ of (13.16) due to the coherent effect is more complicated. It can modify the absorption spectrum significantly. Only a qualitative discussion is presented here; Ref. 7 provides mathematical details. For counterpropagating pump and probe beams, the coherent part of $\rho_{21}(\omega', v_z)$ has three poles, but in the contour integration over v_z the integration path should be closed in the upper plane that contains only one pole. It gives a contribution superimposed on the hole caused by the saturation effect to form a broader but shallower dip in the absorption spectrum. This is seen in Fig. 13.5. It is seen that when I/I_s is large, the coherent effect can drastically change the absorption spectrum seen by the probe beam. However, if $I/I_s \leq 1$, the coherent effect is not very significant, and the saturation effect alone gives a good description of the absorption spectrum. For co-propagating pump and probe beams, the contour integral of the coherent part over v_z should be closed in the lower plane that contains two poles. The coherent effect again broadens the saturation hole. For $I/I_s \ll 1$, the hole is composed of two Lorentzian dips at the same resonant frequency $\omega' = \omega$, one with a halfwidth 2Γ and a depth $\frac{1}{2}I/I_s$ and the other with a halfwidth $1/T_1$ and a depth $\frac{1}{2}I/I_s$. [If the two levels have different lifetimes γ_1^{-1} and γ_2^{-1} in the more general case, then the

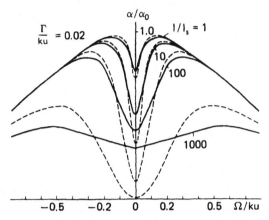

Fig. 13.5 Absorption spectra with a saturation dip seen by a weak probe beam in the presence of a strong counterpropagating pump wave of different strengths I/I_s. The solid curves include the coherent effect, but the dashed curves do not. (After Ref. 7.)

absorption dip is a superposition of three Lorentzian dips, with halfwidths 2Γ, γ_1, γ_2 and depths $\frac{1}{2}I/I_s$, $\frac{1}{2}(I/I_s)\gamma_2/(\gamma_1 + \gamma_2)$, $\frac{1}{2}(I/I_s)\gamma_1/(\gamma_1 + \gamma_2)$, respectively.] By analyzing the shape of the overall dip it is possible to deduce Γ and T_1 (or Γ, γ_1, and γ_2).

As mentioned earlier, the coherent effect is caused by coherent scattering of the pump wave at ω by the population modulation at $(\omega - \omega')$ resulting from the beating of the pump and probe waves at ω and ω'. The coherently scattered output at ω' can constructively or destructively interfere with the incoming probe wave causing a decrease or increase in the absorption of the probe beam. Then it is obvious that if pulsed lasers are used, the coherent effect can be eliminated by delaying the probe pulse from the pump pulse.

The preceding discussion shows that one can deduce the homogeneous linewidth of a transition from the observed saturation dip, but the main purpose of saturation spectroscopy is to resolve the closely spaced lines normally hidden under the inhomogeneously broadened profile. This can not be achieved with co-propagating pump and probe beams since the saturation dip always appears at $\omega' = \omega$. With counterpropagating pump and probe beams, however, the dip appears at $\omega' = 2\omega_{21} - \omega$; different transitions with different resonant frequencies ω_{21} then show up as different dips and can be well resolved as long as the frequency separation is larger than the dip width.

Absorption of the Probe Beam in the Presence of a Counterpropagating Pump Beam of the Same Frequency

The spectroscopic method described above requires two tunable lasers of high monochromaticity, which is seldom affordable. The saturation spectroscopy can, however, be carried out with counterpropagating pump and probe beams of the same frequency. This can be seen from the fact that the counterpropagating pump and probe waves interact with the same velocity group of atoms under the Doppler profile when $\omega - kv_z = \omega + kv_z = \omega_{21}$ or $v_z = 0$, that is, when ω is tuned to the center of the Doppler-broadened transition line.

In the case of a strong pump and a weak probe, the calculation is similar to that in the previous section. In (13.14), for the sake of simplicity, the wavevector dependence in the density matrix components is not explicitly shown. Actually, for counterpropagating waves we have $\rho_{11}(\omega' - \omega, k' + k)$. Then even when $\omega = \omega'$ the coherent effect is still present because the pump and probe waves can interfere and yield a spatial modulation in the population difference, $\Delta\rho(k' + k)$. However, as pointed out earlier, the coherent effect is relatively unimportant when $I/I_s \ll 1$. We therefore assume $I/I_s \ll 1$ and neglect the coherent effect in the following discussion. The absorption coefficient of the weak probe beam is then given by (13.18) with $\omega = \omega'$.

A typical experimental arrangement is seen in Fig. 13.6.[8] The error arising from the beams being not exactly antiparallel is not significant for most applications. As an example, Fig. 13.7 compares the saturation spectrum of the Balmer α-line of atomic deuterium with the emission line profile of a cooled deuterium gas discharge.[8] The Lamb shift is resolved in the saturation spectrum. Note that the spectrum in Fig. 13.7 was obtained by subtraction of the saturation spectrum from the original Doppler-broadened spectrum. Each resonant peak here corresponds to an absorption dip in the saturation spectrum.

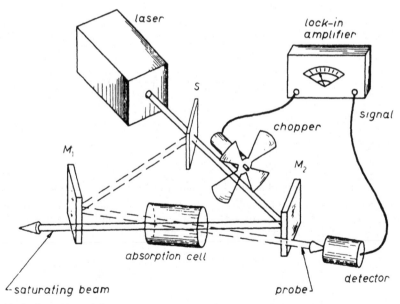

Fig. 13.6 Experimental arrangement of saturation spectroscopy with two counterpropagating waves of the same frequency. (After Ref. 8.)

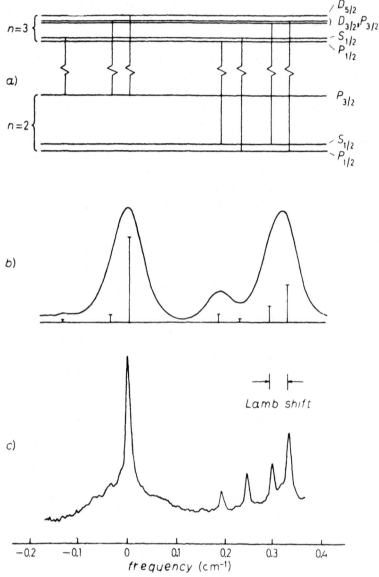

Fig. 13.7 Balmer α spectrum of atomic deuterium. (*a*) Fine structure of the *n* = 2 and *n* = 3 levels. (*b*) Emission line profile of a cooled deuterium gas discharge and theoretical fine-structure lines with relative transition probabilities (*T* = 50 K). (*c*) Observed saturation spectrum with the optical Lamb shift. (After Ref. 8.)

More generally, the pump and probe beams can be of comparable intensities. We consider the special case of two beams of equal intensities and neglect the coherent effect. Under the resonant excitation, the population difference between the two levels of an atom with a velocity component v_z is obtained from the usual saturation formula

$$\Delta\rho(v_z) = \Delta\rho^0(v_z)\left[\frac{1}{1 + \dfrac{\Gamma^2 I/I_s}{(\omega - \omega_{21} + kv_z)^2 + \Gamma^2} + \dfrac{\Gamma^2 I/I_s}{(\omega - \omega_{21} - kv_z)^2 + \Gamma^2}}\right]$$

$$(13.19)$$

where I is the intensity of each beam. The absorption coefficient for each beam is given by

$$\alpha(\omega) = \alpha_0 \int dv_z \left\{\frac{\Delta\rho(v_z)\Gamma}{\pi\left[(\omega - \omega_{21} + kv_z)^2 + \Gamma^2\right]}\right\} \qquad (13.20)$$

where $\alpha_0 = (4\pi\omega'/c)(N|p_{12}|^2/\hbar\Gamma)$. In the weak saturation limit, $I/I_s \ll 1$, we find

$$\alpha(\omega) = \alpha_0(\omega)\left[1 - \frac{1}{2}\left(\frac{I}{I_s}\right)\left(1 + \frac{\Gamma^2}{(\omega - \omega_{21})^2 + \Gamma^2}\right)\right] \qquad (13.21)$$

with

$$\alpha_0(\omega) = \alpha_0\Delta\rho^0\left(v_z = \frac{\omega - \omega_{21}}{k}\right).$$

The result shows that when $|\omega - \omega_{21}| \gg \Gamma$, the absorption coefficient is $\alpha(\omega) = \alpha_0(\omega)(1 - 1/2I_s)$, but when $|\omega - \omega_{21}| \sim \Gamma$, additional absorption appears in the form of a dip, with a depth $(\frac{1}{2}I/I_s)\alpha_0(\omega)$ and a halfwidth equal to the homogeneous halfwidth Γ. This is sketched in Fig. 13.8a.

In practice, the two counterpropagating waves of equal intensities can be provided by the field in a laser cavity. The medium in this case can be the laser medium with a negative $\Delta\rho$ corresponding to an inverted population. A reduction in $|\Delta\rho|$ decreases the gain and hence the laser output. Thus the hole-burning effect appears here as a dip in the gain spectrum or a dip in the laser output spectrum. This was first proposed by Lamb, and is known as the Lamb dip.[9] It was the observation of the Lamb dip that opened the field of high-resolution laser spectroscopy.

One can, of course, also insert an absorbing medium into a laser cavity. In this case, the absorption generally reduces the laser gain and decreases the laser

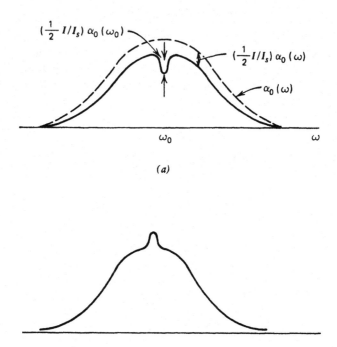

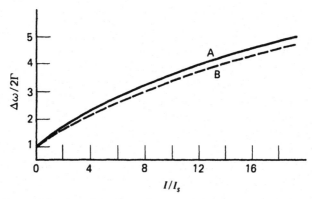

Fig. 13.8 (*a*) Lamb dip and (*b*) inverted Lamb dip in saturation spectroscopy with two counterpropagating waves of the same frequency and intensity.

Fig. 13.9 Dip width as a function of pumping strength I/I_s. Curve A is exact calculation; B is an approximation ignoring the coherent effect. (After Ref. 7.)

output. Then a dip (decrease) in the absorption arising from hole burning should lead to an inverted dip in the laser gain spectrum, as in Fig. 13.8b. This is known as an inverted Lamb dip. The Lamb-dip experiment requires the coincidence of the laser frequency with the transition frequency. It is therefore more limited than the general absorption spectroscopic technique described earlier.

We have neglected the coherent effect here. To see that it is indeed negligible for $I/I_s \ll 1$, Fig. 13.9 shows the dependence of the dip width as a function of I/I_s with and without the coherent contribution.[7] The difference clearly is appreciable only when $I/I_s \gg 1$.

Multilevel Saturation Spectroscopy

The preceding discussion can be extended to a three-level system with two transitions sharing a common level (Fig. 13.10).[10] A strong monochromatic beam with frequency $\omega \simeq |\omega_{10}|$ is used to induce a population change in levels 0 and 1 for a selected group of atoms or molecules. A weak beam at $\omega' \simeq |\omega_{20}|$ is used to probe the induced population change. Since only a selected group of atoms or molecules is seen by the probe beam, the inhomogeneous broadening of the $\langle 0| \rightarrow \langle 2|$ transition is greatly reduced. For this case, the calculation is fairly straightforward. The induced population change has already been derived, so that the absorption or gain spectrum obtained by the probe beam can be calculated readily. The coherent effect is negligible when ω and ω' are very different. When $\omega \sim \omega'$, it is still negligible for counterpropagating pump and probe waves in the weak saturation limit, $I/I_s \ll 1$. The technique is most useful for resolving transitions between two sets of closely spaced multilevels.

If the difference of the two resonant frequencies $|\omega_{10}|$ and $|\omega_{20}|$ is smaller than or comparable to the Doppler width in a gas medium, saturation spectroscopy can also be carried out with two counterpropagating waves of the same frequency ($\omega = \omega'$).[10] They interact with the same velocity group of atoms when ω and v_z satisfy the relations $\omega - kv_z = |\omega_{10}|$ and $\omega + kv_z = |\omega_{20}|$. The saturation dip appears at $\omega = \frac{1}{2}(|\omega_{10}| + |\omega_{20}|)$. The experiment can be performed in either an absorption cavity or a laser cavity as in the Lamb-dip

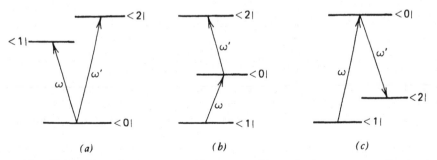

Fig. 13.10 Three-level systems with two transitions sharing a common level.

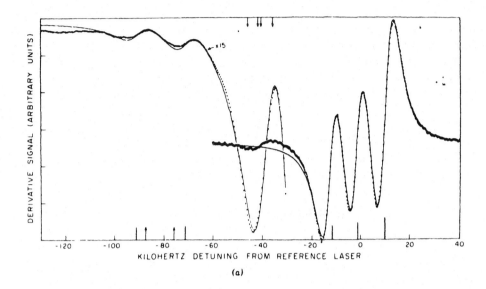

(a)

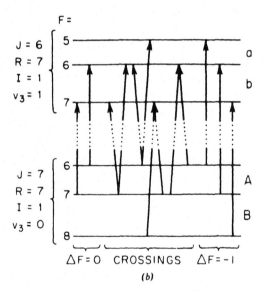

(b)

Fig. 13.11 (a) Magnetic hyperfine structure of the $F_2^{(2)}$ component of the $P(7)$ line in the ν_3 band of CH_4, obtained by the saturation spectroscopic method on methane in a frequency-stabilized He-Ne laser cavity operating at 3.39 μm. (b) Transitions responsible for the structure in (a); the four pairs of transitions labeled "crossings" give rise to the three-level saturation dip structure indicated by the downward arrows in (a). (After Ref. 11.)

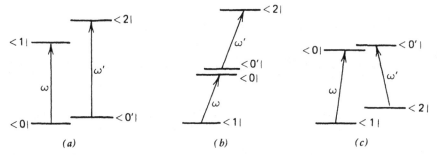

Fig. 13.12 Four-level systems with two transitions involving two coupled levels.

case discussed previously. An example is shown in Fig. 13.11.[11] The three strong structures on the right and two weak ones on the left all result from two-level saturation dips. They correspond to the three $\Delta F = -1$ and two $\Delta F = 0$ single transitions, respectively. Only the middle structure in the spectrum originates from three-level saturation. It results from superposition of the four saturation dips arising from the double transitions labeled "crossings" in the energy level diagram of Fig. 13.11b.

The technique can also be extended to two transitions with no common level, but coupled through atomic or molecular collisions.[12] As seen in Fig. 13.12, collisions may tend to equalize the populations in nearby energy levels for atoms or molecules of the same velocity group. If the saturation pumping induces a population change in $|0\rangle$, there will be a corresponding population change in $|0'\rangle$. A scan of ω' about $\omega_{20'}$ should exhibit a saturation dip at $\omega' = \omega_{20'} \pm k'v_z = \omega_{20'} \pm k'(\omega_{10} - \omega)/k$, where $+$ and $-$ refer to counter- and co-propagating pump and probe beams, respectively.

In three- or four-level saturation spectroscopy, if spontaneous emission exists between $|2\rangle$ and $|0\rangle$, then the emitted light can be used to play the role of the probe beam.[13] The pump beam at ω induces a population change in $\langle 0|$ (and $\langle 0'|$) for a selected velocity group of atoms, and thus modifies the emission between $\langle 0|$ (or $\langle 0'|$) and $\langle 2|$ for that group of atoms. Clearly, the induced emission spectrum is Doppler-free (to the first order). Because the detection of emission often is far more sensitive than the measurement of absorption or amplification, this technique can be used to study gases of very low pressure. The same technique can be extended to condensed matter. Laser-induced fluorescence line-narrowing has recently become a very popular technique for high-resolution spectroscopic study of ions in solids.[14]

13.4 TWO-PHOTON DOPPLER-FREE ABSORPTION SPECTROSCOPY

For gas media, two-photon absorption of two counterpropagating waves of the same frequency can also yield a first-order Doppler-free spectrum.[15] The two-photon transition probability for an atom moving with a velocity $\mathbf{v}$ is, following (12.1) with ω_i replaced by $\omega_i + \mathbf{k}_i \cdot \mathbf{v}$,

$$\frac{dW_{fi}}{d(\hbar\omega_1)} = \frac{dW_{fi}}{d(\hbar\omega_2)} = \frac{8\pi^3 N}{(\varepsilon_1\varepsilon_2)^{1/2}c^2}|M_{fi}|^2 g(\hbar\,\Delta\omega)(\rho_i - \rho_f)I_1 I_2$$

$$|M_{fi}|^2 = \sum_s \left[\frac{\langle f|e\mathbf{r}\cdot\hat{e}_2|s\rangle\langle s|e\mathbf{r}\cdot\hat{e}_1|i\rangle}{\hbar(\omega_1 + \mathbf{k}_1\cdot\mathbf{v} - \omega_{si})} + \frac{\langle f|e\mathbf{r}\cdot\hat{e}_1|s\rangle\langle s|e\mathbf{r}\cdot\hat{e}_2|i\rangle}{\hbar(\omega_2 + \mathbf{k}_2\cdot\mathbf{v} - \omega_{si})} \right]. \tag{13.22}$$

If no intermediate resonance is involved, the $\mathbf{v}$ dependence in $|M_{fi}|^2$ can be neglected. Then the only dependence of W_{fi} on $\mathbf{v}$ is through the lineshape function

$$g(\hbar\,\Delta\omega) = g\left[\hbar(\omega_1 + \mathbf{k}_1\cdot\mathbf{v} + \omega_2 + \mathbf{k}_2\cdot\mathbf{v} - \omega_{fi})\right]. \tag{13.23}$$

It is readily seen that if $\omega_1 = \omega_2$ and $\mathbf{k}_1 = -\mathbf{k}_2$, then the first-order Doppler effect of $g(\hbar\,\Delta\omega)$ is completely eliminated, with $g(\hbar\,\Delta\omega)$ reduced to $g[\hbar(2\omega - \omega_{fi})]$. In the case of a Lorentzian lineshape, we have

$$g(\hbar\,\Delta\omega) = \frac{\Gamma/\pi}{(2\omega - \omega_{fi})^2 + \Gamma^2}. \tag{13.24}$$

With $g(\hbar\,\Delta\omega)$ independent of $\mathbf{v}$ to the first order of $\mathbf{v}$, the two-photon absorption spectrum of a gas medium is then the same as that of stationary atoms or molecules. Here, unlike saturation spectroscopy, all molecules contribute equally to two-photon absorption. The absorption coefficient is there-

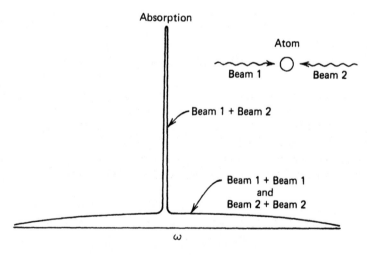

Fig. 13.13 A two-photon absorption line obtained by counterpropagating waves in a gas medium. The sharp peak is the Doppler-free line and the broad background is the Doppler-broadened spectrum.

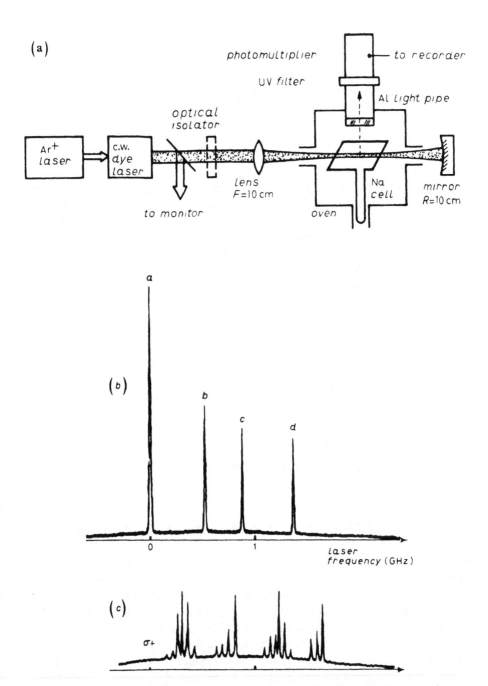

Fig. 13.14 (*a*) A typical experimental arrangement for Doppler-free two-photon spectroscopy with a CW dye laser. (*b*) The 3*s*-4*d* transitions of Na observed by two-photon Doppler-free spectroscopy with circularly polarized light. (*c*) Zeeman splitting of the 3*s*-4*d* transitions at H = 170 G observed by two-photon Doppler-free spectroscopy. (After Ref. 17.)

fore proportional to the density of molecules, but the observed spectral width corresponds to the homogeneous linewidth.

However, with counterpropagating beams, the two participating photons in the two-photon absorption process may come in the manner of either one from each beam or both from the same beam. While the former gives a Doppler-free line, the latter still leads to a Doppler-broadened peak. The spectrum should appear as a sharp line sitting on a broad background, as in Fig. 13.13. If the intensities of the two counterpropagating beams are the same, the integrated strength of the background should be equal to that of the Doppler-free line. Fortunately, the Doppler width is (ku/Γ) times larger than the homogeneous width. In the visible range, (ku/Γ) can be 10^3 to 10^5. Therefore, the background appears to be fairly weak and does not hurt the quality of the Doppler-free spectrum. In some cases the background can also be completely eliminated by using appropriate pump polarizations.

Two-photon Doppler-free spectroscopy is attractive because of its simplicity.[16] An example is shown in Fig. 13.14, where the hyperfine structure and the Zeeman splittings of the $3S$–$4D$ transition in sodium are clearly resolved.[17] The technique allows the use of a single laser to probe a transition at twice the laser frequency. The sensitivity of the technique is good, since all the atoms or molecules in the beam participate in the absorption. It can be further improved with the fluorescence or ionization detection schemes mentioned in Section 12.2. In general, however, two-photon absorption spectroscopy is still less sensitive than saturation spectroscopy because of the lack of an intermediate resonance. Pulsed lasers with high peak power must then be used, and the resolution often is limited by the laser linewidth. If a close intermediate resonance exists, then a CW laser can be intense enough to be used to obtain a high-resolution two-photon absorption spectrum. Figure 13.14 is a good example.

13.5 HIGH-RESOLUTION POLARIZATION SPECTROSCOPY

A polarized pump beam, resonantly exciting a transition in a medium, is expected to induce a dichroism and a corresponding birefringence in the medium through the induced population change. Both the dichroism and the birefringence should exhibit a resonant behavior at transitions involving levels with the induced population change, and can be measured by the polarization variation of a probe beam through the medium. This is the underlying principle of polarization spectroscopy.[18] We consider it here as a modification of the saturation spectroscopy discussed in Section 13.3. Only a qualitative discussion is attempted; quantitative details can be found in Ref. 19.

Consider a circularly polarized monochromatic laser beam which resonantly excites a transition of a selected velocity group of atoms to saturation.[18] This group of atoms will then preferentially absorb the oppositely polarized component of a probe beam probing the saturation. In addition, there is an induced circular birefringence since, as a result of the selective excitation, the refractive indices for the two circularly polarized components are no longer the same. Consequently, in traversing the medium, a linearly polarized probe beam, with frequency in the region of the saturation dip, becomes elliptically polarized along with a rotation of the major polarization axis. If the probe frequency is away from the saturation dip, then the polarization of the probe remains essentially unchanged. The transmitted probe beam can now be analyzed by a crossed analyzer. Only at the saturation dip in the Doppler-broadened peak will the probe light leak through the analyzer. Unlike the absorption saturation spectroscopy, the polarization spectroscopy here yields a Doppler-free spectrum with no inhomogeneous background. This is certainly an advantage since the sensitivity can be greatly improved without the strong background. It allows the use of lower laser power and lower gas pressure. In general, an elliptically polarized pump beam can also be used as long as the analyzer is set correspondingly.[18]

The technique applies to both two- and three-level systems. In the latter case, there should be a common level for the pump and probe transitions as described in Fig. 13.10. With a system of many closely spaced transitions, as in the case of a molecule, the polarization spectroscopy also has the advantage of being able to greatly simplify the spectrum.[20, 21] The polarized pump beam modifies the population of the common level with a particular molecular orientation. The probe beam monitoring transitions from this common level to other levels should experience a dichroism and a birefringence. Then, with a crossed analyzer, these transitions can be easily distinguished from the others. The resulting spectrum seen by the probe beam through the analyzer is usually far simpler and better characterized than the ordinary absorption spectrum. In this technique, the common level is labeled through saturation pumping by the polarized pump beam. It is therefore known as polarization labeling spectroscopy.[20, 21]

Figure 13.15 shows the polarization labeled spectra of Na_2 as an example.[20] The pump laser beam was circularly polarized and tuned to a $X'\Sigma_g^+ \rightarrow B'\Pi_u$ transition of Na_2 near 4825 Å. The first row of spectrum in Fig. 13.15 was obtained with the pump frequency adjusted to the $(v = 0, J = 49) \rightarrow (4, 50)$ transition. Transitions from $(0, 49)$ to the various (v', J') states were monitored by the polarization change of the probe light induced by the population change in $(0, 49)$. For each $v \rightarrow v'$ transition, three spectral lines are expected according to the selection rules $\Delta J = \pm 1, 0$. The signal intensity for $\Delta J = \pm 1$ is, to the first order, proportional to α_0^2, and for $\Delta J = 0$, proportional to $1/J^2$, where α_0 is the absorption coefficient in the absence of pumping. Therefore, for large J, the $\Delta J = 0$ transitions are much weaker and may not show up in the spectrum. We can then identify the successive doublets in the first row of the spectrum in Fig. 13.15 as $(0, 49) \rightarrow [(0, 48), (0, 50)]$, $[(1, 48), (1, 50)]$, $[(2, 48),$ $(2, 50)], \ldots, [(6, 48), (6, 50)]$, respectively. Similarly, the successive rows of spectra in Fig. 13.15 were obtained with pumping transitions $(1, 25) \rightarrow (5, 24)$, $(0, 42) \rightarrow (4, 41)$, $(1, 29) \rightarrow (5, 29)$ and $(1, 33) \rightarrow (5, 34)$, respectively. The $v = 1$

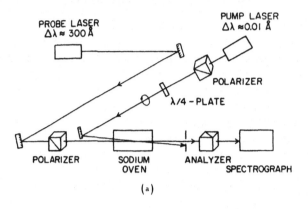

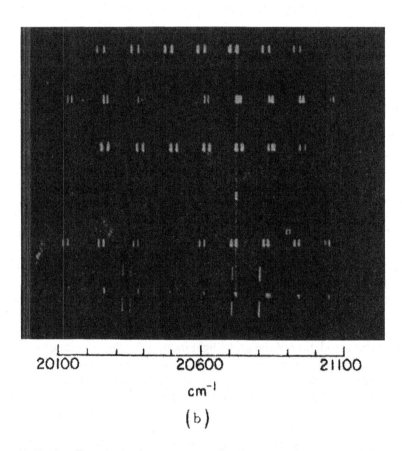

Fig. 13.15 (a) Experimental arrangement of polarization labeling spectroscopy. (b) Polarization-labeled spectra of $Na_2 \, B^1\Pi_u$ band. For the top five spectra, the pump laser was circularly polarized. For the bottom spectrum, the pump was linearly polarized. (After Ref. 20.)

$\rightarrow v' = 3$ transitions in the second and fourth rows are missing because of the small Franck-Condon factors. The upper-state vibrational quantum numbers in the observed transitions can be assigned here easily because each row of spectrum should end up at the low-frequency side with a $v' = 0$ doublet. The spectrum in Fig. 13.15 is much simpler than the ordinary absorption spectrum, where the transitions from many rotational states of the lower $v = 0$ and $v = 1$ vibrational states will create a nearly intangible forest of lines in the spectrum.

The polarization labeling technique can also be extended to the effective four-level systems described in Fig. 13.12. The polarized pumping partially orients the molecules in the $|0\rangle$ level. Through molecular collision, the orientation is transferred from $|0\rangle$ to $|0'\rangle$. The probe beam probing the transition from $|0'\rangle$ to $|2\rangle$ should again experience an induced dichroism and birefringence.[20]

Polarization spectroscopy also can be used to study two-photon transitions. Instead of absorption, the polarization variation of the incoming beams is measured. We consider this in more detail in Chapter 15 under the rubric of Raman-induced Kerr effect, recognizing that Raman transitions are just a special case of two-photon transitions.

13.6   OPTICAL RAMSEY FRINGES

In radio atomic spectroscopy, an ingenious high-resolution technique was invented in the late 1940s by Ramsey using an atomic beam.[22] As seen in Fig. 13.16, an atomic beam interacts with the applied radio fields in two regions. Atoms passing through the first region are coherently excited. If the coherence persists when the excited atoms reach the second region, they may absorb or emit, depending on whether the atomic coherence is in phase or out of phase with the exciting radio field. The absorption in the second region therefore appears as interference fringes as the radio frequency scans through the atomic resonance. These are known as Ramsey fringes. The fringe pattern depends on the coherent lifetime (dephasing time) of the atomic excitation from which the high-resolution spectrum can be deduced.

The technique can be extended to optical spectroscopy.[23] Since the dephasing time of an optical transition is usually very short, a modification of the technique is necessary. Instead of the em fields exciting an atomic beam at two separate spatial points, one can use two laser pulses separated in time to excite the same group of atoms in a vapor cell.[24,25] The effect is the same. The

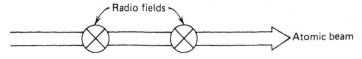

Fig. 13.16 Schematic describing the interaction of an atomic beam with two radio frequency fields leading to the observation of Ramsey fringes.

absorption spectrum seen by the second pulse or the emission spectrum after the second pulse exhibits the Ramsey fringes.

It is interesting to note that the spectroscopic resolution of the Ramsey-fringe technique is not limited by the pulsed laser linewidth, which, by the uncertainty principle, is given by the inverse of the pulsewidth. This can be seen as follows. For two identical coherent pulses separated by time T, the field can be written as

$$E(t) = \int d\omega \left[\mathscr{E}(\omega)\sin \omega t + \mathscr{E}(\omega)\sin(\omega t + \omega T) \right]$$
$$= \int d\omega \left[2\mathscr{E}(\omega)\cos\frac{\omega T}{2}\sin(\omega t + \omega T) \right] \qquad (13.25)$$

with $\omega = \omega_0 + \Delta\omega$. As seen in Fig. 13.17, it has a sawtooth spectrum whose envelope is the spectrum of a single pulse. Then, in tuning the laser over a resonance, it is the sawtooth spectral profile of the laser scanning over the resonance. Therefore, the spectral resolution is now limited by the width of the sawtooth instead of the width of the envelope. In scanning ω_0 over a δ-function resonance, the linear transmission spectrum should reproduce the sawtooth spectrum of the laser excitation. If the resonance has a finite linewidth, then the sawteeth in the transmission spectrum are broadened with a decreased contrast of peaks to background.

The Ramsey fringes depend on the phase correlation of the two pulse excitations. If the pulses are not phase-correlated, no fringes can be seen. Also, if the coherent excitation imposed on the atoms by the first pulse dephases before the second pulse arrives, the fringe pattern will disappear. The dephas-

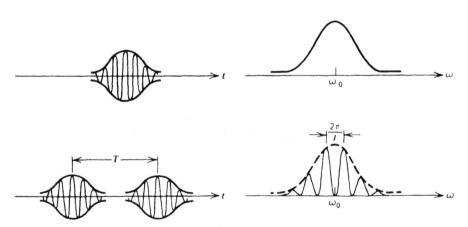

Fig. 13.17 Single and double coherent laser pulses with their corresponding frequency spectra.

ing effect on the observed spectrum is in broadening the sawteeth and decreasing the contrast of the fringes.

In an actual experiment, atomic motion also may cause dephasing. This, however, can be eliminated if two-photon excitation is used.[24] As discussed in Section 13.4, the Doppler-free two-photon absorption is independent of atomic velocity and has all atoms participate in the absorption. In the following discussion, we use this case as an example to give a simple mathematical derivation of the Ramsey fringes.[26]

Consider a two-level system excited via two-photon transitions by two square pulses of duration τ; one is switched on at $t = 0$ and the other at $t = T$. At $t \leq 0$ the system is in state $|1\rangle$, and at $t > 0$ the system is in a coherent state

$$|\psi\rangle = a(t)|1\rangle + b(t)|2\rangle. \tag{13.26}$$

Assume that the perturbation theory is valid. Then $a(t) \cong 1$, and $b(t)$ can be obtained from the Schrödinger equation or

$$i\frac{\partial}{\partial t}b(t) \cong A e^{-i(2\omega - \omega_{21})t} \tag{13.27}$$

where A is a coupling coefficient proportional to the square root of the two-photon transition probability. We find, for $t \geq \tau$,

$$b(t) = \frac{A}{2\omega - \omega_{21}}\left[e^{-i(2\omega - \omega_{21})\tau} - 1\right] \tag{13.28}$$

and for $t \geq T + \tau$,

$$b(\tau) = \frac{A}{2\omega - \omega_{21}}\left[e^{-i(2\omega - \omega_{21})(T+\tau)} - e^{-i(2\omega - \omega_{21})T}\right] + b(t)$$

$$= \frac{A}{2\omega - \omega_{21}}\left[e^{-i(2\omega - \omega_{21})\tau} - 1\right]\left[e^{-i(2\omega - \omega_{21})T} + 1\right]. \tag{13.29}$$

If fluorescence from $\langle 2|$ is detected, the signal should be proportional to $|b(t)|^2$, which, after the second pulse, is given by

$$|b(T + \tau)|^2 = 4A^2\tau^2\left[\frac{\sin\frac{1}{2}(2\omega - \omega_{21})\tau}{\frac{1}{2}(2\omega - \omega_{21})\tau}\right]^2 \cos^2\left(\frac{2\omega - \omega_{21}}{2}T\right). \tag{13.30}$$

The spectrum of $|b(T + \tau)|^2$ versus $(2\omega - \omega_{21})$ is seen to be a fringe pattern shown in Fig. 13.18. The periodic spacing of the fringes is $1/2T$. In the presence of a dephasing rate Γ with $\tau \ll \Gamma^{-1}$, (13.29) should be modified by

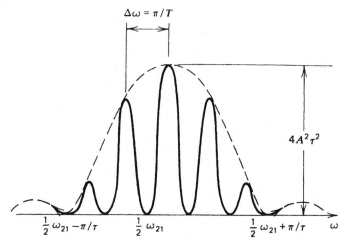

Fig. 13.18 Ramsey fringes resulting from two-photon excitation of a transition by two square pulses of duration τ, one switched on at $t = 0$ and the other at $t = T$.

replacing $b(T)$ by $b(T)e^{-\Gamma T}$. We then find

$$|b(T + \tau)|^2 = \left(\frac{A}{2\omega - \omega_{21}}\right)^2 \left|[e^{-i(2\omega - \omega_{21})\tau} - 1][e^{-i(2\omega - \omega_{21})T} + e^{-\Gamma T}]\right|^2.$$

(13.31)

The fringe contrast is given by

$$C = \frac{|b(T + \tau)|^2_{\max} - |b(T + \tau)|^2_{\min}}{|b(T + \tau)|^2_{\max} + |b(T + \tau)|^2_{\min}}$$

$$= \frac{2e^{-\Gamma T}}{1 + e^{-2\Gamma T}},$$

(13.32)

which is 1 for $\Gamma = 0$ and decreases rapidly as Γ increases.

Figure 13.19 is an example of the optical Ramsey fringes obtained by two-pulse two-photon excitation in sodium together with the experimental arrangement.[26] On each spectral line, the interference fringes are clearly visible. The observed periodic spacing of the fringes is indeed given by $1/2T$ and decreases with increase of T.

An obvious improvement of the Ramsey fringes can be achieved by using a series of many equally spaced pulses.[27] It can be obtained from either a mode-locked laser or a pulse through a resonant cavity with partially transmitting mirrors. If the pulses are phase-correlated, the frequency spectrum will appear as a series of equally spaced spikes, as in Fig. 13.20. The spectral resolution is now limited by the width of each spike. With N pulses coherently exciting the atoms within the dephasing time, the width of the spikes is

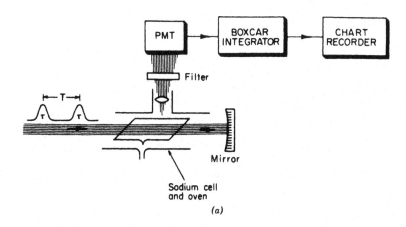

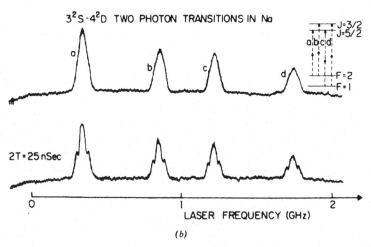

Fig. 13.19 (*a*) Arrangement of the Ramsey fringe experiment in Na. (*b*) Observed spectra of the $3^2S - 4^2D$ two-photon transitions in Na; the lower trace, showing the Ramsey fringes on each peak, was obtained with a delay of 25 nsec between the two excitation pulses. (After Ref. 26.)

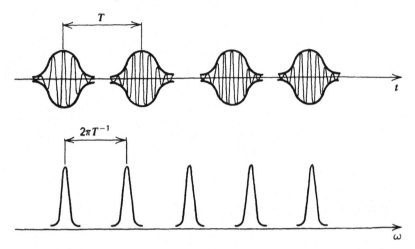

Fig. 13.20 A series of equally spaced laser pulses and the corresponding frequency spectrum.

inversely proportional to N, while the signal intensity is proportional to N^2. This technique has been demonstrated by Teets et al.[27]

13.7 OTHER HIGH-RESOLUTION SPECTROSCOPIC TECHNIQUES

There are a number of other high-resolution spectroscopic techniques. Most of them are variations of the techniques we have already discussed, including multiphoton Doppler-free absorption and multiphoton saturation spectroscopy. Coherent transient spectroscopy and four-wave mixing spectroscopy, however, are unique and deserve a more detailed discussion. The former is discussed in Chapter 21 on coherent transient effects, and the latter in Chapter 15.

REFERENCES

1 J. L. Hall, in M. S. Feld, A. Javan, and N. Kurnit, eds., *Fundamental and Applied Laser Physics* (Wiley, New York, 1973), p. 463; A. Brillet and J. L. Hall, *Phys. Rev. Lett.* **42**, 549 (1979).

2 B. Bleaney and K. W. H. Stevens, *Rep. Prog. Phys.* **16**, 108 (1953).

3 E. B. Aleksandrov, *Opt. Spectrosc.* **17**, 957 (1964); J. N. Dodd, R. D. Kaul, and D. M. Warrington, *Proc. Phys. Soc.* (London) **84**, 176 (1964).

4 S. Haroche, J. A. Paisner, and A. L. Schawlow, *Phys. Rev. Lett.* **30**, 948 (1973).

5 See, for example, N. Bloembergen and Y. R. Shen, *Phys. Rev.* **133**, A37 (1964).

6 W. R. Bennett, *Phys. Rev.* **126**, 580 (1962).

7 V. S. Letokhov and V. P. Chebotayev, *Nonlinear Laser Spectroscopy* (Springer-Verlag, Berlin, 1977), Chap. II.

8   T. W. Hänsch, in N. Bloembergen ed., *Nonlinear Spectroscopy* (North-Holland Publishing Co., Amsterdam, 1977), p. 17.

9   W. E. Lamb, *Phys. Rev.* **134A**, 1429 (1964); R. A. McFarlane, W. R. Bennett, and W. E. Lamb, *Appl. Phys. Lett.* **2**, 189 (1963); A Szöke and A. Javan, *Phys. Rev. Lett.* **10**, 521 (1963).

10 H. R. Schlossberg and A. Javan, *Phys. Rev.* **150**, 267 (1966); erratum: *Phys. Rev.* **A5**, 1974 (1972).

11 J L. Hall and C. Bordé, *Phys. Rev. Lett.* **30**, 1101 (1973).

12 R. G. Brewer, R. L. Shoemaker, and S. Stenholm, *Phys. Rev. Lett.* **33**, 63 (1974)

13 M. S. Feld and A. Javan, *Phys. Rev.* **177**, 540 (1969).

14 A. Szabo, *Phys. Rev. Lett.* **25**, 924 (1970).

15 L. S. Vasilenko, V. P. Chebotagev, and A. V. Shishaev, *JETP Lett.* **12**, 113 (1970).

16 F. Biraben, B Cagnac, and G. Grynberg, *Phys. Rev. Lett.* **32**, 643 (1974); M. D. Levenson and N. Bloembergen, *Phys. Rev. Lett.* **32**, 645 (1974); T W. Hansch, K C. Harvey, G Meisel, and A. L. Schawlow, *Opt. Comm* **11**, 50 (1974).

17 F. Biraben, B. Cagnac, and G. Grynberg, *Phys. Lett.* **48A**, 469 (1974).

18 C. Weiman and T. Hänsch, *Phys. Rev. Lett.* **36**, 1170 (1976).

19 C. Weiman, Ph. D. dissertation Stanford University (1976).

20 R. Teets, R. Feinberg, T. W. Hansch, and A. L. Schawlow, *Phys. Rev. Lett.* **37**, 683 (1976).

21 M. E. Kaminsky, R. T. Hawkins, F. V. Kowalski, and A. L. Schawlow, *Phys. Rev. Lett.* **36**, 671 (1976).

22 N. F. Ramsey, *Phys. Rev.* **76**, 996 (1949).

23 J. C. Gergguist, S. A. Lee, and J. L. Hall, *Phys. Rev. Lett.* **38**, 159 (1977).

24 Ye. V. Baklanov, V. P. Chebotayev, and B. Ta. Dubetsky, *Appl. Phys* **11**, 201 (1976).

25 M. Salour and C. Cohen-Tannoudji, *Phys. Rev. Lett.* **38**, 757 (1977).

26 M. Salour, *Rev. Mod. Phys.* **50**, 667 (1978).

27 R. Teets, J. Eckstein, and T. W. Hansch, *Phys. Rev. Lett.* **38**, 760 (1977).

BIBLIOGRAPHY

Bloembergen, N., ed., *Nonlinear Spectroscopy* (North-Holland Publishing Co., Amsterdam, 1977).

Demtröder, W., *Laser Spectroscopy* (Springer-Verlag, Berlin, 1981).

Hänsch, T. W., *Physics Today* **30**, No. 5, 34 (1977).

Letokhov, V. S.; and V. P. Chabotayev, *Nonlinear Laser Spectroscopy* (Springer-Verlag, Berlin, 1977).

Proceedings of International Laser Spectroscopy Conferences: I. R. G. Brewer and A Mooradian, eds. (Plenum Press, New York, 1973); II. S. Haroche, J C. Pebay-Peyroula, T. W. Hänsch, and S E Harris, eds. (Springer-Verlag, Berlin, 1975); III J. L. Hall and J L. Carlsten, eds (Springer-Verlag, Berlin, 1977); IV. H. Walther and K. W. Rothe, eds. (Springer-Verlag, Berlin, 1979); V. A. W. McKellar, T. Oka, and B P Stoicheff, eds (Springer-Verlag, Berlin, 1981); VI. H. Weber, ed. (Springer-Verlag, Berlin, 1983).

Schawlow, A. L., *Rev. Mod. Phys.* **54**, 697 (1982).

Shimoda, K., ed., *High-Resolution Laser Spectroscopy* (Springer-Verlag, Berlin, 1976).

Shimoda, K., and T. Shimizu, *Prog. Quant. Electron.* **2**, 47 (1972).

Walther, H., ed., *Laser Spectroscopy of Atoms and Molecules* (Springer-Verlag, Berlin, 1976).

14

Four-Wave Mixing

Four-wave mixing refers to the nonlinear process with four interacting electromagnetic waves. In the weak interaction limit, it is a third-order process and is governed by the third-order nonlinear susceptibility. Unlike second-order processes, a third-order process is allowed in all media, with or without inversion symmetry. Yet it is generally much weaker than an allowed second-order process because of disparity in the sucsceptibilities, $|\chi^{(3)}| \ll |\chi^{(2)}|$. With high-intensity lasers, however, it is still easily observable, as first demonstrated by Maker and Terhune.[1] This is particularly true if $|\chi^{(3)}|$ shows resonant enhancement. When more than one tunable laser is used for pumping, even multiple resonances of $\chi^{(3)}$ can be excited.

Being flexible and easily observable in all media, four-wave mixing has many interesting applications. It extends the frequency range of tunable coherent sources to the infrared and ultraviolet.[2] In the degenerate case (i.e., four waves having the same frequency) it is used for wavefront reconstruction in adaptive optics.[3] With resonances, it can be adopted as a powerful spectroscopic and analytical tool for material studies. We consider the fundamentals as well as some applications of four-wave mixing in this chapter, and leave the discussion of four-wave mixing spectroscopy to Chapter 15.

14.1 THIRD-ORDER NONLINEAR SUSCEPTIBILITIES

In media with inversion symmetry, third-order nonlinearity is the lowest-order nonlinearity allowed under electric-dipole approximation.[4] The microscopic expression of a third-order nonlinear susceptibility $\chi^{(3)}$ can be derived from a perturbation calculation using, for example, the diagrammatic technique outlined in Section 2.3. In general, it consists of 48 terms, explicitly shown in Ref. 5. While $\chi^{(3)}$ is governed by the overall symmetry of the bulk medium, each term of $\chi^{(3)}$ is governed by the selection rules on its matrix elements.

Near resonances, a few terms of $\chi^{(3)}$ are resonantly enhanced through the resonant denominators. The resonant part of $\chi^{(3)}$ can be separated from the nonresonant part through resonant dispersion. The former is a complex quantity as the damping coefficients in the resonant denominators become nonnegligible. A few examples are discussed here.

Singly Resonant Cases

Assume three input pump frequencies ω_1, ω_2, and ω_3. Single resonance of $\chi^{(3)}$ occurs when any of the three frequencies or their algebraic sums approach a transition frequency of the medium. Consider as an example the case seen in Fig. 14.1 where $\omega_1 - \omega_2$ is at resonance. The third-order susceptibility can be written as the sum of a resonant part $\chi_R^{(3)}$ and a nonresonant part $\chi_{NR}^{(3)}$

$$\chi^{(3)} = \chi_{NR}^{(3)} + \chi_R^{(3)}. \tag{14.1}$$

The expression of $\chi_R^{(3)}$ can be obtained either from the general expression of $\chi^{(3)}$ or from the derivation in Section 2.3 or 10.4. We find

$$\left[\chi_R^{(3)}(\omega_a = \omega_3 - \omega_2 + \omega_1)\right]_{ijkl} = -\frac{N(M_{g'g}^a)_{ij}^*(M_{g'g}^s)_{kl}(\rho_g - \rho_{g'})}{\hbar(\omega_1 - \omega_2 - \omega_{g'g} - i\Gamma_{g'g})} \tag{14.2}$$

where

$$(M_{g'g}^s)_{kl} = \sum_n \left[\frac{\langle g'|er_k|n\rangle\langle n|er_l|g\rangle}{\hbar(\omega_1 - \omega_{ng})} - \frac{\langle g'|er_l|n\rangle\langle n|er_k|g\rangle}{\hbar(\omega_2 + \omega_{ng})}\right]$$

$$(M_{g'g}^a)_{ij} = \sum_n \left[\frac{\langle g|er_j|n\rangle\langle n|er_i|g\rangle}{\hbar(\omega_a - \omega_{ng})} - \frac{\langle g'|er_i|n\rangle\langle n|er_j|g\rangle}{\hbar(\omega_3 + \omega_{ng})}\right].$$

The resonant $\chi_R^{(3)}$ of other singly resonant cases can be similarly derived.

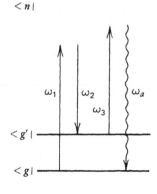

Fig. 14.1 Schematic showing a four-wave mixing process with $\omega_1 - \omega_2$ at resonance.

Doubly Resonant Cases

We assume here two input pump frequencies ω_1 and ω_2, and consider the various doubly resonant cases seen in Fig. 14.2.[6] The expressions for $\chi_R^{(3)}$ can be derived following the diagrammatic technique of Section 2.3. For the cases of Fig. 14.2a–d, the diagrams are shown explicitly, with resonant transitions denoted by the heavy dots. They yield, respectively,

$$
\left[\chi_R^{(3)}(\omega_3 = \omega_1 - \omega_2 + \omega_1)\right]_{ijkl}
$$

$$
= -\frac{Ne^4}{\hbar^3} \sum_m \left[\frac{\langle g|r_i|m\rangle\langle m|r_j|g'\rangle}{2\omega_1 - \omega_2 - \omega_{mg}} + \frac{\langle g|r_j|m\rangle\langle m|r_i|g'\rangle}{2\omega_1 - \omega_2 + \omega_{mg'}}\right]
$$

$$
\times \frac{\langle g'|r_k|n\rangle\langle n|r_l|g\rangle\rho_{gg}^0}{(\omega_1 - \omega_{ng} + i\Gamma_{ng})(\omega_1 - \omega_2 - \omega_{g'g} + i\Gamma_{g'g})}
$$

(14.3a)

+ terms with j and l interchanged

$$
\left[\chi_R^{(3)}(\omega_3 = \omega_1 - \omega_2 + \omega_1)\right]_{ijkl}
$$

$$
= -\frac{Ne^4}{\hbar^3} \sum_m \frac{\langle g|r_i|n'\rangle}{2\omega_1 - \omega_2 - \omega_{n'g} + i\Gamma_{n'g}}
$$

(14.3b)

$$
\times \left[\frac{\langle n'|r_j|m\rangle\langle m|r_k|n\rangle}{\omega_1 - \omega_2 - \omega_{mg}} + \frac{\langle n'|r_k|m\rangle\langle m|r_j|n\rangle}{2\omega_1 - \omega_{mg}}\right]\frac{\langle n|r_l|g\rangle\rho_{gg}^0}{\omega_1 - \omega_{ng} + i\Gamma_{ng}}
$$

+ terms with j and l interchanged

$$
\left[\chi_R^{(3)}(\omega_3 = \omega_1 - \omega_2 + \omega_1)\right]_{ijkl}
$$

$$
= -\frac{Ne^4}{\hbar^3} \sum_m \frac{\langle g|r_i|n'\rangle\langle n'|r_j|g'\rangle}{(2\omega_1 - \omega_2 - \omega_{n'g} + i\Gamma_{n'g})(\omega_1 - \omega_2 - \omega_{g'g} + i\Gamma_{g'g})}
$$

$$
\times \left[\frac{\langle g'|r_k|m\rangle\langle m|r_l|g\rangle}{\omega_1 - \omega_{mg}} + \frac{\langle g'|r_l|m\rangle\langle m|r_k|g\rangle}{-\omega_2 - \omega_{mg}}\right]\rho_{gg}^0
$$

(14.3c)

+ terms with j and l interchanged

$$
\left[\chi_R^{(3)}(\omega_3 = \omega_1 - \omega_2 + \omega_1)\right]_{ijkl}
$$

$$
= -\frac{Ne^4}{\hbar^3} \sum_m \frac{\langle m|r_i|n'\rangle\langle n'|r_j|g\rangle\langle g|r_k|n\rangle\langle n|r_l|m\rangle}{(2\omega_1 - \omega_2 - \omega_{n'm})(\omega_1 - \omega_2 - \omega_{n'n} + i\Gamma_{n'n})}
$$

(14.3d)

$$
\times \left[\frac{1}{\omega_1 - \omega_{n'g} + i\Gamma_{n'g}} - \frac{1}{\omega_2 - \omega_{ng} - i\Gamma_{ng}}\right]\rho_{gg}^0
$$

+ terms with j and l interchanged.

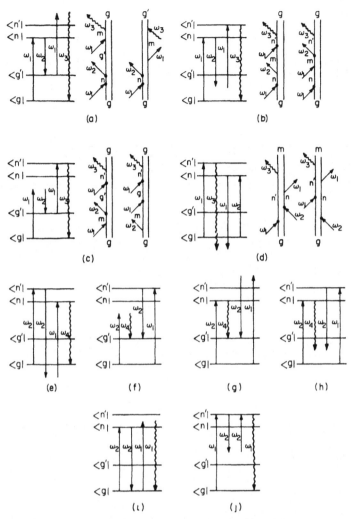

Fig. 14.2 Schematics describing the various cases of doubly resonant four-wave mixing. The double Feymann diagrams for cases (a)–(d) are explicitly shown.

Resonances are explicit shown in these expressions by the frequency denominators with damping constants. Similarly, the expressions of $\chi_R^{(3)}(\omega_4 = \omega_2 - \omega_1 + \omega_2)$ for Fig. 14.2e–h at the output frequency $2\omega_2 - \omega_1$, and those of $\chi_R^{(3)}(\omega_1 = \omega_1 - \omega_2 + \omega_2)$ for Fig. 14.2i and j at the output frequency ω_1 can also be derived.

Triply Resonant Cases

We assume here three input frequencies, ω_1, ω_1', and ω_2, and consider the $\chi^{(3)}$ terms in which all the three frequency denominators are near resonances.[6]

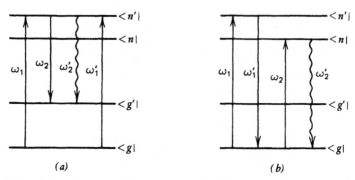

Fig. 14.3 Schematics showing two cases of triply resonant four-wave mixing.

They are schematically shown in Fig. 14.3. The expressions of $\chi_R^{(3)}(\omega_2' = \omega_1 - \omega_1' + \omega_2)$ for the cases of Fig. 14.3a and b are, respectively,

$$\left[\chi_R^{(3)}(\omega_2' = \omega_1 - \omega_1' + \omega_2)\right]_{ijkl}$$

$$= \frac{Ne^4}{\hbar^3}\langle g'|r_i|n'\rangle\langle n'|r_j|g\rangle\langle g|r_k|n'\rangle\langle n'|r_l|g'\rangle\rho_{gg}^0$$

$$\times\Bigg\{\left[\left(\omega_2' - \omega_{n'g'} + i\Gamma_{n'g'}\right)\left(\omega_1' - \omega_2 - \omega_{g'g} - i\Gamma_{g'g}\right)\right.$$

$$\left.\times\left(\omega_1' - \omega_{n'g} - i\Gamma_{n'g}\right)\right]^{-1} + \left(\omega_2' - \omega_{n'g'} + i\Gamma_{n'g'}\right)^{-1} \qquad (14.4a)$$

$$\times\left(\omega_1 - \omega_1' + i/T_{ln'}\right)^{-1}\left[\left(\omega_1 - \omega_{n'g} - i\Gamma_{n'g}\right)^{-1}\right.$$

$$\left.-\left(\omega_1' - \omega_{n'g} + i\Gamma_{n'g}\right)^{-1}\right]\Bigg\}$$

$$\left[\chi_R^{(3)}(\omega_2' = \omega_2 - \omega_1' + \omega_1)\right]_{ijkl}$$

$$= \frac{Ne^4}{\hbar^3}\langle g|r_i|n\rangle\langle n|r_j|g\rangle\langle g|r_k|n'\rangle\langle n'|r_l|g\rangle\rho_{gg}^0$$

$$\times\Bigg\{\left(\omega_2' - \omega_{ng} + i\Gamma_{ng}\right)^{-1}\left(\omega_1' - \omega_2 - \omega_{n'n} - i\Gamma_{n'n}\right)^{-1}$$

$$\times\left[\left(\omega_2 - \omega_{ng} + i\Gamma_{ng}\right)^{-1} - \left(\omega_1' - \omega_{n'g} - i\Gamma_{n'g}\right)^{-1}\right] \qquad (14.4b)$$

$$+ \left(\omega_2' - \omega_{ng} + i\Gamma_{ng}\right)^{-1}\left(\omega_1 - \omega_1' + i/T_{lg}\right)^{-1}$$

$$\times\left[\left(\omega_1 - \omega_{ng} - i\Gamma_{ng}\right)^{-1} - \left(\omega_1' - \omega_{ng} + i\Gamma_{ng}\right)^{-1}\right]\Bigg\}.$$

These expressions are for isolated molecules or ions. As discussed in Section 13.1, the resonant frequencies of molecules or ions often depend on the local environment. The effective $\chi_R^{(3)}$ of an ensemble of molecules or ions should be a weighted average of $\chi_R^{(3)}$ over the distribution of the resonant frequencies.

14.2 GENERAL THEORY OF FOUR-WAVE MIXING

The theory of four-wave mixing follows closely the general theory of optical mixing. We assume here, for simplicity, a cubic or isotropic medium. Three different cases are considered in this section.

Three Pump Fields

Let the pump fields be $\mathbf{E}_m(\omega_m) = \mathcal{E}_m \exp(i\mathbf{k}_m \cdot \mathbf{r} - i\omega_m t)$ with $m = 1, 2, 3$ (Fig. 14.4a). The output field, $\mathbf{E}_s(\omega_s) = \mathcal{E}_s \exp(i\mathbf{k}_s \cdot \mathbf{r} - i\omega_s t)$ with $\omega_s = \omega_1 + \omega_2 + \omega_3$, is governed by the wave equation

$$\left[\nabla^2 + \frac{\omega_s^2}{c^2}\epsilon(\omega_s)\right]\mathbf{E}_s = -\frac{4\pi\omega_s^2}{c^2}\mathbf{P}^{(3)}(\omega_s) \tag{14.5}$$

where $\mathbf{P}^{(3)}(\omega_s) = \chi^{(3)}(\omega_s = \omega_1 + \omega_2 + \omega_3):\mathbf{E}(\omega_1)\mathbf{E}(\omega_2)\mathbf{E}(\omega_3)$. The solution of (14.5) follows that described in Chapter 6. With the usual slowly varying amplitude approximation, negligible pump depletion, and the simplifying boundary condition, it yields

$$\mathcal{E}_{si}(z) = -\frac{2\pi\omega_s^2}{(\Delta\mathbf{k}\cdot\hat{z})k_s c^2}\chi_{ijkl}^{(3)}\mathcal{E}_{1j}\mathcal{E}_{2k}\mathcal{E}_{3l}(1 - e^{i\Delta\mathbf{k}\cdot\mathbf{z}})e^{-\alpha_{si}z} \tag{14.6}$$

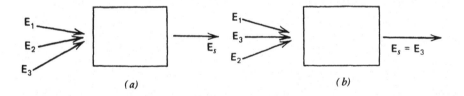

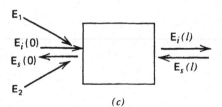

Fig. 14.4 Three different types of four-wave mixing discussed in Section 14.2.

where

$$\Delta \mathbf{k} = \Delta \mathbf{k}' + i\Delta \mathbf{k}'' = \left(\mathbf{k}_1' + \mathbf{k}_2' + \mathbf{k}_3' - \mathbf{k}_s' \right) + i\hat{z}\left(\alpha_{1j} + \alpha_{2k} + \alpha_{3l} - \alpha_{sl} \right),$$

$\Delta \mathbf{k}'$ is the wavevector mismatch, and the α_{ml} are the attenuation coefficients of the waves along $\hat{z}$.

As is common for optical mixing, phase matching ($\Delta \mathbf{k} = 0$) is of prime importance here, since it greatly enhances the signal output. In four-wave mixing, phase matching can be achieved in an infinite number of ways by properly adjusting the directions of propagation of the three pump waves. Which arrangement is preferred often depends on practical considerations, such as optimum beam overlapping length and better spatial discrimination against scattering background.

Output Field in the Same Mode as One of the Input Fields

In this case, we take $E_{sl} = E_{3l}$ (Fig. 14.4b). The input field E_{3l} should then experience gain or loss induced by the nonlinear wave interaction. Since $\omega_s = \omega_3$ and $\mathbf{k}_s = \mathbf{k}_3$ in (14.5), we must have $\omega_1 = -\omega_2$, and $\Delta \mathbf{k} = \mathbf{k}_1(\omega_1) + \mathbf{k}_2(\omega_2)$. With negligible depletion of $\mathbf{E}_1$ and $\mathbf{E}_2$, the solution becomes

$$\mathcal{E}_{sl}(z) = \mathcal{E}_{sl}(0)\exp\left[g_l(z) - \alpha_{sl}z \right],$$

$$g_l(z) = \frac{2\pi\omega_s^2}{(\Delta \mathbf{k} \cdot \hat{z})k_s c^2} \chi_{ijkl}^{(3)} \mathcal{E}_{1j}\mathcal{E}_{2k}\left(1 - e^{i\Delta \mathbf{k}\cdot\hat{z}} \right). \tag{14.7}$$

The real part of $g_l(z)$ represents a gain. For the special case of $E_{1j}(\omega_1) = E_{2k}^*(\omega_2)$ and $\Delta k'' = 0$, we find

$$\text{Re}\left[g_l(z) \right] = \frac{2\pi\omega_s^2}{k_s c^2} \text{Im}\left[\chi_{ijkl}^{(3)} \right] |\mathcal{E}_{1j}|^2 z, \tag{14.8}$$

which can be compared with the result of Raman gain discussed in Section 10.3.

Backward Parametric Amplification and Oscillation

This is a special case of four-wave mixing in which two strong waves act as the pump fields and two counterpropagating weak waves get amplified (Fig. 14.4c). It resembles the parametric amplification case of Section 9.6, except that two pump fields instead of one are used here. The two weak waves are the signal and idler waves, respectively. The solution then is essentially the same as that described in Section 9.6. Assuming perfect phase matching, which can be achieved easily in this case, and negligible pump depletion, we find, following the derivation in Section 9.6, for the signal and idler fields, $\mathbf{E}_s$ and $\mathbf{E}_i$, propagating along $\mp \hat{z}$, respectively,

$$\mathcal{E}_s(z = 0) = \mathcal{E}_s(l) \Big/ \cos\frac{g_0 l}{2} + i\frac{\omega_s}{\omega_i}\left(\frac{k_i}{k_s}\right)^{1/2}\mathcal{E}_i^*(0)\tan\frac{g_0 l}{2},$$

$$\mathcal{E}_i^*(z = l) = -i\frac{\omega_i}{\omega_s}\left(\frac{k_s}{k_i}\right)^{1/2}\mathcal{E}_s(l)\tan\frac{g_0 l}{2} + \mathcal{E}_i^*(0)\Big/\cos\frac{g_0 l}{2}$$

$$(14.9)$$

where

$$\left(\frac{g_0}{2}\right)^2 = \left(\frac{\omega_s^2\omega_i^2}{k_s k_i}\right)|KE_1 E_2|^2,$$

$$K = \frac{2\pi}{c^2}\hat{e}_s \cdot \chi^{(3)}(\omega_s = -\omega_i + \omega_1 + \omega_2) : \hat{e}_i\hat{e}_1\hat{e}_2,$$

$$(14.10)$$

and $\mathbf{E}_1$ and $\mathbf{E}_2$ are the pump fields. As $g_0 l$ approaches π, both $\mathcal{E}_s(0)$ and $\mathcal{E}_i(l)$ diverge according to (14.9). This indicates the onset of oscillation, which yields an output even in the absence of any input, $\mathcal{E}_s(l) = \mathcal{E}_i(0) = 0$.

In the case of sufficiently weak pump fields, $g_0 l \ll 1$, and $\mathcal{E}_i(z) = \mathcal{E}_i(0)$, the signal output should reduce to the expression for an ordinary four-wave mixing process with three pump fields.

14.3 DEGENERATE FOUR-WAVE MIXING

We now consider a special case of four-wave mixing in which all the four waves have the same frequency. The third-order nonlinear polarization governing the process has, in general, three components with different wavevectors:

$$\mathbf{P}_s^{(3)}(\omega) = \mathbf{P}_s^{(3)}(\mathbf{k}_1 + \mathbf{k}_1' - \mathbf{k}_i, \omega) + \mathbf{P}_s^{(3)}(\mathbf{k}_1 - \mathbf{k}_1' + \mathbf{k}_i, \omega)$$
$$+ \mathbf{P}_s^{(3)}(-\mathbf{k}_1 + \mathbf{k}_1' + \mathbf{k}_i, \omega)$$

$$(14.11)$$

where

$$\mathbf{P}_s^{(3)}(\mathbf{k}_1 + \mathbf{k}_1' - \mathbf{k}_i, \omega) = \chi^{(3)}(\omega) : \mathbf{E}_1(\mathbf{k}_1)\mathbf{E}_1'(\mathbf{k}_1')\mathbf{E}_i^*(\mathbf{k}_i),$$

$$\mathbf{P}_s^{(3)}(\mathbf{k}_1 - \mathbf{k}_1' + \mathbf{k}_i, \omega) = \chi^{(3)}(\omega) : \mathbf{E}_1(\mathbf{k}_1)\mathbf{E}_1'^*(\mathbf{k}_1')\mathbf{E}_i(\mathbf{k}_i),$$

$$\mathbf{P}_s^{(3)}(-\mathbf{k}_1 + \mathbf{k}_1' + \mathbf{k}_i, \omega) = \chi^{(3)}(\omega) : \mathbf{E}_1^*(\mathbf{k}_1)\mathbf{E}_1'(\mathbf{k}_1')\mathbf{E}_i(\mathbf{k}_i).$$

$\mathbf{E}_1(\mathbf{k}_1)$, $\mathbf{E}_1'(\mathbf{k}_1')$, and $\mathbf{E}_i(\mathbf{k}_i)$ are the three input fields, and all $\chi^{(3)}(\omega)$ in the above equations are the same under the electric dipole approximation. We note that in this case $\chi^{(3)}(\omega)$ has at least a singly resonant term arising from the two-photon zero-frequency resonance, i.e., a term with $(\omega - \omega + i/T_1)$ in the denominator. It may also have a two-photon resonant term if $\omega + \omega$ is in resonance with a transition of the medium. Finally, $\chi^{(3)}$ can be triply resonant if ω is near a resonance. Because of the strong resonant enhancement, $\chi^{(3)}$ for degenerate four-wave mixing can be very large in some media. As a result, such a third-order process is even observable with CW laser beams.

The output of degenerate four-wave mixing can be calculated using the theory in Section 14.2, but it can also be easily understood from the following physical picture. Two of the three input waves interfere and form either a static grating or a moving grating with an oscillation frequency 2ω; the third input wave is scattered by the grating to yield the output wave. In many cases where 2ω is away from resonance, the contribution from the static gratings should dominate. With three input waves, three different static gratings are formed. The grating formed by the $\mathbf{k}_1$ and $\mathbf{k}_l$ waves scatters the $\mathbf{k}_1'$ wave to yield outputs at $\mathbf{k}_s = \mathbf{k}_1' \pm (\mathbf{k}_1 - \mathbf{k}_l)$. The one formed by the $\mathbf{k}_1'$ and $\mathbf{k}_l$ waves scatters the $\mathbf{k}_1$ wave to yield the outputs at $\mathbf{k}_s = \mathbf{k}_1 \pm (\mathbf{k}_1' - \mathbf{k}_l)$. The one formed by the $\mathbf{k}_1$ and $\mathbf{k}_1'$ waves scatters the $\mathbf{k}_l$ wave to yield the outputs at $\mathbf{k}_s = \mathbf{k}_l \pm (\mathbf{k}_1 - \mathbf{k}_1')$. They are illustrated in Fig. 14.5 for the special case of $\mathbf{k}_1' = -\mathbf{k}_1$. Altogether, three output waves with different wavevectors, $\mathbf{k}_s = \mathbf{k}_1 + \mathbf{k}_1' - \mathbf{k}_l$, $\mathbf{k}_1 - \mathbf{k}_1' + \mathbf{k}_l$, and $-\mathbf{k}_1 + \mathbf{k}_1' + \mathbf{k}_l$, are expected. However, we realize that since $|\mathbf{k}_s|$, in general, is not equal to $\omega \epsilon^{1/2}/c$, the generation of the three output waves may not be all phase matched. Consider, for example, the case with $\mathbf{k}_1' = -\mathbf{k}_1$. The output waves are expected to have $\mathbf{k}_s = -\mathbf{k}_l$ and $\mathbf{k}_l \pm 2\mathbf{k}_1$. While the generation of the output at $\mathbf{k}_s = -\mathbf{k}_l$ is always phase matched, that of the other two is not. Thus usually only the output at $\mathbf{k}_s = -\mathbf{k}_l$ needs to be considered. It is interesting to see the connection between this case and holography. In both cases, the output wave $(\mathbf{k}_s)$, arising from scattering of one of the pump waves $(\mathbf{k}_1$ or $\mathbf{k}_1')$ off the interferogram formed by the object wave $(\mathbf{k}_l)$ and the other pump wave $(\mathbf{k}_1'$ or $\mathbf{k}_1)$, retraces back the path of the object wave $(\mathbf{k}_s = -\mathbf{k}_l)$. Now that an object can be represented by a group of $\mathbf{k}_l$ waves, we see that an image of the object can be reconstructed by the corresponding output waves.

In an isotropic medium, if we require the output of degenerate four-wave mixing not to be in the same mode as one of the pump waves, then for phase-matched output we must have $\mathbf{k}_1' = -\mathbf{k}_1$ and $\mathbf{k}_s = -\mathbf{k}_l$. The effective nonlinear polarization can be written, from symmetry consideration, as

$$\mathbf{P}_s^{(3)}(\mathbf{k}_s = -\mathbf{k}_l, \omega) = \chi^{(3)}(\omega) : \mathbf{E}_1(\mathbf{k}_1)\mathbf{E}_1'(-\mathbf{k}_1)\mathbf{E}_l^*(\mathbf{k}_l)$$

$$= A(\mathbf{E}_1 \cdot \mathbf{E}_l^*)\mathbf{E}_1' + B(\mathbf{E}_1' \cdot \mathbf{E}_l^*)\mathbf{E}_1 + C(\mathbf{E}_1 \cdot \mathbf{E}_1')\mathbf{E}_l^*$$

$$(14.12)$$

where A, B, and C all depend on the angle θ between $\mathbf{E}_1$ and $\mathbf{E}_l$, and $B(\theta) = A(\pi - \theta)$. The brackets $(\mathbf{E}_1 \cdot \mathbf{E}_l^*)$ and $(\mathbf{E}_1' \cdot \mathbf{E}_l^*)$ in the A and B terms in (14.12) describe the static gratings formed by the wave interference, while

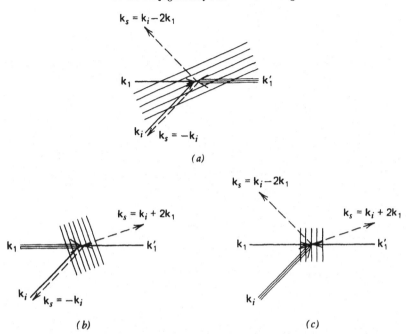

Fig. 14.5 Degenerate four-wave mixing resulting from scattering of an incident wave by the static grating formed by the other two incident waves: (*a*) grating formed by the k_1 and k_i waves, (*b*) grating formed by the k_1' and k_i waves, and (*c*) grating formed by the k_1 and k_1' waves.

$(E_1 \cdot E_1')$ in the C term is a moving grating with an oscillation frequency 2ω. By properly arranging the polarizations of the three incoming waves, it is possible to have only one particular term in (14.12) nonvanishing. The output is polarized along $P_s^{(3)}$. Being in the backward direction with respect to the incoming k_i wave, it can be described by the solution of (14.9).[7] We then notice that with $\mathscr{E}_s(l) = 0$, the output field $\mathscr{E}_s(0)$ has a magnitude proportional to that of the input field $\mathscr{E}_i(0)$, and a phase complex conjugate to that of $\mathscr{E}_i(0)$.

14.4 PHASE CONJUGATION BY FOUR-WAVE MIXING

Phase conjugation[8] is defined as the process in which the phase of the output wave is complex conjugate to the phase of an input wave. In other words, the process reverses the phase of the input. This happens, for example, in difference-frequency generation, parametric amplification, and four-wave mixing. If the phase-conjugated output propagates in the backward direction with respect to the corresponding input wave, then it can be used to correct abberation due to phase distortion experienced by the input wave. As illustrated in Fig. 14.6, the input beam in passing through a medium suffers a

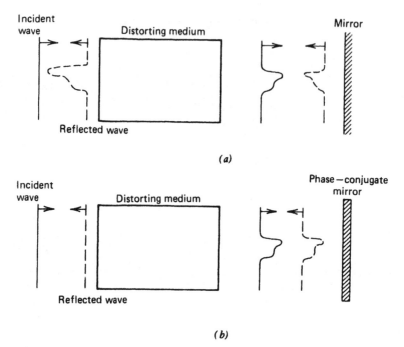

Fig. 14.6 Sketches showing how the wavefront of a beam changes in passing through a distorting medium back and fourth: (*a*) an ordinary mirror is used, and (*b*) a phase-conjugate mirror is used to reflect the beam.

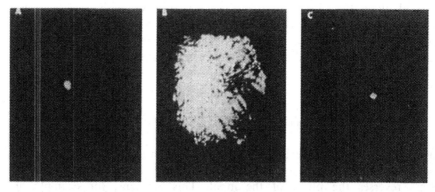

Fig. 14.7 Photographs showing correction of aberration: (*a*) an unperturbed laser beam, (*b*) the same laser beam after passing through an etched glass plate, and (*c*) the same beam after a phase-conjugate reflection and a second pass through the etched plate. (After Ref. 9.)

wavefront distortion. Unlike an ordinary mirror, the phase-conjugate mirror reverses the wavefront distortion of the input beam upon reflection. Then, as the phase-conjugated wave reflects back through the medium again, the wavefront distortion is completely removed (as long as diffraction is negligible). An example is seen in Fig. 14.7.

In the previous section, we saw that the output of degenerate four-wave mixing is a reflected phase-conjugate wave with respect to one of the input waves $[\mathscr{E}_s(0) \propto \mathscr{E}_i^*(0)]$. Thus a nonlinear medium for degenerate four-wave mixing can be used as a phase-conjugate mirror.[3] According to (14.9) with $\mathscr{E}_s(l) = 0$, the phase-conjugated output $|\mathscr{E}_s(0)|^2$ can even be more intense than the input $|\mathscr{E}_i(0)|^2$ if $g_0 l/2 > \pi/4$, and the amplifications approaches the oscillation limit as $g_0 l/2 \rightarrow \pi/2$.

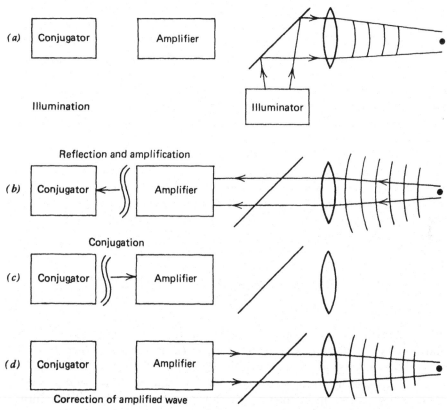

Fig. 14.8 Wavefront reconstruction applied to laser fusion. The sequence of events is (*a*) illumination of the target with a probe beam, (*b*) reflection and amplification resulting in a distorted wave, (*c*) phase conjugation, and (*d*) a second pass through the distorting amplifying medium, producing a correct focusing of the beam on the target. (After Ref. 9.)

There are many interesting and potentially important applications of phase conjugation based on its ability to remove wavefront distortion.[9] One is for correction of distorted images. As an example, consider the amplification of a laser beam in an amplifier. The beam quality may be seriously deteriorated after the beam traverses the amplifying medium. If, however, a phase-conjugate mirror is used to send the beam back through the amplifier once again, then the amplifier output can have its input beam quality restored. This allows the construction of high-power laser systems with beam quality comparable to that of a single-mode oscillator. One can also have a laser oscillator with one of its mirrors replaced by a phase-conjugate mirror, which tends to help in establishing a better beam quality, improve the mode stability, and possibly provide additional gain to the oscillator. In laser fusion work, beam focusing on the target may be impaired by wavefront distortion of the beam in its propagation from the laser to the target. This can be remedied by the scheme of Fig. 14.8. The target is first illuminated; the wave radiated from the target is then amplified by the amplifier, reflected by a phase-conjugate mirror, amplified again, and finally automatically focused onto the target with no net distortion of its wavefront. The scheme can actually be used on any target, not necessary in laser fusion work. In principle, it can also be applied to a moving target at a distance as the beam automatically tracks the target. This may have great potential in military applications. With sufficient gain in the amplifier, it is even possible to have laser oscillation with the target and the phase-conjugate mirror forming the cavity, so that no external illumination on the target is necessary.

Phase conjugation can also be obtained from stimulated scattering of a highly multimode pump beam.[10] In this case, the output is phase conjugated to the input pump field. This has been observed experimentally. Theoretically, however, it is shown that only approximate phase conjugation can be achieved, but the approximation becomes better when the number of pump modes increases.[11] A better physical understanding of the effect is still needed.

14.5 TUNABLE INFRARED AND ULTRAVIOLET GENERATION[2]

Four-wave mixing can be used to extend the frequency range of coherent radiation to infrared and ultraviolet. The process is again governed by the general theory of optical mixing. It is the third-order susceptibility $\chi^{(3)}(\omega_s = \omega_1 + \omega_2 + \omega_3)$ that determines how efficient the frequency conversion process can be with given pump intensities. As shown in Section 7.4, $\chi^{(3)}$ for third harmonic generation can be greatly enhanced by resonances. In alkali vapor, for example, it increases from less than 10^{-33} esu/atom to $\sim 10^{-31}$ esu/atom near a single resonance. Now, in four-wave mixing, ω_1, ω_2, ω_3 are not necessarily equal; it is even possible to have $\chi^{(3)}$ doubly or triply resonant. A few examples are schematically shown in Fig. 14.9 using potassium vapor as

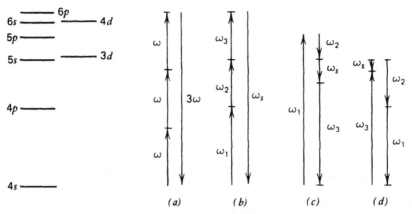

Fig. 14.9 Energy level diagram of a potassium atom and schematics for a number of resonant third-order optical mixing processes: (*a*) third-harmonic generation, (*b*) sum frequency generation with ω_1, $\omega_1 + \omega_2$, and $\omega_1 + \omega_2 + \omega_3$ near resonances, (*c*) infrared generation with ω_1 and $\omega_1 - \omega_2$ near resonances, and (*d*) infrared generation with ω_1 and $\omega_1 + \omega_2$ near resonances.

the nonlinear medium. As one would expect, $\chi^{(3)}$ doubly or triply resonant can be orders of magnitude larger than that of the singly resonant case. We use here the process of Fig. 14.9*c* for illustration.

Let us assume that in Fig. 14.9*c*, ω_1 is near the $4s \rightarrow 5p$ resonance, $\omega_1 - \omega_2$ is exactly on the $4s \rightarrow 5s$ resonance, and ω_3 is not too far from the $4s \rightarrow 4p$ resonance. The dominating resonant term of $\chi^{(3)}$ in this case is

$$\left(\chi_R^{(3)}\right)_{ijkl} \cong \left(\frac{Ne^4}{\hbar^3}\right) \sum_{s,p} A_{sp}\langle 4s|r_i|4p\rangle\langle 4p|r_i|5s\rangle\langle 5s|r_k|5p\rangle\langle 5p|r_l|4s\rangle,$$

$$A_{sp} = \left[\left(\omega_1 - \omega_{5p-4s} + i\Gamma_{5p}\right)\left(\omega_1 - \omega_2 - \omega_{5s-4s} + i\Gamma_{5s}\right)\right.$$

$$\left. \times \left(\omega_3 - \omega_{4p-4s}\right)\right]^{-1} \quad (14.13)$$

where $\sum_{s,p}$ sums over the fine structures in the s and p levels. With $\omega_1 - \omega_{5p-4s} = 50 \text{ cm}^{-1}$, $\Gamma_{5s} = 0.1 \text{ cm}^{-1}$, and $\omega_3 - \omega_{4p-4s} = 5000 \text{ cm}^{-1}$, we find $\chi^{(3)} = 6 \times 10^{-27}$ esu/atom.[12] For an atomic density of 10^{17} atoms/cm³, the value of $\chi^{(3)} = 6 \times 10^{-10}$ esu is already larger than the nonresonant $\chi^{(3)}$ of a typical condensed matter. We notice that since ω_3 is still far away from resonance, $\chi^{(3)}$ does not vary appreciably with ω_3. This, together with the large value of $\chi^{(3)}$, means that efficient generation of tunable output over a broad range of infrared frequency $\omega_s = \omega_1 - \omega_2 - \omega_3$ is possible as long as collinear phase matching can be achieved. If ω_3 is tuned toward the $4s \rightarrow 4p$ resonance, then $\chi^{(3)}$ is further enhanced. For $\omega_3 - \omega_{4p-4s} = 50 \text{ cm}^{-1}$, we have $\chi^{(3)} = 6 \times 10^{-25}$ esu/atom.

Collinear phase matching is essential for high conversion efficiency. In the above infrared generation, it may be achievable using anomalous dispersion of the alkali vapor. The collinear phase matching relation $k_1 = k_2 + k_3 + k_s$ for the process of Fig. 14.9c can be written in the form

$$\omega_1 n_1 = \omega_2 n_2 + \omega_3 n_3 + \omega_s n_s. \qquad (14.14)$$

Since both ω_2 and ω_s are very far from resonances involving the ground level, the normally dispersive refractive indices n_2 and n_s do not vary appreciably with ω_2 and ω_s. For a prescribed output frequency ω_s, the frequency ω_3 is fixed from the relation $\omega_{5s-4s} = \omega_3 + \omega_s = \omega_1 - \omega_2$, and hence both n_s and n_3 are fixed. Now that n_1 has an anomalous dispersion while n_2 is nearly independent of ω_2, it is possible to satisfy (14.14) by adjusting ω_1 and ω_2 properly. For smaller ω_s, we must have ω_1 closer to ω_{5p-4s} to achieve phase matching. This unfortunately increases the absorption of the pump field at ω_1 and decreases drastically the efficiency of infrared generation. The problem may be alleviated by mixing foreign gas into the vapor, for example, sodium in potassium, and utilizing the additional dispersion in the refractive index provided by the foreign gas.[12]

Tunable infrared generation using the process of Fig. 14.9c was actually demonstrated by Sorokin et al.[12,13] In their experiment, only two pump beams, at ω_1 and ω_3, were used, while the ω_2 beam was automatically generated in the cell by stimulated Raman scattering of the ω_1 beam. With peak powers of 1 kW at ω_1 and 10 kW at ω_3 and an active length of 30 cm in potassium vapor, they observed a tunable infrared output from 2 to 25 μm having peak powers of 100 mW at 2 μm and 0.1 mW at 25 μm. Extension to a broader tuning range may be possible. The theoretical analysis of this experiment is not yet complete. Strictly speaking, it is a four-wave parametric amplification process with waves at ω_s and ω_2 being the signal and idler waves. The calculation should be a straightforward extension of that given in Chapter 9, including a Raman resonant transition.

As a variation of the process, the Raman transition $\omega_1 - \omega_2 = \omega_{5s-4s}$ can be replaced by a two-photon transition $\omega_1 + \omega_2 = \omega_{5s-4s}$. The process should be at least equally efficient if ω_1 is also near the ω_{4p-4s} resonance, as in Fig. 14.9d. Then, for infrared generation, ω_3 can be either larger or smaller than ω_{5s-4s} in both Fig. 14.9c and Fig. 14.9d. A reverse four-wave mixing process is, of course, also possible. Thus, by using ω_1, ω_2, and ω_s as pump frequencies in Fig. 14.9d, the ω_3 wave is generated. The process can therefore be used to construct an infrared-to-visible converter.[14]

With laser powers much higher than 10 kW, the conversion efficiency of infrared generation by four-wave mixing can, in principle, be high. It may, however, be limited by other simultaneously occurring nonlinear effects, such as self-focusing, saturation, and ionization. A thorough study of how the conversion efficiency can be improved has not yet been reported.

Four-wave mixing can also be used to generate tunable ultraviolet radiation.[2,15] In comparison with third-harmonic generation, the four-wave mixing process has the advantage that $\chi^{(3)}(\omega_s = \omega_1 + \omega_2 + \omega_3)$ can be greatly enhanced through multiple resonances. An example appears in Fig. 14.9b. Alkali vapor, however, is not very good for vacuum uv generation because its ionization energy is too low and there is no discrete state in the ionization continuum to resonantly enhance $\chi^{(3)}$ at the uv output frequency. Alkali earth vapor appears to be much better. The process of Fig. 14.10 is used in Sr vapor for an example. The pump frequencies ω_1 and ω_2 are adjusted so that ω_1 is near the $(5s)^2 \rightarrow (5s)(5p)$ resonance and $\omega_1 + \omega_2$ is on the $(5s)^2 \rightarrow (5p)^2$ resonance. The other pump frequency ω_3 is near the $(5p)^2 \rightarrow (6s)(6p)$ resonance, and can be tuned to yield tunable uv output at $\omega_s = \omega_1 + \omega_2 + \omega_3$. Note that $(6s)(6p)$ is a discrete autoionization state in the ionization continuum. Because of the multiple resonances, $\chi^{(3)}(\omega_s)$ can be greatly enhanced.

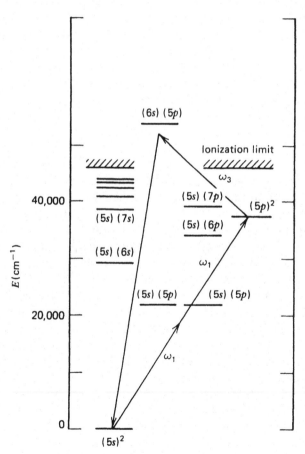

Fig. 14.10 Energy level diagram of Sr with arrows showing a resonant four-wave mixing process for tunable uv generation at $2\omega_1 + \omega_3$.

To avoid strong attenuation of the pump beams, the pump frequencies should be sufficiently far away from resonant absorption. Then, with phase matching, the conversion efficiency for uv generation can be significant.

Molecular gases, such as CO and NO, also can be used as effective media for tunable uv generation by four-wave mixing.[16] Their resonantly enhanced nonlinear susceptibilities are lower in comparison with those of metal vapors because of the weaker resonances, but the gases are much easier to handle. The medium can be in the form of a molecular beam, which, for vacuum uv spectroscopy, has the advantage of requiring no window between the uv source and the sample in the vacuum.[17]

Phase matching for uv generation can be achieved by using the anomalous dispersion of the refractive index as in the case of infrared generation. It can also be achieved by mixing of foreign gas into the vapor. In the actual experiment of Hodgson et al. on Sr,[15] only two pump beams were used with $\omega_1 = \omega_2$, and $2\omega_1$ was tuned to the resonant transition from $(5s)^2$ to an even-parity excited state. By varying ω_3 to have the resonant enhancement of various autoionization states successively coming into play, the uv output was observed to have a tuning range from 1578 to 1957 Å. By using Mg, Hg, and Zn, the tunable uv generation can be extended to the 1060 Å region.[18] With optimal focusing of ~ 1 MW pump beams into a metal vapor of ~ 10 torr, conversion efficiency of ~ 1% is possible but is limited by the usual detrimental effects—self-focusing, saturation, ionization, and so on.

In principle, one can also use condensed matter as the nonlinear medium for efficient infrared and ultraviolet generation by four-wave mixing. In practice, however, a condensed matter is often strongly absorbing in the uv below 2000 Å, and has rather broad absorption bands in the visible and infrared. Therefore, the output cannot be in the vacuum uv or in any of the absorption bands. If all the four frequencies should stay out of the absorption region, then the resonant enhancement may not be strong enough to make $\chi^{(3)}$ larger than the multiply resonant $\chi^{(3)}$ of a metal vapor at ~ 10 torr.

To conclude this section, consider an interesting application of four-wave mixing to time-resolved infrared spectroscopy.[19] Figure 14.11 is the schematic of the experimental arrangement. A pulsed broadband infrared beam (ω_{IR}^B) is first generated by a stimulated electronic Raman process in a metal vapor cell using a pulsed broadband dye laser. This infrared beam, in passing through a sample, carries the spectral information of the sample absorption. It then interacts with a narrow-band laser beam (ω_l) in a second metal vapor cell. The latter generates a narrow-band Stokes beam (ω_s) in the cell, and through four-wave mixing ($\omega_{out}^B = \omega_l - \omega_s + \omega_{IR}^B$), up-converts ω_{IR}^B to a broadband visible output ω_{out}^B, which can be recorded on a spectrograph. With this technique, the infrared absorption spectrum of the sample is displayed as a corresponding visible spectrum, and hence the detection sensitivity is greatly improved. Moreover, since nanosecond or picosecond laser pulses can be used, the technique has a time-resolving capability in the nanosecond or picosecond

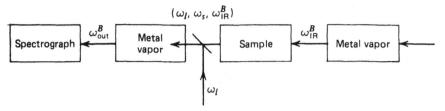

$(\omega_l, \omega_s, \omega_{IR}^B)$

Fig. 14.11 Schematic of an experimental arrangement using four-wave mixing for time-resolved infrared spectroscopy.

regime. It may therefore find important applications in studies of chemical reactions and radical spectra.

14.6   TRANSIENT FOUR-WAVE MIXING

We have thus far considered four-wave mixing only in the steady state. With pulsed resonant excitation, however, transient effect in four-wave mixing may become important. As in the steady-state case, transient four-wave mixing is also governed by the third-order polarization $\mathbf{P}^{(3)}$. The only difference is that $\mathbf{P}^{(3)}$ is now a time-varying function of excitation and relaxation of the medium. Here we discuss the derivation of $\mathbf{P}^{(3)}$, leaving the actual solution of the wave equation with $\mathbf{P}^{(3)}$ as the driving source to Chapter 21.

Using the density matrix formalism of Section 2.1, we can write

$$\mathbf{P}^{(3)} = \mathrm{Tr}\left(-Ne\mathbf{r}\rho^{(3)}\right). \tag{14.15}$$

Then, to find $\mathbf{P}^{(3)}$, it is necessary only to find the density matrix $\rho^{(3)}$. We adopt here again the diagrammatic technique of Yee and Gustafson[20] for the derivation of $\rho^{(3)}$. The notation here follows Section 2.3. Consider the general case where three successive fields, $\mathscr{E}_1(\omega_1, t)\exp[i(\mathbf{k}_1 \cdot \mathbf{r} - \omega_1 t)]$, $\mathscr{E}_2(\omega_2, t)\exp[i(\mathbf{k}_2 \cdot \mathbf{r} - \omega_2 t)]$, and $\mathscr{E}_3(\omega_3, t)\exp[i(\mathbf{k}_3 \cdot \mathbf{r} - \omega_3 t)]$ interact with a material system at t_1, t_2, and t_3 with $t_1 < t_2 < t_3$. In this given time order, $\rho^{(3)}(\omega = \omega_1 + \omega_2 + \omega_3, t)$ has eight terms derived from the eight diagrams in Fig. 14.12. In comparison, $\rho^{(3)}(\omega)$ for the steady-state case has 48 terms from 48 diagrams as mentioned in Section 2.3. The rules for deriving an expression from a diagram are the same as those given in Section 2.3 expect that the propagation from one vertex at t_i to the next at t_j is now represented by the phase factor

$$\langle\langle ab|\hat{A}(t_i - t_j)|ab\rangle\rangle \equiv e^{-(i\omega_{ab} + \gamma_{ab})(t_j - t_i)} \tag{14.16}$$

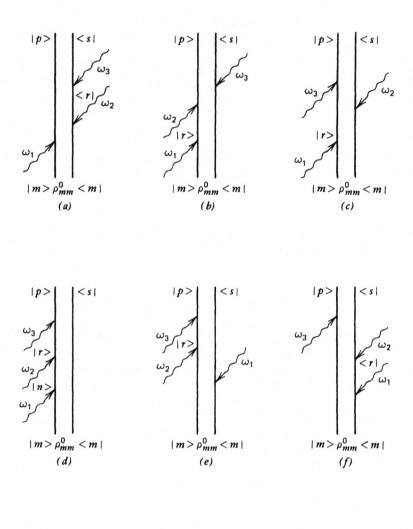

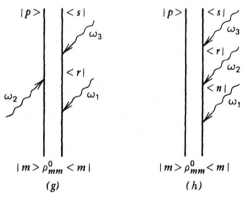

Fig. 14.12 The eight Feymann diagrams representing the eight terms of $\rho^{(3)}(t)$.

where $|a\rangle$ is the ket state on the left between t_i and t_j and $\langle b|$ is the bra state on the right, and γ_{ab} is the damping constant. The final expression has integration over all possible time separations between vertices and summation over all possible initial, intermediate, and final states.

Two terms of $\rho^{(3)}(t)$ corresponding to the diagrams with Figs. 14.12a and 14.12b are used here as examples.[21] From Fig. 14.12a, we have, with the interaction Hamiltonian given by $\mathcal{H}'(\omega) = -\mathbf{p} \cdot \mathcal{E}(\omega)^{i(\mathbf{k} \cdot \mathbf{r} - \omega t)}$,

$$[\rho^{(3)}(t)]_{(a)} = - \sum_{m,r,p,s} \left\{ \int_{-\infty}^{0} d\tau_1 \int_{-\infty}^{0} d\tau_2 \int_{-\infty}^{0} d\tau_3 \left(\frac{-1}{i\hbar} \right)^3 \right.$$

$$\times e^{-(i\omega_{ps}+\phi_{ps})\tau_3 - (i\omega_{pr}+\phi_{pr})\tau_2 - (i\omega_{pm}+\phi_{pm})\tau_1}$$

$$\text{(14.17)}$$

$$\times \langle p|\mathbf{p} \cdot \mathcal{E}_1(t - \tau_1 - \tau_2 - \tau_3)e^{i\mathbf{k}_1 \cdot \mathbf{r} - i\omega_1(t - \tau_1 - \tau_2 - \tau_3)}|m\rangle$$

$$\times \langle m|\mathbf{p} \cdot \mathcal{E}_2(t - \tau_2 - \tau_3)e^{i\mathbf{k}_2 \cdot \mathbf{r} - i\omega_2(t - \tau_2 - \tau_3)}|r\rangle$$

$$\left. \times \langle r|\mathbf{p} \cdot \mathcal{E}_3(t - \tau_3)e^{i\mathbf{k}_3 \cdot \mathbf{r} - i\omega_3(t - \tau_3)}|s\rangle (|p\rangle\rho_{mm}^0\langle s|) \right\}$$

With the substitution of variables $\xi_1 = t - \tau_1 - \tau_2 - \tau_3$, $\xi_2 = t - \tau_2 - \tau_3$, and $\xi_3 = t - \tau_3$, this equation becomes

$$[\rho^{(3)}(t)]_{(a)} = \sum_{m,r,p,s} \left\{ e^{-(i\omega_{ps}+\phi_{ps})t} \left(\frac{1}{i\hbar} \right)^3 (|p\rangle\rho_{mm}^0\langle s|) \right.$$

$$\times \langle p|\mathbf{p} \cdot \hat{e}_1|m\rangle\langle m|\mathbf{p} \cdot \hat{e}_2|r\rangle\langle r|\mathbf{p} \cdot \hat{e}_3|s\rangle$$

$$\text{(14.18)}$$

$$\times \int_{-\infty}^{t} d\xi_3 e^{[(i\omega_{ps}+\phi_{ps}) - (i\omega_{pr}+\phi_{pr}) - i\omega_3]\xi_3} \mathcal{E}_3(\xi_3)e^{i\mathbf{k}_3 \cdot \mathbf{r}}$$

$$\times \int_{-\infty}^{\xi_3} d\xi_2 e^{[(i\omega_{pr}+\phi_{pr}) - (i\omega_{pm}+\phi_{pm}) - i\omega_2]\xi_2} \mathcal{E}_2(\xi_2)e^{i\mathbf{k}_2 \cdot \mathbf{r}}$$

$$\left. \times \int_{-\infty}^{\xi_2} d\xi_1 e^{[(i\omega_{pm}+\phi_{pm}) - i\omega_1]\xi_1} \mathcal{E}_1(\xi_1)e^{i\mathbf{k}_1 \cdot \mathbf{r}} \right\}.$$

Similarly, we find, from Fig. 14.12b,

$$
\left[\rho^{(3)}(t)\right]_{(b)} = -\sum_{m,r,p,s}\left\{ e^{-(\iota\omega_{ps}+\phi_{ps})t}\left(\frac{1}{i\hbar}\right)^3 (|p\rangle\rho^0_{mm}\langle s|) \right.
$$

$$
\times \langle p|\mathbf{p}\cdot\hat{e}_2|r\rangle\langle r|\mathbf{p}\cdot\hat{e}_1|m\rangle\langle m|\mathbf{p}\cdot\hat{e}_3|s\rangle
$$

$$
\times \int_{-\infty}^{t} d\xi_3 e^{[(\iota\omega_{ps}+\phi_{ps})-(\iota\omega_{pm}+\phi_{pm})-\iota\omega_3]\xi_3}\mathscr{E}_3(\xi_3)e^{i\mathbf{k}_3\cdot\mathbf{r}}
$$

$$
\times \int_{-\infty}^{\xi_3} d\xi_2 e^{[(\iota\omega_{pm}+\phi_{pm})-(\iota\omega_{rm}+\phi_{rm})-\iota\omega_2]\xi_2}\mathscr{E}_2(\xi_2)e^{i\mathbf{k}_2\cdot\mathbf{r}}
$$

$$
\left. \times \int_{-\infty}^{\xi_2} d\xi_1 e^{[(\iota\omega_{rm}+\phi_{rm})-\iota\omega_1]\xi_1}\mathscr{E}_1(\xi_1)e^{i\mathbf{k}_1\cdot\mathbf{r}} \right\}. \quad (14.19)
$$

The full expression of $\rho^{(3)}(t)$ is the sum of all eight terms derived from the eight diagrams in Fig. 14.12. In an actual case of transient four-wave mixing, however, resonant or near-resonant excitations are involved, and terms in $\rho^{(3)}$ which are nonresonant can often be neglected. The effective number of terms of $\rho^{(3)}$ then is greatly reduced.

As in the steady-state case, transient four-wave mixing yields a coherent output. It belongs to the category of coherent transient optical effects, which will be discussed in more detail in Chapter 21. Here we consider only some general characteristics of transient four-wave mixing.[21] First, as is explicitly shown in (14.18) and (14.19), the nonlinear polarization $\mathbf{P}^{(3)} \propto \rho^{(3)}$ has a wavevector $\mathbf{k}_s = \mathbf{k}_1 + \mathbf{k}_2 + \mathbf{k}_3$. For the mixing process to be efficient, the phase-matching condition should be satisfied, that is, $\mathbf{k}(\omega) \equiv [\omega\sqrt{\varepsilon(\omega)}/c]\hat{k} = \mathbf{k}_s$. This is the same as in the steady-state case, as one would expect. Second, the transient behavior of $\rho^{(3)}(t)$ arises from resonant or near-resonant excitation of the medium and is governed by the time-dependent phase factors in $\rho^{(3)}$. For illustration, we discuss the case of a molecular gas excited by three resonant pulses.

In a gas, molecules with different velocities should interact with different fields at different times. Let $\mathbf{r}(t)$ be the position of a molecule at time t moving with a velocity $\mathbf{v}$. The field seen by the molecule at an earlier time ξ_i is $\mathscr{E}_i(\xi_i)\exp[i\mathbf{k}_i\cdot\mathbf{r}(\xi_i) - i\omega_i\xi_i]$ at $\mathbf{r}(\xi_i)$. Since $\mathbf{r}(\xi_i) = \mathbf{r}(t) - (t - \xi_i)\mathbf{v}$, we have

$$
\mathscr{E}_i(\xi_i)\exp[i\mathbf{k}_i\cdot\mathbf{r}(\xi_i)] = \mathscr{E}_i(\xi_i)\exp[i\mathbf{k}_i\cdot\mathbf{r}(t) - i\mathbf{k}_i\cdot\mathbf{v}(t - \xi_i)]. \quad (14.20)
$$

As a result, $\rho^{(3)}(t)$ for the molecules is a function of $\mathbf{v}$. For example, from

(14.18) we find

$$\left[\rho^{(3)}(\mathbf{v}, t)\right]_{(a)} = \sum_{m,r,p,s} \left\{ e^{-(i\omega_{ps} + \phi_{ps})t} \left(\frac{1}{i\hbar}\right)^3 (|p\rangle\rho_{mm}^0\langle s|) e^{i\mathbf{k}_s\cdot(\mathbf{r}-\mathbf{v}t)} \right.$$

$$\times \langle p|\mathbf{p}\cdot\hat{e}_1|m\rangle\langle m|\mathbf{p}\cdot\hat{e}_2|r\rangle\langle r|\mathbf{p}\cdot\hat{e}_3|s\rangle$$

$$\times \int_{-\infty}^{t} d\xi_3 e^{[i(\omega_{rs}-\omega_3)+\phi_{ps}-\phi_{pr}+i\mathbf{k}_3\cdot\mathbf{v}]\xi_3}\mathscr{E}_3(\xi_3) \qquad (14.21)$$

$$\times \int_{-\infty}^{\xi_3} d\xi_2 e^{[i(\omega_{mr}-\omega_2)+\phi_{pr}-\phi_{pm}+i\mathbf{k}_2\cdot\mathbf{v}]\xi_2}\mathscr{E}_2(\xi_2)$$

$$\left.\times \int_{-\infty}^{\xi_2} d\xi_1 e^{[i(\omega_{pm}-\omega_1)+\phi_{pm}+i\mathbf{k}_1\cdot\mathbf{v}]\xi_1}\mathscr{E}_1(\xi_1) \right\}.$$

The overall $\rho^{(3)}$ for the gas system is then given by an average of $\rho^{(3)}(\mathbf{v}, t)$ over the velocity distribution $n(\mathbf{v})$:

$$\rho^{(3)}(t) = \int_{-\infty}^{\infty} n(\mathbf{v})\rho^{(3)}(\mathbf{v}, t)d\mathbf{v}. \qquad (14.22)$$

Equation (14.21) shows explicitly that $\rho^{(3)}$ has a wavevector $\mathbf{k}_s \equiv \mathbf{k}_1 + \mathbf{k}_2 + \mathbf{k}_3$. The transient behavior of $\rho^{(3)}(\mathbf{v}, t)$ as governed by the time-dependent phase factor is now also a function of $\mathbf{v}$. If we assume that the exciting pulses are resonant with $\omega_1 = \omega_{pm}$, $\omega_2 = \omega_{mr}$, and $\omega_3 = \omega_{rs}$, and the pulses are short compared to the relaxation time $|\phi_{ij}|^{-1}$ and the velocity dephasing time $|\mathbf{k}_i\cdot\mathbf{v}|^{-1}$, then (14.21) reduces to

$$\left[\rho^{(3)}(\mathbf{v}, t)\right]_{(a)} = \left(\frac{1}{i\hbar}\right)^3 |p\rangle\langle s|e^{-i\omega_{ps}t + i\mathbf{k}_s\cdot\mathbf{r}}$$

$$\times e^{-i\mathbf{v}\cdot[\mathbf{k}_3(t-\xi_{30})+\mathbf{k}_2(t-\xi_{20})+\mathbf{k}_1(t-\xi_{10})]}$$

$$\times e^{-\phi_{ps}(t-\xi_{30})-\phi_{pr}(\xi_{30}-\xi_{20})-\phi_{pm}(\xi_{20}-\xi_{10})} \qquad (14.23)$$

$$\times \langle p|\mathbf{p}\cdot\hat{e}_1|m\rangle\langle m|\mathbf{p}\cdot\hat{e}_2|r\rangle\langle r|\mathbf{p}\cdot\hat{e}_3|s\rangle$$

$$\int_{-\infty}^{t} d\xi_3\mathscr{E}_3(\xi_3)\int_{-\infty}^{\xi_3} d\xi_2\mathscr{E}_2(\xi_2)\int_{-\infty}^{\xi_2} d\xi_1\mathscr{E}_1(\xi_1)\rho_{mm}^0$$

where ξ_{i0} is the time when the center of the ith pulse arrives at $\mathbf{r}(t)$. It is seen that the only phase factor in $\rho^{(3)}(\mathbf{v}, t)$ depending on $\mathbf{v}$ is

$$e^{-i\theta(\mathbf{v})} \equiv e^{-i\mathbf{v}\cdot[\mathbf{k}_3(t-\xi_{30})+\mathbf{k}_2(t-\xi_{20})+\mathbf{k}_1(t-\xi_{10})]}. \qquad (14.24)$$

If $\theta(\mathbf{v}) = 0$ for all $\mathbf{v}$, then molecules with different $\mathbf{v}$ will radiate in phase, and the transient four-wave mixing output will be a maximum. This happens at

$t = t_e$ provided $t_e \geq \xi_{30}$, where

$$t_e = \mathbf{k}_s \cdot [\mathbf{k}_1 \xi_{10} + \mathbf{k}_2 \xi_{20} + \mathbf{k}_3 \xi_{30}] / |k_s|^2. \qquad (14.25)$$

This result shows that the coherent output from transient four-wave mixing is expected to last only for a short duration during which $\theta(v)$ is small for all v. This is generally true for all coherent transient effects. The other factors in (14.23) determine the intensity of the four-wave mixing output, with $\exp[-\phi_{ps}(t - \xi_{30}) - \phi_{pr}(\xi_{30} - \xi_{20}) - \phi_{pm}(\xi_{20} - \xi_{10})]$ describing the decay of molecule excitation due to random perturbation, and hence the decay of the coherent output signal. We do not dwell on the details of transient four-wave mixing here but postpone the discussion to Chapter 21 in conjunction with other coherent transient optical effects.

REFERENCES

1 P. D. Maker and R. W. Terhune, *Phys. Rev. A* **137**, 801 (1965).

2 See, for example, J. J. Wynne and P. P. Sorokin, in Y. R. Shen, ed., *Nonlinear Infrared Generation* (Springer-Verlag, Berlin, 1977), p. 159; D. C. Hanna, M. A. Yuratich, and D. Cotter, *Nonlinear Optics of Free Atoms and Molecules* (Springer-Verlag, Berlin, 1979).

3 B. I. Stepanov, E. V. Ivakin, and A. S. Rubanov, *Sov. Phys. Doklady* **16**, 46 (1971); R. W. Hellwarth, *J. Opt. Soc. Am.* **67**, 1 (1977).

4 R. W. Hellwarth, *Prog. Quant. Electron.* **5**, 1 (1977).

5 N. Bloembergen, H. Lotem, and R. T. Lynch, *Ind. J. Pure Appl. Phys.* **16**, 151 (1978).

6 J-L. Oudar and Y. R. Shen, *Phys. Rev. A* **22**, 1141 (1981).

7 A. Yariv and D. M. Pepper, *Opt. Lett.* **1**, 16 (1977).

8 A. Yariv, *IEEE J. Quant. Electron.* **QE-14**, 650 (1978).

9 C. R. Giuliano, *Phys. Today* **34**, No. 4, 27 (1980), and references therein.

10 B. Ya. Zel'dovich, V. I. Popovicher, V. V. Ragul'skii, and F. S. Faizullov, *Sov. Phys. JETP* **15**, 109 (1972).

11 R. W. Hellwarth, *J. Opt. Soc. Am.* **68**, 1050 (1978).

12 J. J. Wynne, P. P. Sorokin, and J. R. Lankard, in R. G. Brewer and A. Mooradian, eds., *Laser Spectroscopy* (Plenum, New York, 1974), p. 103.

13 P. P. Sorokin, J. J. Wynne, and J. R. Lankard, *Appl. Phys. Lett.* **22**, 342 (1973).

14 D. M. Bloom, J. R. Yardley, J. F. Young, and S. E. Harris, *Appl. Phys. Lett.* **24**, 427 (1974).

15 R. T Hodgson, P. P. Sorokin, and J. J. Wynne, *Phys. Rev. Lett.* **32**, 343 (1974).

16 K. K. Innes, B. P. Stoicheff, and S. C. Wallace, *Appl. Phys. Lett.* **29**, 715 (1976); F. Vallee, S. C. Wallace, and J. Lugasik, *Opt. Commun.* **42**, 148 (1982); F. Vallee and J. Lugasik, *Opt. Commun.* **43**, 287 (1982).

17 A. H. Kung, *Opt. Lett.* **8**, 24 (1983); Technical Digest, Conference on Lasers and Electro-Optics, Baltimore, (1983), paper Tu 13.

18 T. J. McKee, S. C. Wallace, and B. P. Stoicheff, *Opt. Lett.* **3**, 207 (1978); F. S. Tomkins and R. Mahon, *Opt. Lett.* **6**, 179 (1981); J. Bokor, R. R. Freeman, R. L. Panock, and J. C. White, *Opt. Lett.* **6**, 182 (1981); W. Jamroz, P. E. LaRocque, and B. P. Stoicheff, *Opt. Lett.* **7**, 617 (1982).

19 D. S. Bethune, J. R. Lankard, and P. P. Sorokin, *Opt. Lett.* **4**, 103 (1979); Ph. Avouris, D. S.

Bethune, J. R. Lankard, J. Ors, and P. P. Sorokin, *J. Chem. Phys.* **74**, 2304 (1981).

20   T. K. Yee and T. K. Gustafson, *Phys. Rev. A* **18**, 1597 (1978).

21   P. Ye and Y. R. Shen, *Phys. Rev. A* **25**, 2183 (1982).

BIBLIOGRAPHY

Fisher, R. A., ed., *Optical Phase Conjugation*, (Academic Press, New York, 1983).

Hanna, D. C., M. A. Yuratich, and D Cotter, *Nonlinear Optics of Free Atoms and Molecules* (Springer-Verlag, Berlin, 1981).

Hellwarth, R. W., *Prog. Quant. Electron.* **5**, 1 (1977).

Proceedings of the International Workshop on Optical Phase Conjugation and Instabilities, C. Flytzanis et al., eds. (Cargese, France, 1982).

15

Four-Wave Mixing Spectroscopy

Four-wave mixing with resonant excitations can be a versatile spectroscopic technique.[1,2] It has already found important applications in many areas, such as analytical chemistry, combustion, and material studies. Its advantages over other techniques are in the capabilities for high resolution, for elimination of strong fluorescence background, and for time-resolving measurements of ultrafast dynamic properties. In this chapter we discuss four-wave mixing spectroscopy in its various forms. Note that most of the high-resolution nonlinear spectroscopic techniques discussed in Chapter 13 can be regarded as four-wave mixing spectroscopy when the nonlinearity is limited to the third order.

15.1 GENERAL DESCRIPTION

As shown in Chapter 14, the signal in a four-wave mixing process is directly related to the third-order nonlinear susceptibility $\chi^{(3)}$, which exhibits resonances characteristic of the nonlinear medium. Therefore, from $\chi^{(3)}$ as a function of pump frequencies, one can obtain spectroscopic information about the medium.

The expressions of $\chi^{(3)}(\omega = \omega_1 + \omega_2 + \omega_3)$ at the beginning of Section 14.1 show that it can have single, double, or triple resonances. Singly resonant four-wave mixing has the merit of a relatively simple experimental arrangement since only one pump frequency needs to be tuned. The corresponding theoretical analysis is also straightforward. The process can yield information one normally obtains from ordinary one-photon or two-photon transition measurements. Doubly and triply resonant four-wave mixing may require more than one tunable pump frequency and hence a more complicated experimental set-up. It can, however, give more selective spectroscopic information about the medium. Doppler-free spectra, for instance, can be obtained with double or triple resonances. The high-resolution nonlinear spectroscopic techniques discussed in Chapter 13 are good examples.

The two general types of four-wave mixing processes discussed in Section 14.2 can lead to two different kinds of four-wave mixing spectroscopic techniques. In the first case, the output is in a mode different from the input modes. The signal is proportional to $|\chi^{(3)}|^2$ and can be selectively detected through filtering against the transmitted pump beams. The technique is most commonly used in singly resonant cases where the analysis is relatively simple. In the second case, the output is in the same mode as one of the input waves. The signal appears in the form of gain or loss of that particular input wave and is proportional to $\mathrm{Im}(\chi^{(3)})$ (see Section 14.2, "Output Field in the Same Mode as One of the Input Fields"). This then allows a simple interpretation of the observed spectrum.

In the following sections, we discuss the few well-known four-wave mixing spectroscopic methods in detail. While in most cases we consider $\omega_1 - \omega_2$ near resonance (the Raman case), keep in mind that the same description applies to cases with $\omega_1 + \omega_2$ near resonance.

15.2 COHERENT RAMAN SCATTERING SPECTROSCOPY

Coherent Anti-Stokes Raman Scattering (CARS)

As shown schematically in Fig. 15.1, CARS refers to the Raman scattering process in which the material excitational wave at $\omega_1 - \omega_2 \simeq \omega_v$ is first coherently excited by the beating of two incoming waves at ω_1 and ω_2 and then mixed with the wave at ω_1 to yield a coherent output at the anti-Stokes frequency $\omega_a = 2\omega_1 - \omega_2$. It is therefore a coherent version of the spontaneous Raman scattering process. Because of its numerous important applications, CARS is perhaps the best known four-wave mixing spectroscopic method. A large number of review articles already exist on the subject.[2] The theory of CARS is simple, as it follows the general theory of optical mixing discussed earlier in Chapters 6 and 14. With the input pump beams at ω_1 and ω_2, the output field is governed by the wave equation (assuming an isotropic medium)

$$\left[\nabla^2 + \frac{\omega_a^2}{c^2}\epsilon_a(\omega_a)\right]\mathbf{E}_a = -\frac{4\pi\omega_a^2}{c^2}\mathbf{P}^{(3)}(\omega_a) \qquad (15.1)$$

where $\mathbf{P}^{(3)}(\omega_a) = \chi^{(3)}(\omega_a = \omega_1 + \omega_1 - \omega_2):\mathbf{E}_1(\omega_1)\mathbf{E}_1(\omega_1)\mathbf{E}_2^*(\omega_2)$. The solution of (15.1) with negligible pump depletion and damping gives an output intensity I_a from a slab medium of length l:

$$I_a = \frac{c}{2\pi}|E_a|^2 = \frac{2\pi\omega_a^2}{c\epsilon_a}|\chi_{as}^{(3)}|^2|E_1|^4|E_2|^2\frac{\sin^2\frac{1}{2}\Delta k_l l}{\left(\frac{1}{2}\Delta k_l l\right)^2} \qquad (15.2)$$

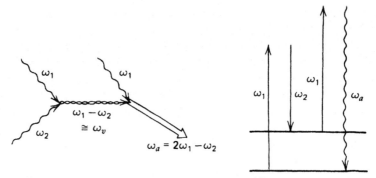

Fig. 15.1 Schematic diagrams describing the CARS process.

where

$$\chi_{as}^{(3)} = \hat{e}_a \cdot \boldsymbol{\chi}^{(3)} : \hat{e}_1 \hat{e}_1 \hat{e}_2$$
$$\Delta k_l = \Delta \mathbf{k} \cdot \hat{l},$$
$$\Delta \mathbf{k} = \mathbf{k}_a - (2\mathbf{k}_1 - \mathbf{k}_2).$$

We note that the anti-Stokes output is proportional to $|\chi_{as}^{(3)}|^2$. If the frequency scan is limited to a sufficiently narrow range, the variation of I_a versus the pump frequencies is entirely dominated by the resonant dispersion of $\chi_{as}^{(3)}$. As discussed in Section 14.1, $\chi_{as}^{(3)}$ can be decomposed into a resonant part $\chi_R^{(3)}$ and a nonresonant part $\chi_{NR}^{(3)}$. Here, $\chi_R^{(3)}$ is singly resonant at $\omega_1 - \omega_2 = \omega_v$ and can be written in the form

$$\chi_R^{(3)} = \frac{a}{(\omega_1 - \omega_2 - \omega_v) + i\Gamma} \tag{15.3}$$

where, from (14.2), $a = -N(M_{g'g}^a)^*(M_{g'g}^s)(\rho_g - \rho_{g'})/\hbar$ is essentially independent of ω_1 and ω_2 as $\omega_1 - \omega_2$ scans over the Raman resonance. Therefore, the anti-Stokes output has a spectrum given by

$$|\chi_{as}^{(3)}|^2 = \left[\chi_{NR}^{(3)} + \frac{a(\omega_1 - \omega_2 - \omega_v)}{(\omega_1 - \omega_2 - \omega_v)^2 + \Gamma^2}\right]^2 + \frac{a^2\Gamma^2}{(\omega_1 - \omega_2 - \omega_v)^2 + \Gamma^2}. \tag{15.4}$$

Figure 15.2 is an example. Because of the presence of the nonresonant part $\chi_{NR}^{(3)}$ in $\chi_{as}^{(3)}$, the spectrum appears asymmetric with respect to the resonant point $\omega_1 - \omega_2 = \omega_v$, and has a peak and a dip at

$$(\omega_1 - \omega_2)_\pm = \omega_v + \frac{1}{2}\left\{-\frac{a}{\chi_{NR}^{(3)}} \pm \left[\left(\frac{a}{\chi_{NR}^{(3)}}\right)^2 + 4\Gamma^2\right]^{1/2}\right\}. \tag{15.5}$$

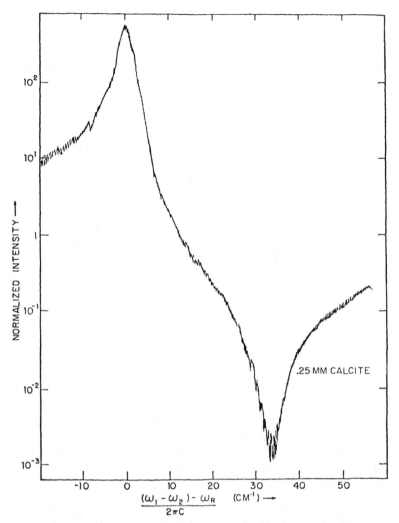

Fig. 15.2 A CARS spectral line around the 1088-cm^{-1} resonance in calcite. [After M. D. Levenson, *IEEE J. Quant. Electron.* **QE-10**, 110 (1974).]

We find

$$(\omega_1 - \omega_2)_+ + (\omega_1 - \omega_2)_- = 2\omega_v - \frac{a}{\chi_{NR}^{(3)}}$$

$$\left[(\omega_1 - \omega_2)_+ - (\omega_1 - \omega_2)_-\right]^2 = \left(\frac{a}{\chi_{NR}^{(3)}}\right)^2 + 4\Gamma^2.$$

(15.6)

Measurements of $(\omega_1 - \omega_2)_\pm$ together with the output intensities at $(\omega_1 - \omega_2)_\pm$

can in principle determine all the four quantities ω_v, Γ, a, and $\chi_{NR}^{(3)}$. In practice, however, the accuracy of intensity measurements is usually poor but that of frequency measurements can be very good. Then, we note from (15.6) that if ω_v or Γ is known from other measurements, $a/\chi_{NR}^{(3)}$ can be obtained simply from the frequency measurement of $(\omega_1 - \omega_2)_+ \pm (\omega_1 - \omega_2)_-$. If a is also known from spontaneous Raman scattering, then $\chi_{NR}^{(3)}$ can be deduced very accurately.[3] This contrasts with the inaccurate determination of $\chi_{NR}^{(3)}$ by the intensity measurements.

Although $(\omega_1 - \omega_2)_+$ alone cannot determine the resonant frequency ω_v, they do give an approximate value for ω_v, and therefore can be used to characterize the resonant medium. In many cases the Raman resonance is strong, while the nonresonant background $\chi_{NR}^{(3)}$ is weak, so that $|a/\chi_{NR}^{(3)}| \gg \Gamma$. This often occurs, for example, in pure gas media, where, one finds $(\omega_1 - \omega_2)_+ \simeq \omega_v$ and $(\omega_1 - \omega_2)_- \simeq \omega_v - a/\chi_{NR}^{(3)}$. The resonant frequency ω_v can be directly obtained from the peak position of the CARS spectrum. In other cases where $|a/\chi_{NR}^{(3)}| \lesssim \Gamma$, both the peak and the dip of the spectrum appear around ω_v within a range of the order of Γ. However, if $|\chi_{NR}^{(3)}|$ is much larger than a/Γ, then the peak and dip will not be prominent in the spectrum. This is clearly a drawback of CARS as a spectroscopic technique because weak Raman resonances are not easily observable. The sensitivity of CARS can be greatly improved if the nonresonant background due to $\chi_{NR}^{(3)}$ can be suppressed. As we shall see later, this can be done by selectively detecting the proper polarization component of the output.

Maker and Terhune[4] first demonstrated the CARS process. Later, Wynne,[5] Bloembergen and his associates,[6] and Akhmanov et al.[7] used the method for spectroscopic studies in a large number of liquids and solids. A typical experimental arrangement is presented in Fig. 15.3. Dye lasers pumped by either a nitrogen laser or a frequency-doubled Nd : YAG laser are often used as tunable pump sources. The signal strength can be estimated from (15.2). Assume $|\chi_R^{(3)}|_{max} = |a/\Gamma| > |\chi_{NR}^{(3)}|$ and perfect phase matching $\Delta k = 0$. Then, if input beams of 10 kW peak power at ~ 5000 Å are focused to a spot of 10^{-5} cm^2 into a medium 1 mm long with $|\chi_R^{(3)}|_{max} \simeq 10^{-12}$ esu, we find that the output can have a peak power of 5 W. If the medium is 2 cm long with $|\chi_R^{(3)}|_{max} \simeq 10^{-15}$ esu, the output can still have a peak power of 2 mW, which is readily detectable. The last example actually corresponds to the case of CARS in a gas medium at nearly the atmospheric pressure. In fact, while initial CARS work was on condensed matter, the more recent applications of CARS to material studies have been on gases. Even CARS from molecules in molecular beams has been observed.[8]

Because the spectral resolution of CARS is basically limited by the laser linewidths, it can be used to obtain high-resolution Raman spectra of gases.[9] Conventional Raman scattering, in comparison, suffers not only poor spectral resolution limited by the spectrometer but also low signal intensity. The high sensitivity of CARS allows the method to be used for detecting trace molecules in gases. In particular, it has found applications in combustion studies as a

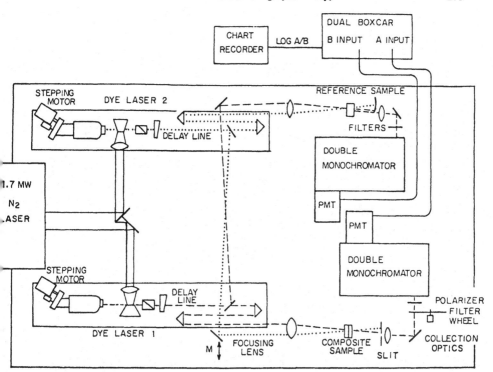

Fig. 15.3 A CARS spectroscopy system. Laser beam 1 (---) and laser beam 2 ($\cdots$) are blocked after the samples, and the output due to frequency mixing is collected into the double monochromators. The signal from the reference arm and the ratioing system are used for normalization in order to reduce the effect from laser fluctuations. [After M. D. Levenson, *IEEE J. Quant. Electron.* **QE-10**, 110 (1974).]

means to monitor the temporal and spatial distributions of various species and to find the internal energy distribution of the species.[10] Because the output is coherent and highly directional, temporal, spatial, and spectral filtering can be used to suppress the strong luminescence background from the combustion process. In fact, its ability to discriminate against luminescence makes coherent Raman spectroscopy in general a unique technique for combustion studies. Similarly, CARS can be used to probe the reaction products and their internal energy distribution from a chemical reaction, and to obtain Raman spectra of fluorescent materials. That CARS is capable of probing materials in an enclosed, hostile environment with good space and time resolution has prompted the suggestion of using CARS to study high-temperature plasmas[11] and to monitor target implosion in laser fusion.[12]

Although the foregoing discussion is focused on Raman resonances, the theoretical treatment applies equally well to two-photon resonances with

$\omega_1 + \omega_2 \simeq \omega_0$, or $2\omega_1 \simeq \omega_0$, or $2\omega_2 \simeq \omega_0$. The coherent output is at either $\omega_0 \pm \omega_1$ or $\omega_0 \pm \omega_2$.

Polarization CARS

It is easy to see from (15.4) that if $|\chi_R^{(3)}|_{max} = |a/\Gamma| \ll \chi_{NR}^{(3)}$, the resonant structure sitting on the nonresonant background in the CARS spectrum will be easily lost in the presence of background fluctuations. With an appropriate polarization arrangement in the detection scheme, however, the nonresonant background can be largely suppressed, and hence the sensitivity of CARS is greatly improved.[13, 14] The basic principle of the polarization CARS is described below.

The solution of (15.1) shows that the output field $\mathbf{E}_a$ is proportional to the nonlinear polarization $\mathbf{P}^{(3)}(\omega_a)$, which can be written as

$$\mathbf{P}^{(3)} = \mathbf{P}_{NR}^{(3)} + \mathbf{P}_R^{(3)} \tag{15.7}$$

with $\mathbf{P}_{NR}^{(3)} = \chi_{NR}^{(3)} : \mathbf{E}_1 \mathbf{E}_1 \mathbf{E}_2^*$ and $\mathbf{P}_R^{(3)} = \chi_R^{(3)} : \mathbf{E}_1 \mathbf{E}_1 \mathbf{E}_2^*$. If $\mathbf{P}_{NR}^{(3)}$ is in the direction $\hat{e}_\mu$, then since $\mathbf{P}_R^{(3)}$ and $\mathbf{P}_{NR}^{(3)}$ are generally not in the same direction, the output field has the form

$$\begin{aligned}
\mathbf{E}_a = \Big[& \hat{e}_\mu A^{NR}\big(\hat{e}_\mu \cdot \chi_{NR}^{(3)} : \mathbf{E}_1 \mathbf{E}_1 \mathbf{E}_2^* \big) + \hat{e}_\mu A^R \big(\hat{e}_\mu \cdot \chi_R^{(3)} \mathbf{E}_1 \mathbf{E}_1 \mathbf{E}_2^* \big) \\
& + \hat{e}_\nu A^R \big(\hat{e}_\nu \cdot \chi_R^{(3)} : \mathbf{E}_1 \mathbf{E}_1 \mathbf{E}_2^* \big) \Big]
\end{aligned} \tag{15.8}$$

where A^{NR} and A^R are coefficients, and $\hat{e}_\nu$ is orthonormal to $\hat{e}_\mu$. With an analyzer in front of the detector to block out the $\hat{e}_\mu$ component, the output signal is then given by

$$I_a \propto |A^R \hat{e}_\nu \cdot \chi_R^{(3)} : \mathbf{E}_1 \mathbf{E}_1 \mathbf{E}_2^*|^2 \tag{15.9}$$

which is proportional to only $|\hat{e}_\nu \cdot \chi_R^{(3)} : \hat{e}_1 \hat{e}_1 \hat{e}_2|^2$. The nonresonant background can in principle be completely suppressed. In practice, because the analyzer is not perfect, a residual nonresonant background still shows up but is certainly greatly reduced from the original magnitude. For spectroscopic studies, one is interested in the dispersion of $\chi_R^{(3)}$ and not in the absolute value of $\chi_R^{(3)}$. Then, with weak resonances such that $|\hat{e}_\mu \cdot \chi_R^{(3)} : \hat{e}_1 \hat{e}_1 \hat{e}_2| \ll |\hat{e}_\mu \cdot \chi_{NR}^{(3)} : \hat{e}_1 \hat{e}_1 \hat{e}_2|$, it is most convenient to record the ratio $|R|^2$ of the $\hat{e}_\mu$ and $\hat{e}_\nu$ components of the output, which is given by

$$|R|^2 = \left| \frac{A_\nu^R \big(\hat{e}_\nu \cdot \chi_R^{(3)} : \hat{e}_1 \hat{e}_1 \hat{e}_2 \big)}{A_\mu^{NR} \big(\hat{e}_\mu \cdot \chi_{NR}^{(3)} : \hat{e}_1 \hat{e}_1 \hat{e}_2 \big)} \right|^2 . \tag{15.10}$$

Note that $|R|^2$ is independent of the pump intensities, and therefore is little affected by laser intensity fluctuations. Since the dispersion of $\chi_{NR}^{(3)}$ is negligible in the resonant region, $|R|^2$ versus $\omega_1 - \omega_2$ essentially reproduces the spectrum of $|\chi_R^{(3)}|^2$.

A possible experimental arrangement for this polarization CARS is schematically shown in Fig. 15.4. The $\hat{e}_\mu$ and $\hat{e}_\nu$ components, corresponding to the transmitted and rejected components from the analyzer, are simultaneously recorded by the two photodetectors, and their ratio is taken by the divider. The polarization CARS spectrum of benzene in CCl_4 is shown in Fig. 15.5 as an example.[14] The same spectral line would hardly be visible in ordinary CARS because of the enormous nonresonant background.

In some cases, one would like to obtain the spectra of $Re(\chi_R^{(3)})$ and $Im(\chi_R^{(3)})$ separately. This can be achieved with a simple modification of the polarization arrangement in Fig. 15.4. If we rotate the analyzer by a small angle θ from the null point, then R is replaced by $R_1 \cong \tan\theta + R$, and the ratio of the two output components by

$$|R_1|^2 = |\tan\theta + R|^2. \tag{15.11}$$

Let us choose $\tan^2\theta \gg |R|^2$. Then the output ratio can be approximated, with $R = R' + iR''$, by

$$|R_1|^2 = \tan^2\theta + 2R'\tan\theta. \tag{15.12}$$

Since $\chi_{NR}^{(3)}$ is generally real, R' is directly proportional to $Re(\hat{e}_\nu \cdot \chi_R^{(3)} : \hat{e}_1\hat{e}_1\hat{e}_2)$. The spectrum of $|R_1|^2$ therefore yields the resonant spectrum of $Re(\hat{e}_\nu \cdot \chi_R^{(3)} : \hat{e}_1\hat{e}_1\hat{e}_2)$ on top of a constant background. The relative strength of the resonant structure versus the background in this case is $2R'/\tan\theta$, which is certainly much larger than that in ordinary CARS. In actual experiments, it is more convenient to use a polarization rotator in front of the analyzer than to rotate the analyzer. This is to avoid problems that may arise because of sensitivity of the photodetection system to the variation of beam polarization.[14]

If, in addition, a quarter wave plate is also inserted before the polarization rotator to change the relative phase of the $\hat{e}_\nu$ component by $\pm90°$, then the

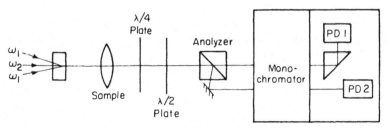

Fig. 15.4 An experimental arrangement for polarization CARS. (After Ref. 15.)

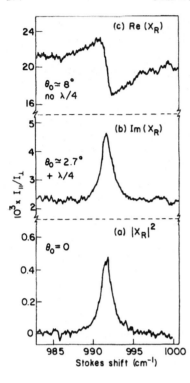

$10^3 \times I_\parallel / I_\perp$

(c) Re (χ_R)

$\theta_0 \simeq 8°$
no $\lambda/4$

(b) Im (χ_R)

$\theta_0 \simeq 2.7°$
+ $\lambda/4$

(a) $|\chi_R|^2$

$\theta_0 = 0$

Stokes shift (cm⁻¹)

Fig. 15.5 Polarization CARS spectra of 0.011 M benzene in carbon tetrachloride: (a) with suppression of the nonresonant background, (b) with Im $\chi_R^{(3)}$ superimposed on a weak nonresonant background, and (c) with Re $\chi_R^{(3)}$ superimposed on a weak nonresonant background. (After Ref. 14.)

ratio of the two output components becomes

$$|R_2|^2 = |\tan\theta \pm iR|^2$$
$$\simeq \tan^2\theta \mp 2R''\tan\theta, \tag{15.13}$$

which yields the resonant spectrum of Im$(\hat{e}_\nu \cdot \chi_R^{(3)} : \hat{e}_1\hat{e}_1\hat{e}_2)$ on top of a constant background. An example of $|R_1|^2$ and $|R_2|^2$ versus $\omega_1 - \omega_2$ is shown in Fig. 15.5 for 0.01 M benzene dissolved in CCl₄.[14] The small value of $|\chi_R^{(3)}/\chi_{NR}^{(3)}|$ due to the small amount of benzene in this case would make the detection of the Raman resonance of benzene very difficult with ordinary CARS. The technique of measuring R' and R'' described here is essentially the same as the well-known heterodyne technique. The uncrossed coherent background signal here plays the role of the local oscillator.

We realize that even in detecting $|R|^2$, the resonant spectrum is not completely free of background because of the finite extinction coefficient of the analyzer. Fluctuations of the background still limit the sensitivity of the polarization CARS. Even though the background in the measurements of R' and R'' is higher than that in the measurement of $|R|^2$, a signal-to-noise ratio analysis shows that the former can be orders of magnitude more sensitive.[15] This is because heterodyning effectively increases the signal-to-background ratio. The greatly improved sensitivity of the polarization CARS makes it particularly useful for probing trace molecules, as in combustion studies.[16]

Other Coherent Raman Spectroscopic Techniques

As a variation of CARS, one can detect, with the same input pump beams, the coherent output at the frequency $2\omega_2 - \omega_1$. This is known as coherent Stokes Raman scattering (CSRS). The theory of CSRS is the same as CARS, except that the values of $\chi_{NR}^{(3)}$ and a are somewhat different because of dispersion. The polarization arrangement can also be used to improve the sensitivity of CSRS.

Raman gain and inverse Raman spectroscopic techniques have already been discussed in Section 10.6. They can be considered four-wave mixing processes with the output in the same mode as one of the inputs. The gain or loss, which is directly proportional to $\mathrm{Im}\chi_R^{(3)}$, shows a resonant spectrum without the nonresonant background. The techniques are particularly suitable for obtaining high-resolution Raman spectra of gases. With sufficiently strong imput laser intensities, the Raman transition can even be saturated. Saturation Raman spectroscopy, similar in principle to saturation spectroscopy discussed in Section 13.3, can be used to obtain sub-Doppler Raman spectra.[17]

Although the discussion in this section has been focused on Raman resonances, the theoretical treatment applies equally well to general two-photon resonances with $\omega_1 + \omega_2 = \omega_0$, or $2\omega_1 = \omega_0$, or $2\omega_2 = \omega_0$, where ω_0 is the resonant frequency. The coherent output can be detected at either $|\omega_1 \pm \omega_0|$ or $|\omega_2 \pm \omega_0|$. Interference may show up if the output exhibits Raman and two-photon resonances at the same time.[18]

15.3 RAMAN-INDUCED KERR EFFECT SPECTROSCOPY (RIKES)

A gain or loss is always accompanied by a corresponding resonant birefringence. Thus for an induced Raman gain or loss, there should be a corresponding induced birefringence seen by the probe beam. This is known as the Raman-induced Kerr effect,[19] where the Kerr effect is used indiscriminantly to denote the field-induced birefringence phenomenon. (The optical Kerr effect is discussed in more detail in Chapter 16.)

We consider here the same experimental geometry as in the Raman gain spectroscopy with a pump beam at ω_1 and a probe beam at ω_2, both propagating along $\hat{z}$. The frequency $\omega_1 - \omega_2$ is scanned over the Raman resonance at ω_v. For simplicity, let the probe beam be linearly polarized along $\hat{x}$. In propagating through the medium, the polarization of the probe beam becomes elliptical as a result of the Raman-induced birefringence. When the birefringence is small, the polarization change can be treated as creation of a new polarization component at ω_2 along $\hat{y}$ through the four-wave mixing process as follows. The incoming fields induce a third-order polarization with a

$\hat{y}$ component

$$P_y^{(3)}(\omega_2) = \hat{y} \cdot \chi^{(3)}(\omega_2 = \omega_1 - \omega_1 + \omega_2) \cdot \hat{e}_1 \hat{e}_1 \hat{x} E_1 E_1^* E_2. \quad (15.14)$$

In a medium of length l, this nonlinear polarization component generates a field component along $\hat{y}$

$$E_y(\omega_2) = \frac{i2\pi\omega_2^2}{c^2 k_2} P_y^{(3)}(\omega_2). \quad (15.15)$$

Measurement of the output $I_y(\omega_2) \propto |E_y(\omega_2)|^2$ versus $\omega_1 - \omega_2$ should therefore yield the spectrum of $|\hat{y} \cdot \chi^{(3)}: \hat{e}_1 \hat{e}_1 \hat{x}|^2$, which exhibits Raman resonance.

In many media, if the collinearly propagating beams are along some symmetry direction, the subindices of $\chi_{ijkl}^{(3)}$ must appear in like pairs from symmetry consideration. We then have

$$P_y^{(3)}(\omega_2) = \chi_{yxyx}^{(3)} E_{1x} E_{1y}^* E_{2x} + \chi_{yyxx}^{(3)} E_{1y} E_{1x}^* E_{2x}. \quad (15.16)$$

With the pump beam linearly polarized in the direction bisecting $\hat{x}$ and $\hat{y}$, so that $|E_{1x}| = |E_{1y}|$, we find

$$I_y(\omega_2) \propto |\chi_{yxyx}^{(3)} + \chi_{yyxx}^{(3)}|^2 |E_{1x}|^4 |E_{2x}|^2. \quad (15.17)$$

The phase-matching condition is automatically satisfied in this case. Since $\chi^{(3)} = \chi_{NR}^{(3)} + \chi_R^{(3)}$, the observed spectrum has the resonant structure superimposed on the nonresonant background, similar to that obtained by CARS. If the pump beam is circularly polarized, then $E_{1x} = \pm i E_{1y}$, and we have

$$I_y(\omega_2) \propto |\chi_{yxyx}^{(3)} - \chi_{yyxx}^{(3)}|^2 |E_1|^4 |E_2|^2. \quad (15.18)$$

In an isotropic medium, the nonresonant part has the symmetry relation $(\chi_{NR}^{(3)})_{yxyx} = (\chi_{NR}^{(3)})_{yyxx}$, but $(\chi_R^{(3)})_{yxyx} \neq (\chi_R^{(3)})_{yyxx}$, and we find

$$I_y(\omega_2) \propto |(\chi_R^{(3)})_{yxyx} - (\chi_R^{(3)})_{yyxx}|^2 |E_1|^4 |E_2|^2. \quad (15.19)$$

The observed spectrum therefore shows no nonresonant background and is similar to that obtained in polarization CARS (Fig. 15.5a). In experiments, $I_y(\omega_2)/I_x(\omega_2)$ should be recorded to minimize the effect of probe beam fluctuations. This can be done by simultaneously recording the transmitted and rejected components of $I(\omega_2)$ from an analyzer in front of the detector.

One can also use the heterodyne scheme to obtain $\mathrm{Re}\chi_R^{(3)}$ and $\mathrm{Im}\chi_R^{(3)}$ separately.[20] Again, by rotating the analyzer a small angle θ away from $\hat{y}$, the ratio of the transmitted and rejected components becomes

$$|R_1|^2 = |\tan\theta + \gamma\{(\chi_R^{(3)})_{yxyx} - (\chi_R^{(3)})_{yyxx}\}|^2 \quad (15.20)$$

where γ is a real coefficient proportional to $|E_1|^2$. If $\tan^2\theta \gg |\gamma\{(\chi_R^{(3)})_{yxyx} - (\chi_R^{(3)})_{yyxx}\}|^2$, then we have

$$|R_1|^2 \simeq \tan^2\theta + 2\gamma(\tan\theta)\text{Re}\{(\chi_R^{(3)})_{yxyx} - (\chi_R^{(3)})_{yyxx}\}, \quad (15.21)$$

which yields a spectrum of $\text{Re}\chi_R^{(3)}$. On the other hand, with a quarter-wave plate inserted in the path, the ratio of the transmitted and rejected components becomes

$$\begin{aligned}|R_2|^2 &= |\tan\theta + i\gamma\{(\chi_R^{(3)})_{yxyx} - (\chi_R^{(3)})_{yyxx}\}|^2 \\ &\simeq \tan^2\theta + 2\gamma(\tan\theta)\text{Im}\{(\chi_R^{(3)})_{yxyx} - (\chi_R^{(3)})_{yyxx}\}\end{aligned} \quad (15.22)$$

which yields a spectrum of $\text{Im}\chi_R^{(3)}$.

Again, while the discussion here is on RIKES, it can be applied equally well to other two-photon resonances with $\omega_1 \pm \omega_2 = \omega_0$, where ω_0 is the resonant frequency.

15.4 MULTIPLY RESONANT FOUR-WAVE MIXING

The resonant nonlinear susceptibility $\chi_R^{(3)}$ can be further enhanced with double and triple resonances. It should give more selective spectroscopic information than in the singly resonant case. In general, multiply resonant four-wave mixing can provide more specific details about a particular one-photon resonant transition. A few interesting applications of such resonant processes are described here.

High-Resolution Doppler-Free Spectroscopy

Resonant four-wave mixing with appropriate input beams can yield Doppler-free spectra. A few examples were discussed in Chapter 13. The Doppler-free saturation and polarization spectroscopy with the lowest-order saturation is a triply resonant four-wave mixing process in which the pump field $E(\omega)$ creates a population change $\rho_{nn}^{(2)}$ proportional to $E(\omega)E^*(\omega)$ and then the probe field $E(\omega')$ detects the change through the nonlinear polarization $P^{(3)}(\omega') \propto E(\omega)E^*(\omega)E(\omega')$ with ω' and ω either on the same or different resonant transitions. The two-photon Doppler-free absorption spectroscopy, on the other hand, is based on a singly resonant four-wave mixing process.

In general, multiply resonant four-wave mixing can yield a Doppler-free spectrum if the damping coefficients of the resonant denominators of $\chi^{(3)}$ satisfy a certain condition. This is seen from the following mathematical derivation.[21]

Consider first a doubly resonant case. The resonant nonlinear susceptibility $\chi_R^{(3)}$ has the general form (see Section 2.2)

$$\chi_R^{(3)} = \frac{A}{(\omega_a - \omega_{ij} \pm i\Gamma_{ij})(\omega_b - \omega_{kl} \pm i\Gamma_{kl})} \tag{15.23}$$

where A is a coefficient with negligible frequency dependence in the present discussion. In a gas medium, $\chi_R^{(3)}$ should be an average over the Doppler profile (see Section 13.1), while ω_{ij} and ω_{kl} should be replaced by $\omega_{ij}^0(1 - \mathbf{v} \cdot \hat{k}_a/c)$ and $\omega_{kl}^0(1 - \mathbf{v} \cdot \hat{k}_b/c)$, respectively:

$$
\langle \chi_R^{(3)} \rangle
$$
$$
= \int_{-\infty}^{\infty} dv \, \frac{Ag(v)}{\left[\omega_a - \omega_{ij}^0(1 - \mathbf{v} \cdot \hat{k}_a/c) \pm i\Gamma_{ij}\right]\left[\omega_b - \omega_{kl}^0(1 - \mathbf{v} \cdot \hat{k}_b/c) \pm i\Gamma_{kl}\right]} \tag{15.24}
$$

where $g(v) = (1/\sqrt{\pi}\,u)\exp(-v^2/u^2)$ and $u = 2kT/m$. Equation (15.24) can be written in the form

$$\langle \chi_R^{(3)} \rangle = \int_{-\infty}^{\infty} dv \, \frac{K\exp(-v^2)}{\sqrt{\pi}\,(v - \xi_a)(v - \xi_b)} \tag{15.25}$$

with $v = v/u$, $\xi_a = [c/(\hat{v} \cdot \hat{k}_a)u](1 - \omega_a/\omega_{ij}^0 \mp i\Gamma_{ij}/\omega_{ij}^0)$, and a similar expression for ξ_b. Then the integral can be evaluated in terms of the plasma dispersion function

$$Z(\xi = \xi' + i\xi'') = \pi^{-1/2}\int_{-\infty}^{\infty}\frac{dv\,e^{-v^2}}{v - \xi} \qquad \text{for } \xi'' > 0 \tag{15.26}$$
$$Z(\xi^*) = -Z^*(-\xi)$$

and $\langle \chi_R^{(3)} \rangle$ behaves differently near the double resonance $\xi_a' = \xi_b'$, depending on the relative signs of ξ_a'' and ξ_b''. If ξ_a'' and ξ_b'' are of the same sign, we have

$$\langle \chi_R^{(3)} \rangle \propto \frac{[Z(\xi_a) - Z(\xi_b)]}{\xi_a - \xi_b}, \tag{15.27}$$

which shows no singularity as $\xi_a' \to \xi_b'$, and as a function of ξ_a' (or ω_a), has a resonant width more than twice the Doppler width. If, however, ξ_a'' and ξ_b'' have opposite signs, we find, as $\xi_a' \to \xi_b'$,

$$\langle \chi_R^{(3)} \rangle \propto \left(\frac{dZ}{d\xi} \pm \frac{i2\sqrt{\pi}\,e^{-\xi_a^{*2}}}{\xi_a - \xi_b} \right) \tag{15.28}$$

where + and − refer to the cases of $\xi_a'' > 0$ and $\xi_a'' < 0$, respectively. The last term in (15.28) has a Lorentzian resonant lineshape with a halfwidth $\xi_a'' + \xi_b''$ $= [c/(\hat{v} \cdot \hat{k})u](\Gamma_{ij}/\omega_{ij}^0 + \Gamma_{kl}/\omega_{kl}^0)$. In other words, as $\omega_a/\omega_{ij}^0 \to \omega_b/\omega_{kl}^0$ at double resonance, $\langle \chi_R^{(3)} \rangle$ versus ω_a/ω_{ij}^0 has a Lorentzian linewidth given by the sum of the normalized natural linewidths, $(\Gamma_{ij}/\omega_{ij}^0 + \Gamma_{kl}/\omega_{kl}^0)$. Therefore, measurement of $\langle \chi_R^{(3)} \rangle$, or $|\langle \chi_R^{(3)} \rangle|^2$ in this case, can yield a Doppler-free spectrum. For the various doubly resonance cases shown in Fig. 14.2, spectra obtained with 14.2d, g, and h are Doppler-free.[15] Note that Fig. 14.2g describes a CSRS process, while d and h are CARS and CSRS, respectively, probing a resonant transition between excited states.

A similar result is obtained for the triply resonant case. We can write

$$\langle \chi_R^{(3)} \rangle \propto \int_{-\infty}^{\infty} d\nu \, \frac{\exp(-\nu^2)}{(\nu - \xi_a)(\nu - \xi_b)(\nu - \xi_c)}. \tag{15.29}$$

At triple resonance, $\xi_a' \to \xi_b' \simeq \xi_c'$, if one of the ξ'' has a different sign from the two others, then (15.29) gives a Doppler-free spectrum. In general, $\langle \chi_R^{(3)} \rangle$ versus ω_a shows two Lorentzian peaks. They merge into one when $\xi_b' \to \xi_c'$. All the triply resonant processes in Fig. 14.3 should yield Doppler-free spectra.

The discussion here can of course be extended to the reduction of inhomogeneous broadening in solids.[15] However, because of the many crystal-field parameters governing the broadening (see Section 13.1), the elimination of crystal-field broadening is never perfect.

Measurement of Longitudinal Relaxation Times

If the relaxation of an excess population in a state can be approximated by

$$\left[\frac{\partial}{\partial t}(\rho_{nn} - \rho_{nn}^0) \right]_{\text{damping}} = -\frac{(\rho_{nn} - \rho_{nn}^0)}{T_{1n}}, \tag{15.30}$$

the resonant four-wave mixing spectroscopy can also be used to deduce the longitudinal relaxation time T_{1n}.[15] The physical idea is fairly simple. Two imput beams at ω_1 and ω_1' near a resonance can beat in the medium and induce a time-varying population change oscillating at the frequency $\omega_1 - \omega_1'$ with an amplitude inversely proportional to the zero-frequency resonant denominator $[(\omega_1 - \omega_1') + i/T_{1n}]$. The induced population change in the particular state is then probed by a resonant transition from that state. The output spectrum of $\langle \chi_R^{(3)} \rangle$ versus $\omega_1 - \omega_1'$ has a halfwidth $1/T_{1n}$, although the halfwidths of the one-photon transitions involved are much larger than $1/T_{1n}$.

Consider the process in Fig. 14.3a described by (14.4a) as an example. For a gas medium, the average $\chi_R^{(3)}$ with $T_1 \gg \Gamma^{-1}$ is

$$
\langle \chi_R^{(3)} \rangle \propto \frac{1}{\omega_1 - \omega_1' + i/T_{1n'}} \int dv \, \frac{g(v)}{\omega_2' + \mathbf{k}_2' \cdot \mathbf{v} - \omega_{n'g'} + i\Gamma_{n'g'}}
$$

$$
\times \left(\frac{1}{\omega_1 + \mathbf{k}_1 \cdot \mathbf{v} - \omega_{n'g} - i\Gamma_{n'g}} - \frac{1}{\omega_1' + \mathbf{k}_1' \cdot \mathbf{v} - \omega_{n'g} + i\Gamma_{n'g}} \right).
$$

$$\tag{15.31}$$

Since $T_{1n'} \gg \Gamma_{n'g}^{-1}, \Gamma_{n'g'}^{-1}$, the integral in (15.31) is practically independent of ω_1' in the range of $|\omega_1 - \omega_1'| \sim T_{1n'}^{-1}$, and therefore we have the spectrum of $\langle \chi_R^{(3)} \rangle$ versus $\omega_1 - \omega_1'$ as

$$
\langle \chi_R^{(3)} \rangle \propto \left(\omega_1 - \omega_1' + i/T_{1n'} \right)^{-1}.
$$

$$\tag{15.32}$$

Yajima et al.[22] proposed and demonstrated a method for measurements of T_1 and T_2 using four-wave mixing with two laser beams. It was also mentioned in Section 13.3 that T_1 and T_2 can be deduced simultaneously from the saturation spectroscopy. Unlike the case described above, however, the lack of a third adjustable input frequency in the latter cases for selective resonant probing makes the interpretation of the results less straight-forward.

Coherent Raman Spectroscopy of Excited States

The doubly resonant case of Fig. 14.2d described by (14.3d) with an output at $2\omega_1 - \omega_2$ shows a resonance at $\omega_1 - \omega_2 = \omega_{n'n}$. Thus the four-wave mixing process can be used to probe a Raman transition between the two excited states $|n\rangle$ and $|n'\rangle$.[23] We, notice however, that the bracket in (14.3d) is

$$
\left(\frac{1}{\omega_1 - \omega_{n'g} + i\Gamma_{n'g}} - \frac{1}{\omega_2 - \omega_{ng} - i\Gamma_{ng}} \right)
$$

$$
= \frac{-(\omega_1 - \omega_2 - \omega_{n'n} + i\Gamma_{n'n}) + i(\Gamma_{n'n} - \Gamma_{n'g} - \Gamma_{ng})}{(\omega_1 - \omega_{n'g} + i\Gamma_{n'g})(\omega_2 - \omega_{ng} - i\Gamma_{ng})}
$$

$$\tag{15.33}$$

which nearly vanishes at $\omega_1 - \omega_2 = \omega_{n'n}$ if ω_1 and ω_2 are far away from resonances so that Γ_{ng} and $\Gamma_{n'g}$ are negligible. Yet, even with the double resonance $\omega_1 \simeq \omega_{n'g}$ and $\omega_2 \simeq \omega_{ng}$, no Raman resonance can be observed if $\Gamma_{n'n} - \Gamma_{n'g} - \Gamma_{ng} = 0$ because the resonant denominator $(\omega_1 - \omega_2 - \omega_{n'n} + i\Gamma_{n'n})$ in $\chi_R^{(3)}$ of (14.3d) is canceled by the numerator in (15.33). This actually happens when the damping is dominated by spontaneous emission, since

$\Gamma_{n'g} + \Gamma_{ng} = \Gamma_{n'n}$.[23] In the presence of collisions, this relation no longer holds, and we usually have $\Gamma_{ij} = \Gamma_{ij}^N + \gamma_{ij} p$, where Γ_{ij}^N is due to spontaneous emission, p is the gas pressure, and γ_{ij} is the coefficient describing the collisional broadening. The relevant frequency factor in $\chi_R^{(3)}$ of (14.3d) then becomes

$$\frac{1}{(\omega_1 - \omega_2 - \omega_{n'n} + i\Gamma_{n'n})} \left(\frac{1}{\omega_1 - \omega_{n'g} + i\Gamma_{n'g}} - \frac{1}{\omega_2 - \omega_{ng} - i\Gamma_{ng}} \right)$$

$$= \frac{1}{(\omega_1 - \omega_{n'g} + i\Gamma_{n'g})(\omega_2 - \omega_{ng} - i\Gamma_{ng})}$$

$$\times \left[-1 + \frac{i(\gamma_{n'n} - \gamma_{n'g} - \gamma_{ng})p}{\omega_1 - \omega_2 - \omega_{n'n} + i(\Gamma_{n'n}^N + \gamma_{n'n} p)} \right]$$

$$(15.34)$$

which clearly shows a Raman resonance between $|n\rangle$ and $|n'\rangle$ with an amplitude proportional to p. This is called the pressure-induced extra resonance in four-wave mixing (PIER-4) by Bloembergen et al.[23] They observed the PIER-4 signal between the $3P_{3/2}$ and $3P_{1/2}$ states of sodium vapor. The signal strength is proportional to

$$|\chi_R^{(3)}|^2 \propto N^2 \left| 1 - \frac{i(\gamma_{n'n} - \gamma_{n'g} - \gamma_{ng})p}{\omega_1 - \omega_2 - \omega_{n'n} + i(\Gamma_{n'n}^N + \gamma_{n'n} p)} \right|^2. \qquad (15.35)$$

With buffer gas, $\gamma_{ij} p$ should be replaced by $\gamma_{ij} p + \gamma'_{ij} p_{\text{buffer}}$. The linewidth of the resonance should increase linearly with the pressures. The experimental results agree with these predictions. It shows that PIER-4 can be used as a spectroscopic method not only to probe the Raman resonance between excited states, but also to study molecular collisions.

The PIER-4 process is, of course, not restricted to gases. In condensed matter, the relation $\Gamma_{n'n} = \Gamma_{n'g} + \Gamma_{ng}$ usually does not hold, so that the PIER-4 signal should be observable.[24] Note that the case of Fig. 14.2d can yield a Doppler-free spectrum. Among the other processes in Fig. 14.2, case h with the output at $2\omega_2 - \omega_1$ also probes the Raman resonance between excited states. The theoretical discussion is essentially the same as case d.

15.5 FORCED LIGHT-SCATTERING SPECTROSCOPY

As described in Section 10.3, a transverse excitation, denoted by the density matrix $\rho_{nn'}$ with $n \neq n'$, can be treated as a material excitational wave governed by a characteristic wave equation. Thus CARS can be considered a process in which the material excitation wave at $\omega_1 - \omega_2 \simeq \omega_{\text{ex}}$ is resonantly excited by optical mixing of two incoming waves at ω_1 and ω_2, and then the material excitational wave coherently scatters the incoming wave at ω_1 and yields the anti-Stokes output. Theoretically, therefore, the process can be described by coupling of light waves at ω_1, ω_2, and $2\omega_1 - \omega_2$ with the material excitational wave at $\omega_1 - \omega_2$. In the steady-state case, elimination of the material excitational wave in the coupled equations leads to the same result given in Section 15.2.

We realize that the material excitational wave is not restricted to vibrational or electronic excitation. It can be any excitation including acoustic wave, entropy wave, spin wave, and charge density wave. Each, however, generally has its own governing wave equation. A CARS process involving low-frequency material excitation is often known as a forced light-scattering process.[25] It differs from the spontaneous anti-Stokes scattering process in the sense that the material excitation is not thermally excited but is coherently driven by beating of two optical waves. (In general, it can of course also be directly excited by an electromagnetic wave with $\omega \simeq \omega_{ex}$ or by other means.) In the discussion of stimulated light scattering in Chapter 11, the wave equations for a number of low-frequency excitations were given. As an example, consider here briefly the case of forced light scattering of concentration variation.

The driven equation of motion for the concentration variation is, following (11.26),

$$\left(\frac{\partial}{\partial t} - D\nabla^2 + \frac{1}{\tau} \right) \Delta C = -A\nabla^2 (\mathbf{E}_1 \cdot \mathbf{E}_2^*) \tag{15.36}$$

where $A = D(\partial\varepsilon/\partial C)_{\rho,T}/8\pi\rho_0(\partial\mu/\partial C)_{\rho,T}$, and we have generalized the equation by including a term $\Delta C/\tau$ to allow relaxation of ΔC. With monochromatic incoming plane waves, $\mathbf{E}_1 = \mathscr{E}_1\exp(i\mathbf{k}_1 \cdot \mathbf{r} - i\omega_1 t)$ and $\mathbf{E}_2 = \mathscr{E}_2\exp(i\mathbf{k}_2 \cdot \mathbf{r} - i\omega_2 t)$, the solution of (15.36) gives

$$\Delta C = \frac{A(\mathbf{k}_1 - \mathbf{k}_2)^2 (\mathscr{E}_1 \cdot \mathscr{E}_2^*)}{(\omega_1 - \omega_2) + i\left[1/\tau + D(\mathbf{k}_1 - \mathbf{k}_2)^2\right]} e^{i[(\mathbf{k}_1 - \mathbf{k}_2) \cdot \mathbf{r} - (\omega_1 - \omega_2)t]}. \tag{15.37}$$

This driven concentration wave then coherently scatters the $\mathbf{E}_1$ wave to yield an anti-Stokes output, which is governed by (15.1) with $-4\pi(\omega_a^2/c^2) \times \mathbf{P}^{(3)}(\omega_a)$ replaced by $-(\omega_a^2/c^2)(\partial\varepsilon/\partial C)_{\rho,T}(\Delta C)\mathbf{E}_1$. From a medium of length l, the anti-Stokes output in the phase-matched direction $\mathbf{k}_a = 2\mathbf{k}_1 - \mathbf{k}_2$ has an intensity

$$I_a = \frac{\omega_a^2}{8\pi\varepsilon C} \left| \left(\frac{\partial\varepsilon}{\partial C} \right)_{\rho,T} (\Delta C)E_1 + 4\pi\chi_{NR}^{(3)} E_1^2 E_2^* \right|^2 \tag{15.38}$$

where the $\chi_{NR}^{(3)}$ term is the nonresonant contribution to the output. The spectrum of I_a versus $\omega_1 - \omega_2$ in this case is dominated by the zero-frequency resonance of the concentration wave. The width of the resonant line has contributions from both relaxation and diffusion of the local excess concentration.

15.6 TRANSIENT FOUR-WAVE MIXING SPECTROSCOPY

With resonant excitation in four-wave mixing, the relaxation of the excitation naturally leads to a time-dependent output signal when pulsed lasers are used for pumping and probing. Thus transient four-wave mixing spectroscopy can be used to study the relaxation behavior of a resonant excitation. In fact, transient CARS was already discussed in Section 10.10. It was shown that the transverse relaxation time of the Raman resonance can be obtained from the transient measurements. Transient forced light scattering is quite similar.[26] Let us consider again the forced concentration scattering as an example.

Assume that the pump pulses are much shorter than the characteristic time for concentration variation, and that at $t = 0$, they induce a concentration grating in the $\hat{x}$ direction, $\Delta C(x, t = 0) = \Delta C_0 \frac{1}{2}(1 + \cos Kx)$, where $K = (\mathbf{k}_1 - \mathbf{k}_2) \cdot \hat{x}$ and $\Delta C_0 \propto \mathscr{E}_1 \mathscr{E}_2$. After the pump pulses are over, the induced concentration grating gradually decays away, following the transient solution of (15.36)

$$\Delta C(x, t) = \Delta C_0 e^{-t\tau} \frac{1}{2} \left[1 + e^{-DK^2 t} \cos Kx \right] \tag{15.39}$$

A short probe pulse with intensity I_3 scattered by ΔC at time t then yields a coherently scattered signal with a wavevector $\mathbf{k}_a = \mathbf{k}_3 \pm (\mathbf{k}_1 - \mathbf{k}_2)$:

$$S_a(t) \propto I_1 I_2 I_3 e^{-2(1/\tau + DK^2)t}. \tag{15.40}$$

From the measured exponential decay of the signal with time and its dependence on the grating spacing $(2\pi/K)$, we can then deduce separately the relaxation time τ and the diffusion constant D. More rigorously, the finite widths of the pump and probe pulses should also be taken into account in the calculation by proper convolution.

Transient four-wave mixing spectroscopy can be more general, involving possible multiple resonances. This more general discussion is postponed to Chapter 21, where transient four-wave mixing is shown to be related to other coherent transient effects such as photon echoes and free-induction decays.

REFERENCES

1 N Bloembergen, in H. Walther and K. W. Rothe, eds , *Laser Spectroscopy IV* (Springer-Verlag, Berlin, 1979), p 340; see also the bibliography.

2 M D Levenson, *Physics Today* **30**, No 5, 44 (1977); A. B Harvey, J R McDonald, and W M Tolles, in *Progress in Analytical Chemistry* (Plenum Press, New York, 1977), p. 211,

S. A. Akhmanov and N. I. Koroteev, *Sov. Phys. Uspekhi* **20**, 899 (1977); S. A. Akhmanov, in N. Bloembergen, ed., *Nonlinear Spectroscopy*, Proceedings of International School of Physics "Enrico Fermi" (North-Holland, Publishing Co. Amsterdam, 1977), p. 217; W. M. Tolles, J. W. Nibler, J. R. McDonald, and G. V. Knighton, in A. Weber, ed., *Topics in Current Physics* (Springer-Verlag, Berlin, 1977); S. Druet and J. P. Taran, in C. B. Moore, ed., *Chemical and Biochemical Applications of Lasers*, (Academic Press, New York, 1979), Vol. IV; M. D. Levenson, in A. Harvey, ed., *Chemical Applications of Nonlinear Optics* (Academic Press, New York, 1980); M. D. Levenson and J. J. Song, in V. Letokhov and M. S. Feld, eds., *Advances in Coherent and Nonlinear Optics* (Springer-Verlag, Berlin, 1980); G. I. Eesley, *Coherent Raman Spectroscopy* (Pergamon, New York, 1981).

3   M. D. Levenson and N. Bloembergen, *J. Chem. Phys.* **60**, 1323 (1974)

4   P. D. Maker and R. W. Terhune, *Phys. Rev.* **137**, A801 (1965).

5   J. J. Wynne, *Phys. Rev. Lett.* **29**, 650 (1972); *Phys. Rev. B* **6**, 534 (1972).

6   M. D. Levenson, C. Flytzanis, and N. Bloembergen, *Phys. Rev. B* **6**, 3462 (1972); E. Yablonovitch, C. Flytzanis, and N. Bloembergen, *Phys. Rev. Lett.* **29**, 865 (1972); M. D. Levenson, *IEEE J. Quant. Electron.* **QE-10**, 110 (1974); M. D. Levenson and N. Bloembergen, *Phys. Rev. B* **10**, 4447 (1974); *J. Chem. Phys.* **60**, 1323 (1974).

7   S. A. Akhmanov, V. G. Dimitriev, A. I. Koverigin, N. I. Koroteev, V. G. Tunkin, and A. I. Kholodnykh, *JETP Lett.* **15**, 425 (1972).

8   H. Huber-Walchli, D. M. Guthals, and J. W. Nibler, *Chem. Phys. Lett.* **67**, 233 (1979); M. D. Duncan, P. Oesterlin, and R. L. Byer, *Opt. Lett.* **6**, 90 (1981); M. D. Duncan, P. Oesterlin, F. Konig, and R. L. Byer, *Chem. Phys. Lett.* **80**, 253 (1981).

9   F. DeMartini, F. Simoni, and E. Santamato, *Opt. Comm.* **9**, 176 (1973); M. A. Henesian, L. Kulevski, and R. L. Byer, *J. Chem. Phys.* **65**, 5530 (1976).

10 P. R. Regnier and J. P. Taran, *Appl. Phys. Lett.* **23**, 240 (1973); J. W. Nibler, J. R. McDonald, and A. B. Harvey, *Opt. Comm.* **18**, 371 (1976).

11 H. C. Praddaude, D. W. Scudder, and B. Lax, *Appl. Phys. Lett.* **35**, 766 (1979).

12 R. L. Byer and E. Gustafson, private communications (to be published).

13 S. A. Akhmanov, A. F. Bunkin, S. G. Ivanov, and N. I. Koroteev, *JETP Lett.* **25**, 46 (1977).

14 J-L. Oudar, R. W. Smith, and Y. R. Shen, *Appl. Phys. Lett.* **34**, 758 (1979).

15 J-L. Oudar and Y. R. Shen, *Phys. Rev. A* **22**, 1141 (1980).

16 L. A. Rahn, L. J. Zych, and P. L. Mattern, *Opt. Comm.* **30**, 249 (1979).

17 A. Owyoung and P. Esherick, *Opt. Lett.* **5**, 421 (1980).

18 S. D. Kramer, F. G. Parsons, and N. Bloembergen, *Phys. Rev B* **9**, 1853 (1974).

19 D. Heiman, R. W. Hellwarth, M. D. Levenson, and G. Martin, *Phys. Rev. Lett* **36**, 189 (1976).

20 G. L. Eesley, M. D. Levenson, and W. M. Tolles, *IEEE J. Quant Electron.* **QE-14**, 45 (1978).

21 S. A. J. Druet, J-P. E. Taran, and Ch. J. Bordé, *J. Phys.* (Paris) **40**, 841 (1979).

22 T. Yajima and H. Souma, *Phys. Rev. A* **17**, 309 (1978); T. Yajima, H. Souma, and Y. Ishiba, *Phys. Rev. A* **17**, 324 (1978).

23 Y. Prior, A. R. Bogdan, M. Dagenais, and N. Bloembergen, *Phys. Rev. Lett.* **46**, 111 (1981); A. R Bogdan, M. Downer, and N. Bloembergen, *Phys. Rev. A* **24**, 623 (1981); N. Bloembergen, A. R. Bogdan, and M. W. Downer, in A. R. W. McKellar, T. Oka, and B. P. Stoicheff, eds., *Laser Spectroscopy V* (Springer-Verlag, Berlin, 1981), p. 157

24 P. L. Decola, J. R. Andrews, R. M. Hochstrasser, and H. P. Trommsdorff, *J. Chem. Phys.* **73**, 4695 (1980).

25 H. Eichler, G. Salje, and H. Stahl, *J. Appl. Phys.* **44**, 5383 (1973); D. Pohl, S. E. Schwarz, and V. Irniger, *Phys. Rev. Lett.* **31**, 32 (1973).

26 D. W. Phillion, D. J. Kuizenga, and A E. Siegman, *Appl. Phys. Lett.* **27**, 85 (1975); J. R. Salcedo, A. E. Siegman, D. D. Dlott, and M. D. Fayer, *Phys. Rev. Lett.* **41**, 131 (1978); J. R. Salcedo and A. E. Siegman, *IEEE J. Quant. Electron.* **QE-15**, 250 (1979).

BIBLIOGRAPHY

Bloembergen, N., ed., *Nonlinear Spectroscopy* (North-Holland, Publishing Co. Amsterdam, 1977).

Laser Spectroscopy I–V (Springer-Verlag, Berlin, 1973, 1975, 1977, 1979, 1981).

Letokhov, V., and M. S. Feld, eds., *Advances in Coherent and Nonlinear Optics* (Springer-Verlag, Berlin, 1980).

Levenson, M. D , *Introduction to Nonlinear Laser Spectroscopy* (Academic Press, New York, 1982).

16

Optical-
Field-Induced
Birefringence

An applied dc electric or magnetic field can effectively modify the refractive index of a medium. Electrooptic and magnetooptic effects were discussed in Chapter 4. The same is possible with an optical field. A sufficiently intense laser beam can induce a significant change in the refractive indices of a medium. The refractive index change in turn affects the beam propagation and leads to a new class of nonlinear optical effects characteristically different from either optical mixing or nonlinear wave attenuation. This chapter describes the various physical mechanisms contributing to the optical-field-induced birefringence, and the resulting effect on the beam polarization. Connection to four-wave mixing is made. Only media with inversion symmetry are considered, to avoid complications caused by the presence of second-order processes.

16.1 GENERAL FORMS OF OPTICAL-FIELD-INDUCED REFRACTIVE INDICES

The nonlinear polarization $\mathbf{P}^{NL}(\omega)$ induced by an intense monochromatic field $E(\omega)$ has a general form

$$\mathbf{P}^{NL}(\omega) = \Delta\chi\left[\omega, E_i(\omega)E_j^*(\omega)\right] \cdot \mathbf{E}(\omega). \qquad (16.1)$$

A similar expression exists for the nonlinear polarization $\mathbf{P}^{NL}(\omega')$ at the probe frequency ω'

$$\mathbf{P}^{NL}(\omega') = \Delta\chi\left[\omega', E_i(\omega)E_j^*(\omega)\right] \cdot \mathbf{E}(\omega'). \qquad (16.2)$$

Here the probe field $\mathbf{E}(\omega')$ is assumed to be sufficiently weak so that only the term linearly proportional to $\mathbf{E}(\omega')$ in $\mathbf{P}^{NL}(\omega')$ needs to be taken into account. The induced susceptibility $\Delta\chi$ is related to the induced refractive index $\Delta\mathbf{n}$ by the simple relationship

$$\Delta\left(n_{ij}^2\right) = \Delta\varepsilon_{ij} = 4\pi\Delta\chi_{ij}. \tag{16.3}$$

We consider here only the third-order nonlinear polarization or $\Delta\chi$ to the lowest order. Then, by the conventions given in Section 2.9, the third-order nonlinear polarization $\mathbf{P}^{(3)}(\omega')$ can be written as

$$
\begin{aligned}
P_i^{(3)}(\omega') &= \sum_{j,k,l} \chi_{ijkl}^{(3)}(\omega' = \omega' + \omega - \omega) E_j(\omega') E_k(\omega) E_l^*(\omega) \\
&= 6 \sum_{j,k,l} C_{ijkl}^{(3)}(\omega' = \omega' + \omega - \omega) E_j(\omega') E_k(\omega) E_l^*(\omega).
\end{aligned} \tag{16.4}
$$

For $\omega = \omega'$, it becomes

$$P_i^{(3)}(\omega) = 3 \sum_{j,k,l} C_{ijkl}^{(3)}(\omega = \omega + \omega - \omega) E_j(\omega) E_k(\omega) E_l^*(\omega). \tag{16.5}$$

A general review of third-order susceptibilities can be found in Ref. 1. In an isotropic medium, the nonzero components of $\chi^{(3)}$ are $\chi_{1111}^{(3)}$, $\chi_{1122}^{(3)}$, $\chi_{1212}^{(3)}$, and $\chi_{1221}^{(3)}$, with $\chi_{1111}^{(3)} = \chi_{1122}^{(3)} + \chi_{1212}^{(3)} + \chi_{1221}^{(3)}$.

16.2 PHYSICAL MECHANISMS

A number of physical mechanisms can contribute to the optical-field-induced refractive indices. We discuss here only the few commonly encountered ones.

Electronic Contribution

The applied optical field can distort the electronic charge distribution in a medium, which will lead to a change in the refractive indices. Microscopically, the electronic contribution to the third-order susceptibility can be derived from the third-order perturbation calculation, as outlined in Chapter 2. For a typical transparent liquid or solid, $\chi^{(3)}$ falls in the range between 10^{-13} and 10^{-15} esu. However, as the optical frequencies approach an absorption band, $\chi^{(3)}$ can be greatly enhanced. This is particularly true if the resonant absorption is sharp. A population redistribution induced by the resonant excitation often accounts for the major part of the enhancement.

Consider, as an example, a monochromatic beam of frequency ω propagating in a gas medium with a transition frequency ω_{ng} close to ω. It is easy to show, from the density matrix formalism, for example, that the beam sees, aside from a nonresonant background, a resonant susceptibility

$$\chi_R(\omega) = N \frac{|\langle n|er_i|g\rangle|^2}{\hbar(\omega - \omega_{ng} + i\Gamma_{ng})} \Delta\rho \qquad (16.6)$$

where $\Delta\rho = \rho_{gg} - \rho_{nn}$ is the population difference between the states $|g\rangle$ and $|n\rangle$, and N is the density of molecules. Clearly, if the beam intensity is sufficiently strong, the resonant excitation will yield a significant population redistribution, with $\Delta\rho$ decreasing as the beam intensity increases. This is known as the saturation effects, as discussed in Section 13.3. Following (13.7) for an effective two-level system, we then have

$$\Delta\rho = \Delta\rho^0 \left[1 - \frac{4p_{ng}^2 T_1 \Gamma_{ng} |E(\omega)|^2 / \hbar^2}{(\omega - \omega_{ng})^2 + \Gamma_{ng}^2 + 4p_{ng}^2 T_1 \Gamma_{ng} |E(\omega)|^2 / \hbar^2} \right] \qquad (16.7)$$

where $p_{ng} = \langle n|er_i|g\rangle$ and $\Delta\rho^0$ is the thermal population difference. In the weak saturation limit, (16.7) reduces to

$$\Delta\rho = \Delta\rho^0 \left[1 - \frac{4p_{ng}^2 T_1 \Gamma_{ng} |E(\omega)|^2 / \hbar^2}{(\omega - \omega_{ng})^2 + \Gamma_{ng}^2} \right]. \qquad (16.8)$$

The resonant susceptibility of (16.6) in the presence of an intense beam can therefore be written in the form

$$\chi_R(\omega) = \chi_R^{(1)} + \Delta\chi_R(\omega, |E(\omega)|^2). \qquad (16.9)$$

Here $\chi_R^{(1)}(\omega)$ is the resonant part of the linear susceptibility independent of the beam intensity, and

$$\Delta\chi_R = -\chi_R^{(1)} \frac{4p_{ng}^2 T_1 \Gamma_{ng} |E(\omega)|^2 / \hbar^2}{(\omega - \omega_{ng})^2 + \Gamma_{ng}^2 + 4p_{ng}^2 T_1 \Gamma_{ng} |E(\omega)|^2 / \hbar^2}, \qquad (16.10)$$

which, in the weak saturation limit, reduces to

$$\Delta\chi_R = \chi_R^{(3)} |E(\omega)|^2$$

with

$$\chi_R^{(3)} = -\chi_R^{(1)} \frac{4p_{ng}^2 T_1 \Gamma_{ng} / \hbar^2}{(\omega - \omega_{ng})^2 + \Gamma_{ng}^2}. \qquad (16.11)$$

It is interesting to see how large $\chi_R^{(3)}$ can be for an atomic vapor system. Assume $N = 10^{15}/cm^3$, $p_{ng} = 5 \times 10^{-17}$ esu (for the $s \rightarrow p$ transition in an alkali vapor, for example), $T_1 \Gamma_{ng} \simeq 1$, and $|\omega - \omega_{ng}| = 1$ cm$^{-1} \gg \Gamma_{ng}$. Then we obtain $\chi_R^{(1)} \simeq 0.01$ esu and $\chi_R^{(3)} \simeq 2.5 \times 10^{-3}$ esu. This shows that even with $|E(\omega)| = 1$ esu only (corresponding to a beam intensity of 250 W/cm^2), the induced susceptibility is as large as $\Delta\chi_R \simeq 2.5 \times 10^{-3}$ esu, or, equivalently, the induced refractive index is $\Delta n_R \simeq 1.5 \times 10^{-2}$ esu. The large value of $\chi_R^{(3)}$ for atomic vapor has prompted researchers to use it as the nonlinear medium in degenerate four-wave mixing work such as phase conjugation[2] and nonlinear optical diffraction.[3]

The saturation effect in semiconductors can also yield a large $\chi_R^{(3)}$. In InSb, for example, $\chi_R^{(3)}$ can reach a value of ~ 1 esu when the optical frequency moves into the direct absorption band.[4] The mechanism responsible for the large $\chi_R^{(3)}$ here is somewhat different from the atomic systems because of the band structure in semiconductors. The resonant optical field pump electrons into the conduction band and leaves holes in the valence band. Because of the fast relaxation rate of carriers within a band, the excited electrons and holes quickly relax to thermal population distributions in the conduction band and the valence band, respectively. The steady-state population distributions of electrons and holes are finally determined by the balance between the excitation and the electron-hole recombination across the band gap. This induced population redistribution (pump saturation) results in a change in the absorption spectrum, which is related to the optical-field-induced refractive index through the Kramers-Kronig relation.

The large $\chi_R^{(3)}$ resulting from the saturation effect is not limited to the cases with electronic resonance. In molecular systems, the same thing can occur near a vibrational transition, although $\chi_R^{(3)}$ is usually not as large as that for the atomic systems because of the weaker oscillator strength of the vibrational transition.

Raman or Two-Photon Contribution

The third-order susceptibility $\chi^{(3)}(\omega' = \omega' + \omega - \omega)$ can also be resonantly enhanced if $|\omega' - \omega|$ approaches a Raman transition. This fact forms the basis of the Raman-induced Kerr effect spectroscopy (RIKES) described in Section 15.3. It is seen there that the presence of the field $E(\omega)$ can induce at the probe frequency ω' a susceptibility change $\Delta\chi(\omega') = \chi^{(3)}(\omega'):E(\omega)E^*(\omega)$, or a corresponding refractive index change $\Delta n(\omega')$. Typically, $\chi^{(3)} \sim 3 \times 10^{-13}$ esu in a liquid with $|\omega - \omega'|$ near a strong Raman resonance. Therefore, with $|E(\omega)| = 100$ esu (corresponding to a beam intensity of 2.5 MW/cm^2), one finds $\Delta n \sim 10^{-8}$.

Similarly, $\chi^{(3)}(\omega' = \omega' + \omega - \omega)$ can be resonantly enhanced if $\omega' + \omega$ approaches a two-photon resonance. The presence of $E(\omega)$ again induces a susceptibility change $\Delta\chi(\omega') = \chi^{(3)}(\omega'):E(\omega)E^*(\omega)$ or a corresponding $\Delta n(\omega')$ at ω'. For both Raman and two-photon cases, $\chi^{(3)}(\omega')$ is further enhanced if ω or ω' is also near a transition to an intermediate state. In sodium vapor, for example, with $\omega_1 + \omega_2$ on the $3s \rightarrow 4d$ transition and ω_1 being 10 cm^{-1} away from the $3s \rightarrow 4p$ transition, one finds $\chi^{(3)} \sim 10^{-25}N$ esu where N is the atomic density. For $N \sim 10^{15}/cm^3$ and $|E| \sim 100$ esu, the induced refractive index change Δn is as large as 10^{-5}.[5] In molecular systems, $\chi^{(3)}(\omega = \omega + \omega - \omega)$ can often be greatly enhanced as 2ω approaches the $v = 0 \rightarrow v = 2$ vibrational transition because ω, at the same time, is near the $v = 0 \rightarrow v = 1$ resonance. The large $\chi^{(3)}$ seen by a 10.6-μm CO_2 laser in SF_6 gas is a typical example.[6]

Electrostriction

Application of a dc electric field to a local region of a medium causes an increase in the density of the medium in that region. This field-induced density redistribution occurs in order to minimize the free energy of the system in the presence of the field, and is known as electrostriction. The same effect is expected with the optical field, since the dc field energy and the optical field energy are equivalent in this case. The induced density change then leads to a change in the susceptibility or refractive index. The essential mathematical derivation has already been given in Section 11.1 in connection with the discussion on stimulated Brillouin scattering.

The induced density variation $\Delta \rho_D$ should obey the driven acoustic equation of (11.2), which appears in the case of monochromatic pump beam as

$$\left[\frac{\partial^2}{\partial t^2} - 2\Gamma_B \frac{\partial}{\partial t} - v^2 \nabla^2 \right] \Delta \rho_D = -\nabla^2 \left(\frac{1}{2\pi} \gamma |E(\omega)|^2 \right). \tag{16.12}$$

If the beam is CW, the time-independent solution is

$$\Delta \rho_D = \frac{\gamma}{2\pi v^2} |E(\omega)|^2. \tag{16.13}$$

The resulting nonlinear polarization seen by a probe beam at ω' is

$$\mathbf{P}^{(3)}(\omega') = \frac{1}{4\pi} \frac{\partial \varepsilon(\omega')}{\partial \rho_D} \Delta \rho_D \mathbf{E}(\omega') \tag{16.14}$$

and hence

$$\Delta \chi(\omega') = \frac{1}{4\pi} \frac{\partial \varepsilon(\omega')}{\partial \rho_D} \Delta \rho_D. \tag{16.15}$$

Here we assume that ω' is very different from ω so that $\Delta \rho_D(\omega' - \omega)$ induced by the beating of $E(\omega)$ and $E(\omega')$ is negligible. Otherwise, an additional term in $P^{(3)}(\omega')$, and hence $\Delta \chi(\omega')$, appears resulting from coherent scattering of

$E(\omega)$ by $\Delta\rho_D(\omega' - \omega)$. Typically, for liquid, $\Delta\chi \sim 10^{-12}$ esu or $\Delta n \sim 10^{-11}$ esu with $|E| \sim 1$ esu.[7] In a transparent medium this electrostrictive contribution often is appreciably larger than the electronic contribution. However, as we discuss later, the response of the density variation to the applied field is slow, and therefore, with short optical pulses, Δn can never reach its full steady-state value.

Molecular Reorientation and Redistribution

Assume a linearly polarized beam propagating in a liquid. If the molecules are anisotropic, the optical field tends to align the molecules through interaction of the field with the induced dipoles on the molecules. Furthermore, because of the presence of induced dipole interaction between molecules, the molecules will spatially redistribute themselves in order to minimize the free energy of the system. Both molecular reorientation and redistribution lead to a change in the refractive index of the medium.

In Section 11.3, optical-field-induced molecular reorientation in an isotropic medium was described at length. We reviewed it here with a more general derivation. We can define an orientational order parameter Q by the macroscopic relation[8]

$$\chi_{xx} = \bar{\chi} + \tfrac{2}{3}\Delta\chi_0 Q \tag{16.16}$$

assuming $\mathbf{E}$ along $\hat{x}$, where $\bar{\chi} = \tfrac{1}{3}(\chi_{xx} + \chi_{yy} + \chi_{zz})$ is the average linear susceptibility of the medium, and $\Delta\chi_0$ is the anisotropy of the medium when all molecules are aligned parallel ($Q = 1$). We also have

$$\chi_{yy} = \chi_{zz} = \bar{\chi} - \tfrac{1}{3}\Delta\chi_0 Q. \tag{16.17}$$

The order parameter Q defined here is a macroscopic quantity while the one in Section 11.3 is microscopic. The two become identical if the molecules are noninteracting.

In the liquid phase, the induced ordering is expected to be small, so that $Q \ll 1$. According to Landau's theory, the free energy of the system can then be expanded into power series of Q:[9]

$$F = F_0 + \tfrac{1}{2}a(T - T^*)Q^2 + \tfrac{1}{3}BQ^3 + \cdots - 2(\bar{\chi} + \tfrac{2}{3}\Delta\chi_0 Q)|E(\omega)|^2 \tag{16.18}$$

where F_0 is independent of Q, T is the temperature, a and B are constant coefficients, and T^* is a fictitious second-order phase transition temperature at which the system would necessarily make a transition to the ordered phase if it had not already done so. In the limit of small Q, terms of orders higher than

Q^2 in F can be neglected. Then minimization of F immediately gives

$$Q = \frac{4\Delta\chi_0|E(\omega)|^2}{3a(T - T^*)}. \tag{16.19}$$

The induced change in the susceptibility tensor seen by a probe beam at ω', from (16.16) and (16.17), appears to be

$$\Delta\chi_{xx} = -2\Delta\chi_{yy} = -2\Delta\chi_{zz} = \tfrac{2}{3}\Delta\chi_0(\omega')Q$$

$$= \frac{8\Delta\chi_0(\omega')\Delta\chi_0(\omega)}{9a(T - T^*)}|E(\omega)|^2. \tag{16.20}$$

Again, we have assumed that $|\omega - \omega'|$ is much larger than the inverse of the relaxation time for molecular reorientation, so that $Q(\omega' - \omega)$ induced by the beating of $E(\omega)$ and $E(\omega')$ is negligible. Otherwise, an additional term should appear in Q, and hence $\Delta\chi$.

In the ideal limit of noninteracting molecules, one would, of course, have $T^* = 0$, and the foregoing results should reduce to those in Section 11.3, namely (11.23) with $E_1 = E_2 = E$ and $\omega_1 = \omega_2$. A direct comparison of the expressions of Q then yields $a = 5Nk_B$ for this ideal case.

We have considered only molecular reorientation induced by a linearly polarized light. What happened if, instead, a circularly polarized light is used? Will there be a circular birefringence induced in an isotropic medium? Physically, it is easy to see that the optical field should reorient the molecules toward the $(\hat{x}$-$\hat{y})$ plane perpendicular to the direction of propagation $(\hat{z})$, but in the $\hat{x}$-$\hat{y}$ plane, the molecules are randomly distributed since they cannot follow the field rotating at the optical frequency. Thus one expects a field-induced linear birefringence between $\hat{z}$ and $\hat{x}$-$\hat{y}$, and no birefringence in the $\hat{x}$-$\hat{y}$ plane. The induced susceptibility changes should be

$$\Delta\chi_+ = \Delta\chi_- = -\tfrac{1}{2}\Delta\chi_{zz} \tag{16.21}$$

where the subindices "+" and "$-$" denote the right and left circular polarizations, respectively. Since $\Delta\chi_{zz}$ should be the same, irrespective of a linearly polarized aligning field along $\hat{x}$ or a circularly polarized field in $\hat{x}$-$\hat{y}$, we have

$$\Delta\chi_+ = \Delta\chi_- = \frac{2\Delta\chi_0(\omega')\Delta\chi_0(\omega)}{9a(T - T^*)}|E(\omega)|^2 \tag{16.22}$$

which is four times less than $\Delta\chi_{xx}$ induced by a linearly polarized field along $\hat{x}$.

In ordinary liquids composed of anisotropic molecules, molecular reorientation can yield a Δn of the order of 10^{-12} to 10^{-11} esu for $|E| = 1$ esu.[7] The response time of molecular orientation is usually of the order of 10 psec.

Therefore, with nanosecond and subnanosecond laser pulses, molecular reorientation is often the dominant mechanism for the observed Δn. For liquid crystalline media in the isotropic phase, however, Δn can be much larger.[9] This is because in such media T^* may be less than 1 K below the isotropic–mesomorphic transition, and according to (16.20), as T approaches the transition temperature, $\Delta \chi$ or Δn increases inversely with $(T - T^*)$. This pretransitional behavior is commonly known as critical divergence. It has been found that Δn can reach a value of $\sim 10^{-9}$ esu for $|E| = 1$ esu even at a temperature 5 K above the transition. The response time of Δn, however, also increases with $(T - T^*)^{-1}$, which is known as the critical slowing-down behavior. At 5 K above the transition, it is of the order of 100 nsec.[9]

Molecules in the isotropic liquid phase are more or less uncorrelated. Their response to the external perturbation therefore is an unconcerted effort and cannot be very large. On the other hand, molecules in a liquid crystal phase are well correlated, at least in orientation. The orientational response of the molecules to the external perturbation is a group effort, and can be expected to be extremely large. Indeed, with $|E| \sim 1$ esu, the observed average Δn in a nematic liquid crystal film can be ~ 0.05.[10] Because of the wall-aligning force on molecules, the induced molecular orientation is not uniform across the film. The induced Δn is usually a strong nonlinear function of $|E|^2$ even at $|E| \sim 1$ esu. Being a correlated response to the applied field, this induced molecular reorientation is very slow, having a response time of the order of 1 sec or larger.

The molecular reorientation can contribute to Δn only if the molecules are anisotropic. However, it has been found that in transparent liquids composed of nearly spherical molecules or atoms, an appreciable Δn can still be induced by laser pulses. The observed Δn is much larger than what one would expect from electronic and electrostrictive contributions and must arise from optical-field-induced redistribution.[7] The theory of molecular redistribution is unfortunately not yet well formulated.

Both molecular reorientation and molecular redistribution can occur only if molecules are free to rotate and move in the medium. With a few exceptions, this is generally not the case in solids. Therefore, they should not contribute to the observed Δn in solids.

Other Mechanisms

There are many other possible mechanisms that may contribute to Δn. In an absorbing medium, the optical-field-induced temperature rise ΔT will certainly lead to a change in the refractive index. The induced concentration variation ΔC in a mixture is another possible mechanism. The expressions for ΔT and ΔC can be obtained following the derivation in Chapter 11. An extremely large Δn can also be obtained in photorefractive materials such as $BaTiO_3$.[11] The optical beam presumably excites and redistributes the charges trapped at various sites in the crystal. The charge redistribution sets up a strong internal electric field, which in turn induces a large Δn via the electrooptical effect. Even with a 10-mW/cm^2 laser beam, the induced Δn can be larger than 10^{-5}, although the response is very slow ($\sim$ 1 sec) at such a low laser intensity.

16.3 OPTICAL KERR EFFECT AND ELLIPSE ROTATION

The optical-field-induced birefringence can change the polarization of a beam propagating in the medium. Conversely, measurements of the change of the beam polarization should allow us to deduce values of the induced birefringence.

Consider first the optical Kerr effect, which usually refers to the phenomenon of linear birefringence induced by a linearly polarized optical field. We assume an isotropic medium here for simplicity. With a pump beam at ω and a probe beam at ω', the third-order polarization at ω' takes the form

$$
\begin{aligned}
P_i^{(3)}(\omega') = \sum_j \Big[&\chi_{1122}^{(3)}(\omega' = \omega' + \omega - \omega)E_i(\omega')E_j(\omega)E_j^*(\omega) \\
+ &\chi_{1212}^{(3)}(\omega' = \omega' + \omega - \omega)E_j(\omega')E_i(\omega)E_j^*(\omega) \quad (16.23) \\
+ &\chi_{1221}^{(3)}(\omega' = \omega' + \omega - \omega)E_j(\omega')E_j(\omega)E_i^*(\omega) \Big].
\end{aligned}
$$

If the two beams are parallel and their polarizations are linear but at 45° with respect to each other, then with the pump field $\mathbf{E}(\omega)$ along $\hat{x}$ propagating along $\hat{z}$, one finds

$$
\begin{aligned}
P_x^{(3)}(\omega') &= \left(\chi_{1122}^{(3)} + \chi_{1212}^{(3)} + \chi_{1221}^{(3)} \right) E_x(\omega')E(\omega)E^*(\omega), \\
P_y^{(3)}(\omega') &= \chi_{1122}^{(3)} E_y(\omega')E(\omega)E^*(\omega).
\end{aligned}
\qquad (16.24)
$$

The field-induced anisotropy in the susceptibility is therefore given by[12]

$$
\begin{aligned}
\delta\chi(\omega') &= \Delta\chi_{xx} - \Delta\chi_{yy} \\
&= \left(\chi_{1212}^{(3)} + \chi_{1221}^{(3)} \right) |E(\omega)|^2
\end{aligned}
\qquad (16.25)
$$

or the induced linear birefringence given by

$$
\begin{aligned}
\delta n_l(\omega') &= \frac{2\pi}{n} \delta\chi(\omega') \\
&= \frac{2\pi}{n} \left(\chi_{1212}^{(3)} + \chi_{1221}^{(3)} \right) |E(\omega)|^2.
\end{aligned}
\qquad (16.26)
$$

Then, in propagating through a medium of length l, the $\hat{x}$ and $\hat{y}$ components of the probe field experience a relative phase difference

$$\delta\phi = (\omega'/c)\delta n_{\parallel}l. \tag{16.27}$$

This phase difference changes the polarization state of the probe beam, and can be measured by an analyzer crossed with the polarization of the incoming probe beam in front of the detector. A typical experimental arrangement is seen in Fig. 16.1.[12] The signal should be proportional to $\sin^2(\delta\phi/2)$, from which $\delta\phi$, and hence $\delta n_{\parallel}$, can be deduced. We note here that the same result with $\delta\phi \ll 1$ can be obtained from the four-wave mixing approach, as described in Section 15.3. The four-wave mixing output corresponds to the generation of an orthogonal polarization component in the probe beam.

As mentioned in the previous section, when ω approaches ω', the beating of $E(\omega)$ and $E(\omega')$ provides an additional contribution to $\Delta\chi$. The degenerate case with $\omega = \omega'$ and $\mathbf{k} = \mathbf{k}'$ is particularly interesting since we now have a single monochromatic beam propagating in the medium. Will the polarization state of the beam vary during propagation as a result of the induced birefringence? The answer is yes, in general, as one can expect from the four-wave mixing picture. We consider here the special case of an isotropic medium. The third-order polarization has the same general expression given in (16.23) with $\omega' = \omega$. In terms of circular polarization components, we can write[12]

$$P_{\pm}^{(3)}(\omega) = \left(\chi_{1122}^{(3)} + \chi_{1212}^{(3)}\right)|E_{\pm}(\omega)|^2 E_{\pm}(\omega)$$
$$+ \left(\chi_{1122}^{(3)} + \chi_{1212}^{(3)} + 2\chi_{1221}^{(3)}\right)|E_{\mp}(\omega)|^2 E_{\pm}(\omega). \tag{16.28}$$

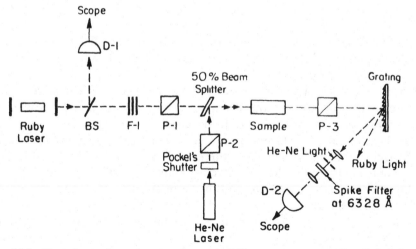

Fig. 16.1 Experimental arrangement for optical Kerr measurement. P-1, P-2, P-3 are polarizers, and D-1, D-2 are detectors. (After Ref. 9.)

For a linearly polarized field, $\mathbf{E}(\omega) = \hat{x}E(\omega)$, (16.28) reduces to

$$\mathbf{P}^{(3)}(\omega) = \hat{x}\chi_{1111}^{(3)}|E(\omega)|^2 E(\omega) \tag{16.29}$$

and for a circularly polarized field, $\mathbf{E} = \hat{e}_{\pm}E(\omega)$, it becomes

$$\mathbf{P}^{(3)}(\omega) = \hat{e}_{\pm}\left(\chi_{1122}^{(3)} + \chi_{1212}^{(3)}\right)|E(\omega)|^2 E(\omega). \tag{16.30}$$

In both cases, the induced $\mathbf{P}^{(3)}(\omega)$ has the same polarization state as the incoming field, and therefore no change in the beam polarization is expected as the beam propagates in the medium. More generally, however, for an elliptically polarized field, $\mathbf{E} = \hat{e}_{+}E_{+} + \hat{e}_{-}E_{-}$, (16.28) shows that the induced refractive indices for the two circularly polarized components are, respectively,

$$\Delta n_{\pm} = \frac{2\pi}{n}\left[\left(\chi_{1122}^{(3)} + \chi_{1212}^{(3)}\right)|E_{\pm}|^2 + \left(\chi_{1122}^{(3)} + \chi_{1212}^{(3)} + 2\chi_{1221}^{(3)}\right)|E_{\mp}|^2\right]. \tag{16.31}$$

The difference between Δn_{+} and Δn_{-} here indicates that the elliptically polarized beam can induce a circular birefringence in the medium

$$\Delta n_c = \Delta n_{+} - \Delta n_{-} = -(2\pi/n)2\chi_{1221}^{(3)}\left(|E_{+}|^2 - |E_{-}|^2\right). \tag{16.32}$$

As discussed in Section 4.2, a circular birefringence renders a rotation of the beam polarization. In the present case, the rotation of the elliptical polarization across the medium of length l is given by[13]

$$\theta = (\omega/2c)\Delta n_c l. \tag{16.33}$$

Note that θ depends on both the beam intensity and the ellipticity according to (16.32). Measurement of this intensity-dependent ellipse rotation θ allows us to deduce Δn_c and hence $\chi_{1221}^{(3)}$.

Figure 16.2 shows a typical experimental arrangement for ellipse rotation measurements.[12] Because of the beam intensity variation in the transverse direction $\hat{r}$, θ actually varies with r. To deduce $\chi_{1221}^{(3)}$ from the measurement of θ, one should limit the detection to a small region in the beam profile where the intensity variation is negligible. Alternatively, one can use a focused beam with a Gaussian beam profile.[12] The average ellipse rotation of the entire beam can be calculated in terms of Δn_c and the beam characteristics, and compared with the measurements.

From (16.26) and (16.32), it is seen that the combined results of the optical Kerr measurement (with $\omega \sim \omega'$) and the ellipse rotation measurement allow us to determine two of the three independent elements of $\chi^{(3)}$, namely $\chi_{1212}^{(3)}$ and $\chi_{1221}^{(3)}$. The third, $\chi_{1122}^{(3)}$, generally can be obtained only by resorting to

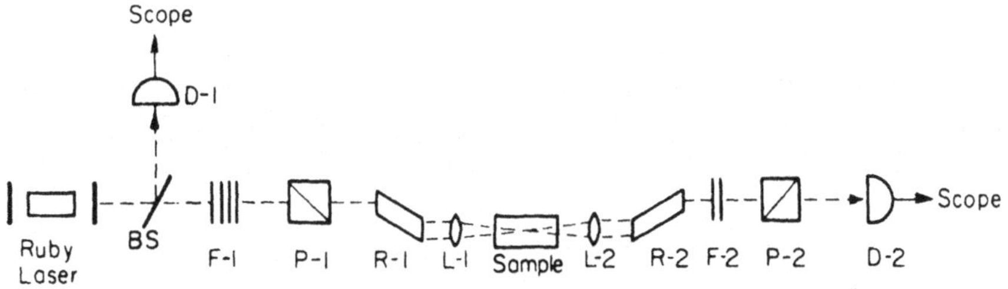

Fig. 16.2 Experimental arrangement for ellipse rotation measurement. P-1, P-2 are polarizers, R-1, R-2, Fresnel Rhombs, and D-1, D-2, detectors. (After Ref. 9.)

another independent experiment, for example, a degenerate four-wave mixing experiment.

The optical-field-induced variation of the beam polarization in an aniso-tropic medium is often complicated, because of the existing linear birefrin-gence even in the absence of an induced Δn. The problem is particularly interesting when Δn is so large that it changes drastically the beam propaga-tion characteristics in the medium. This may happen, for example, in liquid crystals.[14]

16.4 TRANSIENT EFFECTS

The foregoing discussion was limited to the steady-state case. The response of a medium, however, becomes transient if laser pulses with pulsewidths shorter than the response times are used. The response times are different for different physical mechanisms contributing to the induced refractive index. The elec-tronic contribution has a response time on the order of $(\omega - \omega_0)^{-1} \sim 10^{-15}$ sec, if the optical frequency ω is far away from any absorption band, where ω_0 is the position of the major absorption band. On the other hand, for Δn induced by the population redistribution, the response time is determined by the population relaxation.

For the electrostrictive contribution, the transient response is governed by the equation of motion (16.12) for acoustic waves.[7] With the laser pulse given, the time-dependent density variation $\Delta \rho_D(\mathbf{r}, t)$ can be solved from (16.12) and $\Delta \chi(\omega)$ is directly proportional to $\Delta \rho_D(\mathbf{r}, t)$. In this case, the response time is characterized by the inverse of the damping coefficient, Γ_B^{-1}, and the time it takes for the acoustic wave to travel across the beam radius, R/v. If the pulsewidth T_p is much longer than both Γ_B^{-1} and R/v, the response is quasi-steady-state. Yet if $T_p \ll R/v < \Gamma_B^{-1}$, the response becomes transient, with $\Delta \rho_D \sim \Delta \rho_D^{ss}[T_p^2/\Gamma_B^{-1}(R/v)]$ near the end of the pulse, where $\Delta \rho_D^{ss}$ is the steady-state value. If $T_p \ll \Gamma_B^{-1} \ll R/v$, the response is also transient, with $\Delta \rho_D \sim \Delta \rho_D^{ss}[T_p^2/(R/v)^2]$. For $R = 1$ mm, $v = 2 \times 10^5$ cm/sec, $\Gamma_B^{-1} \sim R/v = 5 \times 10^{-7}$ sec, and $T_p = 10^{-8}$ sec, we have $\Delta \rho_D \sim 4 \times 10^{-4} \Delta \rho_D^{ss}$. The corresponding induced refractive index is also 4×10^{-4} times smaller than the steady-state value. This example shows that the electrostrictive contribution to Δn is far less important with nanosecond or picosecond laser pulses than with longer pulses.

For the molecular reorientational contribution, the order parameter Q obeys the driven Debye equation given in (11.22), or more generally, by the following dynamic equation for Q including dissipation:

$$
\begin{aligned}
\nu \frac{\partial Q}{\partial t} &= -\frac{\partial F}{\partial Q} \\
&= -a(T - T^*)Q + \tfrac{4}{3}\Delta\chi_0 |E(\omega)|^2
\end{aligned}
\tag{16.34}
$$

where ν is a viscosity coefficient. The solution of the equation is simply

$$
Q(t) = \int_{-\infty}^{t} \frac{4\Delta\chi_0 |E(\omega)|^2}{3\nu} e^{-(t-t')/\tau} dt'
\tag{16.35}
$$

with the relaxation time τ defined as

$$
\tau = \nu/a(T - T^*).
\tag{16.36}
$$

For ordinary liquids, τ is of the order of 10 psec. Then the response of molecular reorientation to a nanosecond laser pulse is expected to be quasi-steady-state, but to a picosecond pulse, it is still transient. For liquid crystals in the isotropic phase, τ can be much larger, approaching 1 μsec as T approaches the isotropic–mesomorphic transition temperature.[9] In the mesomorphic phase, the collective motion of the molecules slows down the response even more drastically. The response time can be longer than 1 sec.[10]

16.5 APPLICATIONS

A number of applications can be derived from the optical-field-induced refractive index change. In Section 14.3, we saw how degenerate four-wave mixing can be used for phase conjugation and image reconstruction. As described there, degenerate four-wave mixing can result from coherent scattering of a light beam by a refractive index grating induced in the medium by the interference of two pump light waves. Thus a medium with a large optical-field-induced refractive index per unit field intensity is most useful for efficient degenerate four-wave mixing applications.

The optical Kerr effect can be used in optical switching.[15] As shown in Fig. 16.3, the transmission of the weak signal beam passing through the nonlinear medium is normally blocked by the crossed analyzer. In the presence of an intense pump beam, however, the medium becomes birefringent, and the signal beam, experiencing a polarization change in traversing the medium, is no longer completely blocked by the analyzer. Therefore, if a picosecond pump pulse is used, a nonlinear medium with a picosecond response time can act as a fast optical gate for a signal beam with a picosecond on–off time. Such an optical switch is clearly very useful in many applications, especially in dynamic study of physical phenomena and mechanisms.

If a Fabry–Perot interferometer is filled with a nonlinear medium, then because of the optical-field-induced refractive index, the transmission of a monochromatic beam through the interferometer depends on the beam intensity. For example, with the interferometer spacing tuned to a value for peak linear transmission, an intense light beam may find the interferometer badly mistuned because the field-induced refractive index has caused an additional phase change on the beam traversing the interferometer. In general, the transmission of the nonlinear Fabry–Perot interferometer is a highly nonlinear function of the beam intensity, depending on the initial phase setting of the interferometer. Three general forms of transmission characteristics of the interferometer can be obtained, and can be used for intensity limiting, differential amplification, and bistable operation.[16] The last one has been the focus of very active research because of its great potential applications in optical data processing.[17]

Optical bistable operation of a nonlinear Fabry–Perot (FP) interferometer is illustrated in Fig. 16.4. The FP transmission curve T versus ϕ in Fig. 16.4a is described by the equation

$$T = \frac{T_0}{1 + F\sin^2(\phi/2)} \qquad (16.37)$$

where T_0 and F are constants. Because of the field-induced refractive index change in the medium inside the interferometer, the round-trip phase retardation ϕ now depends on the field intensity. Assuming $\Delta n = n_2|E|^2$, we can write

$$\phi = \phi_0 + KI_{\mathrm{in}}T. \qquad (16.38)$$

With ϕ_0 and K being constants, (16.38) describes a straight line in Fig. 16.4a,

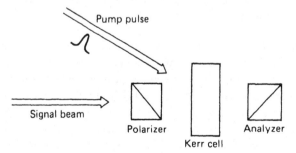

Fig. 16.3 Optical Kerr cell as a fast optical switch.

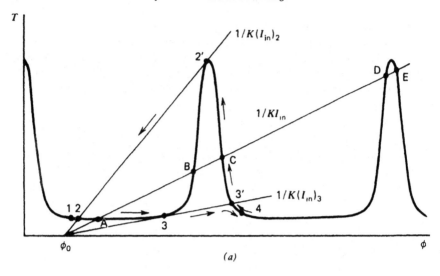

(a)

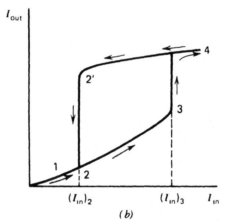

(b)

Fig. 16.4 (a) Graphical method of finding the operating points of a nonlinear Fabry–Perot interferometer. When I_{in} increases, the operating point moves along the path 1–2–A–3–3'–4, and when I_{in} decreases, the operating point moves along the path 4–3'–C'–2'–2–1. (b) I_{out} versus I_{in} in the form of a hysteresis loop corresponding to the operating path described in (a).

the slope of which is inversely proportional to the incoming laser intensity I_{in}. Given I_{in}, the operating point of the interferometer is then determined by the solution of (16.37) and (16.38), corresponding to the crossing point of the straight line with the FP transmission curve. It is seen in Fig. 16.4a that if I_{in} is sufficiently large, more than one operating point can exist. Some of them (such as B and D in Fig. 16.4a) are unstable. Among the stable ones, the real operating point is selected by the operating condition that if the operating point is varied, the interferometer prefers to have it varied smoothly along the curve. Thus, as shown in Fig. 16.4a, when I_{in} increases, the operating point should move along 1-2-A-3-3'-4, but when I_{in} decreases, it should move along 4-3'-C-2'-2-1. As a result, the corresponding I_{out} versus I_{in}, sketched in Fig. 16.4b, takes the form of a hysteresis loop. For I_{in} between $(I_{in})_2$ and $(I_{in})_3$, the output can have either a high value or a low value depending on the operation path. This bistable behavior is the basis of binary switching elements. Therefore, the nonlinear FP interferometer can become an important optical device in optical data processing and all-optical logic and computing systems. Although the nonlinear FP interferometer is commonly used to study optical bistability, we should note that the same phenomenon can arise in many other systems. Away from the operational point of view, a nonlinear FP interferometer is also interesting because its switching action resembles a phase transition,[18] and it is a nonlinear system with positive feedback that may lead to bifurcations and chaos in the output.[19]

Both the optical Kerr effect and the intensity-dependent ellipse rotation can be used for pulse shaping. The polarization state of an intense laser pulse in traversing a nonlinear medium is time-dependent, resulting from the intensity-dependent polarization rotation. Therefore, through the use of an analyzer, the transmitted pulse can be reshaped. The same effect can be used in a laser cavity to reshape the laser pulses.[20]

The optical-field-induced refractive index can also lead to an intensity-dependent distortion of the beam wavefront, resulting in self-focusing and other self-actions of light. This is the subject of discussion in Chapter 17.

REFERENCES

1 R. W. Hellwarth, *Prog. Quant. Electron.* **5**, 1 (1977).

2 D. M Bloom, P. F. Liao, and N. P. Economou, *Opt. Lett.* **2**, 58 (1978).

3 N. Tan-no, K. Ohkawara, and H. Inaba, *Phys. Rev. Lett.* **46**, 1282 (1981); C. V. Heer, *Opt. Lett.* **6**, 549 (1981).

4 D. A. B.-Miller, C. T. Seaton, M. E. Prise, and S. D. Smith, *Phys. Rev. Lett.* **47**, 197 (1981).

5 P. F. Liao and G. C. Bjorklund, *Phys. Rev. Lett.* **36**, 584 (1976).

6 D. G. Steel and L. F. Lam, *Phys. Rev. Lett.* **43**, 1588 (1979).

7 Y. R. Shen, *Phys. Lett.* **20**, 378 (1966).

8 P. G. De Gennes, *Mol. Cryst. Liq. Cryst.* **12**, 193 (1971).

9 G. K. L. Wong and Y. R. Shen, *Phys. Rev. A* **10**, 1277 (1974). In this reference, a tensorial order parameter is used, and the coefficient a is three times less than the one defined here. Also, the field energy is $\epsilon|E|^2/8\pi$ there, while here it is defined as $\epsilon|E|^2/2\pi$.

10 S. D. Durbin, S. M. Arakelian, and Y. R. Shen, *Phys. Rev. Lett.* **47**, 1411 (1981).

11 J. Feinberg, D. Heiman, A. R. Tanguay, and R. W. Hellwarth, *J. Appl. Phys.* **51**, 1297 (1980).

12 See, for example, E. G. Hanson, Y. R. Shen, and G. K. L. Wong, *Phys. Rev.* **14**, 1281 (1976); A. Owyoung, R. W. Hellwarth, and N. George, *Phys Rev. A* **4**, 2342 (1971).

13 P. D. Maker, R. W. Terhune, and C. M. Savage, *Phys. Rev. Lett.* **12**, 507 (1964).

14   S. D. Durbin, S. M. Arakelian, and Y. R. Shen, *Opt. Lett.* **6**, 411 (1981); *Phys. Rev. Lett.* **47**, 1411 (1981).

15   M. A. Duguay and J. W. Hanson, *Opt. Comm.* **1**, 254 (1969).

16   H. Seidal, U.S. Patent No. 3610731; A. Szoke, V. Danean, J. Goldhar, and N. A. Kurnit, *Appl. Phys. Lett.* **15**, 376 (1969); H. M. Gibbs, S. L. McCall, and T. N. C. Venkatesan, *Phys. Rev. Lett.* **36**, 1135 (1976).

17   See C. M. Bowden, M. Ciftan, and H. R. Robl, eds., *Optical Bistability* (Plenum, New York, 1981); C. M. Bowden, H. M. Gibbs, and S. L. McCalls, eds., *Optical Bistability II*, (Plenum, New York, 1983).

18   See, for example, R. Bonafacio and L. A. Lugiato, in H. Haken, ed., *Patent Formation by Dynamic Systems and Pattern Recognition* (Springer-Verlag, Berlin, 1979).

19   K. Ikeda, H. Daido, and O. Akimoto, *Phys. Rev. Lett.* **45**, 709 (1980); **48**, 617 (1982).

20   See, for example, J. Schwartz, W. Weiler, and R. K. Chang, *IEEE J. Quant. Electron.* **QE-6**, 442 (1970); L. Dahlstrom, *Opt. Comm.* **4**, 289 (1971); K. Sala, M. C. Richardson, and N. R. Isenor, *IEEE J. Quant. Electron.* **QE-13**, 915 (1977).

BIBLIOGRAPHY

Chang, T. Y., *Opt. Eng.* **20**, 220 (1981).

Hellwarth, R. W., *Prog. Quant Electron.* **5**, 1 (1977).

17

Self-Focusing

Self-focusing of light has fascinated many researchers in the past. It is typical of the type of nonlinear wave propagation that depends critically on the transverse profile of the beam. Theoretically, the wave equation governing the effect is the prototype of an important class of partial differential equations such as the Landau–Ginzburg equation for type-II superconductors and the Schrödinger equation for particles with self-interactions. Practically, the effect often is responsible for optical damage of transparent materials, is a limiting factor in the design of high-power laser systems, and sometimes plays an important role in the occurrence of other physical processes in a medium.

Although a complete solution of self-focusing and related effects requires extensive numerical calculations, good physical understanding of the problem can still be obtained from solutions with approximations based on experimental findings. This is the emphasis of our discussion in the present chapter. Aside from self-focusing, there also exist a number of other self-action phenomena. Only self-defocusing, self-phase-modulation, and self-steepening are briefly discussed here.

17.1 PHYSICAL DESCRIPTION

We begin with a physical description of the self-focusing phenomenon. Briefly, self-focusing is an induced lens effect. It results from wavefront distortion inflicted on the beam by itself while traversing a nonlinear medium. Consider a single-mode laser beam with a Gaussian transverse profile propagating into a medium with a refractive index n given by $n = n_0 + \Delta n(|E|^2)$, where $\Delta n(|E|^2)$ is the optical-field-induced refractive index change (see Chapter 16). If Δn is positive, the central part of the beam having a higher intensity should experience a larger refractive index than the edge and therefore travel at a slower velocity than the edge. Consequently, as the beam travels in the medium, the original plane wavefront of the beam gets progressively more distorted, as seen

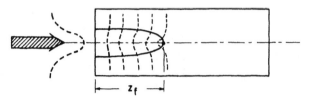

Fig. 17.1 Distortion of the wavefront of a laser beam leading to self-focusing in a nonlinear medium.

in Fig. 17.1. The distortion is similar to that imposed on the beam by a positive lens. Since the optical ray propagation is in the direction perpendicular to the wavefront, the beam appears to focus by itself.

However, a beam with a finite cross section should also diffract. Only when self-focusing is stronger than diffraction will the beam self-focus. Crudely speaking, the self-focusing action is proportional to $\Delta n(|E|^2)$, while the diffraction action is inversely proportional to the square of the beam radius. Therefore, as the beam shrinks on self-focusing, both the self-focusing action and the diffraction action become stronger. If the latter increases faster than the former, then at some point diffraction overcomes self-focusing, and the self-focused beam, after reaching a minimum cross section (the focal point), should diffract. In many cases, however, the field-induced refractive index can be approximated by $\Delta n = n_2|E|^2$, where n_2 is a constant. Then, because $|E|$ is inversely proportional to the beam radius, the self-focusing action is always stronger than the diffraction action if it is initially so. The beam may keep on self-focusing until some other nonlinear optical effect sets in to terminate the process. In such a case, the cumulative action of the nonlinear iterative effect makes the beam shrink sharply and suddenly, as in Fig. 17.1. The focal point and the focal distance (z_f) are then well defined. Which nonlinear effect actually sets in at the focal point to terminate the self-focusing process depends on the medium. It could be, for example, stimulated Raman scattering, stimulated Brillouin scattering, two-photon absorption, or optical breakdown.

A special case of interest occurs when the self-focusing action and the diffraction action on the input beam just balance each other. One would then expect that the beam should propagate in the medium over a long distance without any change in its beam diameter. This is known as self-trapping. In practice, however, self-trapping in the above context is not a stable situation. Any small loss of the laser power due to absorption or scattering can upset the balance between self-focusing and diffraction and cause the beam to diffract.

As we shall see in Section 17.2, for self-focusing to be stronger than diffraction, we must have $\Delta n \gtrsim 1/k^2a^2$, where k is the wavevector and a is the beam radius. Thus, for $a \sim 1$ mm, and $k \sim 2 \times 10^4$ cm^{-1}, self-focusing occurs only if $\Delta n \gtrsim 10^{-7}$ esu. In most media, such a large Δn can be induced only by a laser intensity higher than several megawatts per square centimeter, normally

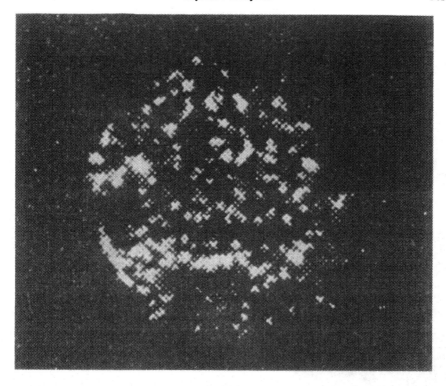

Fig. 17.2 Image of small-scale filaments at the exit windows of a CS_2 cell created by self-focusing of a multimode laser beam. [After S. C. Abbi and H. Mahr, *Phys. Rev. Lett.* **26**, 604 (1971).]

achievable only with pulsed lasers. Self-focusing should then have a time dependence resulting from the amplitude variation of the input laser pulse. Yet if the response of the medium to the field can be considered as instantaneous, the steady-state description of self-focusing is still applicable, except that the focal distance now varies with time in response to the laser intensity variation. This is known as quasi-steady-state self-focusing. If, however, the laser pulse-width is shorter than or comparable to the response time of Δn, then the time variation of Δn also becomes important in self-focusing since propagation of the leading part of the pulse can influence propagation of the lagging part. This is the regime of transient self-focusing. A more detailed discussion on quasi-steady-state and transient self-focusing will be given later.

Askar'yan[1] first suggested the possibility of self-focusing due to $\Delta n(|E|^2)$. Hercher[2] found, in early 1964, that by propagating a Q-switched laser beam of a few megawatts in a solid, one could obtain long threads of damage spots only a few microns in diameter. Chiao et al.[3] soon proposed the self trapping model to explain the observation, assuming that the damage tracks were induced by

Fig. 17.3 Images of a self-focused single-mode laser beam at the exit window of a toluene cell of different cell lengths: (a) short cell length, beam not yet self-focused ($\sim$ 700 μm); (b) cell length close to self-focusing threshold, the self-focused beam at nearly one-tenth of its original size ($\sim$ 50 μm); (c) cell length above the self-focusing threshold, the self-focused beam at its limiting size —the filament (10 μm). [After Y. R. Shen, *Prog. Quant. Electron.* **4**, 3 (1975).]

the self-trapped laser filaments. It was shown much later that the damage tracks were actually due to time-varying self-focusing with moving focal points.[4,5] In the meantime, stimulated Raman scattering was discovered, but it was found that in many solids and liquids, it had a very sharp threshold which could not be explained by the usual theory of stimulated Raman scattering.[6] It was then realized that stimulated Raman emission in such media was actually initiated by self-focusing at the focal point, and the sharp focusing of the self-focusing process was the cause of the sharp onset for stimulated Raman scattering.[7] Self-focusing can also account for many other observed anomalies in stimulated Raman scattering.

Earlier photographs of the self-focused beam in a Kerr liquid showed that the beam shrank upon self-focusing and then broke into many filaments with nearly constant diameters.[8] A typical example is shown in Fig. 17.2. Each filament has a diameter of the order of $10 \ \mu m$, which appears to be a characteristic of the medium. That many filaments could result from self-focusing of an apparently single-mode Gaussian beam was a surprising fact and attracted a great deal of attention. Later, however, it was found that the multiple filaments actually originated from the weak multimode structure in the beam. When a truly single-mode laser was used, self-focusing of the beam did lead to only one single filament, as shown in Fig. 17.3. Then the problem remained interesting because the formation and the characteristics of the filament were not understood. An important question was whether the observed filament was a manifestation of the predicted self-trapping phenomenon.[3] It turned out that the filament was simply the trajectory of the focal spot in the time-varying self-focusing process achieved with a pulsed input.[9,10] We discuss the filament problem in more detail in later sections, but first we consider a more quantitative theory of self-focusing.

17.2 THEORY

The formal theory of self-focusing is fairly simple. It is described by the nonlinear wave equation

$$\nabla^2 E - (\partial^2/c^2\partial t^2)\left[(n_0 + \Delta n)^2 E\right] = 0 \qquad (17.1)$$

assuming that the medium is isotropic, the field is transverse, and the medium response is instantaneous so that $\Delta n(|E|^2)$ does not depend explicitly on t. In this case, the differentiation in the transverse direction is nonnegligible, but since the field amplitude is not expected to vary appreciably over a distance of a wavelength, we can still use the slowly varying amplitude approximation.

Then, for a quasi-monochromatic beam propagation along $\hat{z}$, (17.1) can be reduced to a first-order partial differential equation in z and t. The equation can be further simplified by eliminating $\partial/\partial t$ with the substitution of the reduced time variable $\xi \equiv t - z/v_g$, where v_g is the group velocity. Thus by writing $E = \mathcal{E}(r, z, \xi)\exp(ikz - i\omega t)$, (17.1) becomes

$$\left(i2k\frac{\partial}{\partial z} + \nabla_\perp^2\right)\mathcal{E} = -2k^2\left(\frac{\Delta n}{n_0}\right)\mathcal{E} \tag{17.2}$$

where the beam profile is assumed to be circularly symmetric with r being the radial coordinate. Both the absolute amplitude and the phase of the field are expected to be functions of r, z, and ξ. with $\mathcal{E} \equiv A\exp(i\phi)$, (17.2) can be split into two coupled equations for the absolute amplitude A and the phase ϕ:[11]

$$k\frac{\partial}{\partial z}A^2 = -\nabla_\perp\cdot\left(A^2\nabla_\perp\phi\right) \tag{17.3a}$$

and

$$\frac{\partial}{\partial z}\phi + \frac{1}{2k}(\nabla_\perp\phi)^2 - \frac{k}{2}\left(\frac{\nabla_\perp^2 A}{k^2 A} + 2\frac{\Delta n}{n_0}\right) = 0. \tag{17.3b}$$

Equation (17.3a) is an energy relation, while (17.3b) describes the ray trajectory. Since the phase function $\phi(r, z, \xi)$ actually represents the wavefront of the beam, (17.3b) is a description of how the self-focusing action, represented by $2\Delta n/n_0$, and the diffraction action, represented by $\Delta_\perp^2 A/k^2 A$, distort the wavefront. If at $z = z_0$ there is an exact balance of self-focusing and diffraction such that

$$2\frac{\Delta n}{n_0} + \frac{\nabla_\perp^2 A}{k^2 A} = 0 \tag{17.4}$$

for all r, and if, in addition, the wavefront is flat at z_0 so that $\nabla_\perp\phi = 0$, then (17.3) yields $\partial\phi/\partial z = 0$ and $\partial A/\partial z = 0$ for $z > z_0$. This is the self-trapping case: the wave propagates in the medium with a plane wavefront and a constant transverse profile. The self-trapping solution of (17.4) for $\Delta n = n_2|E|^2$ can be obtained analytically.[3] However, it is rather unstable. A small deviation of $A(r, z)$ from the specific form of the self--trapping solution will cause the beam to either self-focus or diffract, or partly self-focus and partly diffract.

Equation (17.3b) has the same form as the Hamilton–Jacobi equation $\mathcal{H} + \partial S/\partial t = 0$ in classical mechanics,[12] where $\mathcal{H} = p^2/2m + V$ is the Hamiltonian of a particle in a potential well V, and $S(q, p, t)$ is the Hamiltonian's principal function. In our case, $\phi(r, z)$ plays the role of S, and z, r, k

correspond to t, q, m, respectively, with $(\nabla_\perp \phi)^2$ equivalent to p^2 and

$$V = -\frac{k}{2}\left(\frac{\nabla_\perp^2 A}{k^2 A} + 2\frac{\Delta n}{n_0} \right).$$

Then, by knowing that the Hamilton–Jacobi equation should lead to the usual equation of motion, $md^2q/dt^2 = \partial V/\partial q$, for the particle, we obtain similarly, from (17.3b), the equation

$$\frac{d^2 r}{dz^2} = -\frac{1}{k}\frac{\partial V}{\partial r} = \frac{1}{2}\frac{\partial}{\partial r}\left(\frac{\nabla_\perp^2 A}{k^2 A} + 2\frac{\Delta n}{n_0} \right), \qquad (17.5)$$

which governs the optical ray trajectory r as a function of z.

The solution of (17.5) can therefore be perceived from the motion of a particle in the potential well V. However, in our case, V is known only if $A(r, z)$ is known, assuming the function $\Delta n(|E|^2)$ is specified, but $A(r, z)$ can be obtained only by solving the coupled equations in (17.3). As an approximation, one can assume a certain functional form for $A(r, z)$. This approximation is found to be reasonable as long as ray bending during focusing is not significant. For example, we can assume that the central part of the beam retains its Gaussian profile as it propagates, but the beam radius varies with z: this means, for $0 < r \ll a$

$$A(r, z) = A_0 \left[\frac{a_0^2}{a^2(z)} \right] \exp\left[-\frac{r^2}{2a^2(z)} \right], \qquad (17.6)$$

which is known as the paraxial or aberrationless approximation. Each ray follows a trajectory with $r/a = r_0/a_0$, where r_0 and a_0 are the ray coordinate and the beam radius at $z = 0$, respectively.

With $A(r, z)$ given by (17.6), the potential V takes the form

$$V = -k\left(-\frac{1}{k^2 a^2} + \frac{r^2}{2k^2 a^4} + \frac{\Delta n}{n_0} \right).$$

Since $r \ll a$ in the paraxial approximation, we have

$$V(a) \simeq -k\left(-\frac{1}{k^2 a^2} + \frac{\Delta n}{n_0} \right) \qquad (17.7)$$

and (17.5) with $r(z)$ replaced by $a(z)$ can then be solved. One finds

$$\frac{k}{2}\left(\frac{da}{dz} \right)^2 + V(a) = \text{constant}, \qquad (17.8)$$

which is analogous to the energy conservation relation of the particle case. The boundary conditions are $a = a_0$ and $da/dz = (da/dz)_0$ at $z = 0$, and the focal point corresponding to $a = a_{\min}$ and $da/dz = 0$ appears at $z = z_f$. The solution of (17.5) or (17.8) can be written as

$$z = \int_{a_0}^{a} \left\{ \frac{2}{k} [V(a_0) - V(a')] + \left(\frac{da}{dz}\right)_0^2 \right\}^{-1/2} da'. \tag{17.9}$$

The focal length is then given by

$$z_f = \int_{a_0}^{a_{\min}} \left\{ \frac{2}{k} [V(a_0) - V(a)] + \left(\frac{da}{dz}\right)_0^2 \right\}^{-1/2} da. \tag{17.10}$$

We can understand physically how a beam self-focuses and diffracts by resorting to the picture of a particle moving in a potential well. As seen in (17.7), the potential V is positive or negative depending on whether the diffraction term $1/k^2a^2$ or the self-focusing term $\Delta n/n_0$ dominates. If Δn is sufficiently large, but gets saturated at high field intensities, then V may have the shape in Fig. 17.4. It shows that at large beam radius a, and hence relatively weak beam intensity, self-focusing may dominate over diffraction, but as the beam radius shrinks and the beam intensity increases, Δn becomes saturated and diffraction soon dominates. It is then easy to see, from the analogy of a particle trapped in a well, that if the initial beam divergence or convergence $|(da/dz)_0|$ (corresponding to the initial particle velocity) is smaller than $[-V(a_0)/2k]^{1/2}$, the beam will converge and diverge periodically be-

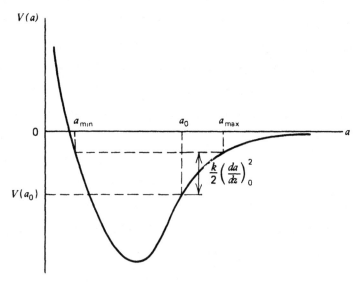

Fig. 17.4 A plot of V as a function of beam radius. This illustrates the analog between self-focusing and motion of a particle in a potential well.

tween a_{max} and a_{min} (Fig. 17.4) as it propagates. The period can be determined from (17.9). If, on the other hand, $|(da/dz)_0| > [-V(a_0)/2k]^{1/2}$, then the beam can never self-focus but will diffract to infinity, although it may first focus if $(da/dz)_0 < 0$.

In real experiments, however, saturation of Δn may not occur even in the focal region. Other nonlinear optical effects often set in to affect self-focusing long before the laser intensity reaches a level to saturate Δn. In fact, the paraxial approximation used in the above derivation also breaks down in the sharp focusing region in the common practical cases. Then, in reality, our calculation here applies only to the prefocusing region with $\Delta n = n_2|E|^2$. In such a case, for a Gaussian beam in the paraxial approximation $r \ll a$, the potential V takes the form

$$
V \cong -k\left(-\frac{1}{k^2a^2} + \frac{n_2 A_0^2}{n_0}\right)
$$
$$
= \frac{1}{ka^2}\left(1 - \frac{P}{P_0}\right) .
$$

(17.11)

where

$$
P = \frac{n_0 c}{2\pi}\int_0^\infty A^2 2\pi r\, dr = \frac{n_0 c a^2 A_0^2}{2}
$$

is the laser power, and $P_0 = c\lambda^2/8\pi^2 n_2$. Using the particle analogy, we immediately see that self-focusing can occur only if $P > P_0$. As long as the initial beam divergence $(k/2)(da/dz)_0^2$ is less than $-V(a_0)$, the self-focusing action is always stronger than the diffraction action, and the beam radius should eventually reduce to zero. In the special case of $P = P_0$, we have $V = 0$, and if $(da/dz)_0 = 0$, the beam should propagate without a change in the beam radius. This corresponds to the self-trapping case. With V given in (17.11) the integral of (17.9) yields

$$
\frac{a^2}{a_0^2} = \left(1 - \frac{P}{P_0}\right)\frac{2z^2}{k^2 a_0^4} + \left[1 + \left(\frac{da}{dz}\right)_0\frac{z}{a_0}\right]^2
$$

(17.12)

which shows how the beam radius reduces as a function of the propagation distance z. The sharp focal point should appear at $z = z_f$ corresponding to $a = 0$. We find

$$
z_f = \frac{ka_0^2/\sqrt{2}}{(P/P_0 - 1)^{1/2} - (ka_0/\sqrt{2})(da/dz)_0} .
$$

(17.13)

If $(da/dz)_0 = 0$, then it becomes

$$z_f = \frac{ka_0^2/\sqrt{2}}{(P/P_0 - 1)^{1/2}}. \tag{17.14}$$

The above solution obtained with the paraxial approximation is of course valid only for ray propagation close to the beam axis. The more rigorous solution of (17.3) can be obtained numerically on computers.[13] It is found that for $\Delta n = n_2|E|^2$ and $P > P_{cr} \simeq (1.22\lambda)^2 c/128 n_2$ (P_{cr} is known as the critical power for self-trapping[3]), the initially plane Gaussian beam can self-focus into a sharp focal spot at[14]

$$z_f = \frac{0.367 ka_0^2}{\left\{\left[(P/P_{cr})^{1/2} - 0.852\right]^2 - 0.0219\right\}^{1/2}}. \tag{17.15}$$

The relation between z_f and $\sqrt{P/P_{cr}}$ is plotted in Fig. 17.5. For $P > 1.2 P_{cr}$,

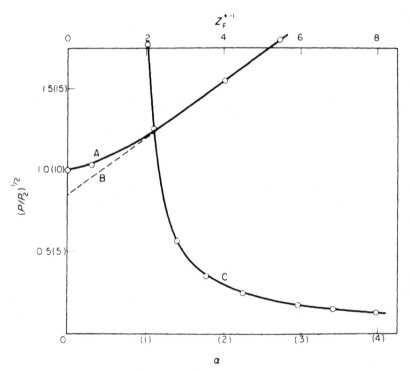

Fig. 17.5 Curve A describes the dependence of the self-focusing distance on input power; curve B is the asymptote of curve A at high powers; curve C describes α in (17.17) as a function of the input power. (After Ref. 14.)

(17.15) can be approximated by the asymptotic form

$$z_f = \frac{K}{\sqrt{P} - 0.852\sqrt{P_{\mathrm{Cr}}}} \tag{17.16}$$

with $K = 0.367ka_0^2\sqrt{P_{cr}}$.

In a Kerr liquid with $n_2 \sim 10^{-11}$ esu, we find $P_{\mathrm{cr}} \sim 8$ kW for a laser beam with $\lambda \sim 5000$ Å. The focal length z_f is proportional to the square of the initial beam radius a_0^2. For $a_0 = 150$ μm and $P/P_{\mathrm{cr}} = 1.5$, we have $z_f = 31$ cm. The on-axis beam intensity as a function of z can also be calculated, and can be approximated by

$$\frac{I(z)}{I(0)} = \left[1 - \left(\frac{z}{z_f}\right)^2 \right]^{-\alpha/2} \tag{17.17}$$

where α is a parameter depending on P as seen in Fig. 17.5.

In the quasi-steady-state case, the field amplitude $\mathscr{E}$, and hence I and P, are also functions of $\xi = t - z/v_g$. Then (17.16) and (17.17) become

$$z_f(t) = \frac{K}{\sqrt{P(\xi)} - 0.852\sqrt{P_{cr}}} \tag{17.18}$$

and

$$\frac{I(z, \xi)}{I(0, \xi)} = \left[1 - \left(\frac{z}{z_f}\right)^2 \right]^{-\alpha/2} \tag{17.19}$$

An immediate consequence of this time-dependent solution is that the focal spot position, given by z_f, should vary with time.[9, 10] This moving focus picture describes the observed results of self-focusing of nanosecond laser pulses in liquids very well, as we shall see in the following section.

17.3 QUASI-STEADY-STATE SELF-FOCUSING

For self-focusing of Q-switched laser pulses in liquids, having the pulsewidth ($\sim$ 10 nsec) much longer than the response time of the medium ($\sim$ 10 psec), the preceding discussion of quasi-steady-state self-focusing should apply. Being a strongly nonlinear effect, self-focusing depends critically on the input beam characteristics. Weak ripples on the otherwise smooth transverse profile may get strongly amplified in the self-focusing process and lead to the breaking of the beam into several independently self-focused sections. With $n = n_0 + n_2|E|^2$, the critical size Λ below which a beam with intensity I becomes unstable against the transverse intensity variation can be estimated from the expression of P_{cr}.[15]

$$\Lambda \simeq \left(\frac{4P_{cr}}{\pi I}\right)^{1/2} = \frac{1.22\lambda c^{1/2}}{(32\pi n_2 I)^{1/2}}. \tag{17.20}$$

For $n_2 \sim 10^{-11}$ esu, $I \sim 50$ MW/cm^2, and $\lambda \sim 5000$Å, we find $\Lambda \sim 10$ μm. The use of laser beams with relatively poor mode quality has led to the observation of beam break-up and multiple filaments. The results then become very difficult to interpret. To compare experiment with theory, therefore, single-mode lasers must be used. We consider here only self-focusing of single-mode laser pulses.

Self-Focusing in the Prefocal Region

Equation (17.19) describes self-focusing in the prefocal region in a medium with $n = n_0 + n_2|E|^2$. It has been confirmed experimentally by measuring the peak intensity on the beam axis as a function of z.[16] The shrinkage of the beam radius due to self-focusing has also been observed (Fig. 17.3).

The polarization dependence of self-focusing is very interesting. It has been found that the output from the focal region is always linearly polarized irrespective of the input polarization. For a circularly polarized input beam, the direction of the output polarization is random. This can be understood as the result of nonlinear coupling between the two circularly polarized field components via the field-induced refractive indices in the medium. [17] As shown in (16.31), the field-induced refractive indices for the two circularly polarized fields can be written as

$$\Delta n_{\pm} = \left(\frac{2\pi}{n}\right)\left[A|E_{\pm}|^2 + B|E_{\pm}|^2\right] \tag{17.21}$$

For ordinary liquids, $A - B = -2\chi^{(3)}_{1221} < 0$. Therefore, if both circularly polarized fields are present, the weaker one sees a larger Δn and hence self-focuses more readily until its intensity equals that of the other component. The output then becomes linearly polarized. The above argument applies even to the case of a circularly polarized input beam, since, in practice, the beam can never be perfectly circularly polarized. The quantitative analysis of this self-focusing problem with mode coupling, however, has not yet been worked out.

Filaments and Moving Foci

The self-focusing threshold for a given medium of length l is determined by the condition $z_f(P_{max}) = l$, where P_{max} is the peak power of the input pulse. Equation (17.18) describing z_f as a function of P_{max} has been experimentally confirmed by measuring the self-focusing threshold powers at different l.[18]

We now consider what happens when the input peak power is above the self-focusing threshold. In early experiments, it was found that after the beam self-focused it broke into a number of intense thin filaments.[8,19] The multiple filaments were the result of the multimode structure in the input beam, as mentioned earlier. It was shown later that a single-mode input laser actually resulted in a single filament on the beam axis. For a given medium, the filament had a diameter constant to within $\pm 20\%$ and lasted over a distance of a few centimerers. The intensity of the filament could be as high as a few tens of gigawatts per square centimeter.

From the picture of quasi-steady-state self-focusing, one can perhaps realize that the filament may correspond to the track of the moving focal spot as it appears on a time-integrated photograph. That this is indeed the case has been confirmed by motion pictures taken with streak camera.[20] The diameter of the filament then corresponds to the diameter of the focal spot, and the filament intensity to the intensity in the focal region.

We can obtain a better perspective of the moving-focus picture from Fig. 17.6. The upper U curve describes the position of the focal spot as a function of time. It is constructed from the input pulse $P(t)$ using (17.18) and assuming K and P_{cr} are known.[10] In practice, K and P_{cr} can be determined from the measurements of $z_f(P_{max})$ versus P_{max}.[18] Experimental determination of the U curve (at least partially) is also possible from some sort of time-of-flight measurements of the moving focus.[21] It has been found that the measured curve agrees very well with the one calculated from (17.18). As seen from Fig. 17.6, the U curve has the following characteristics. If the length of the medium l is sufficiently long, then the focal spot first appears at z_D inside the medium. It then splits into two: one moves backward and then forward after it reaches the minimum self-focusing distance $z_B(P_{max})$ corresponding to the peak of the input pulse; the other moves forward with a velocity faster than light. Both branches of the U curve have their slopes approach light velocity as $z \rightarrow \infty$. Note that the faster-than-light feature does not violate the special theory of relativity because the focal spots appearing at different times actually come from self-focusing of different parts of the input pulse, and therefore the "motion" of the focal spots does not transmit anything real. However, a strong polarization induced by the moving focal spot still appears in the medium and can have an apparent velocity faster than the light velocity. This is similar to the case of Cerenkov radiation, but the problem has not yet been worked out.

The unusual characteristics of the U curve for the moving focus lead to a number of interesting results.[22] First, the focal spot should spend a relatively long time at $a_f(P_{max})$, and hence optical damage is more likely at $z_f(P_{max})$. Indeed, in transparent liquids, laser-induced bubbles have been observed in this region. Second, when $z_f(P_{max})$ is appreciably smaller than l, the light pulse diffracted from the filament within a few centimeters at the end of the cell has a very short pulsewidth, less than 100 psec for a nanosecond input pulse. Third, the high laser intensity (~ 10 GW/cm^2) in the focal region readily initiates other nonlinear optical processes. One is a strong phase modulation

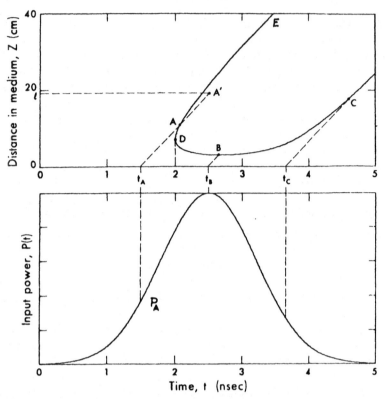

Fig. 17.6 Lower trace describes input power $P(t)$ as a function of time t. Peak power is 42.5 kW and the half-width at the $1/e$ point is 1 nsec. Upper trace calculated from (17.18) describes the position of the focal spot as a function of time. Values of $(0.852)^2 P_{cr}$ and K used are 8 kW and 11.6 cm $(kW)^{1/2}$, respectively, which corresponds roughly to an input beam of 400 μm in diameter propagating in CS_2. The dotted lines, with the slope equal to the light velocity, indicate how light propogates in the medium along the z-axis at various times. (After Ref. 22.)

and a resultant spectral broadening on the light diffracted from the filament region due to the large field-induced refractive index change Δn. We discuss the problem in more detail in a later section. Another is the initiation of stimulated Raman and Brillouin scattering. The stimulated scattering, in turn, drastically affects self-focusing. The interplay between the two, which is most intriguing and interesting, is discussed below.

The sharp stimulated Raman threshold in Kerr liquids (those in which Δn is dominated by molecular reorientation) was a problem that attracted a great deal of attention in the early development of nonlinear optics (See Section 10.6). We now understand that it results from self-focusing. The extremely high laser intensity in the focal region readily initiates stimulated Raman and Brillouin scattering. The sharp stimulated Raman and Brillouin thresholds should therefore nearly coincide with the self-focusing threshold. The buildup of the Raman and Brillouin intensities, however, depends on the characteristics of the two stimulated scattering processes in the particular medium.[22, 23] Stimulated Brillouin scattering has a large steady-state gain but a slow transient response ($\sim$ 10 nsec), whereas stimulated Raman scattering has a much smaller steady-state gain but an almost instantaneous response ($\sim$ 5 psec).[23] As the laser intensity increases upon self-focusing, the Brillouin scattering may or may not appear earlier than the Raman scattering, depending on the medium.

When the laser power is well above the self-focusing threshold, the stimulated Raman and Brillouin generation can be understood with the help of Fig. 17.7.[22] The moving focal spot initiates both forward Raman and backward Raman and Brillouin scattering along the U curve. The backward radiation initiated from the lower branch of the U curve intersects with the incoming laser light in the shaded region and gets effectively amplified. Since the Raman scattering has an instantaneous response, it appears first, and its strong amplification soon depletes the incoming laser power to a level below the self-focusing threshold. Termination of self-focusing then stops the Raman radiation. As a result, the backward Raman output appears in the form of an intense subnanosecond pulse, [24] as seen in Fig. 17.8. With the Raman emission fading out, the incoming laser power recovers from depletion and reaches the self-focusing threshold again. The backward Brillouin radiation then initiated can have a larger transient gain than the Raman scattering. It builds up in intensity and depletes the incoming laser power. Through self-adjustment, the backward Brillouin scattering keeps the transmitted laser power just below the self-focusing threshold. If the transmitted laser power is too high or too low, the Brillouin scattering intensity would increase or decrease accordingly to deplete more or less laser power. This explains the observations in Fig. 17.8 that after the sharp dip afflicted by the Raman generation, the transmitted laser pulse shows a depleted flat top, while the sum of the transmitted laser power and the backward Brillouin power is equal to the incoming laser

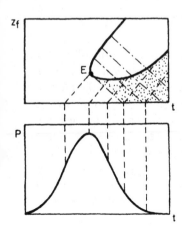

Fig. 17.7 The interaction between backward stimulated scattering and incoming laser radiation. Backward stimulated Raman and Brillouin radiation, initiated along the upper branch of the U-curve, propagates along the dot-dashed lines and interacts with the non-self-focused incoming laser light in the shaded region. [After Y. R. Shen *Prog. Quant. Electron.* **4**, 12 (1975).]

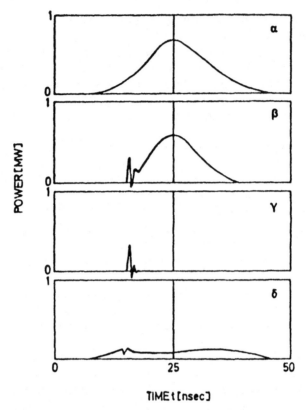

Fig. 17.8 Oscilloscope traces of the incident laser pulse (α), the total stimulated emission in the backward direction (β), the backward stimulated Raman emission alone (γ), and the transmitted laser light (δ). (After Ref. 24.)

power.[24] The depletion of the incoming laser power to a level below the self-focusing threshold also terminates the moving focus or filament. Consequently, the later portion of the lower branch of the U curve can never be observed.

The forward stimulated Raman scattering can also be initiated in the moving focal region. Figure 17.7 shows that its amplification is through interaction with the diffracted laser light after the focal region, and therefore is expected to be much smaller than the backward Raman amplification. This is indeed the experimental observation.[25] As the laser power or the medium length increases further, the focal region becomes longer, and so does the laser–Raman interaction length. As a result, the forward Raman output increases steadily. It can eventually deplete nearly all the laser power in the focal region. Then diffracted Raman, instead of laser light would show up in that focal region, and the laser filament resulting from the moving focus along the upper branch of the U curve would appear effectively terminated.[22]

Other anomalous observations of stimulated Raman and Brillouin scattering with self-focusing can also be successfully explained by the moving focus picture. A more detailed description of the problem is given in Ref. 22. Quantitative solution of the problem can in principle be obtained by solving the coupled nonlinear wave equations for the laser, Raman, and Brillouin fields. However, this is a horrendous job even with computers, and it has not yet been seriously attempted.[26]

Not all experimental observations about the filaments in quasi-steady-state self-focusing are understood. For example, the diameter of the filament or the moving focal spot is presumably determined by the nonlinear processes initiated in the focal region, but how it can be a characteristic of the medium independent of the input laser pulse is not clear. The finite relaxation time of the field-induced refractive index can affect the diffraction of light from the filament especially when the focal spot has a velocity faster than light. The quasi steady-state description of self-focusing is clearly not valid for the focal region. Transient dynamics with complications from other nonlinear optical processes have not yet been worked out. Finally, the observation of Stokes and anti-Stokes rings around the filament[27] is yet to be explained by the moving-focus model.

17.4   TRANSIENT SELF-FOCUSING

When the input laser pulsewidth is shorter than or comparable to the response time of Δn, the time variation of Δn becomes important in the self-focusing process. This is the transient self-focusing regime. We consider here the case where Δn is caused predominantly by field-induced molecular reorientation, and is governed by the Debye relaxation equation[28]

$$\left(\frac{\partial}{\partial t} + \frac{1}{\tau}\right)\Delta n = \frac{1}{\tau}\Delta n_0 \tag{17.22}$$

where Δn_0 is a function of $|E(\mathbf{r}, t)|^2$, and in the lowest order, Δn_0 is proportional to $|E(\mathbf{r}, t)|^2$. Integration of (17.22) with E being a function of $(\mathbf{r}, \xi)$, where $\xi = t - z/v_g$, yields

$$\Delta n(\mathbf{r}, \xi) = \frac{1}{\tau}\int_{-\infty}^{\xi}\Delta n_0\big(|E(\mathbf{r}, \eta)|^2\big)\exp\left[-\frac{(\xi - \eta)}{\tau}\right]d\eta. \tag{17.23}$$

The transient self-focusing dynamics is then governed by the coupled equations (17.2) and (17.23). The solution has been attempted both analytically[29] and numerically.[26, 30] Only a qualitative description is presented here.

Because of the transient response of Δn, the leading part of the laser pulse can affect self-focusing of the trailing part. Figure 17.9 shows how the different parts of the pulse would propagate in the medium.[31] The very first part (a in the figure) of the pulse sees little induced Δn, and diffracts almost linearly as it propagates. The next part (b in the figure) sees somewhat larger Δn, but not large enough to cause self-focusing, and therefore it still diffracts, although not as strongly. Then, the c–f part of the pulse sees a sufficiently large Δn induced by the earlier part of the pulse to be able to self-focus. However Δn is smaller at larger z, so that the beam will eventually diffract. Both self-focusing and diffraction of the c–f part of the pulse are fairly gradual and yield a relatively long focus. The diameter of the focus depends on how large Δn is. The later part of the pulse may see a larger Δn, and therefore self-focus at a shorter distance to a smaller focal diameter, but its diffraction is still gradual because of the slow diffraction of the leading part. In practice, it is likely that the minimum diameter is limited by some other nonlinear process. When this happens, the focal diameter is the same for a finite section of the input pulse, as seen in Fig. 17.9. The above picture is actually an extension of the moving-focus picture, which it reduces to in the quasi-steady-state case.

From Fig. 17.9, we can see how the transverse profile of the pulse gets deformed. As the trailing portion of the pulse shrinks due to self-focusing, the pulse quickly deforms into a horn shape. Because of the slow focusing and diffraction, the horn-shaped pulse soon appears to have reached a stable form, and then propagates on for a long distance without appreciable change in its shape. This stable form of pulse propagation often is known as dynamic trapping. The neck part of the pulse sweeping along the axis should lead to the appearance of a filament on a time-integrated photograph. The horn-shaped pulse gradually expands in the transverse dimension through diffraction and yields a filament with a fading end if the length of the medium is sufficiently long.

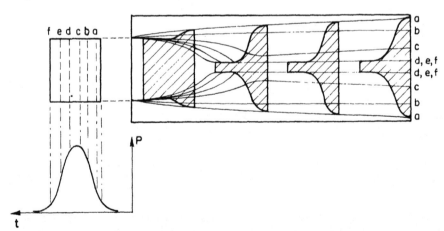

Fig. 17.9 Self-focusing of a picosecond pulse in a Kerr liquid. Different parts (a, b, c, etc.) of the pulse focus and defocus along different ray paths. The pulse first gets deformed into a horn shape and then propogates on without much further change. [After Y. R. Shen, *Prog. Quant. Electron.* **4**, 27 (1975).]

Transient self-focusing with filament formation can be readily observed with picosecond laser pulses in Kerr liquids.[32] Quantitative measurements with picosecond pulses to verify the dynamic trapping picture are, however, difficult because of limitations of picosecond technology. It turns out that in isotropic liquid crystals the response time of Δn can be varied from a few nanoseconds to a few hundred nanoseconds by simply varying the temperature.[28] Then, with the same nanosecond pulses, one can study self-focusing from the transient to the quasi-steady-state limit.[33] In the transient case, it was shown by measuring the on-axis intensity variation of the output that the self-focused laser beam indeed deformed into a horn-shaped pulse.[31] The neck diameter remained almost constant over a fairly long distance. The dynamic trapping model is therefore well proven. In the above case, the limiting neck diameter did not result from stimulated Raman and Brillouin scattering but presumably from two-photon absorption in the focal region.

17.5 SELF-FOCUSING IN A SOLID

Self-focusing can also occur in a solid and leave a damage track.[2] Careful experimental study has shown that the damage has the following characteristics.[34] The damage track appears as a cylindrical region of altered refractive index, a few microns in diameter, straight to within an rms deviation of one wavelength, and several centimeters in length. It starts with a damage star and may terminate before it reaches the exit end of the sample. Track formation is characterized by a flash of white light emission from the track, an increase of divergence of the transmitted laser beam, and a short pulse of backward stimulated Brillouin radiation.

For a Q-switched input laser pulse, electrostriction is the dominant mechanism for the induced Δn in most solids. The equation governing Δn is simply the acoustic wave equation (Section 16.2),

$$\left(\nabla^2 - \frac{1}{v_a^2}\frac{\partial^2}{\partial t^2} + \frac{2\Gamma_B}{v_a^2}\frac{\partial}{\partial t}\right)\Delta n = \frac{1}{v_a^2}\nabla^2\left(\frac{\gamma}{2\pi}|E|^2\right). \qquad (17.24)$$

Because of the slow response of Δn in this case, self-focusing falls into the transient regime. The self-focusing dynamics is given by the solution of the coupled equations (17.2) and (17.24). This has actually been carried out numerically by Kerr.[35] Qualitatively, the transient focusing behavior described in Fig. 17.9 should still be true here. The difference is in the fact that when the intensity of the self-focused beam reaches a certain value at a local spot, optical damage occurs there and effectively terminates self-focusing of the beam beyond that spot. The self-focused beam is strongly diffracted from the damage spot, and therefore the horn-shaped pulse seen in Fig. 17.9 cannot be formed. The picture then resembles more the moving-focus picture, except that the damage spot now plays the role of the moving focus. The trajectory of this well-defined damage spot can be determined if the axial intensity variation with time can be calculated. As seen from Fig. 17.9, the damage spot resulting from transient self-focusing is expected to move in the backward direction.

The above qualitative features have been confirmed by experiment.[36] The damage spot marked by the emission of white light can be followed with a streak camera. An example is seen in Fig. 17.10. The damage spot indeed moves in the backward direction. It reaches the end of the track around the peak of the input pulse, and stays there for a relatively long time, with the laser input continuously feeding energy into this region and creating a damage star.

Although the understanding of self-focusing in solids appears to be satisfactory, experimental investigation so far has not been extensive. For example, self-focusing in semiconductors has seldom been studied. In order for self-focusing to be strong enough to cause optical damage in a solid, the solid must be longer than a threshold length. The thickness of an optical window frequently is smaller than this length. Optical damages in thin windows, in

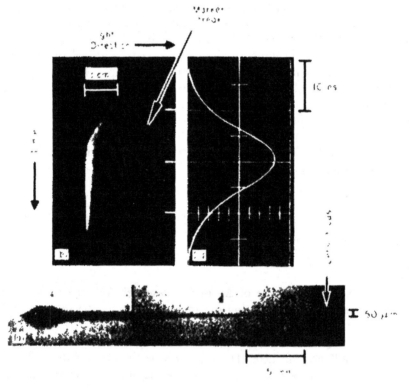

Fig. 17.10 Typical example of (a) damage filament, (b) streak photograph, and (c) oscilloscope trace for a smooth incident pulse. (After Ref. 36.)

absorbing materials, and on surfaces are generally not related to self-focusing. (See Section 27.3)

17.6 OTHER SELF-FOCUSING PHENOMENA

Self-focusing in gases is also observable if the incoming laser beam has a frequency slightly above an absorption line so that Δn is positive and sufficiently large.[37] This has been seen, for example, with a Raman-shifted pulsed ruby laser in potassium vapor,[37] and a pulsed CO_2 laser in SF_6.[38] Even self-focusing of a CW laser beam has been observed in potassium vapor. Under suitable conditions, the self-focused beam is found to funnel into a trapped filament.[39] This particular case of steady-state self-focusing is different from the previous cases we discussed, since Δn is now dependent on atomic diffusion; also, the optical energy in the self-focused beam is partially lost through atomic excitation, relaxation, and diffusion. The theory of such a CW self-focusing case has not yet been worked out.

A similar CW self-focusing has been observed in a liquid with colloidal suspension of submicron spheres.[40] The induced Δn in this case arises from the field-induced density increase of the colloidal spheres. This case is somewhat similar to electrostrictive self-focusing and is different from the atomic case in the sense that no optical energy of the self-focused beam is lost into the medium.

Another CW self-focusing case of similar behavior is in a slightly absorbing solid.[41] This happens when a positive Δn arises from heating of the medium through optical absorption

$$\Delta n = \frac{\partial n}{\partial T}\Delta T + \left(\frac{\partial n}{\partial \rho_D}\right)_T\left(\frac{\partial \rho_D}{\partial T}\right)\Delta T. \qquad (17.25)$$

The first term can be positive if the temperature rise shifts the absorption bands in such a way as to increase Δn. The second term, however, is always negative with increasing ΔT. If the resultant Δn of (17.25) is still positive, then self-focusing can occur. This is known as thermal self-focusing. Actually, both transient and steady-state thermal self-focusing have been observed. An example of CW thermal self-focusing with the formation of a trapped filament is seen in Fig. 17.11.[41]

Optical damage caused by self-focusing in a laser amplifier is now well recognized as a limiting factor in the design of high-power laser systems. Yet it is ironic that so far, little research has been done on self-focusing in an amplifying medium. Fleck and Layne[42] performed a numerical calculation on the problem. Quantitative experimental results, however, are not yet available to check the theory.

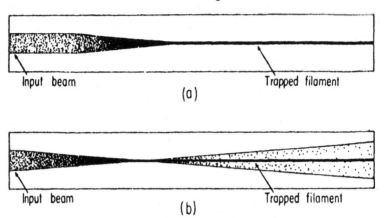

Fig. 17.11 Thermal self-focusing of an argon laser beam inside a lead glass rod ($l = 35$ cm) with input power (a) $P_0 = 3$ W and (b) $P_0 = 8$ W. Part of the self-focused beam funnels into a trapped filament. (After Ref. 41.)

17.7   SELF-PHASE MODULATION

In quasi-steady-state and transient self-focusing, a very interesting observation which we have not yet discussed is the appearance of strong spectral broadening of light emitted from the filament regions.[43] With a nonosecond input pulse, the broadening can be several tens of inverse centimeters, while with a picosecond input pulse, it can be more than several thousand. An example is shown in Fig. 17.12. The result is at least partially due to self-induced phase modulation on the self-focused beam.

Let us first consider the effect with a simple model.[44] Assume a laser pulse $|E(t)|^2$ propagating in a self-trapped filament of length l. If Δn in the filament has an instantaneous response $\Delta n(t) = n_2|E(t)|^2$, then the output from the filament has a self-phase modulation $\Delta\phi(t) = (\omega/c)\Delta n(t)l = (\omega/c)n_2|E(t)|^2 l$ and a corresponding frequency modulation $\Delta\omega(t) = -\partial(\Delta\phi)/\partial t$, which appears as a broadened spectrum. More rigorously, the output spectrum is given by the Fourier transform

$$|E(\omega)|^2 = \left|\int_0^\infty \mathscr{E}(t)e^{-i\omega_0 t + i\Delta\phi(t)}\, dt\right|^2 \qquad (17.26)$$

which in the slowly varying approximation can be evaluated with $\mathscr{E}(t)$ outside

Fig. 17.12 Observed spectral broadening in a mixture of CS_2 and benzene with a multimode Q-switched ruby laser pulse. [After T. K. Gustafson, J. P. E. Taran, H. A. Haus, J. R. Lifsitz, and P. L. Kelley, *Phys. Rev.* **177**, 306 (1969).]

the integral. If $\Delta\phi(t) \propto |E(t)|^2$ is an ordinary bell-shaped pulse as in Fig. 17.13, then, qualitatively, the output spectrum can be expected to have the following characteristics. First, since $\Delta\phi(t)$ is symmetric, the power spectrum is also symmetric with respect to the incoming laser frequency ω_0. Second, the maximum of the frequency spread is given approximately by $|\Delta\omega|_{max} \simeq |\partial(\Delta\phi)/\partial t|_{max}$, which appears at the inflection points of the $\Delta\phi(t)$ curve. Third, there generally exist two points of the same slope on the $\Delta\phi(t)$ curve. These two points, crudely speaking, represent two waves of the same frequency but different phases. They will interfere constructively or destructively depending on the phase difference between them. The output spectrum should therefore show a semiperiodic structure with clear peaks and valleys. The farthest peaks on the two sides, arising from the inflection point on the $\Delta\phi(t)$ curve, are the strongest. The number of peaks on either side is approximately given by the integer closest to but smaller than $|\partial(\Delta\phi)/\partial t|_{max}/2\pi$. In Fig. 17.13a, such a spectrum is shown corresponding to the $\Delta\phi(t)$ curve given. If Δn has a relaxation time comparable to the pulsewidth, then the transient response of Δn yields a $\Delta\phi(t)$ with a long tail (Fig. 17. 13b). Consequently, the spectral broadening on the anti-Stokes side is greatly reduced.

Thus it is clear that the kind of phase modulation shown in Fig. 17.13 can lead to semiperiodic spectral broadening. Self-trapping of a laser pulse, however, is only an ideal case. We must show that the output from the moving focus can be similarly phase-modulated in order to explain the observed spectral broadening in quasi-steady-state self-focusing.[45] Let us assume that the medium length l is much larger than the minimum self-focusing distance $z_f(P_{max})$. As shown in Fig. 17.14, the beam entering the medium at t_A self-focused sharply at A and leaves the medium at A'. If we know how the beam self-focuses at various times, then $|E(z, t)|^2$ can be calculated. In practice, although the detailed shape of $|E(z, t)|^2$ at a given z is not known without a real calculation, we know that the pulsewidth of $|E(z, t)|^2$ must be of the order of the relaxation time τ for Δn. It cannot be much shorter because the observed Δn in the focal region is not much less than the steady-state value $\Delta n_{max} = n_2 |E(z, t)|^2_{max}$. It cannot be much longer since otherwise the nearly steady-state response of Δn would lead to a sharper focusing and hence a pulsewidth smaller than τ, contrary to the assumption. In Fig. 17.14 the hatched area indicates the region of large $|E|^2$, where one also expects a large Δn, which can be calculated from (17.22) assuming $|E(z, t)|^2$ is known. Note that τ for molecular reorientation in a liquid is of the order of 10 psec. It is then easily seen from Fig. 17.14 that the output from the filament region at the end of the medium is strongly phase-modulated, because the beams entering the medium at different times cross different sections of the hatched area.

The phase increment of the self-focused beam traversing the medium can be written as

$$\Delta\phi\left(t = t_A + \frac{ln}{c}\right) = \int_0^l \left(\frac{\omega_0}{c}\right)\Delta n\left(z, t' = t_A + \frac{zn}{c}\right) dz \qquad (17.27)$$

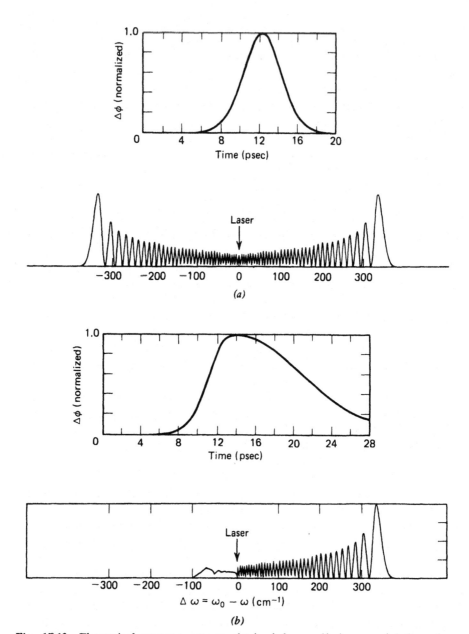

Fig. 17.13 Theoretical power spectrum obtained from self-phase modulation of a pulse propagating in a nonlinear medium without any change of its shape. (*a*) Spectrum corresponding to a phase modulation $\Delta\phi$ symmetric in rise and fall; (*b*) spectrum corresponding to a phase modulation $\Delta\phi$ with a much sharper leading than falling edge.

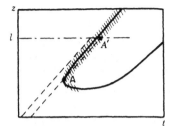

Fig. 17.14 A U-curve describing the moving focus. The refractive index change Δn is appreciable in the hatched region, which has a width of about a few relaxation times. Light traversing the cell along the dashed line acquires a phase increment $\Delta \phi$, which varies with time t. [After Y. R. Shen, *Prog. Quant. Electron.* **4**, 18 (1975).

where for simplicity, we neglect the diffraction contribution to $\Delta \phi$. Qualitatively, $\Delta \phi(t)$ increases to its maximum during the presence of the pulse $|E(l, t)|^2$ in a time of the order of τ, and then decays away much more slowly, similar to the curve in Fig. 17.13*b*. The corresponding asymmetrically broadened spectrum is indeed what was observed experimentally. The maximum broadening on the Stokes side can be obtained analytically with a simple approximation as follows. We approximate the last portion of the U curve toward the end of the medium in Fig. 17.14 by a straight line with a slope $v > c/n$. Then, the light emitted from the hatched section at $z = l$ has the phase increment.

$$\Delta \phi \simeq \left(\frac{\omega_0}{c} \right) \left(\frac{n}{c} - \frac{1}{v} \right)^{-1} \int_{t_0}^{t} \Delta n(l, t') \, dt' \qquad (17.28)$$

where t_0 is a time before which $\Delta n(l, t_0)$ is negligible. From (17.16) we can find

$$\left(\frac{n}{c} - \frac{1}{v} \right)^{-1} = \left(\frac{l^2}{K} \right) \left[\frac{1}{\sqrt{P}} \frac{dP(t)}{dt} \right]_{P = P(z_f = l)} \qquad (17.29)$$

The maximum spectral broadening on the Stokes side then is given by

$$\Delta \omega_{max} = -\left(\frac{\partial \Delta \phi}{\partial t} \right)_{max}$$

$$= -\left(\frac{\omega_0}{c} \right) \left(\frac{n}{c} - \frac{1}{v} \right)^{-1} \Delta n_{max} \qquad (17.30)$$

As an example,[46] consider an input Gaussian pulse with a full width of 1.2 nsec at the $1/e$ points, a beam diameter of 300 μm, and a peak power of 28 kW self-focused in a 20-cm cell of CS_2. The trajectory of the moving focus is described by (17.16) with $K = 5.6$ cm(kW)$^{1/2}$ and $P_{cr} = 8.65$ kw. The part of the input pulse that self-focuses at the end of the cell has the instantaneous input power $P(z_f = l) = 9.8$ kW. The diameter of the filament in CS_2 is 5 μm, which leads to a $\Delta n_{max} \sim 2.5 \times 10^{-3}$ esu in the focal region at the end of the cell. Then, from (17.29) and (17.30), we immediately find $\Delta \omega_{max}/\omega_0 \simeq 0.0076$, or $\Delta\omega_{max} \simeq 110$ cm^{-1} for a ruby laser input. Equation (17.30) has been experimentally verified.[46]

For the case of transient self-focusing of picosecond pulses in a Kerr liquid, the dynamic trapping picture of Fig. 17.9 should be used. Again, it is clear from the picture of ray propagation in Fig. 17.9 that the laser light emerging from the axial region at the end of the medium must be strongly phase-modulated. Assuming a given intensity distribution of the horn-shaped pulse, we can calculate $\Delta n(z, t)$ and hence $\Delta\phi(t)$ of the emitted light. In this case, $\Delta\phi_{max}$ can be very large because of a long dynamic trapping length, while the rise and fall of $\Delta\phi(t)$ are still on the picosecond time scale. The spectral broadening can therefore extend over several hundred or perhaps even a few thousand cm^{-1}. It is also semiperiodic and may have a strong broadening on the anti-Stokes side if the fall time of $\Delta\phi(t)$ is short.

In many cases, however, focusing of a high-intensity picosecond pulse into a liquid or solid with little molecular reorientational contribution to Δn can still yield an output with a huge spectral broadening extending more than a few thousand cm^{-1} on both the Stokes and anti-Stokes sides.[47] Self-phase modulation could also be the cause of this spectral broadening, as photopreionization could lead to a fast and strong self-phase modulation.[48] However, it has also been suggested that wave mixing[49] could be responsible for the broadening. The broadband emission is in the form of a picosecond pulse. It can therefore be used in picosecond spectroscopy as a tunable picosecond source. In another application, an ultrashort pulse can be self-phase-modulated in a nonlinear medium. It can then be reflected from a pair of gratings acting as a pulse compressor for the phase-modulated light to yield a much shorter pulse. Compression of a 90-femtosecond pulse to a 30-fsec pulse has been demonstrated.[50]

Self-phase-modulation in space on the transverse profile of a beam is also possible. It appears as a distortion on the wavefront and leads to self-focusing if the medium is sufficiently long. In a thin medium, strong self-phase modulation can still occur, but physical shrinkage of the beam inside the medium due to self-focusing is hardly visible. The case is then analogous to that of self-phase-modulation in time. For a beam with a Gaussian-like transverse profile, the phase increment $\Delta\phi(r)$ across the beam profile has a bell shape centered at $r = 0$. If $[\Delta\phi(r)]_{max}$ is much larger than 2π, then the output power spectrum in the transverse wavevector $k_\perp$ space should show peaks and valleys resulting from constructive and destructive interferences. They appear in the form of interference rings on a projection screen. The number of bright rings is approximately given by the integer closest to but smaller than $[\Delta\phi(r)]_{max}/2\pi$, and the diameter of the outermost ring is determined from the maximum slope of $\Delta\phi(r)$ at the inflection point. Such an effect has actually been demonstrated in a nematic liquid crystal film.[51] A large Δn can be induced in this kind of media with a CW laser beam of several hundred watts per square centimeter (see Section 16.2). A maximum of $\Delta\phi$ of several tens of 2π radians can be readily obtained in a nematic film several hundred micrometers thick. Inter-

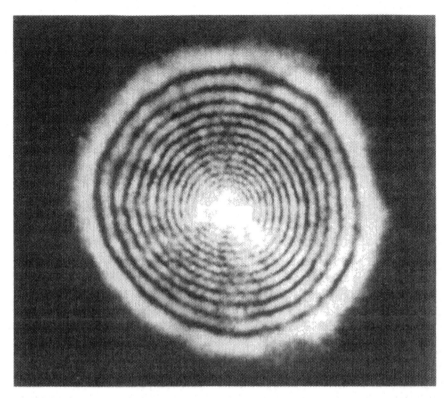

Fig. 17.15 Diffraction ring pattern of a CW Ar^+ laser beam after passing through a 300-μm nematic liquid crystal film. (After Ref. 51.)

ference rings up to ~ 100 have in fact been observed. An example is shown in Fig. 17.15.

17.8 SELF-STEEPENING AND SELF-DEFOCUSING

This section considers only a physical description of the self-steepening and self-defocusing effects. Readers are referred to the literature for details on these subjects.

Self-steepening of a pulse occurs when the group velocity of light depends on the light intensity through the induced Δn.[52] If Δn is positive and has an instantaneous response, then a light velocity that decreases with laser intensity can lead to the formation of a steep front in the trailing edge of the pulse, resembling the usual shock wave formation. If Δn is negative, the steep front can be developed in the leading edge of the pulse. This type of self-steepening relying only on the intensity-dependent Δn has never been observed, however.

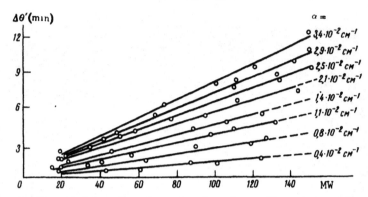

Fig. 17.16 Thermal defocusing of an argon laser beam. Dependence of beam divergence on laser power in a water cell, $l = 44$ cm, with different attenuation constants α. (After Ref. 57.)

It turns out that if the group velocity also has a linear dispersion, for example, when the laser frequency is near an absorption band, self-steepening can be expected to occur in a much shorter propagation distance. This is because self-phase modulation also occurs in the medium and modifies the frequency spectrum of the pulse. The pulse then reshapes according to the linear dispersion of the group velocity. Self-steepening has indeed been observed in Rb vapor with a pulsed dye laser at a frequency slightly below the $s \rightarrow p$ transition. The results agree well with the theoretical prediction.[53]

Self-defocusing can occur if Δn decreases with increase of laser intensity since the wavefront distortion is now opposite to that in the self-focusing case.[54] This happens when the incoming laser frequency is somewhat below a

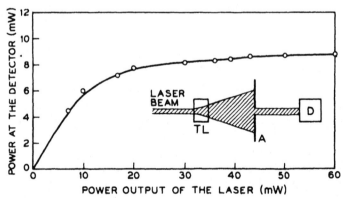

Fig. 17.17 Output power of a thermally defocused beam through an aperture as a function of the input power from an Ar^+ laser at 4880 Å. The inset shows the power-limiting device. TL is the thermal lens, A is the aperture, and D is the detector. (After Ref. 59.)

resonant absorption line. More commonly, it happens in absorbing media with $dn/dT < 0$. Self-defocusing from a laser-induced thermal effect is also known as thermal defocusing. Physically self-defocusing is similar to self-focusing except that the wavefront distortion is inverted. In the more quantitative discussion, however, one finds that unlike the self-focusing case, geometric optics can be used to describe self-defocusing, since the beam diffraction is always fairly gradual.[55] The nonlinear medium acts as a thin or thick divergent lens depending on whether the absorption is weak or strong. Because the phase change across a Gaussian-like beam profile is also a bell-shaped curve centered at $r = 0$, interference rings can also be observed if $|\Delta\phi(r)|_{max}$ is much larger than 2π.[56]

Thermal defocusing is readily detectable even in a very weakly absorbing medium. Figure 17.16 shows an example of how the beam divergence depends on the input power and absorption coefficient.[57] The high sensitivity of the beam divergence to small absorption in the medium has prompted the use of thermal self-defocusing as a spectroscopy method.[58] Absorption coefficients smaller than 10^{-6} can be routinely measured. Another application of thermal defocusing is the construction of a laser power limiter,[59] as seen in Fig. 17.17.

REFERENCES

1 G. A. Askar'yan, *Sov. Phys. JETP* **15**, 1088, 1161 (1962).

2 M. Hercher, *J. Opt. Soc. Am.* **54**, 563 (1964).

3 R Y. Chiao, E. Garmire, and C. H. Townes, *Phys. Rev. Lett.* **13**, 479 (1964) [Erratum, **14**, 1056 (1965)].

4 G. M. Zerev, E. K. Maldutis, and V. A. Pashkov, *JETP Lett.* **9**, 61 (1969); G. M. Zerev and V. A. Pashkov, *Sov. Phys. JETP* **30**, 616 (1970).

5 C. R. Giuliano and J. H. Marburger, *Phys. Rev. Lett.* **27**, 905 (1971).

6 See Chapter 11, and the review article by Y. R. Shen, in M. Cardona, ed , *Light Scattering in Solids* (Springer-Verlag, Berlin, 1975), p. 275.

7 Y. R. Shen and Y. J. Shaham, *Phys. Rev. Lett.* **15**, 1008 (1965); P. Lallemand and N. Bloembergen, *Phys. Rev. Lett.* **15**, 1010 (1965); G. Hauchecorne and G. Mayer, *Compt. Rend.* **261**, 4014 (1965); N. F. Pilipetskii and A. R. Rustamov, *JETP Lett.* **2**, 55 (1965).

8 E. Garmire, R. Y. Chiao, and C. H. Townes, *Phys. Rev. Lett.* **16**, 347 (1966); R. Y. Chiao, M. A. Johnson, S. Krinsky, H. A. Smith, C. H. Townes, and E. Garmire, *IEEE J. Quant. Electron.* **QE-2**, 467 (1966).

9 V. N. Lugovoi and A. M. Prokhorov, *JETP Lett.* **7**, 117 (1968).

10 M. M. T. Loy and Y. R. Shen, *Phys. Rev. Lett.* **22**, 994 (1969).

11 See, for example, S. A. Akhmanov, A. P. Sukhorukov, and R. V. Khokhlov, *Sov. Phys. Uspekhi* **93**, 609 (1968).

12 W. G. Wagner, H. A. Haus, and J. H. Marburger, *Phys. Rev.* **175**, 256 (1968). For the Hamilton–Jacobi theory, see, for example, H. Goldstein, *Classical Mechanics* (McGraw-Hill, New York, 1950), p. 273.

13 P. L. Kelley, *Phys. Rev. Lett.* **15**, 1005 (1965); V. I. Talanov, *JETP Lett.* **2**, 138 (1965).

14 J. H. Marburger and E L. Dawes, *Phys. Rev. Lett.* **21**, 556 (1968); E L. Dawes and J. H. Marburger, *Phys. Rev.* **179**, 862 (1969).

15 V. I. Bespalov and V. I. Talanov, *JETP Lett.* **3**, 307 (1966).

16 G. McAllister, J. Marburger, and L. De Shazer, *Phys. Rev. Lett.* **21**, 1648 (1968).

17 D. H. Close, C. R. Giuliano, R. W. Hellwarth, L. D. Hess, F. J. McClung, and W. G. Wagner, *IEEE J. Quant. Electron* **QE-2**, 553 (1966).

18 C. C. Wang, *Phys. Rev. Lett.* **16**, 344 (1964).

19 R. G Brewer, J. R. Lifsitz, E. Garmire, R. Y. Chiao, and C. H. Townes, *Phys. Rev.* **166**, 326 (1968).

20 V. V. Korobkin, A M. Prokhorov, R. V. Serov, and M. Ya Shchelev, *JETP Lett.* **11**, 94 (1970).

21 M. M. T. Loy and Y. R. Shen, *Phys. Rev. Lett.* **25**, 1333 (1970); *Appl. Phys. Lett.* **19**, 285 (1970).

22 M. M. T. Loy and Y. R Shen, *IEEE J. Quant. Electron* **QE-9**, 409 (1973).

23 M. Maier, G. Wendl, and W. Kaiser, *Phys. Rev. Lett* **24**, 352 (1970).

24 M. Maier, W Kaiser, and J. A. Giordmaine, *Phys. Rev. Lett.* **17**, 1275 (1966); *Phys. Rev.* **177**, 580 (1969)

25 Y. R. Shen and Y. J. Shaham, *Phys. Rev.* **163**, 224 (1967).

26 F. Shimizu, *IBM J. Res. Develop.* **17**, 286 (1973).

27 E. Garmire, *Phys Lett.* **17**, 251 (1965)

28 See Section 16.2.

29 S A Akhmanov, A. P. Sukhorukov, and R. V. Khokhlov, *Sov Phys. JETP* **24**, 198 (1966); V. A. Alashkevich, S. A. Akhmanov, A. P. Sukhorukov, and A. M. Khachatryan, *JETP Lett.* **13**, 36 (1971).

30 J A. Fleck and P. L. Kelley, *Appl. Phys. Lett.* **15**, 313 (1969); R. L. Carman and J. A. Fleck, *Appl. Phys. Lett.* **20**, 290 (1972).

31 G. K. L. Wong and Y. R. Shen, *Phys. Rev. Lett.* **32**, 527 (1973).

32 R G. Brewer and C. H. Lee, *Phys. Rev. Lett.* **21**, 267 (1968).

33 E. G. Hanson, Y. R. Shen, and G. K. L. Wong, *Appl. Phys.* **14**, 65 (1977).

34 G. N. Steinberg, *Phys. Rev. A* **4**, 1182 (1971).

35 E. L. Kerr, *Phys Rev. A* **4**, 1195 (1971); **6**, 1162 (1972).

36 C. R. Giuliano and J. H. Marburger, *Phys. Rev. Lett.* **27**, 905 (1971)

37 D. Grischkowsky, *Phys. Rev. Lett.* **24**, 866 (1970).

38 A. V Nowak and D. O. Ham, *Opt. Lett* **6**, 185 (1981).

39 J. E Bjorkholm and A. Ashkin, *Phys. Rev. Lett.* **32**, 129 (1974)

40 A. Ashkin, J. M. Dziedzic, and P. W. Smith, *Opt Lett.* **7**, 276 (1982).

41 F. W. Dabby and J. R. Whinnery, *Appl Phys. Lett.* **13**, 284 (1968)

42 J. A Fleck and C Layne, *Appl. Phys. Lett.* **22**, 467 (1973).

43 N. Bloembergen and P. Lallemand, *Phys. Rev. Lett.* **16**, 81 (1966); R. G. Brewer, *Phys. Rev. Lett.* **19**, 8 (1967).

44 F. Shimizu, *Phys. Rev. Lett.* **19**, 1097 (1967).

45 Y. R. Shen and M. M. T. Loy, *Phys. Rev. A* **3**, 2099 (1971).

46 G. K. L. Wong and Y. R. Shen, *Appl. Phys. Lett.* **21**, 163 (1972).

47 R. R Alfano and S. L. Shapiro, *Phys. Rev. Lett.* **24**, 1217 (1970); D. K. Sharma and R. W. Yip, *Opt. Commun.* **30**, 113 (1979).

48 W. L. Smith, P. Liu, and N. Bloembergen, *Phys. Rev. A* **15**, 2396 (1977).

49 A. Penzkofer, A. Laubereau, and W. Kaiser, *Phys. Rev. Lett.* **31**, 863 (1973); A. Penzkofer, A. Seilmeier, and W. Kaiser, *Opt. Commun.* **14**, 363 (1975).

50 C.V. Shank, R. L. Fork, R. Yen, R. H. Stolen, and W. J. Tomlinson, *Appl. Phys. Lett.* **40**, 761 (1982)

51 S. D. Durbin, S. M. Arakelian, and Y. R. Shen, *Opt. Lett.* **6**, 411 (1981).

52 L. A. Ostrovsky, *Sov. Phys. JETP* **24**, 797 (1967); F. DeMartini, C. H Townes, T K Gustafson, and P. L. Kelley, *Phys. Rev.* **164**, 312 (1967); T. K. Gustafson, J. P Taran, H. A Haus, J. R. Lifstiz, and P. L. Kelley, *Phys. Rev.* **177**, 306 (1969).

53 D. Grischkowsky, E. Courtens, and J. A. Armstrong, *Phys. Rev. Lett.* **31**, 422 (1973).

54 R.C. C Leite, R. S. Moore, and J. R. Whinnery, *Appl. Phys. Lett.* **5**, 141 (1964).

55 J. R. Whinnery, D. T. Miller, and F. Dabby, *IEEE J. Quant. Electron.* **3**, 382 (1967); S. A Akhmanov, D. P. Krindach, A. V. Migulin, A. P. Sukhorukov, and R. V Khokhlov, *IEEE J. Quant. Electron.* **4**, 568 (1968).

56 F. W. Dabby, T. K. Gustafson, J. R. Whinnery, Y. Kohanzadeh, and P. L. Kelley, *Appl. Phys. Lett.* **16**, 362 (1970).

57 S. A. Akhmanov, D. P. Krindach, A. P. Sukhorukov, and R. V. Khokhlov, *JETP Lett.* **6**, 38 (1967).

58 See, for example, A. C. Albrecht, in A. H. Zewail, ed., *Advances in Laser Chemistry* (Springer-Verlag, Berlin, 1978), p. 235.

59 R. C. C. Leite, S. P. S. Porto, and T. C. Damen, *Appl. Phys. Lett.* **10**, 100 (1967).

BIBLIOGRAPHY

Akhmanov, S. A., R. V. Khokhlov, and A. P. Sukhorukov, in F. T. Arrechi and E. O. Schulz-Dubois, eds., *Laser HandBook II* (North-Holland Publishing co., Amsterdam, 1972)

Shen, Y. R., *Prog. Quant. Electron.* **4**, 1 (1975).

Marburger, J. H., *Prog. Quant. Electron.* **4**, 35 (1975)

18

Multiphoton Spectroscopy

Among the many laser spectroscopic techniques that have revolutionized the optical spectroscopy field, multiphoton spectroscopy is certainly one of the most important. It allows one to probe excited states that cannot be reached by one-photon transitions. Through multiphoton processes, transitions between excited states can also be studied. In previous chapters, two-photon and Raman spectroscopies have been discussed. Being special cases of multiphoton spectroscopy, they share many of the general features of multiphoton spectroscopy. In this chapter, aside from a general discussion of the technique, applications of multiphoton spectroscopy are emphasized.

18.1 GENERAL CONSIDERATIONS

Multiphoton spectroscopy is based on the fact that with high-intensity lasers available, multiphoton transitions can be induced with high probabilities and are readily detectable. Let the transition probability for the n-photon transition from $|g\rangle$ to $|f\rangle$ in Fig. 18.1 be

$$W^{(n)} = \sigma^{(n)} I_1(\omega_1) \cdots I_n(\omega_n)/\hbar^n \omega_1 \cdots \omega_n \tag{18.1}$$

where $\sigma^{(n)}$ is the cross section [we assume, for simplicity, that $\sigma^{(n)}$ is scalar] and $I_i(\omega_i)$ the laser intensity at ω_i. The population excitation into the excited state, $\rho_{ff} - \rho_{ff}^0$, is then governed by the equation

$$\left(\frac{\partial}{\partial t} + \frac{1}{T_1}\right)(\rho_{ff} - \rho_{ff}^0) = W^{(n)}(\rho_{gg} - \rho_{ff}) \tag{18.2}$$

the solution of which is straightforward if $W^{(n)}$ is small so that the population difference $(\rho_{gg} - \rho_{ff})$ on the right-hand side of (18.2) can be approximated by

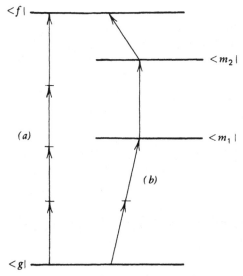

Fig. 18.1 Multiphoton transition from $\langle g|$ to $\langle f|$. (*a*) A single-step 4-photon process; (*b*) a 3-step 4-photon process.

the thermal equilibrium value $\rho_{gg}^0 - \rho_{ff}^0$. In the steady-state case,

$$\begin{aligned}\rho_{ff} - \rho_{ff}^0 &= W^{(n)}T_1\left(\rho_{gg}^0 - \rho_{ff}^0\right) \\ &= \sigma^{(n)}T_1I_1 \cdots I_n\left(\rho_{gg}^0 - \rho_{ff}^0\right)/\hbar^n\omega_1 \cdots \omega_n.\end{aligned} \tag{18.3}$$

The excitation can certainly be large if $\sigma^{(n)}$ and $I_1 \cdots I_n$ are large.

The expression for the multiphoton excitation cross section $\sigma^{(n)}$ can be obtained from the nth-order perturbation calculation. We only describe it qualitatively here. It is proportional to $|A|^2$ with A containing many terms. Each term has, in the numerator, n matrix elements connecting the initial and final states through a number of immediate states and, in the denominator, $(n-1)$ appropriate frequency factors. One can easily show that, off resonance,

$$|\sigma^{(n+1)}/\sigma^{(n)}| \sim (2\pi/c)\hbar\omega|\chi^{(2n+1)}/\chi^{(2n-1)}|.$$

Then, in the visible range, with $|\chi^{(2n+1)}/\chi^{(2n-1)}| \sim 10^{-12}$, we find $|\sigma^{(n+1)}/\sigma^{(n)}| \sim 10^{-34}$. The $(n+1)$-photon excitation is 10^{-6} times smaller than the n-photon excitation if $\sigma^{(n+1)}I/\hbar\omega \sim 10^{-6}\sigma^{(n)}$. This would require a laser intensity $I \sim 1$ GW/cm^2. The above estimate shows why high-intensity lasers are often needed for multiphoton excitations. The cross section $\sigma^{(n)}$ can be strongly enhanced by resonances with intermediate states. Resonant enhancement by many orders of magnitude is possible in atoms and molecules. It then greatly reduces the laser intensity requirement for multiphoton excitations.

Equation (18.2) is actually valid only for an n-photon transition sufficiently far away from intermediate resonances. Such a process is often known as a single-step n-photon transition (process a in Fig. 18.1). On the other hand, if m intermediate resonances occur in the n-photon process (process b in Fig. 18.1), then the population excitation into the final state depends also on the populations pumped into the intermediate states. It is governed by the set of equations

$$
\left(\frac{\partial}{\partial t} + \frac{1}{T_{1f}} \right) \left(\rho_{ff} - \rho_{ff}^0 \right) = W_{m,f}^{(n_{m+1})} \left(\rho_{mm} - \rho_{ff} \right),
$$

$$
\left(\frac{\partial}{\partial t} + \frac{1}{T_{1m}} \right) \left(\rho_{mm} - \rho_{mm}^0 \right) = W_{m-1,m}^{(n_m)} \left(\rho_{m-1,m-1} - \rho_{mm} \right)
$$

$$
- W_{m,f}^{(n_{m+1})} \left(\rho_{mm} - \rho_{ff} \right), \tag{18.4}
$$

$$
\cdots
$$

$$
\left(\frac{\partial}{\partial t} \right) \left(\rho_{11} - \rho_{11}^0 \right) \cong W_{g,1}^{(n_1)} \left(\rho_{gg}^0 - \rho_{11} \right) - W_{1,2}^{(n_2)} \left(\rho_{11} - \rho_{22} \right)
$$

assuming that the population relaxation for each level can be described by a single longitudinal relaxation time T_{1i}. Here, ρ_{ii} with $i = 1, \ldots, m$ is the population in the ith resonant intermediate state, $W_{i,j}^{(n_\alpha)}$ is the n_α-photon transition probability from $|i\rangle$ to $|j\rangle$, and $n_1 + n_2 + \cdots + n_{m+1} = n$. Such an n-photon process with m intermediate resonances is often known as an $(m + 1)$-step n-photon transition. In general, the stepwise excitation greatly increases the population excitation into the final state. This is especially true if the relaxation times of the intermediate states are long. The excitation is usually a complex function of the input laser intensities. With pulsed lasers, the transient response complicates the matter even further.

For multiphoton spectroscopy, however, the quantitative dependence of excitation on laser intensities is often not so important. We are more interested in the resonant feature of $(\rho_{ff} - \rho_{ff}^0)$ as a function of the laser frequencies. The experimental problem involved is mainly how $(\rho_{ff} - \rho_{ff}^0)$ can be detected.

In this chapter we concentrate on atomic and molecular systems. Multiphoton transitions with $n > 2$ seldom have been studied in solids because the high laser intensity needed for multiphoton excitation tends to optically damage the medium. On the other hand, it also seems to be true that not much new information about the electronic properties of a solid can be expected from multiphoton spectroscopy with $n \geq 3$.

An interesting aspect of multiphoton spectroscopy applied to gas systems is its ability to yield Doppler-free spectra. This has already been discussed in detail in Section 13.4 for the two-photon case. The discussion there can be extended to the multiphoton case in general: a Doppler-free n-photon spectrum can be observed if the wavevectors of the pump waves obey the relation $\mathbf{k}_1 + \cdots + \mathbf{k}_n = 0$.

18.2 EXPERIMENTAL TECHNIQUES

Multiphoton spectroscopy requires one or more high-intensity tunable lasers, CW or pulsed, as the excitation sources. Its resolution, assuming Doppler-free, is usually limited by the laser linewidths. Detection of multiphoton excitation is most important in multiphoton spectroscopy. Since the excitation is usually weak, the detection method must be extremely sensitive. A number of such methods have been developed, and they were discussed briefly in Chapter 12. Among them, the most commonly used are the fluorescence and ionization techniques.

Multiphoton-Induced Fluorescence Spectroscopy

The fluorescence yields of atoms and molecules can be very high because, without collisions, an excited atom or molecule would only decay to lower-energy states by emission of photons. If the fluorescence is in the visible, the sensitivity of detection can also be high. A photomultiplier can easily detect a few photons per second in the CW case or per pulse in the pulsed case. Assume that the fluorescence quantum yield of a gas system is close to 1. Then, with a photomultiplier collecting the fluorescence emitted over a π steradian from a local excitation region, roughly 10–20 excited atoms or molecules in that region can be detected without much difficulty. The sensitivity is n times better if each atom or molecule emits n fluorescent photons during the single-pulse excitation or per second in the CW case. This therefore shows that fluorescence detection can be a very sensitive technique for probing multiphoton excitation (Fig. 18.2).

Aside from its sensitivity, the method can also yield a fluorescence spectrum initiating from the multiphoton-pumped excited state. One can again use the polarization-labeling technique discussed in Section 13.5 to selectively pump the excited state and hence obtain a greatly simplified fluorescence spectrum. The properties of various excited states can then be learned by analyzing the spectrum in accordance with the transitions between the excited states. The detection of n-photon-induced fluorescence therefore allows us to probe states that can be reached only by $(n + 1)$-photon transitions.

Multiphoton Ionization Spectroscopy

A more sensitive method to detect multiphoton excitation, when applicable, is the ionization method, because the detection of electrons and ions can be extremely sensitive. The noise of an ion detector can easily be less than 1 ion per minute. Therefore, if all the multiphoton-excited atoms or molecules can be ionized before they relax to lower-energy states, and if all the ions can be collected by the detector, then the sensitivity of the method is limited only by the detector noise, namely, $\sim$ 1 excited atom or molecule per minute.

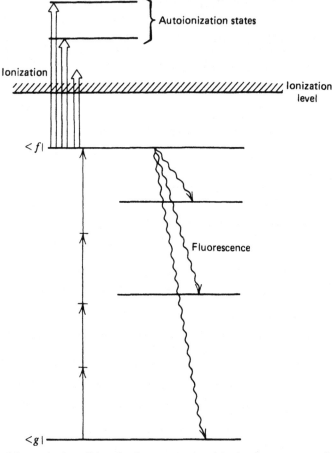

Fig. 18.2 Schematic describing the flourescence and ionization processes for detection of a multiphoton transition.

Several techniques are commonly used for ionization of the excited atoms (or molecules). One is the photoionization method, where the excited atom is ionized by a laser beam with photon energy sufficiently large to pump the atom above the ionization level (Fig. 18.2). The pump intensity has to be strong for the ionization rate from the excited state to exceed the decay rate. Excitation to discrete states instead of continuum above the ionization level can greatly enhance the ionization probability. This method is frequently used for ionization of excited molecules. It is most convenient when one of the strong laser beams used for mutliphoton excitation can also be used for photoionization.

The dc field ionization method also is frequently used. It is well known that in the presence of an external dc electric field, the potential curve seen by the electrons in an atom (or molecule) is distorted; the ionization energy is lowered, and ionization through electron tunneling out of the atom becomes possible. For a sufficiently highly excited atom in a sufficiently strong dc field, the ionization rate can be much higher than the decay rate of the excited atom. The ionization probability can then be close to 1. This method is most useful for detecting atoms in high Rydberg states.

Excited atoms or molecules can also be ionized through collisions. The ionization rate depends on the gas pressure, the excited state and its position relative to the ionization level, the fraction of excited atoms, and so on. In general, for collision ionization to be effective, the gas pressure has to be sufficiently high. This method is useful for ionization of atoms in a gas cell.

Ion detectors can be simple or sophisticated depending on the experimental requirement. Electron multipliers and proportional counters are the two well known examples. Figure 18.3 shows a very simple, but sensitive arrangement for detection of ions in a gas cell.[1] The ionization probe in the cell consists of a metal wire negatively biased with respect to the grounded metal cell wall. Thermionic emission from the metal wire causes the formation of a space-charge region around the wire, and leads to a space-charge-limited current, as in the case of space-charge-limited thermionic diode tube. The ions created in the cell drift toward the metal wire. In an attempt to neutralize the ions, electrons in the space-charge region are attracted toward the ions. They induce a current change many orders of magnitude ($> 10^5$) larger than the one expected from the ion flow. This large current amplification results in a high sensitivity of ion detection. The signal appears as a voltage drop across the load resistance. The bias voltage needed in this case is only ~ 1 V. To avoid the dc Stark effect on the atomic or molecular spectrum, the ionization region can be shielded from the space-charge region by a grid which makes the ionization region field-free but allows the ions to drift into the space-charge region. Because of its simplicity, this ion detection scheme has now been widely adopted for multiphoton spectroscopy in a gas cell.

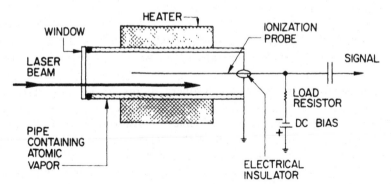

Fig. 18.3 A simple experimental setup for multiphoton ionization spectroscopy. [After P. Esherick, J. J. Wynne, and J. A. Armstrong, in J. L. Hall and J. L. Carlsten, eds., *Laser Spectroscopy III* (Springer-Verlag, Berlin, 1977), p.170.]

18.3   SPECTROSCOPIC APPLICATIONS

Multiphoton spectroscopy enables us to study excited states and transitions between excited states which cannot be reached by single-photon excitations. For example, in alkali atoms, the excited states, $(n's)$, $(n'd)$, $(n'f)$, etc., can be probed by multiphoton excitations, and so can the excited states $(ns)(n's)$, $(ns)(n'd)$, $(np)(n'd)$, etc., of the alkali earth atoms. The results are important for further development of the quantum theory of these simple atoms. In the case of alkali earth atoms, the experimental data have provided a stringent test on the multichannel quantum defect theory. Multiphoton spectra of other multielectron atoms are also interesting, although the theory for such atoms is less well developed. In the field of molecular spectroscopy, multiphoton spectroscopy also forms a new branch. It yields most valuable information about the energy level structures of molecules and the properties of molecules in the excited states.

In this section, instead of general discussion on the countless applications of multiphoton spectroscopy, we concentrate on applications to the studies of Rydberg atoms and autoionization states. These examples can help to illustrate the power and usefulness of the multiphoton spectroscopic technique.

Rydberg Atoms

Rydberg atoms are defined here as atoms in highly excited Rydberg states.[2] They have very different characteristics from normal atoms. As shown in Table 18.1, the excited electron in a Rydberg atom has an orbital radius roughly equal to $n^2 a_0$ where n is the principal quantum number and a_0 is the Bohr radius. For hydrogen with $n = 50$, this radius is $2500 \, a_0 \sim 1000$ Å, which is almost a macroscopic size. The corresponding geometric cross section is 6×10^6 times larger than the ground-state hydrogen atom. The excited electron of the Rydberg atom is then only very weakly bound to the core, and can be easily perturbed by external fields. The radiative lifetime of the highly excited electron (which is also the lifetime of the particular Rydberg atom), however, varies as n^3. Because of the large excited electron orbits, the transition probability between two neighboring Rydberg states of large n is very high. Interaction between Rydberg atoms is also expected to be extremely strong.

The extraordinary properties of Rydberg atoms render the studies of Rydberg atoms most interesting. From the fundamental physics point of view, accurate measurements of energies, lifetimes, ionization probabilities, Stark and Zeeman effects, and so on, allow the determination of many atomic parameters such as core polarizability, configuration interactions, and fine structure splittings. These parameters can be calculated with good approximations, and therefore the experimental measurements provide a meaningful test of the theories. On the other hand, it is easy to have a Rydberg atom subject to an external perturbation which is stronger than its electron binding energy.

Table 18.1
Properties of Rydberg Atoms[a]

Property	n dependence	Na(10d)
Binding energy	n^{-2}	0.14 eV
Energy between adjacent n states	n^{-3}	0.023 eV
Orbital radius	n^2	147 a_0
Geometric cross section	n^4	68000 a_0^2
Dipole moment $\langle nd|r|nf \rangle$	n^2	143 a_0
Polarizability	n^7	0.21 MHz/(v/cm)^2
Radiative lifetime	n^3	1.0 μsec
Fine structure interval	n^{-3}	$\sim$ 92 MHz

[a]After Ref. 2.

This then leads to a new class of interesting problems involving nonperturbative atom–field interactions that do not exist under normal conditions.

Single-photon spectroscopy is certainly also applicable to the study of Rydberg atoms, but the excited states that can be reached are limited. For example, through the $ns \rightarrow n'p$ single-photon excitation, only the p state of the n' Rydberg atom can be reached. Multiphoton spectroscopy, however, allows other states of the Rydberg atom to be probed. Furthermore, the highly excited Rydberg states are very closely spaced. The resolution of these states requires a Doppler-free spectrum. In a gas cell, this can be achieved only with multiphoton spectroscopy.

Alkali atoms have often been the object of research in Rydberg atoms. We discuss here a few selective studies on the subject.

Two-photon Doppler-free spectroscopy has been used to study the highly excited Rydberg states for a number of alkali atoms. In Rb, for example, the Rydberg states have been probed up to $n = 116$, using the detection scheme of Fig. 18.3 and a 50-mW CW narrowband dye laser as the pump source.[3] The detection sensitivity in this case can be further enhanced by the lock-in amplification technique. At $n \sim 100$, the spacing between successive Rydberg states is only a few tens of megahertz. For lower n, the (ns) and (nd) states, and also the spin-orbit split states can be resolved. The latter allows the determination of the fine structure splittings.[4] For $n = 4$ to 55, the fine structure splittings between $^2D_{3/2}$ and $^2D_{5/2}$ of the nd states of Rb have been found to obey the relation $An_{\text{eff}}^{-3} + Bn_{\text{eff}}^{-5}$, where A and B are constants and n_{eff} is the effective quantum number defined by $T_n = -Rn_{\text{eff}}^{-2}$, R being the Rydberg constant and T_n the term value for the principal quantum number n. The results provide a good test on the various theoretical calculations. Pressure shifts and pressure broadenings of Rydberg states can also be investigated.[5] For low n values, they can be detected readily even in the mTorr range. Strong oscillations in the linewidth versus the principal quantum number have been observed but not yet explained quantitatively.

The Rydberg states of the alkali earth atoms, including those with doubly excited electrons, have also been carefully measured.[6] Pulsed dye lasers were often used in such studies. The results are most helpful in establishing the multichannel defect theory developed by Fano and associates.[7]

The Stark and Zeeman effects of the Rydberg states are large because of the large electron orbits.[8,9] In a moderate field, the splittings can be larger than the spacings between Rydberg states. Level crossing and anticrossing effects should be readily observable. The results can be used as a sensitive test of the atomic theory involving field-induced state mixing. When the field perturbation becomes comparable to the Coulomb binding energy of the excited electron, the perturbation theory breaks down and an exact theory is needed. The measurements are therefore most valuable for theoretical exploration in this new regime. We consider here the diamagnetic effect of Na Rydberg states as an example.[9]

The ratio of the diamagnetic energy to the binding energy of a Rydberg atom varies as $n^6 B^2$, where B is the magnetic field. For it to be close to one for an atomic ground state, the field required is $\sim 10^9$ G. Such a field is clearly not achievable in the laboratory. This strong-field regime may exist in neutron stars and for hydrogenlike systems in solids, but the problem can never be as simple as in Rydberg atoms. Because of the n^6 dependence, a Rydberg atom with $n \sim 30$ only requires a magnetic field of several tens of kilogauss to get into the strong field regime. While the nonperturbative theory for such cases is still being developed, the experimental measurements are fairly straightforward.

A representative experimental setup is shown in Fig. 18.4. To avoid Doppler broadening and collisional effects, an atomic beam was used with the exciting laser radiation crossing it at 90° in the magnetic field region. Two 5-nsec pulsed tunable dye lasers were employed to excite the atoms stepwise to a Rydberg state. During the excitation, the dc electric field was absent, but a few kilovolts per centimeter dc electric field was applied 1 μsec later to ionize the Rydberg atom. The electrons from ionization were then accelerated to 10 kV and detected by the detector. The detection scheme could be nearly 100%

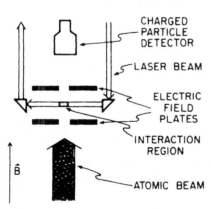

Fig. 18.4 An apparatus for spectroscopic studies of Rydberg atoms. The region shown is located in the center of a superconducting solenoid. The field plates float at −10 kV with respect to the detector. The pulsed field is applied between the plates. (After Ref. 9.)

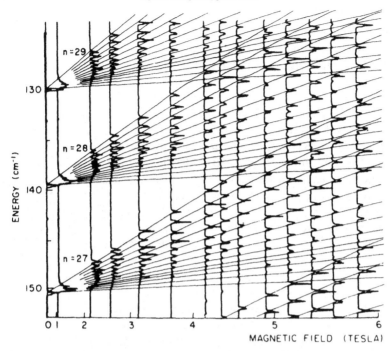

Fig. 18.5 Diamagnetic structure of Na. Experimental excitation curves for even-parity levels, $m_l = 1$, $m_s = \frac{1}{2}$, in the vicinity of $n = 28$. A tunable laser was scanned across the energy range displayed. The zero of energy is the ionization limit. Signals generated by ionizing the excited atoms appear as horizontal peaks. The horizontal scale is quadratic in field. Calculated levels are overlaid in light lines. Some discrepancies are present due to nonlinearity of the laser scan. (After Ref. 9.)

efficient, insensitive to the applied magnetic field, and linear over a wide dynamic range. To study the diamagnetic effect, $\Delta m_l = 0$ transitions from $3P_{3/2}$ to the Rydberg states around $n = 28$ with $m_l = 1$ were chosen so that the magnetic interaction Hamiltonian linear in the field would not contribute. A typical set of spectra is shown in Fig. 18.5. The display gives a clear picture of the diamagnetic shifts of the various lines. Crossing and anticrossing of levels are clearly seen. In this case, a straightforward diagonalization of the Hamiltonian can explain the observed spectra very well, but the procedure used is already near its limit of applicability. For higher n or larger field, new theoretical approaches will be needed.

The large electron orbits of high Rydberg states greatly enhance the transition probabilities between neighboring Rydberg states. For $n \gtrsim 30$, these transitions are in the radio to microwave range. The single-photon transitions $ns \to n'p$, $nd \to n'p$, etc., can be saturated by an intensity $\lesssim 10^{-6}$ W/cm^2, and the double-photon transitions, $ns \to n's$, $nd \to n'd$, etc., by $\lesssim 10^{-2}$ W/cm^2. Since radio and microwave measurements are intrinsically more accurate than the optical measurements, studies of transitions between Rydberg states allow us to deduce more accurately the fine structure splittings, quantum defects, polarizabilities, and so on.[10]

The experimental setup is similar to that shown in Fig. 18.4, except that additional radio or microwave field is now applied to the optically excited atoms. Spectral resolution of microwave transitions can be better than 1 MHz if the residue dc electric field in the interaction region can be eliminated. Detection of the transitions is facilitated by the fact that the dc ionization threshold field of the Rydberg state is inversely proportional to n^4. By applying a sawtooth potential (increasing linearly with time) on the plate electrodes, the signal should appear as pulses shown in Fig. 18.6. Each pulse, labeled by the potential, corresponds to the signal obtained from ionization of atoms in a certain Rydberg state; the pulse strength corresponds to the population in that state. This dc ionization arrangement acts as a Rydberg spectrometer. Then, without microwave excitation of the laser-excited Rydberg atoms only one pulse is observed; with resonant microwave excitation, two pulses should appear, with their relative strengths indicating the relative populations in the two Rydberg states involved in the transition.

For lower n values, transitions between Rydberg states are in the infrared. They can again be detected with high sensitivity using the method just discussed. As a possible application, the Rydberg atoms can be used as a far-infrared detector.[11] With Stark-field tuning, the transition frequency can be continuously varied. It has been found that such a detector can have a noise

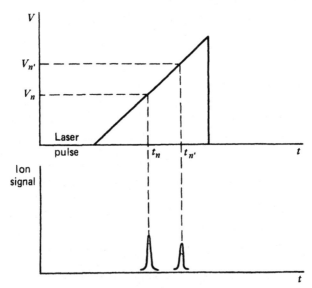

Fig. 18.6 Ion signal detected by a Rydberg spectrometer in response to a linear sweep of an applied dc potential. The two signal pulses at V_n and $V_{n'}$ after the laser pulse excitation are measures of the populations in the Rydberg states $\langle n|$ and $\langle n'|$.

equivalent power of 5×10^{-15} W/Hz$^{1/2}$ at 496 μm, comparable to the best existing far-infrared detector. It has a bandwidth of ~ 1 MHz, a large angular collection aperture, and a wide tuning range extending possibly up to the mid-infrared.

Atoms in the high Rydberg states can radiate and make downward transitions. The transition probability is higher for larger n and smaller transition frequency. Amplified spontaneous emission (or superradiance) and maser (or laser) action can occur when the inverted population is above threshold.[12] The threshold condition is given by

$$N_{nn'} > \gamma_n/\mu_{nn'}\gamma_{nn'}$$

where $N_{nn'}$ is the population difference between the n and n' states, γ_n and $\gamma_{nn'}$ are the total emission rate from $\langle n|$ and the partial emission rate from $\langle n|$ to $\langle n'|$, respectively, and $\mu_{nn'}$ is a form factor. For a cylinder of Rydberg atoms in free space,

$$\mu_{nn'} = 3\pi c^2/2\omega_{nn'}^2 a^2$$

with a being the diameter of the cylinder. If the cylinder is enclosed by a resonant cavity, then $\mu_{nn'}$ should be replaced by $\mu_{nn'}\mathscr{F}$, with $\mu_{nn'}$ now being the cavity-filling factor and $\mathscr{F}$ the finesse. For the $25s \to 24p$ transition of Na, for example, we have $\gamma_n \sim 10^5$/sec, $\gamma_{nn'} \sim 10^3$/sec, $\lambda_{nn'} = 2\pi c/\omega_{nn'} \sim 1$ mm. Then, if $a \sim 1$ mm, $\mu_{nn'} \sim 10^{-2}$, and $\mathscr{F} \sim 200$, the threshold population difference is $N_{nn'} \sim 10^5$ for amplified spontaneous emission, and $N_{nn'} \sim 500$ for maser action. This example shows that with Rydberg atoms, maser action can be observed with a very small number of atoms in the cavity; the number is closer to 1 for higher n. Maser action in such a small system is a subject of great theoretical interest.[12]

Amplified spontaneous emission and maser action in an optically excited Rydberg atomic beam have been observed. The experimental arrangement is again similar to the one shown in Fig. 18.4. In the maser experiment, a microwave cavity is built around the laser-atomic beam interaction region. The microwave emission, in the form of an ~ 1-μsec pulse, has a peak power in the 10^{-11} to 10^{-13} W range. Such a small power is of course difficult to detect directly. It can, however, be detected indirectly using the Rydberg spectrometer described earlier. The number of microwave photons emitted is monitored by the number of atoms appearing in the lower state after a certain time delay (~ 1 μsec) following the laser excitation. Below the masing threshold, little population will end up in this lower state by spontaneous emission after such a short time delay. Above the threshold, appreciable increase of population in the lower state can result from stimulated emission. Figure 18.7 shows how experimentally the maser action from $27S$ to $26P$ populates the lower state with the microwave cavity tuned on resonance and slightly off resonance, respectively.

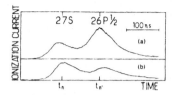

Fig. 18.7 Time-resolved Na ion-signal recordings exhibiting maser effect. (*a*) The cavity is turned on the $27S \rightarrow 26P_{1/2}$ transition. (*b*) The cavity is 40 MHz off resonance. (After Ref. 12.)

Autoionization Spectroscopy

For multielectron atoms and molecules, discrete states coexist with the continuum states above the ionization level. They are known as autoionization states,[13] and have been the subject of immense interest in atomic and molecular spectroscopy. The positions and characteristics of these states, and their interactions with the continuum, which can all be deduced from spectroscopic measurements, are important for the test of theories. Here again multiphoton spectroscopy has the advantage of being able to probe many autoionization states that cannot be reached by single-photon spectroscopy.

The experimental setup in either Fig. 18.3 or Fig. 18.4 can be used for multiphoton autoionization spectroscopic studies. As an example, the autoionization spectra of Sr obtained by three-photon stepwise excitations $(5s)^2 \rightarrow (5s)(5p) \rightarrow (5s)(ns)$ or $(5s)(nd) \rightarrow (5p_{1/2})(ns)$ or $(5p_{1/2})(nd)$ are shown in Fig. 18.8.[14] Stepwise excitations yield high overall transition probabilities. The laser powers required are therefore low, and the background signal due to nonresonant photoionization is consequently weak. With the help of dc field mixing of states, the method can populate many autoionization states normally forbidden by dipole selection rules. From the positions of the spectral lines, quantum defects for a number of Rydberg states $(5s)(ns)$ and $(5s)(nd)$ can be deduced. The lineshapes result from the interactions between the discrete states and the continuum. They also depend on the configuration mixing of the states. The linewidths reflect the lifetimes of the autoionization states in accordance with the uncertainty principle. It is found, for example, the lifetimes of the $(5p_{1/2})(ns)$ states vary as $(n_{eff})^3$, and the lifetimes of the $(5p_{1/2})(16l)$ increase dramaticlly with l. Polarization arrangement of the pump beams can help the assignment of the observed lines. Applied dc electric and magnetic fields can be used to study the Stark and Zeeman splittings of the autoionization states.

Autoionization spectra of rare earth atoms are particularly interesting.[15] Extremely strong, yet narrow (~ 0.05 cm^{-1}), autoionization lines have been observed in Gd and Yb. One may wonder if the sharp lines are characteristic of the rare earth atoms due to excitation of electrons in the shielded but unfilled $4f$ shell. Theoretical calculation, however, suggests that they are the result of excitation of the valence electrons. It will be interesting to see whether

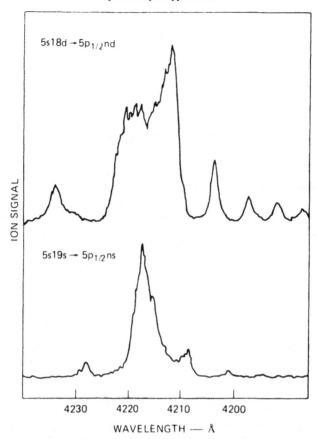

Fig. 18.8 Excitation spectra for the $5p_{1/2}$, nd states (upper curve) and the $5p_{1/2}$, ns states (lower curve) of Sr. In these spectra, the linewidths are artificially broadened beyond the autoionization linewidth by instrumental effects. (After Ref. 14.)

atoms in the actinide group, which has the analogous $5f$ shell, exhibit similar sharp autoionization lines.

Autoionization spectra of molecules are, of course, much more complicated than those of atoms. Nevertheless, they are interesting from the molecular spectroscopic point of view. They are also useful in the sense that excitations to these states increase the ionization probability of the molecule. They may also be used as steppingstones to reach higher autoionization states.

We conclude this section by remarking that multiphoton spectroscopy, having the freedom to vary the individual input beams separately and having an inherently high sensitivity, is an extremely valuable technique. Many variations of the technique can be developed, depending on our imagination and on the system to be investigated. The growth of the field will probably also depend on the advances in the theories of atoms and molecules.

REFERENCES

1   K. H. Kingdon, *Phys. Rev.* **21**, 408 (1923); P. Esherick, J. J. Wynne, and J. A. Armstrong, in J. L. Hall and J. L. Carlsten, eds., *Laser Spectroscopy III* (Springer-Verlag, Berlin, 1977), p. 170.

2   S. A. Edelstein and T. F. Gallagher, in D. R. Bates and B. Bederson, eds., *Advances in Atomic and Molecular Physics* (Academic Press, New York, 1978), Vol. 14, p. 365.

3   B. P. Stoicheff and E. Weinberger, *Can. J. Phys.* **57**, 2143 (1979).

4   K. C Harvey and B. P. Stoicheff, *Phys. Rev. Lett.* **38**, 537 (1977).

5   B. P. Stoicheff and E. Weinberger, *Phys. Rev. Lett.* **44**, 733 (1980).

6   See, for example, P. Esherick, J. A. Armstrong, R. W. Dreyfus, and J. J. Wynne, *Phys. Rev. Lett.* **36**, 1296 (1976); P. Esherick, *Phys. Rev. A* **15**, 1920 (1977).

7   K. T. Lu and U. Fano, *Phys. Rev. A* **2**, 81 (1970); U. Fano, *J. Opt. Soc. Am.* **65**, 979 (1975), and references therein; J. A. Armstrong, P. Esherick, and J. J. Wynne, *Phys. Rev. A* **15**, 180 (1977).

8   M. G. Littman, M. L. Zimmerman, T. W. Ducas, R. R. Freeman, and D. Kleppner, *Phys. Rev. Lett.* **36**, 788 (1976).

9   M. L. Zimmerman, J. C. Castro, and D. Kleppner, *Phys. Rev. Lett.* **40**, 1083 (1978).

10   C. Fabre, S. Haroche, and P. Goy, *Phys. Rev. A* **18**, 229 (1978).

11   T. F. Gallagher and W. E. Cooke, *Appl. Phys. Lett.* **34**, 369 (1979); T. W. Ducas, W. P. Spencer, A. G. Vaidyanathan, W. H. Hamilton, and D. Kleppner, *Appl. Phys. Lett.* **35**, 382 (1979).

12   M. Gross, P. Goy, C. Fabre, S. Haroche, and J. M. Raimond, *Phys. Rev. Lett.* **43**, 343 (1979); J. M. Raimond, P. Goy, M. Gross, C. Fabre, and S. Haroche, *Phys. Rev. Lett.* **49**, 117; 1924 (1982); P. Goy, J. M Raimond, M. Gross, and S. Haroche, *Phys. Rev Lett.* **50**, 1903 (1983).

13   U. Fano, *Phys. Rev.* **124**, 1866 (1961).

14   W. E. Cooke, T. F. Gallagher, S. A. Edelstein, and R. M. Hill, *Phys. Rev. Lett.* **40**, 178 (1978).

15   G. I. Bekov, V. S. Letokhov, O. I. Matveev, and V. I. Mishin, *JETP Lett.* **28**, 283 (1978); G. I. Bekov, E. P. Vidolova-Angelova, L. N. Ivanov, V. S. Letokhov, and V. I. Mishin, *Opt. Commun.* **35**, 194 (1980).

BIBLIOGRAPHY

Laser Spectroscopy III, IV, V (Springer-Verlag, Berlin, 1977, 1979, 1981).

Letokhov, V. S., and V. P. Chebotayev, *Nonlinear Laser Spectroscopy* (Springer-Verlag, Berlin, 1977)

Parker, D. H., J. O. Berg, and M. A. El-Sayed, in A. H. Zewail, ed., *Advances in Laser Chemistry* (Springer-Verlag, Berlin, 1978).

19

Detection of Rare Atoms
and Molecules

Understanding of physical and chemical processes at the atomic and molecular level often is impeded by the lack of sensitive techniques to detect and probe very small numbers of atoms and molecules. This is particularly true when the atoms or molecules to be detected appear only as impurity or trace particles in a host medium. In Chapter 18 laser-induced fluorescence and photoionization as detection methods are seen to be extremely sensitive. They have, in fact, the sensitivity of detecting single atoms and molecules, and therefore hold the promise of becoming a very important experimental tool. With these methods, many interesting problems in physics and chemistry that were hitherto untouchable can now be studied. This chapter describes these methods in some detail, emphasizing their detection capabilities. Possible applications in various fields are discussed.

19.1   BASIC THEORY

The basic idea of laser-selective detection of atoms and molecules is simple. It involves two essential steps: laser labeling of the atoms (or molecules) and laser detection of the labeled atoms. Every atomic (or molecular) species has a characteristic spectrum, acting as its own fingerprint. A monochromatic light can selectively excite a particular kind of atom from a specific ground state to a specific excited state. The excitation puts a label on this particular set of atoms. Following the excitation, if the detection can selectively detect atoms in the particular excited state, then it means that we have succeeded in selectively detecting those labeled atoms originally in the specific ground state.

Clearly, to be able to detect a very small number of atoms in a specific ground state, the labeling should apply to as many of these selected atoms as possible. This means that the light intensity should be sufficiently high to saturate the selected excitation. Then the labeling should also be so selective that as few wrong atoms as possible are excited. Both requirements suggest that intense monochromatic lasers are needed for the selective excitation. Detection, on the other hand, must be very sensitive. It should have the sensitivity to detect a large fraction of the atoms in the selected excited state. The detection also should be highly selective, since it can then discriminate against the wrong atoms that have been excited. In Chapter 18 we saw that laser-induced fluorescence and photoionization are extremely sensitive. In principle, they have the sensitivity to detect single selected atoms. In practice, however, the ultimate sensitivity depends on the background noise arising from the presence of wrong atoms and from the noise in the detection system.

We consider here a simple model illustrating the requirement for detection of single atoms and molecules (Fig. 19.1). Let the selective excitation rate of the atom from $\langle g|$ to $\langle f|$ be W_{ex} and the detection rate of the excited atom be F. We assume two simple cases. First, the detection removes the atom from the excited state and does not put it back into the ground state. Second, all the excited atoms after being detected are put back into the ground state. In Fig. 19.1 we have Γ as the relaxation rate from $\langle f|$ to $\langle g|$ and β as the loss rate of the excited atoms through relaxation into traps, metastable states, and so on.

In the first case, the rate equations for the numbers of atoms, n_g and n_f, in $\langle g|$ and $\langle f|$, and for the number of excited atoms removed by detection, n_D, are

$$\frac{dn_g}{dt} = -W_{ex}(n_g - n_f) + \Gamma n_f,$$

$$\frac{dn_f}{dt} = W_{ex}(n_g - n_f) - (\Gamma + \beta + F)n_f,$$

(19.1)

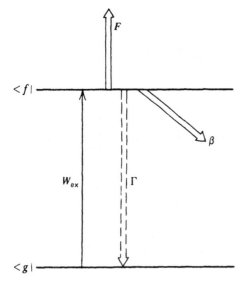

Fig. 19.1 A two-level system under selective excitation W_{ex}, detection F, relaxation Γ, and loss β.

and

$$\frac{dn_D}{dt} = Fn_f$$

assuming $n_f = 0$ at thermal equilibrium. If W_{ex} and F are step functions turned on at $t = 0$, and the initial conditions are $n_g = n_{g0}$ and $n_f = 0$ at $t = 0$, then the solution of (19.1) is

$$n_g = \frac{n_{g0}}{b - a} \left[(b - W_{ex})e^{-at} - (a - W_{ex})e^{-bt} \right],$$

$$n_f = \frac{(a - W_{ex})(b - W_{ex})}{(W_{ex} + \Gamma)(b - a)} n_{g0} [e^{-bt} - e^{-at}], \qquad (19.2)$$

and

$$n_D = \int_0^t Fn_f(t') \, dt'$$

where

$$a = \tfrac{1}{2}x_2 + \left[\left(\tfrac{1}{2}x_2 \right)^2 - x_1^2 \right]^{1/2},$$

$$b = \tfrac{1}{2}x_2 - \left[\left(\tfrac{1}{2}x_2 \right)^2 - x_1^2 \right]^{1/2},$$

$$x_1^2 = W_{ex}(\beta + F),$$

$$x_2 = 2W_{ex} + \beta + F + \Gamma.$$

This result looks complicated. However, we are interested mainly in having a maximum detection sensitivity, which, from physical argument, is expected to occur when $W_{ex} \gg \Gamma$ and $F \gg \beta$. Under these conditions and for the usual simple case of $W_{ex} \gg F$, we have $x_2 \gg x_1$ and $a \gg b$. The solution reduces to

$$n_g \simeq \tfrac{1}{2}n_{g0}e^{-bt},$$

$$n_f \simeq \tfrac{1}{2}n_{g0}e^{-bt}, \qquad (19.3)$$

and

$$n_D \simeq n_{g0}(1 - e^{-bt})$$

with $b \simeq \tfrac{1}{2}F$. If, on the other hand, $F \gg W_{ex}$, we still have $x_2 \gg x_1$, and $a \gg b$ and the solution becomes

$$n_g \simeq n_{g0}e^{-bt},$$

$$n_f \simeq (W_{ex}/F)n_{g0}e^{-bt}, \qquad (19.4)$$

and

$$n_D \simeq n_{g0}(1 - e^{-bt})$$

with $b \simeq W_{ex}$. Then, for square excitation and detection pulses with a pulse-width T, and $bT \gg 1$, we have $n_D \simeq n_{g0}$ at the end of the pulse. This means that all the atoms originally in the specific ground state are detected. It is therefore clear that in order to be able to detect nearly all the selected atoms, we must satisfy the conditions

$$W_{ex} \gg \Gamma, \qquad F \gg \beta,$$
$$FT \gg 1 \text{ if } W_{ex} \gg F, \tag{19.5}$$
$$W_{ex}T \gg 1 \text{ if } F \gg W_{ex}.$$

A more rigorous theory taking into account the coherent nature of the excitation can be found in Ref. 1.

We frequently are interested in detecting single atoms in the presence of a huge number of wrong atoms. The signal counts from the right atoms must then be larger than the background counts from the wrong atoms. If we also use (19.1) to describe the excitation of the wrong atoms (denoted by prime in the notations), then because of lack of selectivity, $W'_{ex} \ll \Gamma'$ and $F' \ll \beta'$. It follows that the fraction of wrong atoms appears as the background signal is, assuming $W'_{ex}T \ll 1$,

$$\frac{n'_D}{n'_{g0}} \simeq \frac{F'W'_{ex}T}{\beta' + \Gamma'}. \tag{19.6}$$

In comparison with n_D in (19.3) or (19.4), the ratio of signal to background for a system of n_{g0} right atoms and n'_{g0} wrong atoms is given by

$$\frac{S}{B} \simeq \frac{n_D}{n'_D} \sim \left(\frac{\beta' + \Gamma'}{F'W'_{ex}T} \right) \left(\frac{n_{g0}}{n'_{g0}} \right) \tag{19.7}$$

for $bT \gtrsim 1$. If $W_{ex}T \sim 1$, the ratio can be written as

$$\frac{S}{B} \sim \left(\frac{\beta' + \Gamma'}{F'} \right) \left(\frac{W_{ex}}{W'_{ex}} \right) \left(\frac{n_{g0}}{n'_{g0}} \right). \tag{19.8}$$

Suppose W_{ex} corresponds to a single-step resonant excitation, while W'_{ex} for wrong atoms is far from resonance. We have $W_{ex}/W'_{ex} \sim |(M/M')(\Delta\omega/\gamma)|^2$, where M and M' are the transition matrix elements in W_{ex} and W'_{ex}, respectively, $\Delta\omega$ is the frequency offset of W'_{ex} from resonance, and γ is the resonant linewidth of W_{ex}. For $M \sim M'$, and $\Delta\omega/\gamma \sim 10^4$, we already have $W_{ex}/W'_{ex} \sim 10^8$, and if $(\beta' + \Gamma')/F' \sim 100$, we would find $S/B \sim 10^{10}(n_{g0}/n'_{g0})$. In other words, the discrimination factor of detecting the right atoms against the wrong atoms is 10^{10}. In the case of discrimination against isotopes, however, $\Delta\omega/\gamma$ is much smaller; to increase the discrimination factor, one should try to make $(\beta' + \Gamma')/F'$ as large as possible through a properly arranged detection scheme. The above result suggests that in order to have a large S/B, the resonant linewidth γ should be very narrow, the detection scheme should be strongly selective, and the laser pulsewidth should be around $T \sim 1/W_{ex}$, but not too long.

In the second case in which all the excited atoms after being detected return to the ground state, (19.1) becomes

$$\frac{dn_g}{dt} = -W_{ex}(n_g - n_f) + (\Gamma + F)n_f,$$

$$\frac{dn_f}{dt} = W_{ex}(n_g - n_f) - (\Gamma + F)n_f, \tag{19.9}$$

$$\frac{dn_D}{dt} = Fn_f$$

assuming $\beta = 0$ for simplicity. The solution with the initial conditions $n_g = n_{g0}$ and $n_f = 0$ at $t = 0$ is

$$n_g = n_{g0} - \frac{W_{ex}}{2W_{ex} + \Gamma + F}n_{g0}[1 - e^{-(2W_{ex} + \Gamma + F)t}],$$

$$n_f = \frac{W_{ex}}{2W_{ex} + \Gamma + F}n_{g0}[1 - e^{-(2W_{ex} + \Gamma + F)t}], \tag{19.10}$$

$$n_D = \frac{W_{ex}F}{2W_{ex} + \Gamma + F}n_{g0}t - \frac{W_{ex}Fn_{g0}}{(2W_{ex} + \Gamma + F)^2}[1 - e^{-(2W_{ex} + \Gamma + F)t}].$$

If $W_{ex} \gg F, \Gamma$, and $W_{ex}T \gg 1$, then we have

$$n_D \simeq \tfrac{1}{2}FTn_{g0}, \tag{19.11}$$

which can be much larger than n_{g0} if $FT \gg 1$. Thus the conditions for high detection sensitivity stated in (19.5) are still true here. The background signal due to excitation and detection of wrong atoms in this case, assuming $\Gamma' \gg W'_{ex}$, F' and $\Gamma'T \gg 1$, is

$$n'_D \simeq (W'_{ex}/\Gamma')F'Tn'_{g0}. \tag{19.12}$$

Therefore, the discrimination factor for detection of the right atoms against the wrong atoms is

$$\frac{S}{B} \sim \frac{1}{2}\left(\frac{\Gamma'}{W'_{ex}}\right)\left(\frac{F}{F'}\right)\left(\frac{n_{g0}}{n'_{g0}}\right) \tag{19.13}$$

for $W_{ex} \gtrsim \Gamma$. If $\Gamma \sim \Gamma'$ and $W_{ex} \sim \Gamma$, then we have

$$\frac{S}{B} \sim \frac{1}{2} \left(\frac{W_{ex}}{W'_{ex}} \right) \left(\frac{F}{F'} \right) \left(\frac{n_{g0}}{n'_{g0}} \right) \tag{19.14}$$

which can be $> 10^{10}(n_{g0}/n'_{g0})$ for $(W_{ex}/W'_{ex})(F/F') > 10^{10}$. The results in (19.13) and (19.14) again show that in order to have a large S/B, the resonant excitation should be very sharp with a large $\Delta\omega/\gamma$, and the detection scheme should be highly selective. The laser intensity should be strong enough to make $W_{ex}\Gamma^{-1} \sim 1$, but not excessively strong so as to reduce Γ'/W'_{ex}.

Through focusing of the exciting laser beams, spatial distribution of rare atoms or molecules in the selected quantum state can be measured by the laser detection methods. That pulsed lasers can be used in these methods suggest also the possibility of time-resolved measurements. The latter is particularly useful for detection of radicals and transient or unstable species.

19.2 EXPERIMENTAL TECHNIQUES

From what we have discussed, it is clear that narrowband tunable lasers, CW or pulsed, are needed for selective excitation of selected species. Either single-photon or multiphoton excitation can be used. To increase the selectivity of the right atoms (or molecules) against the wrong atoms, multistep multiphoton excitation is preferred because the overall selectivity is determined by the product of the resonant features of the successive steps.[2] For example, if $W_{ex}^{(n)}/W_{ex}'^{(n)} \sim [W_{ex}^{(1)}/W_{ex}'^{(1)}]^n$, where $W_{ex}^{(n)}/W_{ex}'^{(n)}$ is the ratio of the n-step excitation rate for the right atoms to that for the wrong atoms, then with $W_{ex}^{(1)}/W_{ex}'^{(1)} \sim 10^8$, the discrimination factor for a two-step excitation to detect the right atoms against the wrong atoms, as estimated from (19.8) or (19.14), can be larger than 10^{16}.

While selective excitation as a labeling process is crucial in discriminating the right atoms (or molecules) against the wrong atoms, it is the high sensitivity of the detection schemes that make the detection of single atoms possible. Among the various methods for detection of excited atoms (or molecules), fluorescence and ionization are most attractive. They have already been discussed in some detail in Chapter 18. Here we concentrate our discussion of these methods on their abilities and possible layout to detect single atoms and molecules.

Laser-Induced Fluorescence

The three main schemes of laser-induced fluorescence are shown in Fig. 19.2, where W_{ex} can be either single-photon or multiphoton excitation. That laser-induced fluorescence has the sensitivity to detect single atoms[3] can be seen from the following example. Consider a single atom with velocity v traversing l mm distance across a CW exciting laser beam. Assume that the atom after excitation will return to the ground state by fluorescence through one of the pathways in Fig. 19.2 in an average lifetime τ. If the excitation is strong enough to saturate the transition ($W_{ex}\tau \gg 1$), then immediately after the atom returns to the ground state, there is a 50% chance that the atom will be reexcited. The number of excitation–fluorescence cycles the atom can experience during the atom–laser interaction time l/v is $l/v\tau$. The number of photons emitted by the atom is therefore $l/2v\tau$. This result is the same as that given by (19.11) with $n_{g0} = 1$, $T = l/v$, and $F = 1/\tau$. For alkali atoms, for example, τ is typically ~ 10 nsec. With $l \sim 2$ mm and $v \sim 10^4$ cm/sec, a single atom is expected to yield 10^3 fluorescent photons, which should be easily detectable if the background noise is sufficiently low.

A typical experimental arrangement using the scheme of Fig. 19.2a is seen in Fig. 19.3.[4] The resonant fluorescence photons emitted by atoms at one focus of the ellipsoidal reflector are collected by the photodetector at the other focus. The overall photon-counting efficiency of 5.5% has been obtained for the case

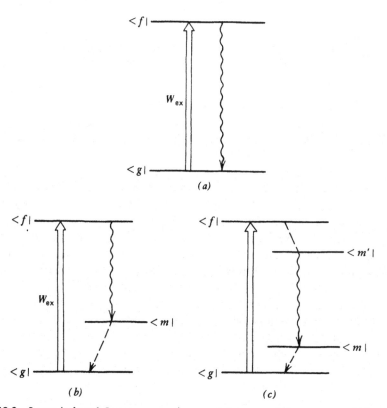

Fig. 19.2 Laser-induced fluorescence schemes: (a) resonant fluorescence; (b) and (c) fluorescence with frequency different from the exciting laser frequency.

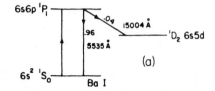

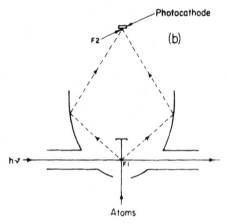

Fig. 19.3 Typical experimental arrangement for single-atom detection by laser-induced fluorescence: (*a*) barium energy levels; (*b*) schematic of apparatus. (After Ref. 4.)

of Ba atoms,[4] with a resonant fluorescent efficiency of ~ 10%. Somewhat more than one photon can be detected when a single Ba atom crosses a laser beam of 1-mm diameter. Limited by noise, a few as 10 atoms/sec are detectable by this technique.

Background noise due to stray laser light and Rayleigh scattering from wrong atoms is often the limiting factor of the detection sensitivity for scheme *a* of Fig. 19.2. The reason is that the background and the signal, having the same wavelength, cannot be easily separated by an ordinary detection method. One way to reduce the background is to use the coincidence-counting technique (Fig. 19.4).[5] The resonant fluorescence signal is monitored by two independent photodetectors. Only when both photodetectors register photons within a certain time interval (~ 1 μsec) will these photons be counted as the signal. This discriminates against the background due to stray light and dark current in the photodetectors which are more random in character. The coincidence-counting technique can also be applied with the two photodetectors monitoring atoms in an atomic beam at two positions along the beam. The measured time delay of the delayed coincident photon pulses from the two detectors readily yields the velocity of the atoms.[6]

Another way to reduce the background is to use scheme *b* and *c* in Fig. 19.2. The fluorescence which is at a different wavelength from that of the exciting laser can be easily detected against the scattered laser light by the use of a

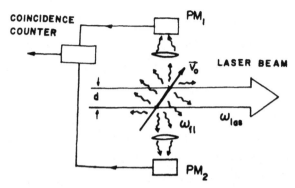

Fig. 19.4 Laser-induced fluorescence of an individual atom detected by the coincidence-counting technique. [After V. S. Letokhov, *Comments Atomic Molec. Phys.* **7**, 93 (1977).]

monochromator or color filter.[7] The ability to detect the right atoms in the presence of overwhelming wrong atoms is then greatly enhanced, as seen from (19.13) or (19.14) with $F \gg F'$. It is, however, important in this method that the relaxations from $\langle f|$ to $\langle m|$ and $\langle m'|$ to $\langle g|$ in Fig. 19.2 are fast, so that they do not appreciably lengthen the overall relaxation time from $\langle f|$ to $\langle g|$; otherwise, the total number of photons emitted by the atom during the excitation will be reduced. The fast relaxations can usually be achieved through collisions. The technique is therefore most useful for detecting rare atoms in a host gas medium. As few as 10 rare atoms/cm^3 (or 10^{-2} atom in 1-mm^3 laser probing region) in a gas of atmospheric pressure can be detected.

An even better way to reduce the background, if applicable, is to use two-step two-photon excitation in scheme a of Fig. 19.2. As mentioned earlier, two-step resonant excitation greatly enhances selectivity. Then the fluorescent frequency is also far away from the exciting laser frequencies, making selective detection very easy. The signal-to-background ratio can therefore be extremely high. The only disadvantage is that much higher laser power is needed for saturation of excitation ($W \geq \Gamma$ in Fig. 19.1).

Photoionization

The detection efficiency of a fluorescent photon emitted by an atom usually is less than 10%. In comparison, the detection of ionization of an atom (or molecule) can be $\sim 100\%$. This makes single-atom (or molecule) detection via selective excitation-ionization most attractive, especially when high-intensity pulsed lasers are required anyway for the excitation. Some experimental details on multiphoton ionization were given in Section 18.2. Detection of either electrons or ions can be used to monitor the ionization.[1] While ionization of the selectively excited atoms can be achieved by different methods, namely, photoionization, dc field ionization, and collisional ionization, we concentrate our discussion here on photoionization.

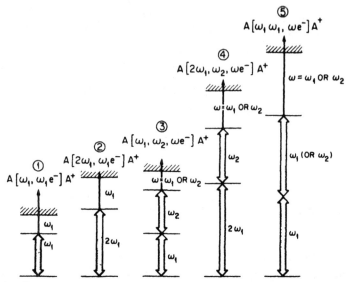

Fig. 19.5 Classification of selective photoionization schemes for sensitive detection of the elements. With these five schemes, all of the elements except He, F, Ne, and possibly Ar can be detected. (After Ref. 1.)

In Fig. 19.5, the various schemes for single-atom detection via photoionization are shown.[1] Note that scheme ⑤ is different from scheme ② or ③ only in the sense that ω_1 and ω_2 are in the uv in the former case. With these five schemes in Fig. 19.5 involving only one- or two-step resonant excitation, all elements, except ground-state He, F, Ne, and possibly Ar, can be selectively detected with currently available laser sources. The possible scheme for the detection of each element in the periodic table is given in Fig. 19.6.

The simple theory described by (19.1) to (19.8) is applicable to the present case, with W_{ex} denoting the one- or two-step resonant excitation. For high detection sensitivity, the excitation should be saturated, and the photoionization rate should be large so that nearly all the excited atoms are ionized. This means that the required tunable laser powers may be very high, especially if multiphoton transitions are involved. To increase the photoionization rate, excitation to discrete autoionization states is preferred.

One possible drawback of the photoionization method for single-atom detection often is the lack of good selectivity in the ionization process, that is, the photoionization step may not be very discriminative against the wrong atoms. This limits the signal-to-background ratio. Several schemes can be used to improve the selectivity of ionization detection. One is to photoionize through selective excitation into an autoionization state of the right atoms. The other is to incorporate a mass spectrometer in the ion detection system to discriminate the wrong atoms which have been ionized. This scheme is particu-

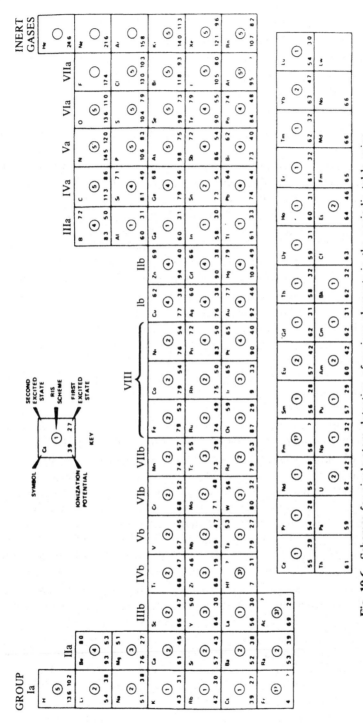

Fig. 19.6 Schemes for single-atom detection of various elements in the periodic table using selective excitation followed by photoionization. (After Ref. 1.)

larly useful for detection of selected molecules[8] and can also be used to study the dissociation of a molecule after photoionization.[9]

19.3 DEMONSTRATIONS OF DETECTION OF RARE ATOMS AND MOLECULES

A few examples of single-atom (or molecule) detection are briefly described here.

Alkali and alkali earth atoms can be detected easily by the laser-induced fluorescence technique. Sodium is a representative case. The $3s(^2S_{1/2}) \rightarrow 3p(^2P_{3/2})$ transition has an absorption cross section $\sigma_a \sim 1.6 \times 10^{-9}$ cm^2. (With Doppler broadening, however, the average σ_a is much lower). The lifetime of the $3p$ state is $\tau = 1/\Gamma \sim 16$ nsec. For saturating excitation, $W_{ex}\tau = \sigma_a(I/\hbar\omega)\tau \sim 1$, the laser intensity needed is $I \sim 10$ mW/cm^2, which can be easily obtained from a CW dye laser. Under such CW excitation, the number of photons emitted by a single atom is $n_D \sim 1/2\tau \sim 3 \times 10^7$/sec. In a cell with low-density sodium, if the laser probing region is ~ 10 mm^3 and each atom spends only $\sim 10^{-4}$ sec in the region, then the average number of photons emitted by each atom is ~ 3000. With a photon detection efficiency of 5%, and a background noise of 10 counts/sec in the photodetector, a density of ~ 10 atoms/cm^3 is detectable. This was indeed the sensitivity limit Fairbank et al.[3] nearly obtained in their first experiment (see Fig. 19.7). With the coincidence-counting technique to reduce the background, and using a sodium atomic beam, Balykin et al.[5] obtained a sensitivity of ~ 10 atoms/sec or 10^{-4} atoms in the probing region. Detection of low-density sodium atoms in the presence of high-density buffer gases has also been demonstrated. Sodium has two $3p$ levels, $^2P_{1/2}$ and $^2P_{3/2}$, separated by 6 Å. It is possible to selectively excite the atoms to $^2P_{1/2}$, and utilize collisions to transfer the excitation to $^2P_{3/2}$. The fluorescence from $^2P_{3/2}$ to the ground $3s$ level is then at a frequency different from the exciting laser frequency, and can be detected through a monochromator with an excellent discrimination against the stray or Rayleigh scattered laser light. Using this scheme, a detection sensitivity of ~ 10 atoms/cm^3 in an argon buffer gas of $\sim 10^{18}$ atoms/cm^3 can be achieved with CW laser excitation.[7]

The detection of molecules by laser-induced fluorescence is generally less sensitive for several reasons. The oscillator strengths of transitions in molecules are relatively weak ($\sigma_a \sim 10^{-18}$ cm^2 for molecules as compared to 10^{-12} cm^2 for atoms). The laser intensity required for saturation of excitation is therefore much higher. Absorption and luminescence via electronic transitions of molecules are often in the form of broad bands. They make selective excitation and detection more difficult and reduce the discrimination factor against the wrong molecules. In addition, with a limited detection bandwidth, the measured fluorescent efficiency of the molecules is often much less than 1. Experimentally, 5×10^4 molecules per cubic centimeter of BaO in a specific rotational-

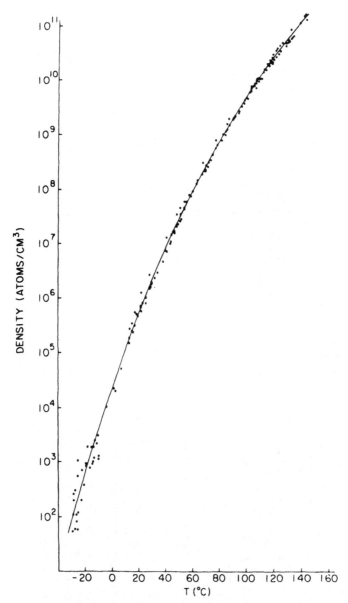

Fig. 19.7 Sodium-vapor density measurements using the laser-induced fluorescence technique. The solid line is a thermodynamically derived curve, using $\Delta H_0^\circ = 25600$ cal/mole. (After Ref. 3.)

Table 19.1
Table of Some RIS Experiments[a]

Process	Experiment	Reference
$H[\omega_1, \omega_2, \omega_1 e^-]H^+$	Nd : Yag pumped dye laser system; $\lambda_1 = 2660$ Å; $\lambda_2 \sim 2240$ Å; saturated a small volume	G. C. Bjorklund, C P Ausschnitt, R R Freeman, and R H. Storz, *Appl Phys Lett* **33**, 54 (1978)
$He(2^1, {}^3S)[\omega_1, \omega_1 e^-]He^+$	Nitrogen pumped dye laser at various wavelengths produce 3^3P, 4^1P, 5^1P, 3^3P, 4^3P, and 5^3P atoms	F. B. Dunning and R F Stebbings, *Phys Rev Lett* **32**, 1286 (1974)
$Li[\omega_1, \omega_2, \omega e^-]Li^+$	Coaxial flashlamp dye lasers; $\lambda_1 = 6709$ Å, $\lambda_2 = 6105$ Å, saturated a large volume	S D Kramer, J P Young, G S. Hurst, and M G Payne, *Opt. Commun* **30**, 47 (1979)
$Ca({}^3P_2)[\omega_1, \omega_2 e^-]Ca^+$	Argon ion pumped dye laser (CW); $\lambda_1 = 6162$ Å, $\lambda_2 = 4880$ Å and other autoionization peaks	U Brinkmann, W Hartig, H Telle, and H Walther, *Appl. Phys.* **5**, 109 (1974)
$Rb[\omega_1, \omega_2 e^-]Rb^+$	Ruby laser, dye laser system; $\lambda_1 = 7948$ Å, $\lambda_2 = 3471$ Å	R V Ambartsumyan, V N. Kalinin, and V. S. Letokhov. *JETP Lett* **13**, 217 (1971), R V Ambartsumyan and V S Letokhov. *Appl Opt* **11**, 354 (1972)
$Cs[\omega_1, \omega_1 e^-]Cs^+$	Linear flashlamp dye laser; $\lambda_1 = 4555$ or 4593 Å; saturated a small volume, demonstrated one-atom detection	G S. Hurst, M H Nayfeh, and J. P. Young, *Phys Rev A* **15**, 2283 (1977)
$Cs[\omega_1, \omega_1 e^-]Cs^+$	Coaxial flashlamp dye laser; $\lambda_1 = 4555$ Å, saturated large volume, absolute measurement of number of photodissociated atoms	L W Grossman, G S Hurst, S. D. Kramer, M G. Payne, and J P Young, *Chem Phys. Lett* **50**, 207 (1977)
$Cs[\omega_1, \omega_1 e^-]Cs^+$	Coaxial flashlamp dye laser, $\lambda_1 = 4555$ or 4593 Å saturated large volume; measured Cs atoms from ^{252}Cf fission	S D. Kramer, C E Bemis, J. P Young, and G S Hurst, *Opt Lett* **3**, 16 (1978)
$Y[\omega_1, \omega_2, \omega_3, Ee]Y^+$	Nitrogen pumped dye laser system, $\lambda_1 = 5556$ Å, $\lambda_2 = 6800$ Å, $\lambda_1 = 5950-5770$ Å	G I Bekov, V S. Letokhov, O I Matveev, and V. I. Mishin. *Opt. Lett.* **3**, 159, (1978).

[a]After Ref. 1.

vibrational level have been detected with a pulsed dye laser,[10] and 5×10^6 molecules/cm^3 of I_2 in a specific rotational-vibrational level have been observed with a CW laser.[11] The selectivity can be greatly improved by using a two-step resonant excitation: the first step is a vibrational transition which can be highly selective because of its narrow linewidth, and the second step is a less selective electronic transition for inducing fluorescence in the visible. High laser intensities are usually required for such a two-step excitation, and hence pulsed lasers are needed for this scheme.[12]

Detection of small numbers of atoms by photoionization has been demonstrated in many cases. A list of some of them is given in Table 19.1. As an example, we consider the detection of He in the long-lived excited state 2^1S.[13] Scheme ① in Fig. 19.5 can be used in this case. The laser at 5015 Å excites the 2^1S to 3^1P with an absorption cross section $\sigma_a \sim 5 \times 10^{-16}$ cm^2. (More rigorously, the degeneracy factors of the two levels should be taken into account in the calculation, and then emission and absorption cross sections between the two levels are generally not the same.)[13] The ionization cross section from the 3^1P level with the same laser frequency is $\sigma_I \sim 1 \times 10^{-17}$ cm^2. Thus $W_{ex} = \sigma_a(I/\hbar\omega) \sim 10^3 \, I/\sec$ and $F = \sigma_I(I/\hbar\omega) \sim 20 \, I/\sec$, where I is in watts per square centimeter. If the laser excitation is a pulsed one with a peak intensity $I \sim 1$ MW/cm^2 and a pulsewidth $T \sim 0.5$ μsec, then we have $W_{ex} \sim 10^9/\sec \sim \Gamma \sim 6 \times 10^8/\sec$, $W_{ex} \gg F \sim 2 \times 10^7/\sec \gg \beta \sim 3 \times 10^5/\sec$, and $FT \sim 10$. From (19.4), we expect that with such a pulsed laser excitation, nearly all He (2^1S) atoms can be detected. This is roughly what was actually observed.[13] A single atom appearing in the probing region during the laser pulse could be detected.

Detection of selected molecules by photoionization has also been repeatedly demonstrated.[14] We take here NO_2 as an example. A tunable dye laser in the range 4470–4970 Å can be used to excite NO_2 from $^2A_2$ to $^2B_1$. Then a H_2 vacuum uv laser (~ 1600 Å) is used to photoionize the molecule from the $^2B_1$ state. Detection of a single molecule in a given rotation-vibrational state is possible if the conditions in (19.5) can be satisfied.[9]

19.4 APPLICATIONS

Applications of rare atom or molecule detection are numerous. An obvious one is to study properties of excited-state atoms or molecules, radicals, ions, and other rare or transient species. For example, laser techniques can be used to probe the number of the excited-state atoms and to measure the photoionization cross sections from this excited state.[13, 15] The latter measurement is possible as σ_I can be deduced from the laser fluence (IT) dependence of the photoionization signal following (19.3) with $bt = \frac{1}{2}\sigma_I(IT/\hbar\omega)$. Spectroscopy of these rare or transient species is an interesting field which has hardly been explored because of lack of sensitive tools. With laser techniques, the field is expected to bloom rapidly, limited only by the power and tuning range of the available lasers. As an example, the $1^3S_1 \rightarrow 2^3S_1$ transition of positronium has recently been measured by the multiphoton ionization technique with good resolution.[16] The result is important as a stringent test on the theory of quantum electrodynamics.

The ability to detect single atoms or molecules in a given state also makes the laser techniques most useful for studies of photodissociation and chemical reactions.[17] Detection allows direct measurements of the dissociation and reaction products before they collide or react with molecules or walls. The measurements enable one to learn not only the velocity and angular distributions of the products but also the internal energy distribution in the products. Again, the limitation of this technique lies in the availability of tunable high-power lasers over a wide range.

The single-atom detection technique can also find important applications in nuclear physics. This technique can be used to monitor fission products[18] or to study the reaction of heavy elements at very low pressures. The sensitivity to detect a single atom in a specific volume at a specific time also permits the detection of exotic or unstable nuclei, rare isotopes, radioactive nuclei, solar neutrinos, and so on. Detection is achieved through monitoring of the daughter atoms released by these particles as they are arranged to undergo a nuclear reaction. Laser techniques have also been proposed for use in the search for quarks.

Single-atom detection can also be used to explore some basic questions in statistical mechanics.[1] One is on the Brownian motion of specific atoms in a gas. Is the assumption of randomness in the theory justified? Another is on the ergodic hypothesis. As suggested by Einstein,[19] a time-resolved diffusion experiment, in which the diffusion equation can be checked by both "time summation" and "space summation," can provide a test on the assumption of an ergodic system. Finally, the atomic fluctuation phenomena can be directly observed.[20] Since the detectable number of atoms N in a specific volume can be small, the normalized number fluctuations $\Delta N / \overline{N}$ can be made quite large and can be accurately measured. The change from a Gaussian distribution of N to a Poisson distribution as the average number $\overline{N}$ increases (from 1 to 20) has been demonstrated experimentally.[20]

Another possible application of single-atom detection is on dating. The more conventional techniques for dating are less sensitive, and require a fair amount of the material to be dated in the analytical process. The single-atom detection technique certainly has the advantage of being able to cut down the minimum amount of material needed for dating.

REFERENCES

1 G. S. Hurst, M. G. Payne, S. D. Kramer, and J. P. Young, *Rev. Mod. Phys.* **51**, 767 (1979).

2 V. S. Letokhov and V. I. Mishin, *Opt. Commun.* **29**, 168 (1979).

3 W. M. Fairbank, T. W. Hansch, and A. L. Schawlow, *J. Opt. Soc. Am.* **65**, 199 (1975).

4   G. W. Greenless, D. L. Clark, S. L. Kaufman, D. A. Lewis, J F. Tonn, and J. H. Broadhurst, *Opt. Commun.* **23**, 236 (1977).

5   V. I. Balykin, V. S. Letokhov, V. I. Mishin, and V. A. Semchishen, *JETP Lett.* **26**, 357 (1977).

6   C Y. She, W. M. Fairbank, and K. M. Billman, *Opt. Lett.* **2**, 30 (1978); J. V. Prodan, C. Y. She, and W. M. Fairbank, *Opt. Commun.* **43**, 215 (1982).

7   J. A. Gelbwachs, C. F. Klein, and J. E. Wessel, *IEEE J. Quant. Electron.* **QE-14**, 121 (1978).

8   M Klewer, M. J. M. Beerlage, J. Los, and M. J. Van der Wiel, *J. Phys.* **B10**, 2809 (1977); J. D. Reilly and K. L Kompa, *Adv. Mass Spectrosc.* **8**, 1800 (1979).

9   L. Zandee, R. B. Bernstein, and D. A. Lichtin, *J. Chem. Phys.* **69**, 3427 (1978); L. Zandee and R. B. Bernstein, *J. Chem. Phys.* **70**, 2574 (1979); **71**, 1359 (1979).

10 A. Schulz, H. W. Cruse, and R. N. Zare, *J. Chem. Phys.* **57**, 1354 (1972).

11 V. I. Balykin, V. I. Mishin, and V. A. Semchishen, *Sov. J. Quant. Electron.* **7**, 879 (1977)

12 A. Laubereau, A. Seilmeier, and W. Kaiser, *Chem. Phys. Lett.* **36**, 232 (1975).

13 G. S. Hurst, M. G. Payne, M. H. Nayfeh, J. P. Judish, and E. B. Wagner, *Phys. Rev. Lett* **35**, 82 (1975).

14 P. M. Johnson, M. R. Berman, and D. Zakheim, *J. Chem. Phys.* **62**, 2500 (1975); G. Petty, C. Tai, and F. W. Dalby, *Phys. Rev. Lett.* **34**, 1207 (1975).

15 R. V. Ambartzumian, A. M. Apatin, V. S. Letokhov, A. A. Makarov, V. I. Mishin, A. A Puretskii, and N. P. Furzikov, *JETP* **43**, 866 (1976); M. H. Nayfeh, G. S. Hurst, M. G. Payne, and J. P. Young, *Phys. Rev. Lett.* **39**, 604 (1977).

16 S. Chu and A. P. Mills, *Phys. Rev. Lett.* **48**, 1333 (1982); S. Chu and A. P. Mills, in H. Weber and W. Lüthy, eds., *Laser Spectroscopy VI* (Springer-Verlag, Berlin, 1983).

17 See, for example, J. Stephenson and D. S. King, *J. Chem. Phys.* **69**, 1485 (1977).

18 S. D. Kramer, C. E. Bemis, J. P. Young, and G. S. Hurst, *Opt. Lett.* **3**, 16 (1978).

19 A. Einstein, in R. Furth, ed., *Investigations on the Theory of Brownian Movement* (Dover, New York, 1956).

20 L W. Grossman, G. S. Hurst, M. G. Payne, and S. L. Allman, *Chem. Phys. Lett.* **50**, 70 (1977).

BIBLIOGRAPHY

Antonov, V. S., and V. S. Letokhov, *Appl. Phys.* **24**, 89 (1981).

Balykin, V. I., G I. Bekov, V. S. Letokhov, and V. I. Mishin, *Phys. Rep.* (to be published).

Bekov, G I., and V. S. Letokhov, *Appl. Phys. B* **30**, 161 (1983)

Hurst, G. S, M. G. Payne, S. D. Kramer, and J. P. Young, *Rev. Mod. Phys.* **51**, 767 (1979).

Letokhov, V S., *Comments Atomic Mol. Phys.* **7**, 93 (1977); **7**, 107 (1977); **10**, 257 (1981).

Parker, D. H., J O. Berg, and M. A. El-Sayed, in A. H. Zewail, ed., *Advances in Laser Chemistry* (Springer-Verlag, Berlin, 1978).

Payne, M. G., C. H. Chen, G. S. Hurst, and G. W. Foltz, in D. Bates and B. Bederson *Advances in Atomic and Molecular Physics*, vol. 17 (Academic Press, New York, 1982).

20

Laser Manipulation
of Particles

Radiation forces from a laser beam can be very strong because of the beam's coherent and highly directional properties. A focused beam of 10^8 W/cm^2 can exert an acceleration of $10^5 g$ on a 1-μm dielectric sphere with a 10% reflectivity. It is therefore possible to use laser beams to accelerate, decelerate, steer, manipulate, cool, or trap small particles including atoms, molecules, and ions. This has opened another fascinating area of laser physics research. Many interesting applications in various fields can be anticipated. Studies of single particle properties and controlled reaction or interaction between particles are the obvious examples. Laser manipulation of particles is actually a subject outside the scope of nonlinear optics, but we include it here as a special topic of laser physics worth knowing.

20.1 RADIATION FORCES

In a dielectric medium, the radiation force per unit volume is given by[1]

$$\mathbf{f} = \nabla \cdot \boldsymbol{\sigma} - \frac{\partial \mathbf{G}}{\partial t}. \tag{20.1}$$

Here, $\boldsymbol{\sigma}$ is the Maxwell stress tensor, and $\mathbf{G}$ is the electromagnetic momentum density in vacuum. Then

$$\boldsymbol{\sigma} = \left[-p - \frac{1}{8\pi} E^2 \left(\varepsilon - \rho \frac{\partial \varepsilon}{\partial \rho} \right) - \frac{1}{8\pi} B^2 \right] \mathbf{1} + \frac{1}{4\pi} (\mathbf{ED} + \mathbf{BB}),$$

$$\mathbf{G} = \frac{1}{4\pi c} (\mathbf{E} \times \mathbf{B}) = \frac{\mathbf{S}}{c^2} \tag{20.2}$$

where p is the pressure in the medium, ρ is the density of the medium, and $\mathbf{S}$ is the Poynting vector. With the help of the Maxwell equations, substitution of (20.2) into (20.1) yields

$$\mathbf{f} = -\nabla p - \frac{E^2}{8\pi}\nabla\varepsilon + \nabla\left[\rho\left(\frac{\partial\varepsilon}{\partial\rho}\right)\frac{E^2}{8\pi}\right] - (\varepsilon - 1)\frac{\partial\mathbf{G}}{\partial t}, \qquad (20.3)$$

which in a uniform medium reduces to

$$\mathbf{f} = -\nabla p + \rho\left(\frac{\partial\varepsilon}{\partial\rho}\right)\frac{1}{8\pi}\nabla E^2 - (\varepsilon - 1)\frac{\partial\mathbf{G}}{\partial t}. \qquad (20.4)$$

The second term in (20.4) is simply the electrostrictive force (see Section 11.1), while the third term arises from the change of the electromagnetic momentum density.

We are usually interested only in the time-averaged $\langle\mathbf{f}\rangle$. Then, in the steady state, although $\langle\partial\mathbf{G}/\partial t\rangle = 0$ inside a medium, it exerts a finite pressure on the boundary surface where reflection and refraction occur. The total force on a macroscopic dielectric object immersed in a fluid is given by

$$\mathbf{F}_{\text{total}} = \int_v \langle\mathbf{f}\rangle \, dv, \qquad (20.5)$$

which from the momentum conservation relation should be equal to the surface integral $\int_s (\Delta\langle(\partial/\partial t)\mathbf{K}\rangle \cdot d\mathbf{s})\hat{n}_s$ around the object where $\hat{n}_s$ is the surface normal, $\Delta\langle(\partial/\partial t)\mathbf{K}\rangle$ is the time-averaged electromagnetic momentum density transferred to the object per unit time at the boundary surface between the two media, with $\mathbf{K} = \varepsilon\mathbf{G}$ being the electromagnetic momentum (or pseudo-momentum) density in a medium with a dielectric constant ε. This can actually be shown with the expression of f in (20.4).[2]

Consider a transparent dielectric sphere with $\varepsilon = \varepsilon_H$ immersed in a fluid with $\varepsilon = \varepsilon_L$ and located off-axis from a laser beam as shown in Fig. 20.1.[3] Assume that the ray approximation is valid. Then the radiation force per unit area on the sphere exerted by the incoming ray along a at the input side is

$$\mathbf{F}' = c\left[\frac{\langle\mathbf{K}'\rangle}{\sqrt{\varepsilon_L}} - \left(\frac{\langle\mathbf{K}'_R\rangle}{\sqrt{\varepsilon_L}} + \frac{\langle\mathbf{K}'_D\rangle}{\sqrt{\varepsilon_H}}\right)\right] \qquad (20.6)$$

where $\mathbf{K}'$, $\mathbf{K}'_R$, and $\mathbf{K}'_D$ are the momentum densities of the incoming, reflected, and refracted waves, respectively. We can write $\langle\mathbf{K}'\rangle = \hat{k}_t(\langle K'_1\rangle + \langle K'_2\rangle)$ with $\langle K'_1\rangle = \langle K'_R\rangle$, and hence

$$\mathbf{F}' = \mathbf{F}'_R + \mathbf{F}'_D \qquad (20.7)$$

with

$$\mathbf{F}'_R = \frac{(\langle K'_1\rangle - \langle K'_R\rangle)c}{\sqrt{\varepsilon_L}} \quad \text{and} \quad \mathbf{F}'_D = \left(\frac{\langle K'_2\rangle}{\sqrt{\varepsilon_L}} - \frac{\langle K'_D\rangle}{\sqrt{\varepsilon_H}}\right)c.$$

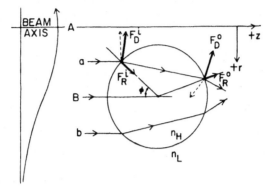

Fig. 20.1 A dielectric sphere situated off the axis A of a TEM_{00}-mode beam with a pair of symmetric rays a and b. The forces due to a are shown for $\varepsilon_H > \varepsilon_L$. The sphere is pushed toward $+z$ and $-r$. (After Ref. 3.)

Then, $\mathbf{F}_R^i$ is along the inward normal of the sphere, and $\mathbf{F}_D^i$ has a component along $+\hat{z}$ and a component toward the beam axis if $\varepsilon_H > \varepsilon_L$ (as shown in Fig. 20.1), or away from the beam axis if $\varepsilon_H < \varepsilon_L$. The radiation pressure force on the sphere due to reflection and refraction of the ray along a at the output surface can be similarly calculated. Again, we can write

$$\mathbf{F}^o = \mathbf{F}_R^o + \mathbf{F}_D^o \tag{20.8}$$

with

$$\mathbf{F}_R^o = \frac{(\langle \mathbf{K}_1^o \rangle - \langle \mathbf{K}_R^o \rangle)c}{\sqrt{\varepsilon_H}},$$

$$\mathbf{F}_D^o = \left(\frac{\langle \mathbf{K}_2^o \rangle}{\sqrt{\varepsilon_H}} - \frac{\langle \mathbf{K}_D^o \rangle}{\sqrt{\varepsilon_L}} \right)c,$$

$$\langle \mathbf{K}^o \rangle = (\langle K_1^o \rangle + \langle K_2^o \rangle)\hat{k}_0, \quad \text{and} \quad \langle K_1^o \rangle = \langle K_R^o \rangle.$$

$\mathbf{F}_R^o$ is along the outward normal of the sphere, and $\mathbf{F}_D^o$ has a component along $+\hat{z}$ and a component toward the beam axis if $\varepsilon_H > \varepsilon_L$, or away from the beam axis if $\varepsilon_H < \varepsilon_L$. The sum of $\mathbf{F}^i$ and $\mathbf{F}^o$ is a net force along $+\hat{z}$ and a net force toward the beam axis if $\varepsilon_H > \varepsilon_L$. A similar analysis shows that the ray along b, which is symmetric to a with respect to the axis of the sphere as shown in Fig. 20.1, gives the sphere a net force along $+\hat{z}$ and a net outward force away from the beam axis if $\varepsilon_H > \varepsilon_L$. Since the wave along a is more intense than that along b, the former exerts a stronger force than the latter. Consequently, the total radiation force on the sphere, obtained by integrating over the sphere, is along $+\hat{z}$ and toward the beam axis if $\varepsilon_H > \varepsilon_L$ (away from the beam axis if $\varepsilon_H < \varepsilon_L$). This result can be easily understood from the physical picture that the dielectric sphere, aside from being pushed forward by the photon flux, should tend to move to a position which minimizes the free energy of the system. Take the case of a latex ball ($\sqrt{\varepsilon_H} = 1.58$) immersed in water ($\sqrt{\varepsilon_L} = 1.33$) as an example.[3] Assume that the ball has a radius r and is sitting on the axis of a focused argon laser beam. The total radiation force on the sphere along $\hat{z}$, obtained by an integration of $\mathbf{F}$ over the sphere, is $(F_{total})_z \simeq 4q|\mathscr{P}|r^2/w_0^2$, where $\mathscr{P}$ is the laser power, w_0 is the e^{-2} radius of the focused beam, and $q \simeq 0.6$. With $\mathscr{P} = 1$w, and $r \simeq w_0/\sqrt{2} \simeq \lambda = 5140$ Å, the total force on the sphere is found to be $\sim 4 \times 10^{-5}$ dyne, and the corresponding acceleration of the sphere of density $\rho \simeq 1$ g/cm^3 is $\sim 10^8$ cm/sec^2 or 10^5 g. This shows that the laser radiation pressure can indeed be used to manipulate particles of micron size.

That the total radiation force on a macroscopic particle can be calculated from the momentum conservation relation is generally true. It can also be used to calculate the radiation force on absorbing objects. By the same token, the conservation of angular momentum can be used to find the torque exerted on a particle by a circularly or elliptically polarized laser beam. Rotation of a particle induced by light is an interesting problem not yet fully explored.[4]

We now consider radiation forces on atoms or molecules. Let the induced dipole on an atom in an electromagnetic field be $\mathbf{p}$. Then the force on the atom is simply the Lorentz force on the dipole

$$\mathbf{f}_{atom} = \mathbf{p} \cdot \nabla \mathbf{E} + \frac{1}{c} \frac{\partial \mathbf{p}}{\partial t} \times \mathbf{B}. \tag{20.9}$$

With $\mathbf{p} = \alpha\mathbf{E}$, $\nabla \times \mathbf{E} = -(1/c)(\partial\mathbf{B}/\partial t)$, and the identity $\mathbf{E} \cdot \nabla\mathbf{E} = \frac{1}{2}\nabla E^2 - \mathbf{E} \times (\nabla \times \mathbf{E})$, (20.9) becomes[2]

$$\mathbf{f}_{atom} = \alpha\left[\frac{1}{2}\nabla E^2 + \frac{1}{c}\frac{1}{\partial t}(\mathbf{E} \times \mathbf{B})\right]. \tag{20.10}$$

One immediately recognizes that (20.10) is nothing but the microscopic counterpart of (20.4), since in the atomic case $p = 0$ and $\varepsilon = 1 + 4\pi\rho\alpha$.

The first term in (20.10) is equivalent to the electrostrictive force in a macroscopic medium. It is called the dipole force when only the real part of α ($= \alpha' + i\alpha''$) is considered.

$$\mathbf{f}_{dip} = \left(\frac{\alpha'}{2}\right)\nabla E^2. \tag{20.11}$$

The direct proportion of $\mathbf{f}_{dip}$ to α' makes $\mathbf{f}_{dip}$ most significant and strongly dispersive near resonances. With $\alpha' > 0$ (immediately below a strong resonance), the force pulls the atom toward regions of higher intensities; with $\alpha' < 0$ (above a strong resonance), it pushes the atom toward regions of lower intensities. If the atom can be treated as an effective two-level system, the polarizability can be derived from the density matrix formalism (Section 2.1) in the nonperturbative limit:

$$\alpha = \frac{|\langle 1|er|2\rangle|^2}{\hbar(\omega - \omega_{21} + i\Gamma)}(\rho_{11} - \rho_{22})$$

$$\alpha' = \frac{|\langle 1|er|2\rangle|^2(\omega - \omega_{21})}{\hbar[(\omega - \omega_{21})^2 + \Gamma^2]}(\rho_{11} - \rho_{22}).$$

(20.12)

In (20.12), the population difference between the two levels $|1\rangle$ and $|2\rangle$ is given by

$$\rho_{11} - \rho_{22} = \frac{\Delta\rho_0}{1 + \Gamma g(\omega)I/I_s} \qquad (20.13)$$

where $I_s = c\hbar^2\Gamma/8\pi|\langle 1|er|2\rangle|^2 T_1$ is the saturation intensity, I is the laser intensity, $\Delta\rho_0$ is the population difference at thermal equilibrium, and $g(\omega) = \Gamma/\pi[(\omega - \omega_{21})^2 + \Gamma^2]$ is the unsaturated resonance lineshape. The polarizability can be written in the form

$$\alpha' = \frac{\alpha_0'(\omega)}{1 + \Gamma g(\omega)I/I_s} \qquad (20.14)$$

with $\alpha_0'(\omega)$ denoting the real part of the polarizability without saturation. Then, from (20.11), the dipole force has the time-averaged expression[5]

$$\langle \mathbf{f}_{dip}\rangle = \frac{\alpha_0'(\omega)}{2[1 + \Gamma g(\omega)I/I_s]}\nabla\langle E^2\rangle. \qquad (20.15)$$

or

$$\langle \mathbf{f}_{dip}\rangle = \frac{\hbar(\omega - \omega_{21})}{1 + \Gamma g(\omega)I/I_s}\left[\frac{g(\omega)}{2I_s T_1}\right]\nabla I\Delta\rho_0. \qquad (20.16)$$

We notice that in the limit of strong saturation, although $\alpha' \to 0$, $\langle \mathbf{f}_{dip}\rangle$ remains finite and increases with the detuning $(\omega - \omega_{21})$. With $\omega < \omega_{21}$ and hence $\alpha' > 0$, a TEM_{00} beam tends to attract the atom radially inward and trap it on the beam axis through the dipole force. The trapping energy $\int_0^\infty \langle \mathbf{f}_{dip}\rangle \cdot d\mathbf{r}$ in the radial direction increases with the laser power even in the saturation limit. Aside from the dipole force, no other terms in (20.10) contribute to the time-averaged $\langle \mathbf{f}_{atom}\rangle$.

Radiation force can also arise from momentum change due to absorption and emission of photons by an atom. It is sometimes called the scattering force.[2,3] In absorbing one photon, the atom receives a linear momentum $\hbar k$. In the ensuing spontaneous emission process, the photon is emitted with equal probabilities in all directions. Therefore, on average, no momentum change results from the spontaneous emission. If the atom can be treated as a two-level system, the scattering force is given by the number of photons absorbed per unit time multiplied by $\hbar \mathbf{k}$:

$$\langle \mathbf{f}_{scatt} \rangle = \frac{\omega \alpha'' \langle E^2 \rangle}{\hbar \omega} \hbar \mathbf{k}$$

$$= \frac{|\langle 1|er|2\rangle|^2 \Gamma}{\hbar \left[(\omega - \omega_{21})^2 + \Gamma^2 \right]} (\rho_{11} - \rho_{22}) \langle E^2 \rangle \mathbf{k} \qquad (20.17)$$

which reduces to

$$\langle \mathbf{f}_{scatt} \rangle = \frac{\frac{1}{2} \hbar \mathbf{k}}{T_1} \qquad (20.18)$$

in the limit of strong saturation, as one would expect from physical argument. The scattering force is in the direction of beam propagation and appears to push the atom along the beam.

On molecules, the scattering force is much smaller because of the much weaker resonant transitions due to spread of oscillator strengths among many vibration-rotational lines and the longer lifetime T_1. The dipole force, however, can still be significant as it does not depend so critically on resonant enhancement.

20.2 OPTICAL LEVITATION OF MACROSCOPIC PARTICLES

We now consider how the radiation force of a laser beam can be used to manipulate a macroscopic particle. The example given in the previous section shows that a 10-mW CW visible laser beam can yield an acceleration of 1 g on a ≤ 10-μm latex or glass ball. With a vertically directed laser beam, it is then possible to levitate such a macroscopic particle in air or liquid against the gravitational force.[6] A typical experimental setup is seen in Fig. 20.2. As actually demonstrated by Ashkin and Dziedzic,[4,6] the particle can indeed be levitated to a height where the radiation pressure force from the focused laser beam just balances the gravitational force. The particle remains on the beam axis, but because of the beam fluctuations, the vertical position of the particle may fluctuate. This can be eliminated using the feedback scheme sketched in Fig. 20.2. There, the vertical position of the particle is monitored by a height sensor via the scattered radiation from the particle. Deviation from a preset height generates an error signal to increase or decrease the laser intensity so as to bring the particle to the correct position.

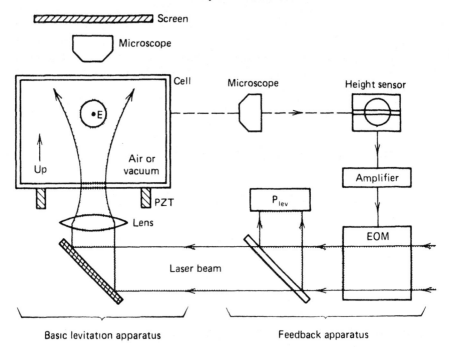

Fig. 20.2 Apparatus for levitating a dielectric sphere with a feedback stabilization scheme. PZT denotes piezoelectric transducer, and EOM, electrooptical modulator. [After A. Ashkin, *Science* **210**, 1081 (1980).]

Since stable levitation of particles is possible, a number of interesting experiments can be imagined. First, the spectroscopy of a single levitated particle can be studied. For example, the surface wave resonances of a spherical particle,[7] which arise when the surface waves run around the sphere and close head-to-tail on themselves, can now be detected either from the fluorescence spectrum or from the variation of the levitation spectrum[8] (laser power required for levitation versus laser frequency). These surface resonances can be very sharp (< 0.25 Å), with their resonant frequencies depending critically on the size of the sphere. They can be employed for size measurement with an accuracy two to three orders of magnitude better than other methods. The accuracy for relative size measurements can be as good as 1 part in 10^6 and can therefore be used to study minute changes of the sphere caused by, for example, evaporation, condensation, or external perturbation. The levitation method can also be used to accurately sort out spherical particles of different sizes. The optical properties of nonspherical particles are also interesting, and can be most conveniently studied by the levitation method.

Levitation of particles can also be used to study interaction between two macroscopic particles, or interaction between a particle and a surface.[9] This can be done by levitating the particles to prescribed positions and observing the force change. Study of melting and crystallization is another possible interesting problem. In the more practical applications, optical levitation can be used to support the small targets in laser-fusion experiments and to map out the gas flow pattern around a particle in various chambers.

The radiation forces can also be used to steer, accelerate, and manipulate a macroscopic particle. This then allows the studies of dynamic properties of particles in a medium, collisions between particles, fusion of particles, and so on.[9]

In a fluid with many suspended dielectric spheres of higher refractive indices, the radiation forces of a laser beam can increase the density of particles in the region of higher intensities. Consequently, the refractive index of the medium becomes higher in the region of high intensities. This is a nonlinear optical effect since the refractive index now depends on the laser intensity. Degenerate four-wave mixing and self-focusing[10] have actually been observed in such a medium.

20.3 LASER STEERING OF ATOMIC BEAMS

Radiation forces can be used to manipulate atoms. Because the forces have strong resonant enhancement, they are very selective in exerting forces on different atomic species. The scattering force following (20.17) or (20.18) can be very strong. In the saturation limit, which requires only ~ 10 mW/cm^2 for the $s \to p$ transition in alkali atoms, an acceleration of $\mathbf{a} \sim \frac{1}{2}\hbar\mathbf{k}/MT_1$ is obtained, where M is the mass of the atom. For a sodium atom with $T_1 \simeq 16$ nsec, a saturating laser beam resonant with the $3s \to 3p$ transition yields an acceleration of $\sim 10^8$ cm/sec^2. This means that an atom with an initial velocity $v_0 \sim 10^5$ cm/sec can be stopped by a head-on laser beam in 1 msec over a distance of 50 cm, or can be deflected by 5 mrad by a perpendicular beam over a 1-cm interaction length. The dipole force following (20.15) or (20.16) can also be appreciable. In the saturation limit, a Gaussian beam, having $I = I_0\exp(-2r^2/w^2)$ with $w = 100\ \mu$m and $\omega - \omega_{21} = -2$ GHz, yields a transverse force $\langle f_{\text{dip}} \rangle \sim 5 \times 10^{-15}$ dyne on the atom at $r = w$. (The equivalent acceleration on a sodium atom is $a \sim 6 \times 10^7$ cm/sec^2.) The radially inward transverse dipole force forms an effective negative potential well around the beam with a minimum on the beam axis. In the above example, atoms with transverse velocities less than $\sim 10^3$ cm/sec can be expected to get trapped in a 1-W laser beam.

We now describe a few experiments on laser manipulation of atomic beams. Observation of atomic beam deflection by a resonantly exciting laser beam crossing at $90°$ provides direct evidence of the existence of the scattering force.[11] The observed transverse deflection of a well-collimated atomic beam agrees with the value predicted. Because it is directly proportional to resonant absorption in the unsaturated limit [see (20.17)], the process has been suggested as a high-resolution atomic spectroscopy method.[12] It can also be used for isotope separation based on the finite isotope shift in the resonant frequencies of different isotopes.[13]

As another demonstration of the scattering force, a resonant laser beam has been used to decelerate atoms in a counterpropagating atomic beam. In the experiment, when the atoms are slowed down, their transition frequency can be Doppler-shifted out of resonance with the laser frequency. To decelerate the atoms appreciably, either the laser frequency should be continuously tuned to resonance with the decelerated atoms[14] or the Doppler-shifted atomic transitions should be continuously tuned by an external field to be always in resonance with the fixed laser frequency.[15] Indeed, with a properly adjusted axially varying magnetic field to Zeeman-tune the $3S_{1/2}(F = 2, M_F = 2) \rightarrow 3P_{3/2}(F = 3, M_F = 3)$ transition of Na, deceleration of sodium atoms to 4% of the initial thermal velocity by a 50-mW CW laser beam with a 10-MHz linewidth has been observed. The temperature characterizing the relative motion of the atoms was reduced to 70 mK.

The existence of the dipole force has been demonstrated by the observation that a Gaussian laser beam can be used to transversely confine and focus an atomic beam.[16] The experimental arrangement is shown in Fig. 20.3. When the dye laser is tuned below the atomic resonance, the inward transverse dipole force decreases the outward transverse velocity component of the atoms, and leads to the focusing of the atomic beam. This is seen in Fig. 20.4a for the case of a sodium atomic beam. If the laser is tuned above resonance, then the transverse dipole force is radially outward, and the atomic beam becomes defocused, as in Fig. 20.4b. The confining and focusing capability of the dipole force makes optical steering of atomic beams possible, as this can be achieved by simple moving the guiding laser beam at a slow enough rate. The process can also be used for isotope separation or for cleaning up dirty atomic beams by confining only the desired species of atoms.

Atomic beam deflection by a transient dipole force has also been observed,[17] and has been considered as a method for isotope separation. The transient

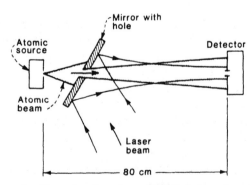

Fig. 20.3 Apparatus for observing focusing and defocusing of an atomic beam by the dipole force of a nearly resonant laser beam. [After A. Ashkin, *Science* **210** (1980).]

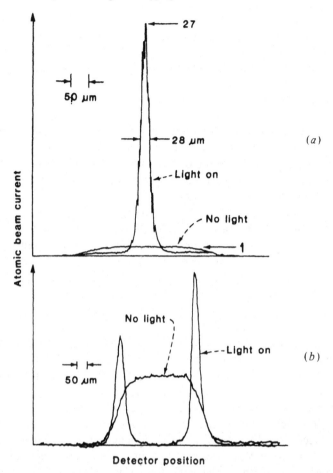

Fig. 20.4 (a) Focusing of an atomic beam by light tuned below resonance. (b) Defocusing of an atomic beam by light tuned above resonance. The detector was displaced along a line passing through the center 0. [After A. Ashkin, *Science* **210**, 1081 (1980).]

behavior of the radiation forces on an atom is itself a subject of theoretical interest.[18]

20.4   OPTICAL COOLING AND TRAPPING OF ATOMS AND IONS

Laser deceleration of atoms is a form of optical cooling for atoms in an atomic beam. The absorption of a photon at $\omega \simeq \omega_{21} - kv$ by a counterpropagating atom changes the atomic velocity from v to $v - \Delta v$ with $\Delta v = \hbar k/M$. Here only the first-order Doppler shift is taken into account. Subsequently, in the reverse transition, a photon is emitted at $\omega' = \omega_{21} + \mathbf{k}' \cdot (\mathbf{v} - \Delta\mathbf{v})$ with $\mathbf{k}'$ being the wavevector. This emitted photon frequency is always higher than the absorbed photon frequency. Therefore, the atom is cooled by losing its kinetic energy to the radiation field. This absorption-emission process has also been suggested for cooling of atoms in a cell.[19] By tuning the laser to the low-frequency side of the Doppler-broadened absorption line, atoms that absorb the laser photons have a velocity component antiparallel to the laser beam. These atoms lose their kinetic energies in reradiation. If, through atomic collisions, thermal equilibrium exists in the atomic gas, then the equilibrium temperature is lowered as the total energy of the atoms decreases continuously through absorption and reradiation.

While the scattering force can be used to cool atoms, the dipole force can be used to trap atoms.[20] We saw in Section 20.3 how the atoms with a transverse velocity $v_\perp < v_{\perp,\max}$ can be trapped in the axial region of a Gaussian beam by the transverse negative potential of the dipole force. This can, of course, be extended to the three-dimensional case. If the Gaussian beam is strongly focused, then the intensity variation along the beam axis can be appreciable (see Fig. 20.5). The corresponding axial dipole force is directed toward the focus and forms a negative potential well around the focus in the axial direction. This together with the transverse negative potential well sets up a three-dimensional local trap for atoms around the focus. The maximum kinetic energy of the atoms that can be confined in the trap is defined by the depth of the potential well. Other laser beams geometries, using possibly more than one laser beam, can also be used to form optical traps. The basic idea is always to find an optical field configuration with a stable equilibrium point such that an atom displaced from this point should experience a restoring force.

Atoms with finite velocities trapped in a potential well will, of course, oscillate back and forth in the well if there is no damping on their motion. They can, however, be optically cooled by the scattering force that damps the motion. For effective cooling and trapping, the scattering force should be provided by a laser beam tuned below but close to an absorption peak, while

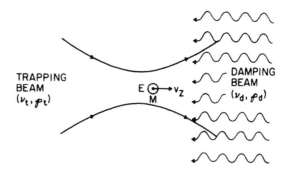

Fig. 20.5 Atom of mass M and velocity v_z located at the equilibrium point E of a Gaussian-beam trap with a plane wave acting to damp the atomic motion. (After Ref. 21.)

the dipole force for trapping is from a separate focused laser beam tuned below but farther away from the resonance (Fig. 20.5).[21] According to the calculations, the trapped atoms can be cooled down to 10^{-3}K by this means. The limit here is set by the quantum fluctuations of both the scattering and the dipole forces.[22]

Fluctuations of the scattering force are anticipated because absorption and emission of photons are random processes. Thus, for example, with N photons both absorbed and emitted by an atom, the momentum of the atom should fluctuate with a root mean square value $[\overline{(Mv)^2}]^{1/2} = \sqrt{2N}\,\hbar k$, and the corresponding average kinetic energy is $N(\hbar k)^2/M$. Fluctuations of the dipole force are more complicated and difficult to understand. Reference 22 provides a detailed discussion of the subject. Force fluctuations have been found to be the limiting mechanism on the focal spot size of the optically focused atomic beam.[23] Because of force fluctuations, even atoms with zero velocity sitting at the equilibrium point of the trap are expected to be heated up with ever increasing energy until they eventually escape from the trap. To keep the atoms in the trap at a low temperature, optical cooling must be used. The equilibrium temperature of the trapped atoms is determined by the balance between heating by fluctuations and optical cooling.

Optical trapping of neutral atoms in a local region has not yet been observed, although no insurmountable experimental difficulties are anticipated. If atoms can be trapped, optical cooling of the trapped atoms should be straightforward. Indeed, optical cooling of trapped ions has already been observed.[24] As charged particles, the ions can be initially trapped by an rf quadrupole field. Cooling of the trapped ions is then achieved with a tunable laser beam. By using optical cooling and trapping, many previously unthinkable experiments are no longer impossible. If atoms or ions are at rest at a local point, one can make observations on single atoms or ions over long periods of time. Then high-resolution spectroscopy of atoms or ions with the Doppler effects completely eliminated could become a reality. Toschek and co-workers[25] have been able to detect and photograph single ions in an rf quadrupole trap from their resonant fluorescence emission. Various fundamental properties of single atoms or ions can then be studied. With the possibility of using lasers to manipulate atoms or ions in a trap, studies of forces between atoms, between atoms and molecules, or between atoms and surfaces, and their influences on radiation lifetimes, formation of molecules, chemical reactions, and so on, can also be conceived. These applications, if possible, will certainly revolutionize the field of atomic physics.

REFERENCES

1   See, for example, L. D. Landau and E. M. Lifshitz, *Electrodynamics of Continuous Media* (Addison-Wesley, Reading, Mass., 1968), p 242.

2   J P. Gordon, *Phys. Rev. A* **8**, 14 (1973).

3   A. Ashkin, *Phys. Rev. Lett.* **24**, 156 (1970).

4 A. Ashkin and J. M. Dziedzic, *Appl. Phys. Lett.* **28**, 333 (1976); **30**, 202 (1977).

5 A. Ashkin, *Phys. Rev. Lett.* **40**, 729 (1978).

6 A. Ashkin and J. M. Dziedzic, *Appl. Phys. Lett.* **19**, 283 (1971).

7 R. E. Benner, P. W. Barber, J. F. Owen, and R. K. Chang, *Phys. Rev. Lett.* **44**, 475 (1980).

8 A. Ashkin and J. M. Dziedzic, *Phys. Rev. Lett.* **38**, 1351 (1977).

9 A. Ashkin, *Science* **210**, 1081 (1980).

10 P. W. Smith, A. Ashkin, and W. J. Tomlinson, *Opt. Lett.* **6**, 284 (1981); A. Ashkin, P. W. Smith, and J. M. Dziedzic, *Appl. Phys.* **28**, 142 (1982).

11 R. Frisch, *Z. Phys.* **86**, 42 (1933).

12 P. Jacquinot, S. Liberman, J. L. Pique, and J. Pinard, *Opt. Commun.* **8**, 163 (1973).

13 A. F. Bernhardt, *Appl. Phys.* **9**, 19 (1976).

14 V. I. Balykin, V. S. Letokhov, and V. I. Mushin, *JETP Lett.* **29**, 614 (1979); *JETP* **51**, 692 (1980).

15 W. D. Phillips and H. Metcalf, *Phys. Rev. Lett.* **48**, 596 (1982); J. V. Prodan, W. D. Phillips, and H. Metcalf, *Phys. Rev. Lett.* **49**, 1149 (1982).

16 J. E. Bjorkholm, R. R. Freeman, A. A. Ashkin, and D. B. Pearson, *Phys. Rev. Lett.* **41**, 1361 (1978); D. B. Pearson, R. R. Freeman, J. E. Bjorkholm, and A. Ashkin, *Appl. Phys. Lett.* **36**, 99 (1980).

17 E. Arimondo, H. Lew, and T. Oka, *Phys. Rev. Lett.* **43**, 753 (1979).

18 A. P. Kazantsev, *JETP* **36**, 81 (1973); **39**, 783 (1974); R. J. Cook, *Phys. Rev. Lett.* **41**, 1788 (1978).

19 T. W. Hansch and A. L. Schawlow, *Opt. Commun.* **13**, 68 (1975).

20 V. S. Letokhov, V. G. Minogin, and B. D. Pavlik, *JETP* **45**, 698 (1977); V. S. Letokhov and V. G. Minogin, *Appl. Phys.* **17**, 99 (1978); A. Ashkin, *Phys. Rev. Lett.* **40**, 729 (1978).

21 A. Ashkin and J. P. Gordon, *Opt. Lett.* **4**, 161 (1979).

22 J. P. Gordon and A. Ashkin, *Phys. Rev. A* **21**, 1606 (1980).

23 J. E. Bjorkholm, R. R. Freeman, A. Ashkin, and D. B. Pearson, *Opt. Lett.* **5**, 111 (1980).

24 D. J. Wineland, R. E. Drullinger, and F. L. Walls, *Phys. Rev. Lett.* **40**, 1639 (1978); W. Neuhauser, M. Hohenstatt, P. Toschek, and H. Dehmelt, *Phys. Rev. Lett.* **41**, 233 (1978).

25 W. Neuhauser, M. Hohenstatt, and P. E. Toschek, *Appl. Phys.* **17**, 123 (1978).

BIBLIOGRAPHY

Ashkin, A., *Sci. Am* **226**, 63 (Feb. 1972).

Ashkin, A., *Science* **210**, 1081 (1980).

Kazantsev, A. P., *Sov. Phys. Uspekhi* **21**, 56 (1978).

21

Transient Coherent Optical Effects

Pulse propagation in a resonant medium can yield many interesting phenomena. Most of them originate from the transient response of the medium to the coherent pulsed excitations. Transient coherent effects are among the most fascinating subjects of resonant wave interaction with matter. They were studied extensively in magnetic resonance before the laser era. With the advent of lasers, extension of these studies to the optical region becomes possible. The necessity of including the wave propagation effects makes transient coherent optical phenomena more interesting and colorful than their counterparts in magnetic resonance. Aside from general theoretical interest, transient coherent effects have also found useful applications in material studies.

21.1 BLOCH EQUATION FOR A TWO-LEVEL SYSTEM

We consider in this chapter mainly transient coherent effects in an effective two-level system, which is often a good approximation for a real system under a quasi-monochromatic resonant excitation. Then, as one may expect, a close analog should exist between the excitation of an optical resonance and that of a magnetic resonance. Indeed, Feynman et al.[1] have shown explicitly that any two-level system is equivalent to a spin-$\frac{1}{2}$ system as far as the resonant excitation is concerned. The proof is fairly simple: One shows that the dynamic response of a two-level system to a resonant excitation obeys a Bloch-type equation of motion[2] as that of a magnetic spin-$\frac{1}{2}$ system does.

Let us consider the two-level system seen in Fig. 21.1. The two levels are eigenstates of the unperturbed Hamiltonian $\mathscr{H}_0$:

$$\mathscr{H}_0|+\rangle = \tfrac{1}{2}\hbar\omega_0|+\rangle \quad \text{and} \quad \mathscr{H}_0|-\rangle = -\tfrac{1}{2}\hbar\omega_0|-\rangle. \qquad (21.1)$$

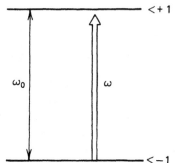

Fig. 21.1 A two-level system under a near-resonant excitation.

The nonvanishing dipole matrix elements are assumed to be

$$\gamma \equiv \langle + |\mu_+| - \rangle = \langle - |\mu_-| + \rangle \quad \text{and} \quad \mu_\pm \equiv \frac{\mu_x \pm i\mu_y}{\sqrt{2}} \qquad (21.2)$$

so that only the circularly polarized fields can induce transitions between the two levels. The interaction Hamiltonian of the system in the presence of a field $\mathbf{E} = \hat{x}E_x(t) + \hat{y}E_y(t)$ is

$$\begin{aligned}
\mathscr{H}_{\text{int}} &= -(\mu_x E_x + \mu_y E_y) \\
&= -(\mu_+ E_- + \mu_- E_+)
\end{aligned} \qquad (21.3)$$

where $E_\pm = (E_x \pm iE_y)/\sqrt{2}$ are the left and right circularly polarized fields.

The dynamic response of the two-level system to the applied field is now governed by the Liouville equation for the density matrix (see Section 2.1):

$$i\hbar \frac{\partial \rho}{\partial t} = [\mathscr{H}_0 + \mathscr{H}_{\text{int}}, \rho] + i\hbar \left(\frac{\partial \rho}{\partial t}\right)_{\text{damping}}, \qquad (21.4)$$

which can be written explicitly as

$$i\hbar \frac{\partial \rho_{+-}}{\partial t} = \hbar\omega_0 \rho_{+-} + \gamma E_-(\rho_{++} - \rho_{--}) - \frac{i\hbar}{T_2}\rho_{+-},$$

$$i\hbar \frac{\partial \rho_{-+}}{\partial t} = -\hbar\omega_0 \rho_{-+} - \gamma E_+(\rho_{++} - \rho_{--}) - \frac{i\hbar}{T_2}\rho_{-+},$$

$$\qquad (21.5)$$

$$i\hbar \frac{\partial}{\partial t}(\rho_{--} - \rho_{++}) = \gamma(E_-\rho_{-+} - E_+\rho_{+-})$$

$$- \frac{i\hbar}{T_1}\left[(\rho_{--} - \rho_{++}) - (\rho^0_{--} - \rho^0_{++})\right].$$

The expectation values of the electric dipole components are then given by

$$\langle \mu_x \rangle = \frac{\langle \mu_+ \rangle + \langle \mu_- \rangle}{\sqrt{2}}$$

$$= \frac{\gamma(\rho_{-+} + \rho_{+-})}{\sqrt{2}}$$

and $\hspace{10cm}$ (21.6)

$$\langle \mu_y \rangle = \frac{\langle \mu_+ \rangle - \langle \mu_- \rangle}{i\sqrt{2}}$$

$$= \frac{\gamma(\rho_{-+} - \rho_{+-})}{i\sqrt{2}}$$

If we now define a pseudo-dipole as

$$\langle \mu \rangle = \hat{x}\langle \mu_x \rangle + \hat{y}\langle \mu_y \rangle + \hat{z}\langle \mu_z \rangle$$

with

$$\langle \mu_z \rangle \equiv \gamma(\rho_{++} - \rho_{--}) \hspace{4cm} (21.7)$$

and an effective electric field as

$$\mathbf{E}_{\text{eff}} = \hat{x}E_x + \hat{y}E_y + \hat{z}(E_z)_{\text{eff}}$$

with a dc component

$$(E_z)_{\text{eff}} \equiv -\hbar\omega_0/\gamma, \hspace{3cm} (21.8)$$

then from (21.5) we find

$$\frac{\partial}{\partial t}\langle \mu_x \rangle = -\frac{\gamma}{\hbar}\left[E_y\langle \mu_z \rangle - (E_z)_{\text{eff}}\langle \mu_y \rangle \right] - \frac{\langle \mu_x \rangle}{T_2},$$

$$\frac{\partial}{\partial t}\langle \mu_y \rangle = -\frac{\gamma}{\hbar}\left[(E_z)_{\text{eff}}\langle \mu_x \rangle - E_x\langle \mu_z \rangle \right] - \frac{\langle \mu_y \rangle}{T_2}, \hspace{1cm} (21.9)$$

$$\frac{\partial}{\partial t}\langle \mu_z \rangle = -\frac{\gamma}{\hbar}\left[E_x\langle \mu_y \rangle - E_y\langle \mu_x \rangle \right] - \frac{\langle \mu_z \rangle - \langle \mu_z^0 \rangle}{T_1}.$$

In the vectorial form, (21.9) becomes

$$\frac{\partial}{\partial t}\langle \mu \rangle = -\frac{\gamma}{\hbar}\mathbf{E}_{\text{eff}} \times \langle \mu \rangle - \frac{1}{T_2}\left(\hat{x}\langle \mu_x \rangle + \hat{y}\langle \mu_y \rangle \right) - \frac{1}{T_1}\hat{z}(\langle \mu_z \rangle - \langle \mu_z^0 \rangle).$$

$$\hspace{11cm} (21.10)$$

This equation is identical to the Bloch equation in magnetic resonance if we take μ to be the magnetic dipole and replace $\mathbf{E}_{\text{eff}}$ by the magnetic field $\mathbf{H}$. We have therefore proved the statement that any two-level system can be treated as a pseudo-spin-$\frac{1}{2}$ system.

Equation (21.10) actually resembles the classical equation of motion for a precessing dipole μ driven by the torque $-(\gamma/\hbar)\mathbf{E}_{\text{eff}} \times \mu$. Neglecting the relaxation terms for simplicity, we can write (21.10) as

$$\frac{\partial}{\partial t}\langle\mu\rangle = \Omega \times \langle\mu\rangle \tag{21.11}$$

where $\Omega = -(\gamma/\hbar)\mathbf{E}_{\text{eff}}$ is the precessing angular frequency. Without an applied field, $\mathbf{E}_{\text{eff}} = \hat{z}(E_z)_{\text{eff}} = -\hat{z}\hbar\omega_0/\gamma$; if the dipole is initially tilted away from $\hat{z}$, it will precess around $\hat{z}$ with the angular frequency $\Omega = \omega_0\hat{z}$. With a time-dependent applied field $\mathbf{E}(t)$, Ω becomes time-dependent. We consider here the special case of a near-resonant excitation by a circularly polarized quasi-monochromatic field $\mathbf{E} = [(\hat{x} + i\hat{y})/\sqrt{2}]\mathscr{E}(t)e^{ikz-i\omega t} +$ complex conjugate with $\omega \sim \omega_0$. As a resonant driving field, $\mathbf{E}$ rotates nearly in synchronization with the dipole precession around $\hat{z}$. The physical picture is more transparent in a coordinate frame rotating with ω around $\hat{z}$. In the rotating frame,[3] (21.11) becomes

$$\frac{\partial}{\partial t}\langle\mu^*\rangle = (\Omega_R - \omega\hat{z}) \times \langle\mu^*\rangle$$
$$= \Omega^* \times \langle\mu^*\rangle \tag{21.12}$$

where $\Omega_R = -(\gamma/\hbar)[(E_z)_{\text{eff}}\hat{z} + 2\mathscr{E}\hat{x}']$ is related to Ω by the rotational transformation, and $\hat{x}'$ is along $\mathbf{E}$. (Here, $2\mathscr{E}$ instead of $\mathscr{E}$ appears in Ω_R because according to our definition of $\mathscr{E}$, the field amplitude of a sinusoidal wave is given by $2\mathscr{E}$.) Thus we have

$$\Omega^* = (\omega_0 - \omega)\hat{z} - \left(\frac{\gamma}{\hbar}\right)2\mathscr{E}\hat{x}'. \tag{21.13}$$

Then if we neglect the slow amplitude variation of $\mathscr{E}$, the dipole sees effectively a stationary field

$$\mathbf{E}^*_{\text{eff}} = -\left(\frac{\hbar}{\gamma}\right)(\omega_0 - \omega)\hat{z} + 2\mathscr{E}\hat{x}'. \tag{21.14}$$

The dynamic response of the dipole, following (21.12), can therefore be described by its precession around $\mathbf{E}^*_{\text{eff}}$ in the rotating frame.

Equation (21.11) or (21.12) governs the response of the medium to the field. To find how the medium in turn affects the field, we must solve the wave equation with the time-varying dipoles as the driving source:

$$\left(\nabla^2 - \frac{\varepsilon}{c^2} \frac{\partial^2}{\partial t^2} \right) \mathbf{E} = \frac{4\pi}{c^2} \frac{\partial^2}{\partial t^2} N\langle \mathbf{p} \rangle. \qquad (21.15)$$

Here, $\langle \mathbf{p} \rangle = \hat{x} \langle \mu_x \rangle + \hat{y} \langle \mu_y \rangle$, and we have assumed a cubic or isotropic medium having N dipoles per unit volume; the local field correction is neglected. In fact, the complete solution of the problem should be found by solving the coupled equations (21.11) and (21.15) together. A number of interesting transient phenomena arise from such solutions, depending on the resonant excitation pulses and the characteristics of the medium. We discuss some of them in the following sections. The emphasis is on physical understanding; hence no rigorous mathematical derivations are attempted. They can be found together with other details in many books and review articles on the subject (see Bibliography).

21.2   TRANSIENT NUTATION AND FREE INDUCTION DECAY

Consider first the case where the initial population of the two-level system is all in the ground state, so that the pseudo-dipole $\langle \mu \rangle = \hat{z} \langle \mu_z \rangle$ of (21.7) is pointing along $-\hat{z}$, as in Fig. 21.2. At $t = 0^+$, a near-resonant circularly polarized field is switched on. Then, $\langle \mu^* \rangle$ in the rotating frame sees a

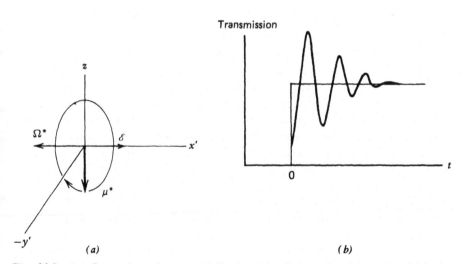

Fig. 21.2 (a) Precession of a pseudo-dipole μ^* in the rotating frame, in which the applied field $\mathscr{E}$ is stationary. (b) Transient nutation signal induced by the switch-on of a resonant excitation.

stationary effective field $\mathbf{E}_{eff}^*$ given by (21.14), and begins the counterclockwise precession around $\mathbf{E}_{eff}^*$ with a frequency.

$$\Omega^* = \left(\frac{\gamma}{\hbar}\right)|\mathbf{E}_{eff}|$$
$$= \left[(\omega - \omega_0)^2 + \left(\frac{\gamma}{\hbar}2\mathscr{E}\right)^2\right]^{1/2}, \qquad (21.16)$$

which is generally known as the Rabi frequency.[4] The precession leads to a sinusoidal oscillation in the amplitudes of $\langle\mu_{y'}^*\rangle$ and $\langle\mu_z^*\rangle = \langle\mu_z\rangle$. The latter corresponds physically to an oscillation in the population difference between the two levels. Since the absorption coefficient of the medium is directly proportional to the population difference, the transmission of the exciting light should experience a sinusoidal intensity modulation at the Rabi frequency. The precession of $\langle\mu^*\rangle$ around $\mathbf{E}_{eff}^*$ will eventually be damped out as $\langle\mu^*\rangle$ relaxes toward its final steady-state value $\langle\mu^*\rangle_\infty$ with

$$\langle\mu_{x'}^*\rangle_\infty = \frac{(\gamma/\hbar)2\mathscr{E}(\omega_0 - \omega)T_2^2}{D},$$

$$\langle\mu_{y'}^*\rangle_\infty = \frac{(\gamma/\hbar)2\mathscr{E}T_2}{D}, \qquad (21.17)$$

$$\langle\mu_z^*\rangle_\infty = -\frac{\left[1 + (\omega_0 - \omega)^2 T_2^2\right]}{D},$$

and

$$D = 1 + (\omega_0 - \omega)^2 T_2^2 + (\gamma/\hbar)^2 4\mathscr{E}^2 T_1 T_2$$

obtained from (21.9). This happens with a time constant $\tau \sim T_2$ if $\mathscr{E}$ is sufficiently small; more generally, τ depends on T_1, T_2, $|\omega_0 - \omega|$, and $\mathscr{E}$.[5] Therefore, following the switch-on of the excitation, the transmission of the exciting light should approach the steady-state value through a damped modulation period as shown in Fig. 21.2. This phenomenon is known as transient nutation.[5] Alternatively, the modulation can be explained by the amplitude-modulated precession of $\langle\mu_{y'}^*\rangle$ around $\hat{z}$ in the lab frame. The radiation from these precessing dipoles superimposed on the incoming light gives rise to the modulation of the transmitted light.

To observe transient nutation with a few cycles of oscillations before it decays away, we must have $\Omega^* > 1/\tau$, or $(\gamma/\hbar)2\mathscr{E} > 1/\tau$ if $\omega = \omega_0$. For the $s \to p$ transitions of alkali atoms, for example, $(\gamma/\hbar) \sim 5 \times 10^9$ esu, and $\tau \sim T_2 \sim 10^{-8}$ sec, we find $\mathscr{E} > 0.01$ esu, corresponding to a laser intensity $I > 0.03$ W/cm^2. For the vibrational transitions of molecules, one may have $(\gamma/\hbar) \sim 5 \times 10^5$ esu, and $\tau \sim 10^{-6}$ sec; the observation of transient nutation then requires $\mathscr{E} > 1$ esu or $I > 250$ W/cm^2. From the theoretical fit of the observed damped oscillation one can deduce γ, and hence the oscillator strength of the transition, and the dephasing time T_2.

Optical transient nutation was first predicted and observed by Tang and co-workers as an analog to the magnetic resonance case.[6] Pulsed lasers with a sufficiently sharp leading edge were needed for the experiments. Brewer and Shoemaker[7] have, however, introduced a pulsed Stark-shift technique that allows the observations of a variety of transient effects, including transient nutation, with the use of a CW laser. The basic setup is sketched in Fig. 21.3. The sample is in a Stark cell. An applied dc electric field on the sample can Stark-shift the atoms or molecules in or out of resonance with the CW exciting laser beam. This is then equivalent to switching on or off of the resonant exciting field. Figure 21.4 shows an example of transient nutation observed by this technique.[7] In this case, the CW laser has a linewidth much narrower than the Doppler width (or inhomogeneous linewidth) of the transition. It is initially in resonance with a group of molecules within the Doppler profile. The sudden application of the electric field shifts the resonance to a different group of molecules, assuming that the Stark shift is within the Doppler profile. This new group of molecules now begins to adsorb and gives rise to the observed transient nutation following the switch-on of the field. Then, if the applied dc electric field is suddenly turned off, the resonant excitation is shifted back to the original group of molecules. These molecules begin to absorb again, and yield another transient nutation signal as seen in Fig. 21.4. Of

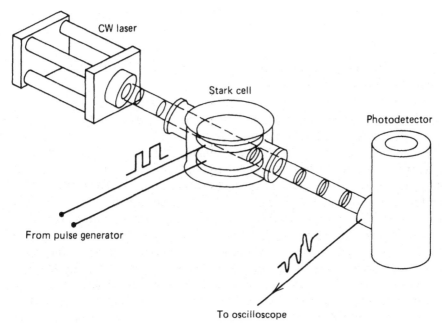

Fig. 21.3 Schematic of the Stark switching apparatus that is used for observing optical transients. [After R. G. Brewer, *Physics Today* **30**, 50 (1977).]

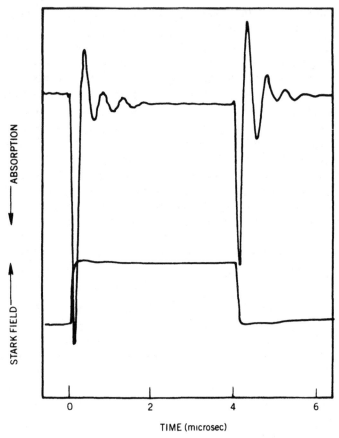

Fig. 21.4 The optical nutation effect in methyl fluoride $C^{13}H_3F$, irradiated by a carbon-dioxide laser at 9.7 μ. (After Ref. 7.)

course, the same experiments can also be performed by shifting the laser frequency instead of the material resonance.

We now consider the case where a resonant exciting laser beam initially in equilibrium with the two-level system is suddenly switched off. The stored energy in the two-level system is expected to radiate out. As shown in Fig. 21.5, the dipole $\langle \mu \rangle$ initially described by (21.17) sees an effective field $\mathbf{E}_{\text{eff}} = (E_z)_{\text{eff}}\hat{z}$ after the laser beam is turned off. It begins to precess around $\hat{z}$ in the lab frame and radiates. A collection of such dipoles radiates coherently until they get out of phase with one another. Dephasing occurs because atoms or molecules under the inhomogeneously broadened profile have different resonant frequencies ω_0, and hence the corresponding dipoles have different precessing frequencies. As a result, the coherent reradiation from the sample should decay away with a time constant τ_F equal to the dephasing time. This decaying coherent reradiation is known as optical free induction decay,[7] which also has an analog in magnetic resonance.[8] If the laser linewidth is much less than $1/T_2$, then

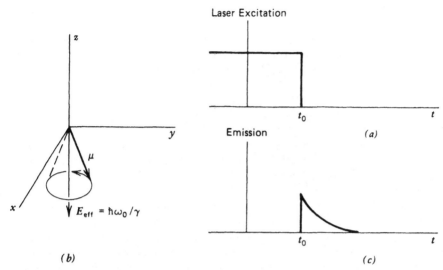

Fig. 21.5 Schematic drawings describing free-induction decay. (a) Laser excitation switched off at t_0; (b) pseudo-dipole μ precessing around $\hat{z}$ after t_0; (c) free-induction decay signal for $t > t_0$.

$\tau_F^{-1} = T_2^{-1} + \Gamma_p$ with $\Gamma_p = [1 + (\gamma/\hbar)^2 4\mathscr{E}^2 T_1 T_2]^{1/2}/T_2$ being the power-broadened width of the excitation. If the laser linewidth is much larger than the Doppler width $\Delta\omega_D \equiv 1/T_2^*$, then $\tau_F = T_2^*$.[†]

To observe the optical free induction decay, it is most convenient to use the heterodyne technique by beating the coherent reradiation with an incoming beam of slightly different frequency and detecting the beat signal. The Stark-shift technique of Brewer and Shoemaker is ideal for such experiments.[7] When a group of molecules initially on resonance with the incoming CW laser is suddenly shifted off resonance by an amount $\delta\omega$, the subsequent free induction decay radiation from these molecules can mix with the laser radiation and yield a damped beat signal at the beat frequency $\delta\omega$, as seen in Fig. 21.6.[9] From the decay of the beat signal, the dephasing time of the transition can be deduced. A homogeneous linewidth ($1/T_2$) as low as ~ 1 kHz has been measured.[10] Therefore, free induction decay can be used as a spectroscopy method of very high resolution. If the shift $\delta\omega$ is smaller than the inhomogeneous linewidth, and $(\gamma/\hbar)2\mathscr{E} > 1/T_2$, the frequency switching should also induce simultaneously the transient nutation process described earlier. The free induction signal then appears to be superimposed on the nutation signal. Since $\tau_F \approx \hbar/(\gamma 2\mathscr{E}) = 1/\Omega^*(\omega = \omega_0) \ll T_2$, the free induction decay is essentially limited to the first half period of the transient nutation.

[†]R. G. DeVoe and R. G. Brewer in *Phys. Rev. Lett.* **50**, 1269 (1983), have shown that in the high-power limit, $(4\gamma^2\mathscr{E}^2/\hbar^2)T_1 T_2 \gg 1$, the dephasing time should be $\tau_F = (1/T_2 + 2\gamma\mathscr{E}/\hbar)^{-1}$, and in the low-power limit, $\tau_F = T_2/2$, as suggested by A. G. Redfield in *Phys. Rev.* **98**, 1787 (1955).

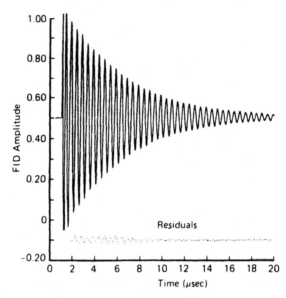

Fig. 21.6 Optical free-induction decay in $LaF_3 : Pr^{3+}$ (0.1%) at 1.6 K, as observed by the Stark-shift heterodyne technique. The residuals lead to an uncertainty of less than 1% in the measured dephasing time of 5.1 μsec. (After Ref. 9.)

More generally, if there exist several closely spaced transitions which can be simultaneously excited by the laser beam, then the free induction decay is the superposition of the induction decay signals from all these excited transitions. Fourier transform of the free induction decay signal should yield the spectrum of these transitions. This is similar to quantum beat spectroscopy (Section 13.2) except that the output in the latter case is the nondirectional fluorescent emission.

21.3   PHOTON ECHOES

Atoms or molecules in different environments have different resonant frequencies, leading to the inhomogeneous broadening of a spectral line. In the pseudo-dipole picture, this means that the dipoles should precess with different frequencies. If initially, through coherent excitation, the precessing dipoles are arranged in phase, they should then emit coherently as in the free induction decay case. However, because of their different precessing frequencies, the dipoles soon run out of phase with one another in a time $\sim T_2^* = 1/\Delta\omega_D$, where $\Delta\omega_D$ is the inhomogeneous broadening width. As a result, the coherent emission would die away. Yet if by some means the dephased dipoles could be rephased, then coherent emission would reappear. That dipole rephasing is indeed possible was first discovered in magnetic resonance by Hahn in a phenomenon known as spin echoes[11] and is a demonstration of the reversibility of some type of statistical dynamic processes. The optical analog of spin echoes, known as photon echoes, was later predicted and observed by Hartmann et al.[12]

Consider a collection of two-level systems with a distribution of resonant frequencies. The populations initially are in the ground level, so that the pseudo-dipoles μ^* are all pointing downward, as in Fig. 21.7a. At $0 \leq t \leq t_1$, a narrow square pulse $\mathbf{E} = [(\hat{x} + i\hat{y})\mathscr{E}/\sqrt{2}]e^{ikz - i\omega t}$ is applied to the systems. If $(\gamma/\hbar)2\mathscr{E} \gg |\omega - \omega_0|$, then with this excitation, all μ^* in the rotating frame where $\mathbf{E}$ is stationary should rotate around $\mathscr{E}$ by an angle.

$$\theta = \int_0^{t_1} \left(\frac{\gamma}{\hbar}\right) 2\mathscr{E} dt. \tag{21.18}$$

Assuming $\theta = \pi/2$, all dipoles end up together in the $\hat{x}-\hat{y}$ plane at the end of the pulse (Fig. 21.7b). They then appear in the lab frame precessing around $\hat{z}$, giving rise to coherent emission in the form of free induction decay, and because of dephasing due to inhomogeneous broadening, the emission signal decays away in a time $T_2^* = 1/\Delta\omega_D$. This is seen by the fanning out of the dipoles in Fig. 21.7c. At $t_2 \leq t \leq t_3$, another narrow square pulse (assuming $t_3 - t_2 \ll t_2 - t_1$) with $(\gamma/\hbar)2\mathscr{E} \gg |\omega - \omega_0|$ and $\theta = \pi$ is applied to the system. In the rotating frame, this makes the dipoles rotate around $\mathscr{E}$ by $180°$. The result is that all μ^* undergo a mirror reflection about the $\mathscr{E} - \hat{z}$ plane. Immediately after t_3, the dipoles are still badly dephased, and no coherent emission or free induction should appear. Yet, because of the mirror reflection of the dipoles caused by the π-pulse excitation, the dipoles, in the rotating frame, should now precess back (Fig. 21.7d) and begin to fan it. It takes the same amount of time for the dipoles to fan in as it takes them to fan out. Therefore, at $t = t_4$ with $t_4 - t_3 \simeq t_2 - t_1$, the dipoles are expected to be back in phase (Fig. 21.7e), with the resultant emission of a coherent pulse. As the dipoles fan out again, the emission signal decays away.

This discussion describes how the photon echo, in the form of a coherent emission pulse, appears from two-level systems. More generally, photon echoes refer to the appearance of coherent emission pulses following successive applications of resonant excitation pulses. Their existence depends on the reversibility of dipole dephasing due to inhomogeneous broadening. However, the dipoles should also experience an intrinsic dephasing process with a dephasing time T_2 related to the homogeneous broadening. This intrinsic dephasing is not reversible. Photon echoes can appear only before the intrinsic phase coherence of the dipoles is destroyed. Therefore, to observe a photon echo, the time delay between the first excitation pulse and the photon echo should not be much longer than T_2. The amplitude of the echo actually decay exponentially with the time delay, from which one can deduce the time constant T_2.

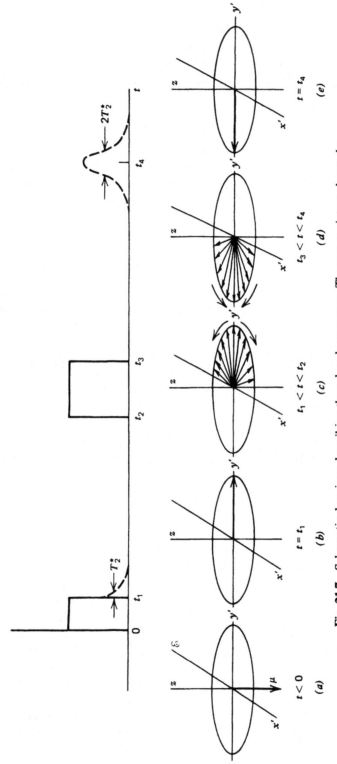

Fig. 21.7 Schematic drawings describing the echo phenomenon. The upper picture shows the pulse excitation sequence. The lower picture depicts the precession of the pseudo-dipoles in the rotating frame at various times.

In photon echoes, the wave propagation effect is also important.[12] Consider a light pulse detected by a detector at, say, $r = 0$ and time t, the actual interaction of the pulse with the atoms at r occurs at the retarded time $t - \mathbf{k} \cdot \mathbf{r}/\omega$. Thus, more correctly, retarded times should be used in this discussion. The photon echo should then appear when

$$\left(t_4 - \mathbf{k}_4 \cdot \frac{\mathbf{r}}{\omega}\right) - \left(t_3 - \mathbf{k}_2 \cdot \frac{\mathbf{r}}{\omega}\right) = \left(t_2 - \mathbf{k}_2 \cdot \frac{\mathbf{r}}{\omega}\right) - \left(t_1 - \mathbf{k}_1 \cdot \frac{\mathbf{r}}{\omega}\right) \quad (21.19)$$

where $\mathbf{k}_1$, $\mathbf{k}_2$, and $\mathbf{k}_4$ are, respectively, the wavevectors of the $\pi/2$ pulse, the π-pulse, and the photon echo. The above equation yields the conditions for the photon echo:

$$t_4 - t_3 = t_2 - t_1 \quad \text{and} \quad \mathbf{k}_4 = 2\mathbf{k}_2 - \mathbf{k}_1. \quad (21.20)$$

The second equation in (21.20) is actually the phase-matching requirement for the generation of the photon echo. It defines the direction of the echo propagation. However, the equation may not be fully satisfied if $\mathbf{k}_1$ and $\mathbf{k}_2$ are not properly chosen. In that case, the echo intensity is expected to reduce because of the phase mismatch.

The more quantitative analysis of photon echoes often starts with the description of the pseudo-dipole in the rotating frame. In that frame, after the $\pi/2$ and π pulse excitations, the precessing dipole with resonant frequency ω_0 is readily shown to have the expression[12]

$$\mu^* = \left(\frac{\hat{x}' + i\hat{y}'}{\sqrt{2}}\right)\mu_z^0 e^{i(\mathbf{k}_1 - 2\mathbf{k}_2)\cdot\mathbf{r} - i\Delta\omega[(t - t_3) - (t_2 - t_1)]} \quad (21.21)$$

where $\Delta\omega = \omega_0 - \omega$, and the phase change during the excitation pulses is neglected. In lab coordinates, this becomes

$$\mu = \left(\frac{\hat{x} + i\hat{y}}{\sqrt{2}}\right)\mu_z^0 e^{i(\mathbf{k}_1 - 2\mathbf{k}_2)\cdot\mathbf{r} - i\Delta\omega[(t - t_3) - (t_2 - t_1)] - i\omega t}. \quad (21.22)$$

For a collection of such dipoles with a resonant frequency distribution $g(\Delta\omega)$ and a density N, the resultant polarization is given by

$$N\langle\mu(\mathbf{r}, t)\rangle = \int_{-\infty}^{\infty} N\mu g(\Delta\omega)\, d(\Delta\omega). \quad (21.23)$$

This polarization acts as the source of coherent radiation governed by (21.15). In the slowly varying amplitude approximation, the solution of (21.15) gives a coherent emission field

$$\mathbf{E}_{em}(t) \propto \left[\int_V d^3r' \int_{-\infty}^{\infty} d(\Delta\omega) N\mu(\mathbf{r}', t) g(\Delta\omega) e^{-i\mathbf{k}_4\cdot\mathbf{r}'}\right] e^{i\mathbf{k}_4\cdot\mathbf{r}} \quad (21.24)$$

where $k_4 = \omega\sqrt{\varepsilon(\omega)}/c$ has its direction determined by $(k_4)_\perp = (2k_2)_\perp - (k_1)_\perp$. With μ given by (21.22), it is readily seen that the emission with the wavevector $\mathbf{k}_4$ is a maximum when the phase factor in the integral vanishes, that is, when (21.20) are satisfied. As t deviates away from t_4 given in (21.20), $\mathbf{E}_{em}$ quickly reduces in amplitude. The coherent emission therefore appears in the form of a pulse centered at t_4. It is circularly polarized in the same sense as the exciting pulses.

Actually, for the generation of photon echoes, the two excitation pulses can have any values of θ. They were chosen to be $\pi/2$ and π in this discussion only for convenience of illustration. The dephasing and rephasing processes are still operative if θ are different from $\pi/2$ and π. In the next section we see that photon echoes can also arise when θ are small, in which case the perturbative transient wave mixing approach can be used to describe transient phenomena. We also see that photon echoes are not restricted to two-level systems. In fact, they are more interesting and intriguing in a multilevel system with multipulse excitations.

The experimental arrangement for photon echo studies is fairly simple. One can use either laser pulses to successively excite the medium or amplitude or frequency switching to shift a CW laser excitation on and off in the form of successive pulses. The latter is most convenient for measuring photon echoes with long time delays. An example of a two-level photon echo is shown in Fig. 21.8.[12] The exponential decay of the echo intensity as a function of the time delay between pulses gives a direct measurement of the dephasing time T_2. A

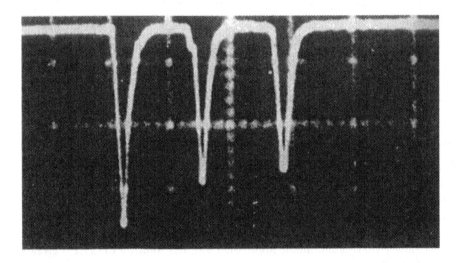

Fig. 21.8 Oscilloscope trace (100 nsec/division) showing the photon echo from a ruby sample at 4.2 K. The echo appears as the third pulse following the two exciting pulses from a ruby laser. (After Ref. 12.)

T_2 as long as a few hundred microseconds, corresponding to a homogeneous linewidth of a few hundred hertz, has been found in transitions of rare earth ions in solids by the photon echo technique.[13] Study of the T_2 dependence on various parameters in gases and condensed matter enables us to have a much better understanding of the various transverse relaxation (dephasing) mechanisms.

21.4   TRANSIENT FOUR-WAVE MIXING

From the microscopic point of view, transient coherent effects result from the fact that the material system can retain for some time the definite phase of a coherent excitation (coherent mixing of two states). In two-level systems, the coherent excitation can be described pictorially by the precession of the transverse dipole components around $\hat{z}$. More generally, however, it is described by the nonvanishing off-diagonal components of the density matrix. Therefore, at least formally, one can find the coherent transient responses of a medium to the applied field from the time-dependent solutions of the equation of motion for the density matrix. This approach is used here to study coherent transient effects in four-wave mixing in the perturbation limit and to show that it can lead to the more general photon echo and free induction phenomena.[14]

We first generalize the diagrammatic technique of Section 2.3 to the time-dependent case.[15] Consider Fig. 21.9, where three successive pulses $\mathbf{E}(\omega_1)$, $\mathbf{E}(\omega_2)$, $\mathbf{E}(\omega_3)$ resonantly excite the transitions $|m\rangle \rightarrow |p\rangle$, $\langle m| \rightarrow \langle r|, \langle r| \rightarrow \langle s|$ at t_1, t_2, t_3, respectively, with $t_1 < t_2 < t_3$. The rules of writing down the time-varying density matrix $\rho^{(3)}(t)$ are the same as in Section 2.3, except that the propagator π_j for propagation from the jth vertex to the $(j + 1)$th vertex

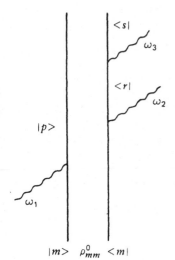

Fig. 21.9 Double Feynmann diagram describing a transient four-wave mixing process.

along the lines $|l\rangle$ and $\langle k|$ should be replaced by the phase factor

$$\langle\langle lk|\hat{A}(t_{j+1} - t_j)|lk\rangle\rangle = e^{-(\iota\omega_{lk} + \Gamma_{lk})(t_{j+1} - t_j)},$$

and the product of all factors in $\rho^{(3)}(t)$ is integrated over all possible time separations between vertices. We assume here that even the population excitation $(l = k)$ in a given state decays exponentially with time. Following these rules, we can find, from Fig. 21.9,

$$\rho^{(3)}(t) = \int_{-\infty}^{0} d\tau_1 \int_{-\infty}^{0} d\tau_2 \int_{-\infty}^{0} d\tau_3 e^{-(\iota\omega_{ps} + \Gamma_{ps})\tau_3 - (\iota\omega_{pr} + \Gamma_{pr})\tau_2 - (\iota\omega_{pm} + \Gamma_{pm})\tau_1}$$

$$\times\langle p|\frac{-1}{i\hbar}\mu\cdot\mathscr{E}_1(t - \tau_1 - \tau_2 - \tau_3)e^{-\iota\omega_1(t - \tau_1 - \tau_2 - \tau_3)}|m\rangle$$

$$\times\langle m|\frac{-1}{-i\hbar}\mu\cdot\mathscr{E}_2(t - \tau_2 - \tau_3)e^{-\iota\omega_2(t - \tau_2 - \tau_3)}|r\rangle \tag{21.25}$$

$$\times\langle r|\frac{-1}{-i\hbar}\mu\cdot\mathscr{E}_3(t - \tau_3)e^{-\iota\omega_3(t - \tau_3)}|s\rangle(|p\rangle\rho^0_{mm}\langle s|).$$

With the substitution of variables $\xi_1 = t - \tau_1 - \tau_2 - \tau_3$, $\xi_2 = t - \tau_2 - \tau_3$, and $\xi_3 = t - \tau_3$, (21.25) becomes

$$\rho^{(3)}(t) = -e^{-(\iota\omega_{ps} + \Gamma_{ps})t}\left(\frac{1}{i\hbar}\right)^3(|p\rangle\rho^0_{mm}\langle s|)$$

$$\times\langle p|\mu\cdot\hat{e}_1|m\rangle\langle m|\mu\cdot\hat{e}_2|r\rangle\langle r|\mu\cdot\hat{e}_3|s\rangle$$

$$\times\int_{-\infty}^{t} d\xi_3\, e^{[(\iota\omega_{ps} + \Gamma_{ps}) - (\iota\omega_{pr} + \Gamma_{pr}) - \iota\omega_3]\xi_3}\mathscr{E}_3(\xi_3) \tag{21.26}$$

$$\times\int_{-\infty}^{\xi_3} d\xi_2\, e^{[(\iota\omega_{pr} + \Gamma_{pr}) - (\iota\omega_{pm} + \Gamma_{pm}) - \iota\omega_2]\xi_2}\mathscr{E}_2(\xi_2)$$

$$\times\int_{-\infty}^{\xi_2} d\xi_1\, e^{[(\iota\omega_{pm} + \Gamma_{pm}) - \iota\omega_1]\xi_1}\mathscr{E}_1(\xi_1).$$

In a gaseous system, $\rho^{(3)}(t)$ is also a function of molecular velocity $\mathbf{v}$. A field $\mathscr{E}_i(\xi_j)$ seen by a molecule with $\mathbf{v}$ at $\mathbf{r}(\xi_j) = \mathbf{r}(t) - (t - \xi_j)\mathbf{v}$ is

$$\mathscr{E}_i(\xi_j) = A_i(\xi_j)e^{\iota\mathbf{k}_i\cdot\mathbf{r}(\xi_j)}$$

$$= A_i(\xi_j)e^{\iota\mathbf{k}_i\cdot\mathbf{r}(t) - \iota(t - \xi_j)(\mathbf{k}_i\cdot\mathbf{v})}. \tag{21.27}$$

We then have

$$
\begin{aligned}
\rho^{(3)}(\mathbf{v}, t) = &-e^{-(i\omega_{ps}+\Gamma_{ps})t}\left(\frac{1}{i\hbar}\right)^3 |p\rangle\rho^0_{mm}\langle s| \\
&\times\langle p|\boldsymbol{\mu}\cdot\hat{e}_1|m\rangle\langle m|\boldsymbol{\mu}\cdot\hat{e}_2|r\rangle\langle r|\boldsymbol{\mu}\cdot\hat{e}_3|s\rangle \\
&\times e^{i(\mathbf{k}_1+\mathbf{k}_2+\mathbf{k}_3)\cdot[\mathbf{r}(t)-\mathbf{v}t]} \\
&\times\int_{-\infty}^{t} d\xi_3\, e^{[i(\omega_{rs}-\omega_3)+\Gamma_{ps}-\Gamma_{pr}+i\mathbf{k}_3\cdot\mathbf{v}]\xi_3}A_3(\xi_3) \qquad (21.28)\\
&\times\int_{-\infty}^{\xi_3} d\xi_2\, e^{[i(\omega_{mr}-\omega_2)+\Gamma_{pr}-\Gamma_{pm}+i\mathbf{k}_2\cdot\mathbf{v}]\xi_2}A_2(\xi_2) \\
&\times\int_{-\infty}^{\xi_2} d\xi_1\, e^{[i(\omega_{pm}-\omega_1)+\Gamma_{pm}+i\mathbf{k}_1\cdot\mathbf{v}]\xi_1}A_1(\xi_1).
\end{aligned}
$$

We consider here the case where all resonant excitation pulses are short enough so that we can write

$$
\begin{aligned}
\int_{-\infty}^{\xi} d\xi'\, e^{[i(\omega_{ij}-\omega_n)+\Gamma_{lm}+i\mathbf{k}_n\cdot\mathbf{v}]\xi'}A(\xi') \\
\simeq e^{[\Gamma_{lm}+i\mathbf{k}_n\cdot\mathbf{v}]\xi_0}\int_{-\infty}^{\xi} d\xi'\, e^{i(\omega_{ij}-\omega_n)\xi'}A(\xi')
\end{aligned}
\qquad (21.29)
$$

$$
\begin{aligned}
\rho^{(3)}(\mathbf{v}, t) = &-\left(\frac{1}{i\hbar}\right)^3 |p\rangle\langle s|e^{-i\omega_{ps}t+i(\mathbf{k}_1+\mathbf{k}_2+\mathbf{k}_3)\cdot\mathbf{r}} \\
&\times e^{-i\mathbf{v}\cdot[\mathbf{k}_3(t-\xi_{30})+\mathbf{k}_2(t-\xi_{20})+\mathbf{k}_1(t-\xi_{10})]} \\
&\times e^{-\Gamma_{ps}(t-\xi_{30})-\Gamma_{pr}(\xi_{30}-\xi_{20})-\Gamma_{pm}(\xi_{20}-\xi_{10})} \\
&\times\langle p|\boldsymbol{\mu}\cdot\hat{e}_1|m\rangle\langle m|\boldsymbol{\mu}\cdot\hat{e}_2|r\rangle\langle r|\boldsymbol{\mu}\cdot\hat{e}_3|s\rangle \\
&\times\int_{-\infty}^{t} d\xi_3\, e^{i(\omega_{rs}-\omega_3)\xi_3}A_3(\xi_3) \qquad (21.30)\\
&\times\int_{-\infty}^{\xi_3} d\xi_2\, e^{i(\omega_{mr}-\omega_2)\xi_2}A_2(\xi_2) \\
&\times\int_{-\infty}^{\xi_2} d\xi_1\, e^{i(\omega_{pm}-\omega_1)\xi_1}A_1(\xi_1)\rho^0_{mm}.
\end{aligned}
$$

The overall density matrix for the Doppler-broadened molecular system has the form

$$
\rho^{(3)}(t) = \int_{-\infty}^{\infty} g(\mathbf{v})\rho^{(3)}(\mathbf{v}, t)\, d\mathbf{v} \qquad (21.31)
$$

where $g(\mathbf{v})$ is the normalized velocity distribution function. The nonlinear polarization is then given by

$$P^{(3)}(t) = N\langle\mu(t)\rangle \tag{21.32}$$
$$= N\mathrm{Tr}\big[\mu\rho^{(3)}(t)\big].$$

We find

$$p^{(3)}(t) = Ce^{i(k_1+k_2+k_3)\cdot r - i\omega_{ps}t}\int_{-\infty}^{\infty} g(v)e^{-i\theta(v,\,t)}\,dv \tag{21.33}$$
$$\times e^{-\Gamma_{ps}(t-\xi_{30})-\Gamma_{pr}(\xi_{30}-\xi_{20})-\Gamma_{pm}(\xi_{20}-\xi_{10})}$$

where

$$\theta(v,\,t) = v\cdot\big[k_3(t-\xi_{30}) + k_2(t-\xi_{20}) + k_1(t-\xi_{10})\big] \tag{21.34}$$

and C is a proportional constant. This shows that the nonlinear polarization has a wavevector $k_s = k_1 + k_2 + k_3$ and a frequency ω_{ps}. As is common in wave mixing problems, for $P^{(3)}$ to radiate efficiently, the radiating wavevector k_4 must satisfy the phase matching condition $k_4 = k_s$. Also, if $\theta(v) = 0$ at $t = t_e$, the integral in (21.33) becomes unity and, correspondingly, $P^{(3)}$ has the maximum amplitude. Both are conditions for the appearance of a photon echo, as already seen in the previous section for two-level systems. The last factor in (21.33) describes the decay of the coherent radiation resulting from damping of the excitations in the various time intervals. Finally, the proportional constant C in (21.33) is responsible for the intensity of the photon echo. Similar results can be obtained for other types of inhomogeneously broadened systems from a similar derivation.

The preceding formalism can be illustrated by considering the case of a two-level system under the resonant excitations of three successive pulses with $\omega_1 = \omega_2 = \omega_3 \simeq \omega_{10}$ but different wavevectors, as in Fig. 21.10. Because of the degenerate frequencies, there are four diagrams contributing to $\rho^{(3)}(t)$, leading to the expression

$$\rho^{(3)}(t) = -\left(\frac{1}{i\hbar}\right)^3 e^{i\omega_{10}t}\big(|1\rangle\rho_{00}\langle 0|\big)e^{-\Gamma_{10}(t-\xi_{30}+\xi_{20}-\xi_{10})}$$
$$\times\big[e^{-\Gamma_{00}(\xi_{30}-\xi_{20})} + e^{-\Gamma_{11}(\xi_{30}-\xi_{20})}\big]$$
$$\times\Big\{\langle 1|\mu\cdot\hat{e}_3|0\rangle\langle 0|\mu\cdot\hat{e}_2|1\rangle\langle 1|\mu\cdot\hat{e}_1|0\rangle e^{i(k_3-k_2+k_1)\cdot r}$$
$$\times\int_{-\infty}^{t} d\xi_3\, A_3(\xi_3)e^{i(\omega_{10}-\omega)\xi_3}\int_{-\infty}^{\xi_3} d\xi_2\, A_2^*(\xi_2)e^{-i(\omega_{10}-\omega)\xi_2}$$
$$\times\int_{-\infty}^{\xi_2} d\xi_1\, A_1(\xi_1)e^{i(\omega_{10}-\omega)\xi_1}\int_{-\infty}^{\infty} dv\, g(v)e^{-i\theta_a(v)} \tag{21.35}$$
$$+ \langle 0|\mu\cdot\hat{e}_1|1\rangle\langle 1|\mu\cdot\hat{e}_2|0\rangle\langle 0|\mu\cdot\hat{e}_3|1\rangle e^{i(k_3+k_2-k_1)\cdot r}$$
$$\times\int_{-\infty}^{t} d\xi_3\, A_3(\xi_3)e^{i(\omega_{10}-\omega)\xi_3}\int_{-\infty}^{\xi_3} d\xi_2\, A_2(\xi_2)e^{i(\omega_{10}-\omega)\xi_2}$$
$$\times\int_{-\infty}^{\xi_2} d\xi_1\, A_1^*(\xi_1)e^{-i(\omega_{10}-\omega)\xi_1}\int_{-\infty}^{\infty} dv\, g(v)e^{-i\theta_\beta(v)}\Big\}.$$

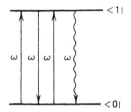

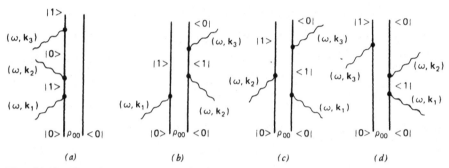

(a) (b) (c) (d)

Fig. 21.10 Transient four-wave mixing in a two-level system resulting from three successive resonant pulse excitations of the same frequency but different wavevectors. The process is generally described by the four separate diagrams in (a), (b), (c), and (d).

The first term in the brackets comes from diagrams a and b, and the second term from c and d in Fig. 21.10. They have different wavevector combinations, and hence different $\theta(\mathbf{v})$:

$$\theta_\alpha(\mathbf{v}) = \left[\mathbf{k}_3(t - \xi_{30}) - \mathbf{k}_2(t - \xi_{20}) + \mathbf{k}_1(t - \xi_{10})\right]\cdot\mathbf{v}$$

and (21.36)

$$\theta_\beta(\mathbf{v}) = \left[\mathbf{k}_3(t - \xi_{30}) + \mathbf{k}_2(t - \xi_{20}) - \mathbf{k}_1(t - \xi_{10})\right]\cdot\mathbf{v}.$$

The phase matching conditions for coherent radiation from the two terms are, respectively,

$$\mathbf{k}_4 = \mathbf{k}_s = \mathbf{k}_3 - \mathbf{k}_2 + \mathbf{k}_1$$

and (21.37)

$$\mathbf{k}_4 = \mathbf{k}_s = \mathbf{k}_3 + \mathbf{k}_2 - \mathbf{k}_1.$$

Figure 21.11 presents the possible arrangements of the $\mathbf{k}$'s to satisfy (21.37).

It is interesting to see that (21.35) actually describes a number of different photon echo phenomena in two-level systems. To have a photon echo appear

$$k_s = k_3 - k_2 + k_1$$

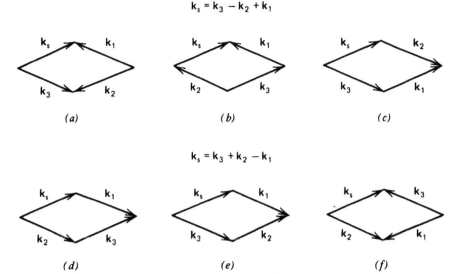

$$k_s = k_3 + k_2 - k_1$$

Fig. 21.11 Possible phase-matching arrangements for the degenerate transient four-wave mixing process described in Fig. 21.10.

after the pulsed excitations, we must have $\theta_\alpha(\mathbf{v})$ or $\theta_\beta(\mathbf{v})$ vanishing at $t = t_e \geq \xi_{30}$. [In general, it is possible to have a photon echo with a reduced amplitude if $\theta(\mathbf{v})$ is a minimum instead of zero at $t = t_m \geq \xi_{30}$,[16] but we do not discuss such a case here.] From (21.36), we have for $\theta_\alpha(\mathbf{v}) = 0$,

$$t_e = \mathbf{k}_s \cdot \frac{(\mathbf{k}_3\xi_{30} - \mathbf{k}_2\xi_{20} + \mathbf{k}_1\xi_{10})}{k_s^2} \qquad (21.38)$$

and for $\theta_\beta(\mathbf{v}) = 0$,

$$t_e = \mathbf{k}_s \cdot \frac{(\mathbf{k}_3\xi_{30} + \mathbf{k}_2\xi_{20} - \mathbf{k}_1\xi_{10})}{k_s^2}. \qquad (21.39)$$

The condition $t_e \geq \xi_{30}$ can be satisfied only for the phase matching arrangements a, b, d, e, in Fig. 21.11. The echo created at $t = t_e$ after the successive pulse excitations at ξ_{10}, ξ_{20}, and ξ_{30} is known as a three-pulse stimulated echo. The nonperturbative derivation of the stimulated echoes has been worked out by Fujita et al.[17] It yields the same result given here in the perturbation limit. As seen in Fig. 21.11, cases a and b generate a backward stimulated echo ($\mathbf{k}_s$ in a direction more or less opposite to $\mathbf{k}_1$ and $\mathbf{k}_2$), and cases d and e generate a forward echo. In a condensed matter, the echo condition is somewhat different. Cases a, b, and c are not allowed, but cases d, e, and f are.[17]

We can let the second and third excitation pulses merge into one with $\xi_{20} = \xi_{30}$, $\mathbf{k}_2 = \mathbf{k}_3$, and $\mathscr{E}_2 = \mathscr{E}_3$, and obtain the two-pulse echo described in the previous section. We find from (21.38) and (21.39) that the echo exists only for case d or e in Fig. 21.11 with $\mathbf{k}_s = 2\mathbf{k}_2 - \mathbf{k}_1$, and it appears at $t_e - \xi_{30} = \xi_{20} - \xi_{10}$ in the forward direction. From (21.35), the echo intensity, proportional to $|\rho^{(3)}(t_e)|^2$, decays with $\exp[-2\Gamma_{10}(t_e - \xi_{10})]$. (Note that $\Gamma_{10} = 1/T_2$.) These are the same results obtained in the previous section for the two-pulse photon echo in a two-level system with the $\pi/2$ and π pulse excitations. This proves that the $\pi/2$ and π excitation pulses are not necessary for the observation of the photon echo.

We can also imagine that the first and second pulses in the three-pulse excitation sequence merge into one to form a two-pulse excitation. With $\xi_{10} = \xi_{20}$ and $\mathbf{k}_1 = \mathbf{k}_2$ in (21.38) and (21.39), it is seen that the coherent emission occurs at $t_e = \xi_{30}$. The output therefore overlaps with and modifies the input pulse at ξ_{30}. From (21.35), one can see that it is in the form of a free induction decay with the emission field proportional to

$$\left[e^{-\Gamma_{00}(\xi_{30} - \xi_{20})} + e^{-\Gamma_{11}(\xi_{30} - \xi_{20})} \right] \times \left[e^{-\Gamma_{10}(t - \xi_{30})} \times \int_{-\infty}^{\infty} d\mathbf{v}\, g(\mathbf{v}) e^{i\mathbf{k}_3 \cdot \mathbf{v}(t - \xi_{30})} \right] e^{-i\omega_{10}t}.$$

The first bracket describes the amplitude decay with the pulse separation $(\xi_{30} - \xi_{20})$, while the second gives the time dependence of the free induction decay for $t \geq \xi_{30}$. The result here shows that, in general, a free induction decay signal is expected after the second excitation pulse in Fig. 21.7. One can also let the three excitation pulses merge into one with $\xi_{10} = \xi_{20} = \xi_{30}$ and $\mathbf{k}_1 = \mathbf{k}_2 = \mathbf{k}_3$. The coherent emission in this case should be the free induction decay appearing at the end of the excitation pulse.

The approach here can be easily extended to transient four-wave mixing in three- and four-level systems. The results should yield many transient coherent phenomena including free induction decays and various kinds of photon echoes (trilevel echoes, grating echoes, Raman echoes, etc.).[18] The derivation can even be extended to the more general case of coherent transient effects in n-level systems with m excitation pulses (transient m-wave mixing). Since the basic principle is the same, we do not dwell further on the problem, but refer the readers to Ref. 14.

The main application of free induction decays, photon echoes, and transient four-wave mixing is to measure the longitudinal (population) relaxation of a prescribed state and the transverse (dephasing) relaxation between a pair of states. That these relaxations depend sensitively on the interactions of the excited system with its surroundings can be used to investigate the interaction mechanisms on the microscopic scale[16, 19] as in the magnetic resonance case. With the help of tunable lasers and frequency-switching methods, such optical transient measurements have become increasingly common.

21.5 ADIABATIC FOLLOWING

We now come back to the two-level systems with the pseudo-dipole description, and discuss the adiabatic following phenomenon which also has an

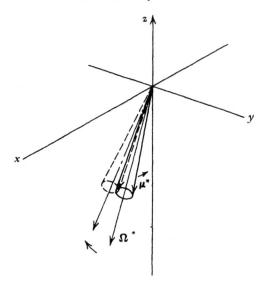

Fig. 21.12 Schematic describing the adiabatic following process. As $\Omega^* = -(\hbar/\gamma)\mathbf{E}^*_{\text{eff}}$ in the rotating coordinates varies slowly, the pseudo-dipole μ^* precessing around Ω^* follows it.

analog in magnetic resonance. We assume in this case that the dipole precessing frequency Ω^* in the rotating frame given by (21.13) is much larger than R, the rate of change of E_{eff}. We also assume $R^{-1} \leq T_1, T_2$ so that damping is ineffective in destroying the coherent effects. Then, as $\mathscr{E}$ varies in time, and Ω^* changes accordingly, the precessing dipole μ^* should follow Ω^* adiabatically, as sketched in Fig. 21.12. The variation of Ω^* can be effected by varying either $\mathscr{E}$ or $\omega_0 - \omega$.

Let us consider the case where the exciting field is initially far from resonance with $\omega_0 - \omega < 0$ and $|\omega_0 - \omega| \gg (\gamma/\hbar)|2\mathscr{E}|$. In this case, the Ω^* vector points downward, making a very small angle $\alpha \simeq |(\gamma/\hbar)2\mathscr{E}/(\omega_0 - \omega)|$ with the $-\hat{z}$-axis, and the pseudo-dipole μ^* precesses around Ω^* with the cone angle α. Now, if $\omega_0 - \omega$ is gradually increased from the initial negative value to a positive value far off resonance, Ω^* changes accordingly from its nearly downward position to a nearly upward position, carrying the precessing μ^* with it. Since μ_z is directly proportional to the population difference between the two levels, the inversion of μ^* corresponds physically to an inversion of the population to the upper level. This is known as adiabatic inversion. With this process, practically all the population in the lower level can be excited into the upper level. Adiabatic inversion is a well-known method in magnetic resonance to create an inverted population. In the optical case,[20] it has also been demonstrated through the detection of a transient stimulated gain effected by the inverted population.[21]

The detailed calculation of adiabatic following from the solution of the Bloch equation has been worked out by Crisp.[22] An approximate expression for μ^* can, however, be obtained from Fig. 21.12. Let the three normalized components of μ^* be

$$u = \frac{\mu_{x'}^*}{\gamma}, \quad v = \frac{\mu_{y'}^*}{\gamma}, \quad \text{and} \quad w = \frac{\mu_z}{\gamma} \qquad (21.40)$$

with $u^2 + v^2 + w^2 = 1$. Then, in the limit of a negligibly small angle $\alpha^* = \alpha$ between μ^* and Ω^*, μ^* is parallel to Ω^*, and we find

$$u = -\frac{(\gamma/\hbar)2\mathscr{E}}{\left[(\omega - \omega_0)^2 + (\gamma/\hbar)^2 4\mathscr{E}^2\right]^{1/2}},$$

$$v = 0,$$

and $\qquad\qquad\qquad\qquad\qquad\qquad\qquad\qquad\qquad\qquad (21.41)$

$$w = \frac{(\omega - \omega_0)}{\left[(\omega - \omega_0)^2 + (\gamma/\hbar)^2 4\mathscr{E}^2\right]^{1/2}}.$$

However, α^* cannot be exactly zero, because then $d\mu^*/dt$ would vanish identically. We can insert the solution of (21.41) into the Bloch equation and use iteration to find the next-order approximation. The corrections on u and w are negligible, but the one on v is important because $v = 0$ in the first approximation. As one would expect, v will appear to be proportional to $[d(\omega - \omega_0)/dt]/[(\omega - \omega_0)^2 + (\gamma/\hbar)^2 4\mathscr{E}^2]$ if $(\omega - \omega_0)$ is varied, or proportional to $[(\gamma/\hbar)2d\mathscr{E}/dt]/[(\omega - \omega_0)^2 + (\gamma/\hbar)^2 4\mathscr{E}^2]$ if $\mathscr{E}$ is varied. The adiabatic following picture has been used by Grischkowsky et al. to discuss a number of coherent near-resonant phenomena.[23]

21.6 SELF-INDUCED TRANSPARENCY

We have so far assumed in the discussion that the exciting pulse propagation in the medium is not affected by the transient response of the medium to the field. This is a good approximation if the medium is "thin" such that no appreciable distortion of the exciting pulse can happen in propagating through the medium. In a "thick" medium, however, the pulse deformation can be appreciable. McCall and Hahn[24] found that if the pulse has an area $\theta \equiv \int_{-\infty}^{\infty}(\gamma/\hbar)2\mathscr{E}dt$ equal to $2n\pi$ with n being an integer, and has a certain definite pulse shape, then it can propagate through the resonant (ordinarily absorbing) medium without any attenuation and change of pulse shape, as long as the T_1 and T_2 relaxations are negligible. This is called self-induced transparency. Since it comes out of the pulse propagation effect, it has no analog in magnetic resonance.

The basic idea of self-induced transparency can be seen from the pseudo-dipole picture. With a $2n\pi$ pulse excitation, μ will precess around $\mathscr{E}$ over full circles and end up in its original position. Therefore, since the medium is the same before and after the pulse, it absorbs no net energy from the pulse. However, during the pulse, it does absorb and emit photons and redistributes energy in the pulse. Consequently, the transmitted pulse appears to be altered in shape unless it already has the proper shape. As we shall see in the following discussion, the proper field envelope for a 2π pulse is of a hyperbolic secant form. In propagating through the medium, the pulse is apparently delayed because the medium absorbs energy from the leading part of the pulse and deposits it back to the tail part.

Formally, self-induced transparency is described by the wave equation (21.15) coupled with the Bloch equation (21.11), in which we have neglected the T_1 and T_2 relaxation terms. In terms of u, v, w defined in (21.40), the Bloch equation becomes

$$\frac{\partial u}{\partial t} = (\Delta\omega)v, \quad \frac{\partial v}{\partial t} = -(\Delta\omega)u + \left(\frac{\gamma}{\hbar}\right)2\mathscr{E}w, \quad \text{and} \quad \frac{\partial w}{\partial t} = -\left(\frac{\gamma}{\hbar}\right)2\mathscr{E}v$$

$$(21.42)$$

where $\Delta\omega = \omega - \omega_0$. With the slowly varying amplitude approximation, the wave equation (21.15) for the circularly polarized wave $\mathbf{E} = \hat{e}_+\mathscr{E}(z,t) \times \exp(ikz - i\omega t)$ reduces to an amplitude equation

$$\frac{\partial\mathscr{E}}{\partial z} + \frac{1}{c}\frac{\partial\mathscr{E}}{\partial t} = \frac{2\pi i\omega}{c}N\langle p_+(z,t)\rangle. \tag{21.43}$$

Assume that the resonance is inhomogeneously broadened with a frequency distribution function $g(\Omega - \omega_0)$, and let $g(\Omega - \omega_0)$ be symmetric and $\omega = \omega_0$. Then we can write

$$\langle p_+(z,t)\rangle = \int_{-\infty}^{\infty} g(\Delta\omega)[u + iv]d(\Delta\omega). \tag{21.44}$$

We consider here the solution with real $\mathscr{E}$. Equation (21.43) then becomes

$$\frac{\partial\mathscr{E}}{\partial z} + \frac{1}{c}\frac{\partial\mathscr{E}}{\partial t} = -\frac{2\pi\omega N}{c}\int_{-\infty}^{\infty} g(\Delta\omega)vd(\Delta\omega). \tag{21.45}$$

The coherent pulse propagation effect is described by (21.45) coupled with (21.42).

We first ask the question whether a pulse can propagate through the resonant medium without change in shape. In this case, if V_p is the pulse velocity, $\partial\mathscr{E}/\partial z = -V_p^{-1}\partial\mathscr{E}/\partial t$, then, from (21.45), we have

$$\frac{\partial}{\partial t}\mathscr{E}\left(t - \frac{z}{V_p}\right) = \frac{2\pi\omega N/c}{V_p^{-1} - c^{-1}}\int_{-\infty}^{\infty} g(\Delta\omega)v\left(\Delta\omega, t - \frac{z}{V_p}\right)d(\Delta\omega).$$

$$(21.46)$$

It can be shown in general that $v(\Delta\omega, t - z/V_p) = v(0, t - z/V_p)f(\Delta\omega)$, so that[24]

$$\int_{-\infty}^{\infty} g(\Delta\omega)v\left(\Delta\omega, t - \frac{z}{V_p}\right)d(\Delta\omega) = v\left(0, t - \frac{z}{V_p}\right)S \qquad (21.47)$$

where

$$S = \int_{-\infty}^{\infty} g(\Delta\omega)f(\Delta\omega)d(\Delta\omega).$$

For our purpose here, we can imagine that (21.47) is obtained by assuming $g(\Delta\omega)$ to be effectively a δ function. From the precessing pseudo-dipole picture, we have

$$v\left(0, t - \frac{z}{V_p}\right) = \sin\theta \qquad (21.48)$$

where θ is a function of $(t - z/V_p)$, with

$$\theta = \int_{-\infty}^{t} \left(\frac{\gamma}{\hbar}\right)2\mathcal{E}\left(t' - \frac{z}{V_p}\right)dt' \quad \text{and} \quad \frac{\partial\theta}{\partial t} = \left(\frac{\gamma}{\hbar}\right)2\mathcal{E}.$$

Equation (21.46) becomes

$$\frac{\partial\mathcal{E}}{\partial t} = \left(\frac{\hbar}{2\gamma}\right)\frac{1}{\tau^2}\sin\theta \quad \text{or} \quad \frac{\partial^2\theta}{\partial t^2} = \frac{1}{\tau^2}\sin\theta \qquad (21.49)$$

with $\tau^2 = (V_p^{-1} - c^{-1})/(4\pi\omega N\gamma\, S/c\hbar)$. This is in the form of the well-known pendulum equation in mechanics, the solution of which can be found as follows. The first equation in (21.49) can be transformed into

$$\frac{\partial\mathcal{E}}{\partial t} = \left(\frac{\gamma}{\hbar}\right)\frac{\partial\mathcal{E}^2}{\partial\theta} = \left(\frac{\hbar}{2\gamma}\right)\frac{1}{\tau^2}\sin\theta. \qquad (21.50)$$

With the initial condition $\mathcal{E}^2 = 0$ at $\theta = 0$, (21.50) has the solution

$$\mathcal{E} = \left(\frac{\hbar}{\gamma\tau}\right)\sin\frac{\theta}{2}. \qquad (21.51)$$

This allows (21.50) to be rewritten in the form

$$\frac{\partial\mathcal{E}}{\partial t} = \frac{1}{\tau}\mathcal{E}\left[1 - \left(\frac{\gamma\tau}{\hbar}\mathcal{E}\right)^2\right]^{1/2}, \qquad (21.52)$$

which can be readily integrated to yield

$$\mathscr{E}\left(t - \frac{z}{V_p}\right) = \frac{\hbar}{\gamma\tau}\,\mathrm{sech}\left[\frac{1}{\tau}\left(t - \frac{z}{V_p}\right)\right]. \tag{21.53}$$

The corresponding field envelope has an area

$$A = \theta(t \to \infty)$$
$$= \int_{-\infty}^{\infty}\left(\frac{\gamma}{\hbar}\right)2\mathscr{E}\,dt = 2\pi. \tag{21.54}$$

This solution shows that a 2π hyperbolic secant pulse can propagate in a resonant medium without any attenuation and change of pulse shape.

We next consider how an arbitrary pulse gets deformed in propagating through a resonant medium. With

$$A = \lim_{t \to \infty}\int_{-\infty}^{t}\left(\frac{\gamma}{\hbar}\right)2\mathscr{E}\,dt', \quad \mathscr{E}(t \to \pm\infty) = 0, \quad \text{and} \quad \int_{-\infty}^{\infty}\left(\frac{\partial\mathscr{E}}{\partial t'}\right)dt' = 0,$$

the integration of (21.45) from $t = -\infty$ to $t \to \infty$ yields

$$\frac{\partial A}{\partial z} = -\frac{4\pi N\omega}{c}\left(\frac{\gamma}{\hbar}\right)\lim_{t \to \infty}\int_{-\infty}^{t}dt'\int_{-\infty}^{\infty}d(\Delta\omega)g(\Delta\omega)v(\Delta\omega, z, t')$$
$$= -\frac{4\pi N\omega}{c}\left(\frac{\gamma}{\hbar}\right)\lim_{t \to \infty}\int_{-\infty}^{\infty}d(\Delta\omega)g(\Delta\omega)\frac{u(\Delta\omega, z, t)}{\Delta\omega}. \tag{21.55}$$

To solve (21.55), the first step is to carry out the integration. We notice that at $t = \infty$, the pulse is over, and the pseudo-dipole must be precessing around $\hat{z}$. Therefore, $u(\Delta\omega, z, t)$ should have a sinusoidal variation with frequency $\Delta\omega$, and the integral can get a contribution from the integrand only in the region $\Delta\omega \sim 0$. Realizing that the pseudo-dipole μ^* for $\Delta\omega \sim 0$ is tilted away from the $-\hat{z}$ axis by an angle A at the end of the pulse (say, $t = t_0$) with $v(t_0) = \sin A$, and $u(t_0) = 0$, we find, from (21.42), $u(\Delta\omega, z, t) = \sin A \sin[\Delta\omega(t - t_0)]$ for $t \geq t_0$. We then have

$$\lim_{t \to \infty}\int_{-\infty}^{\infty}d(\Delta\omega)g(\Delta\omega)\frac{u(\Delta\omega, z, t)}{\Delta\omega}$$
$$\simeq g(0)\sin A \lim_{t \to \infty}\int_{-\infty}^{\infty}d(\Delta\omega)\frac{\sin[\Delta\omega(t - t_0)]}{\Delta\omega} \tag{21.56}$$
$$= \pi g(0)\sin A.$$

Equation (21.55) now reduces to the simple form

$$\frac{\partial A}{\partial z} = -\frac{\alpha}{2}\sin A \tag{21.57}$$

with $\alpha = 4\pi^2 N\omega\gamma g(0)/c\hbar$, the solution of which is

$$A(z) = 2\tan^{-1}\left\{\left[\tan\tfrac{1}{2}A(0)\right]e^{-\alpha z/2}\right\}. \qquad (21.58)$$

This describes how the pulse area changes as the pulse propagates in the medium. This is plotted in Fig. 21.13a. The curve shows that as $z \to \infty$, the pulse area A approaches $2n\pi$ if $(2n - 1)\pi < A(0) < (2n + 1)\pi$, where n is an integer. This means that through exchange of energy with the medium, the input pulse gets deformed and stabilized into a pulse area which is a multiple of 2π. Indeed, numerical calculations show that for $\pi < A(0) < 3\pi$, as z increases, the pulse is gradually deformed into a 2π hyperbolic secant field envelope predicted in (21.53). An example is seen in Fig. 21.13b. For $(2n - 1)\pi < A(0) < (2n + 1)\pi$, the pulse splits and stabilizes into n 2π-pulses after propagating over a sufficiently long distance. For $A(0) < \pi$, the pulse simply decays away in the propagation. In fact, for small $A(0)$, (21.57) and (21.58) can be shown to reduce to the forms for linear propagation as they should, and α can be identified as the linear absorption coefficient of the medium.

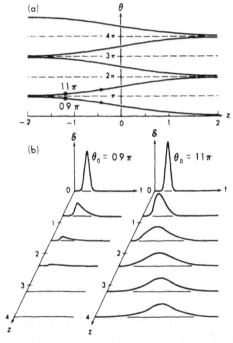

Fig. 21.13 Pulse area plots of self-induced transparency illustrating the area theorem. (a) $\alpha > 0$, the pulse area evolves in the direction of increasing distance z toward the nearest even multiple of π. The entry face of the medium may be at any value of z. (b) Computer plots of evolution of input $A(0) = 0.9\pi$ and $A(0) = 1.1\pi$ pulses with distance. (After Ref. 24.)

As seen from the foregoing discussion, self-induced transparency is characterized by reduced absorption, pulse delay ($V_p < c$), pulse deformation, and pulse splitting. As an example,[25] for the $5s \rightarrow 5p$ transition in Rb, $\gamma \sim 4 \times 10^{-18}$ esu; to form a 2π pulse with a pulsewidth of 10 nsec only requires a field amplitude of $\mathscr{E} \sim 0.1$ esu or a peak intensity of $I \sim 2$ W/cm^2. Using the expressions of τ^2 and α in (21.49) and (21.57), the pulse velocity can be written as

$$V_p = \left[1 + \frac{S\alpha c}{\pi g(0)} \tau^2 \right]^{-1} c. \qquad (21.59)$$

If $\alpha \sim 10^{-2}$ cm^{-1}, $S \sim 1$, $g(0) = 1/\Delta\omega \sim 10^{-7}$ sec and $\tau \sim 10^{-8}$ sec, we find

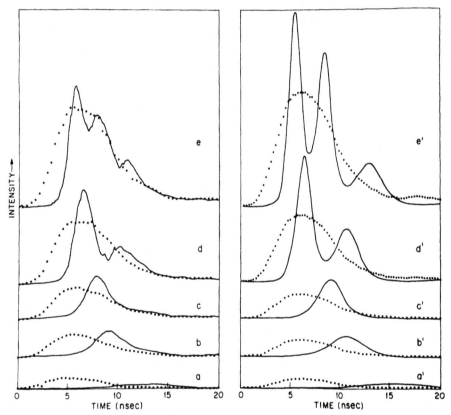

Fig. 21.14 Input and output pulses observed in a self-induced transparency experiment (curves on the left) and calculated from theory (curves on the right). The dotted curves depict input pulses and the full curves depict the corresponding output pulses after propagation through a length of $5/\alpha$. Curves a through e denote pulses with areas of slightly less than π, 2π, between 2π and 3π, slightly less than 5π, and approximately 6π, respectively. Pulse breakup of the pulses with areas above 3π, and the absence of breakup of the pulses with smaller areas, is in excellent agreement with the predictions of the theory. (After Ref. 25.)

$V_p \sim c/10^3$. In passing through a medium 3 cm long, the pulse undergoing self-induced transparency will take 10^{-7} sec, while a normal pulse takes only 10^{-10} sec. The special features of self-induced transparency have actually been experimentally demonstrated, although the experiments are usually complicated by the existence of degenerate states and transverse variation of the laser intensity. The pulse delay and pulse breakup are often the more convincing evidence of the presence of self-induced transparency. An example is seen in Fig. 21.14. While the effect of self-induced transparency is. certainly very intriguing, its applications to either science or technology have not yet been seriously explored.

21.7 SUPERFLUORESCENCE (SUPERRADIANCE)

We saw that a collection of dipoles oscillating in phase should radiate coherently. This is described in the pseudo-dipole picture for effective two-level systems by a set of coinciding pseudo-dipoles μ precessing together around the $\hat{z}$ axis, as in Fig. 21.15. The coherent radiation dies away when the dipole dephasing sets in. It can be seen from (21.15) for example, that the coherent output is proportional to the square of the number of dipoles per unit volume, N^2, while the incoherent emission is proportional to N. Therefore, for $N \gg 1$, the coherent radiation is much stronger than its incoherent counterpart and is sometimes called superradiation.[26]

In the coherent transient effects discussed earlier, the coherent radiation comes from the collective dipole oscillation initially set up by coherent excitation, that is, the pseudo-dipoles are initially tilted away from the $\hat{z}$ axis.

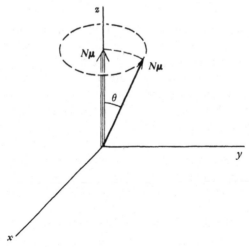

Fig. 21.15 Source of coherent radiation: precessing giant dipole $N\mu$ formed by N dipoles precessing in phase.

Both the population in the excited state (longitudinal excitation) and the average oscillating dipole moment (transverse excitation) are nonvanishing. The radiation output can then be calculated classically from (21.15) with $\langle \mathbf{p} \rangle$ given. Here, however, we consider a different case. We assume a collection of two-level systems whose initial population distribution is completely inverted, as described by the pseudo-dipoles pointing upward along $+\hat{z}$ in Fig. 21.15. In the absence of an external field, these two-level systems are initially uncorrelated and only radiate spontaneously. Subsequently, however, through interaction of radiation with the two-level systems, the latter may become correlated and radiate together coherently. In the pseudo-dipole picture of Fig. 21.15, we can imagine that because of quantum fluctuations the pseudo-dipoles are not strictly along $+\hat{z}$. They can then precess around $\hat{z}$ and the emission tilts the pseudo-dipole further away from $+\hat{z}$. In the meantime, correlation is being established among the dipoles through the radiation field, which acts to align the precessing pseudo-dipoles in phase. Eventually the pseudo-dipoles are completely aligned and tilted significantly from $+\hat{z}$. In that state, they radiate together coherently in the classical sense. The problem that has attracted a great deal of interest is how, in a more precise way, the radiation from such a collection of two-level systems with inverted population changes from the initial spontaneous emission to the final coherent or superradiant emission. The process is now known as superfluorescence.

Dicke first studied the problem assuming a collection of spins ($s = \frac{1}{2}$) in a volume with dimensions much smaller than the wavelength of the field as is in the case of magnetic resonance.[26] Using the quantum description for an N spin-$\frac{1}{2}$ system, he found that when the system is in the $|r, m\rangle$ state, the emission rate is

$$I = (r + m)(r - m + 1)I_0 \tag{21.60}$$

where r is the spin quantum number of the total spin, m is the corresponding magnetic quantum number, and I_0 is the spontaneous emission rate of a single spin. It is seen that for $r = m = r_{max} = \frac{1}{2}N$, we have $I = NI_0$. This indicates that when all spins are inverted, they radiate incoherently (or spontaneously). If, through radiation, the system drops to the $|r = \frac{1}{2}N, m\rangle$ state with $|m| \ll r$, then the emission rate becomes $I \simeq \frac{1}{4}N^2 I_0$, which clearly shows that the emission has become coherent.

Superfluorescence is characterized by a number of special features. First, there is a time delay between the initial set-up of the inverted population and the appearance of superradiation. This time delay corresponds to the time it takes for the system to establish correlation between atoms (or two-level systems). Then, superradiation should appear in the form of a pulse as the emission is over when all the stored energy in the system is extracted. Finally, because of the much faster superradiant emission rate (proportional to N^2 instead of N), the effective emission lifetime of the system becomes much shorter (proportional to $1/N$). Establishment of correlation between atoms through emission can be visualized as follows. Radiation from an atom can induce dipole oscillation on the neighboring atoms. These induced dipoles in turn create a reaction field on the original atom and influence the radiation from that atom.

Superfluorescence was first observed in magnetic resonance by Bloembergen and Pound.[27] In optics, the same effect was also observed by Skribanowitz et al.,[28] and later studied more carefully and extensively by Gibbs et al.[29] The optical case is actually more complicated than described here. The sample dimensions are always much larger than the wavelength, so that the propagation effect must be taken into account. Then the transverse intensity distribution of the beam also affects the output and makes the analysis more complex. The angular distribution of optical superfluorescence is determined by the geometry of the active medium. In the case of a long cylinder, the output appears predominantly in the forward and backward directions with equal intensities. This is sketched in Fig. 21.16. In the same figure, the time variations of the excitation, the incoherent output, and the superradiant pulse, as actually observed in the experiment of Skribanowitz et al. on HF,[28] are also shown.

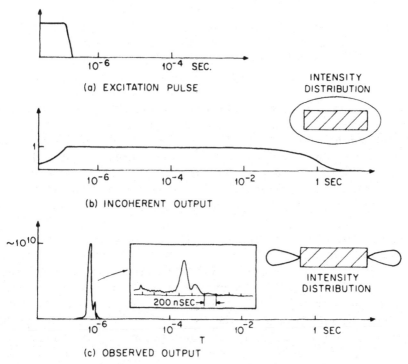

Fig. 21.16 Time variations of (a) the excitation pulse preparing the system of HF with a nearly complete population inversion between two adjacent rotational levels in the $v = 1$ state, (b) the spontaneous emission, and (c) the coherent superradiation. The insets describe the angular distributions of (b) and (c). (After Ref. 28.)

To explain their observation, Skribanowitz et al.[28,30] used the classical theory of radiation by assuming that the polarization fluctuations could initially lead to a finite tilt angle θ_0 for the pseudo-dipole in Fig. 21.15. Superradiation including the propagation effect was calculated by solving the wave equation (21.15) together with the Bloch equation (21.9). Agreement between theory and experiment, however, was limited by the complications in the actual experiment and simplifications in the theoretical calculation. Bonifacio and Lugiato[31] discussed superfluorescence from the fully quantum mechanical point of view using a mean-field approach, and set up explicitly the conditions for superfluorescence. However, they neglected the propagation effect and hence the spatial variation of the field envelope. Glauber and Haake[32] and Polder et al.[33] later constructed the more correct theory of superfluorescence including the propagation effect. Their theories provide a quantitative discussion on the quantum fluctuations and initiation of super-fluorescence. The only limitation of the theories is that they are one-dimensional and neglect the transverse variation of the field and polarization.

It can be imagined that quantum fluctuations and spontaneous emission first set up an initial tilt angle θ_0 for the pseudo-dipole, and then superradiation is built up in the classical manner. The value of θ_0 determines the time delay t_D between the initial set-up of complete population inversion and the peak of the superradiant pulse. A larger θ_0 leads to a shorter t_D. Being originated from quantum fluctuations, θ_0 should also fluctuate, and so should t_D. The mean value of θ_0 reflects the average strength of the quantum fluctuations. It is inversely proportional to $\sqrt{N}$ since, in the more dense medium, correlation between atoms is more easily established and a smaller θ_0 is needed to change from the quantum to the classical emission regime. The average value of θ_0 can be measured experimentally by injecting a pulse with an area $\theta = \int_{-\infty}^{\infty} (\gamma/\hbar) 2\mathscr{E} \, dt$ into the sample.[34] If $\theta < \theta_0$, the injection will have no effect because the initiation of superradiation is still dominated by quantum fluctuations in the medium. If $\theta > \theta_0$, the injection will help the initiation of superradiation, and the time delay t_D will become shorter. The experiment on Cs vapor with a density of $N = 2 \times 10^8$ cm^{-3} gives an average value of $\theta_0 = 5 \times 10^{-4}$ rad.[34]

Theoretical calculations of superfluorescence generally assume a number of simplifications. To test the theories quantitatively, experiments must be designed to simulate as well as possible the conditions laid out in the theories. Gibbs et al carried out such an experiment using Cs vapor as the sample.[29] Optical pumping from $6^2S_{1/2}$ to $7^2P_{3/2}$ by a 2-nsec laser pulse set up the nearly complete population inversion between $7^2P_{3/2}$ and $7^2S_{1/2}$. A 2.8-kOe magnetic field removed the degeneracy of the transition and made the inverted atomic system a good effective two-level system. The atomic density of the vapor and the sample length were selected so that the inequalities between various characteristic times necessary for clear observation of superfluorescence were satisfied. Then, the experimental results indeed showed fairly good agreement with the theoretical predictions. More details on the experiments and the relevant theories can be found in the review articles in Ref. 35.

REFERENCES

1   R. P. Feynman, F. L. Vernon, and R. W. Hellwarth, *J. Appl. Phys.* **28**, 49 (1957).

2   F. Bloch, *Phys. Rev.* **70**, 460 (1946).

3   See, for example, H. Goldstein, *Classical Mechanics* (Addison Wesley, Reading, MA), 2nd ed., p. 132.

4   I. I. Rabi, *Phys. Rev.* **51**, 652 (1937).

5   H. C. Torrey, *Phys. Rev.* **76**, 1059 (1949).

6   C. L. Tang and H. Statz, *Appl. Phys. Lett.* **10**, 145 (1968); G. B. Hocker and C. L. Tang, *Phys. Rev. Lett.* **21**, 591 (1969).

7   R. G. Brewer and R. L. Shoemaker, *Phys. Rev. Lett.* **27**, 631 (1971); *Phys. Rev. A* **6**, 2001 (1972).

8   E. L. Hahn, *Phys. Rev.* **77**, 297 (1950).

9   R. G. DeVoe and R. G. Brewer, *Phys. Rev. Lett.* **50**, 1269 (1983).

10 S. C. Rand, A. Wokaun, R. G. DeVoe, and R. G. Brewer, *Phys. Rev. Lett.* **43**, 1868 (1979).

11 E. L. Hahn, *Phys. Rev.* **80**, 580 (1950).

12 N. A. Kurnit, I. D. Abella, and S. R. Hartmann, *Phys. Rev. Lett.* **13**, 567 (1964); I. D. Abella, N. A. Kurnit, and S. R. Hartmann, *Phys. Rev.* **141**, 391 (1966); I. D. Abella, in E. Wolf, ed., *Progress in Optics* (North-Holland Publishing Co., Amsterdam, 1969), Vol. 7; S. R. Hartmann, in R. Glauber, ed., *Proceedings of the International School of Physics, Enrico-Fermi Course XLII* (Academic Press, New York, 1969), p. 532.

13 R. M. MacFarlane and R. M. Shelby, *Opt. Commun.* **39**, 169 (1981).

14 P. X. Ye and Y. R. Shen, *Phys. Rev. A* **25**, 2083 (1982).

15 T. K. Yee and T. K. Gustafson, *Phys. Rev. A* **18**, 1597 (1978).

16 T. M. Mossberg, R. Kachru, S. R. Hartmann, and A. M. Flusberg, *Phys. Rev. A* **20**, 1976 (1979).

17 M. Fujita, H. Nakatsuka, H. Nakanishi, and M. Matsuoka, *Phys. Rev. Lett.* **42**, 974 (1979).

18 Hartmann and co-workers have also developed other diagrammatic techniques to predict and discuss various types of photon echoes. T. M. Mossberg and S. R. Hartmann, *Phys. Rev. A* **23**, 1271 (1981); R. Beach and S. R. Hartmann, *Phys. Rev.* **25**, 2658 (1982).

19 P. R. Berman, J. M. Levy, and R. G. Brewer, *Phys. Rev. A* **11**, 1668 (1975).

20 E. B. Treacy, *Phys. Lett.* **27A**, 421 (1968).

21 M. M. T. Loy, *Phys. Rev. Lett.* **32**, 814 (1974).

22 M. D. Crisp, *Phys. Rev. A* **8**, 2128 (1973).

23 D. Grischkowsky, *Phys. Rev. Lett.* **24**, 866 (1970); D. Grischkowsky and J. A. Armstrong, *Phys. Rev. A* **6**, 1566 (1972); D. Grischkowsky, *Phys. Rev. A* **7**, 2096 (1973); D. Grischkowsky, E. Courtens, and J. A. Armstrong, *Phys. Rev. Lett.* **31**, 422 (1973).

24 S. L. McCall and E. L. Hahn, *Phys. Rev. Lett.* **18**, 908 (1967); *Phys. Rev.* **183**, 457 (1969).

25 R. E. Slusher and H. M. Gibbs, *Phys. Rev. A* **5**, 1634 (1972); [erratum, **6**, 1255 (1972)].

26 R. H. Dicke, *Phys. Rev.* **93**, 99 (1954).

27 N. Bloembergen and R. V. Pound, *Phys. Rev.* **95**, 8 (1954).

28 N. Skribanowitz, I. P. Herman, J. C. MacGillivray, and M. S. Feld, *Phys. Rev. Lett.* **30**, 309 (1973); I. P. Herman, J. C. MacGillivray, N. Skribanowitz, and M. S. Feld, in R. G. Brewer and A. Mooradian, eds., *Laser Spectroscopy* (Plenum, 1974), p. 379.

29 H. M. Gibbs, Q. H. F. Vrehen, and H. M. J. Hikspoors, *Phys. Rev. Lett.* **39**, 547 (1977).

30 J. C. MacGillivray and M. S. Feld, *Phys. Rev. A* **14**, 1169 (1976).

31 R. Bonifacio and L. A. Lugiato, *Phys. Rev. A* **11**, 1507 (1975); **12**, 587 (1975)

32   R. Glauber and F. Haake, *Phys. Lett.* **68A**, 29 (1978); F. Haake, H. King, G. Schröder, J. Haus, R. Glauber, and F. Hopf, *Phys. Rev. Lett.* **42**, 1740 (1979).

33   D. Polder, M. F. H. Schuurmans, and Q. H. F. Vrehen, *Phys. Rev. A* **19**, 1192 (1979).

34   Q. H. F. Vrehen and M. F. H. Schuurmans, *Phys. Rev. Lett.* **42**, 224 (1979).

35   R. Bonifacio and L. A. Lugiato, in R. Bonifacio, ed., *Dissipative Systems in Quantum Optics* (Springer-Verlag, Berlin, 1982); Q. H. F. Vrehen and H. M. Gibbs, ibid.

BIBLIOGRAPHY

Abragam, A., *The Principles of Nuclear Magnetism* (Oxford University Press, London, 1961).

Allen, L., and J. H. Eberly, *Optical Resonance and Two-Level Atoms* (Wiley, New York, 1975).

Brewer, R. G., in R. Balian, S. Haroche, and S. Liberman, eds., *Frontiers in Laser Spectroscopy*, Vol. 1, (North-Holland Publishing Co., Amsterdam, 1977).

Brewer, R. G , *Physics Today* **30**, 50 (1977).

Shoemaker, R. L., in J. I. Steinfeld, ed., *Laser and Coherence Spectroscopy* (Plenum, New York, 1978), p. 197.

Slichter, C. P., *Principles of Magnetic Resonance* (Springer-Verlag, Berlin, 1978).

22

Strong Interaction
of Light with Atoms

Strong interaction of light with matter generally refers to the case where light and matter form a tightly coupled unit, and the usual pictures derived from the perturbation approach break down. The material system can have its properties changed drastically not only because of light-induced population redistribution but also because of light-induced changes of the energy levels and eigenfunctions. This subject is an interesting subarea of nonlinear optics from both theoretical and practical points of view. In this chapter, we discuss only the case of strong interaction of light with atoms, which under resonant excitations can be regarded as a simple system with, effectively, only a few discrete levels. Part of the discussion has already been presented in Chapter 13 as the basis of some high-resolution nonlinear spectroscopic techniques.

22.1 GENERAL DESCRIPTION

By definition, strong interaction of light with matter occurs when the interaction is so strong that it cannot be treated as a small perturbation. As seen in the microscopic calculation (see Chapter 2), this happens when the matrix elements of the Hamiltonian $|\mathcal{H}_{\text{int}}|_{nn'}$ are comparable with or larger than the frequency denominator $\hbar|\omega - \omega_{nn'} + i\Gamma_{nn'}|$, where ω is the laser frequency, $\omega_{nn'}$ is the resonant transition frequency, and $\Gamma_{nn'}$ is the corresponding damping constant. Therefore, far away from resonance, a very strong optical field is required to get into the strong interaction regime, but near resonance, a weak optical field is often sufficient. For example, in alkali vapor, one can observe strong interaction with a beam intensity not much larger than a few tens of milliwatts per square centimeter at the $ns \rightarrow np$ resonance.

The strong dependence of the interaction strength on resonance allows, in many cases, the simplification of a real material system to an effective system consisting of only a few levels connected by resonant excitations. This is often the case for atoms where nearby levels are sufficiently far apart so that the nonresonant transitions can be truly neglected. In this respect, one deals with an effective two-level system if only two atomic levels are involved in resonant excitation and probe, or with an effective three-level system if three atomic levels are involved, and so on. For large molecules and solids, such simplification may be unrealistic because of the many closely spaced transitions, since the optical-field-induced level shifts and broadening can be comparable to the separation between nearby states. The calculation then appears much more complicated, and in fact, the full theory has not yet been developed. This discussion thus is limited to simple atomic or molecular cases. While the formalism in general applies to any effective n-level system, we discuss mainly systems with effective two or three levels. Analytical solutions are usually possible only for $n \leq 3$.

The problem of strong interaction of light with an n-level system has attracted the attention of physicists ever since the birth of quantum mechanics, and ways of attacking the problem have been discovered and rediscovered by people working in various areas of physics: microwave spectroscopy, magnetic resonance, and optical spectroscopy. There are two usual approaches to the problem. One is the bare-atom approach in which the noninteracting atom-field eigenstates are chosen as the basis in the calculation.[1] The other is the dressed-atom approach, in which all or part of the atom-field interaction is solved exactly, and the resultant atom-field eigenstates are used as the basis for further calculation.[2] The former is perhaps more straightforward in mathematical derivation, while the latter is physically more transparent. The amount of actual calculation in solving a problem is, however, practically the same in the two approaches. In either approach, the fields can be treated as classical if they are intense enough. We use in the following the semiclassical description to illustrate the bare-atom approach, and the full quantum-mechanical description to illustrate the dressed-atom approach.

The immediate consequence of strong optical excitations one would expect is the optical Stark shift and broadening (see Sections 5.3 and 13.3 on saturation in excitation), which is an extension of the picture from the weak interaction regime. Actually, the picture of strong interaction is more complicated. For example, one can indeed find an optical-Stark-broadened (or saturation-broadened) line if the absorption of a strong pump field versus frequency around a resonant transition is measured (see Section 13.3 on saturation in excitation). However, if the frequency of the strong pump field is fixed near resonance, and the absorption of a weak probe beam versus frequency around the resonance is measured, three resonant peaks can be observed in the spectrum (see Section 13.3 on absorption of a weak probe in the presence of a strong pump). (The actual spectrum shows only one absorption line and one amplification line situated symmetrically on the two sides of the pump frequency for reasons to be discussed later.) This can be understood as follows. With the pump frequency ω close to the transition frequency ω_{21}

between the atomic states $|1\rangle$ and $|2\rangle$, the composite states $|1, n\hbar\omega\rangle$ and $|2, n\hbar\omega\rangle$ of the atom-field system are nearly degenerate with the states $|2, (n - 1)\hbar\omega\rangle$ and $|1, (n + 1)\hbar\omega\rangle$, respectively, in the absence of atom-field interaction, but from the selection rules, only the transition between $|1, n\hbar\omega\rangle$ and $|2, n\hbar\omega\rangle$ is allowed. In the presence of strong interaction, however, the nearly degenerate states become mixed and shifted, and all transitions between the two sets of states $\{|2, n\hbar\omega\rangle, |1, (n + 1)\hbar\omega\rangle\}$ and $\{|2, (n - 1)\hbar\omega\rangle, |1, n\hbar\omega\rangle\}$ are allowed. A total of three absorption lines is expected because the energy separation of the two states in each set is the same. More generally, a monochromatic field interacting strongly with n levels (such that $|\langle i|\mathscr{H}_{\text{int}}|j\rangle| \gtrsim \hbar|\omega - \omega_{ij} + i\Gamma_{ij}|$) can lead to the splitting of each level into n levels. The description here essentially follows the dressed-atom picture, although the same results should come out of the bare-atom approach. We discuss in the following sections the bare-atom and dressed-atom approaches in some more detail with a few concrete examples.

22.2 BARE-ATOM APPROACH

We treat the fields as classical, and use the density matrix formalism (Section 2.1) in this section. Assume an effective n-level system in strong interaction with m monochromatic pump fields. There are altogether n^2 density matrix elements ρ_{ij} with $i, j = 1, \ldots, n$ describing the system. Each ρ_{ij} has a set of frequency components which are the linear combinations of the m pump frequencies. In the bare-atom approach, one solves, more or less exactly, the Liouville equation for ρ_{ij} in response to the strong pump fields. Many of the frequency components of ρ_{ij} can be neglected because they are far off resonance. We now assume that a weak field is used to probe the absorption spectrum of transitions either within the n levels or from the n levels to other levels. The next-step calculation is then to find $\rho^{(1)}(\omega_{\text{probe}})$ to the linear order of the probe field in terms of ρ_{ij} already found to all orders of the pump fields. This is done by using the atom-probe interaction as a perturbation Hamiltonian in the Liouville equation. Finally, from the expectation value of the induced dipole $\langle \mathbf{p}(\omega_{\text{probe}})\rangle = \text{Tr}[\rho^{(1)}(\omega_{\text{probe}})\mathbf{p}]$, the probe absorption can be calculated.

The approach here emphasizes the response of the atomic system to the applied fields. Therefore, only the particular solution to the Liouville equation is sought, and hence the density matrix ρ shows no new frequency component other than the linear combinations of the applied optical frequencies. In this respect, the eigenfrequencies of the composite atom-field system are not apparent in the solution since they should only show up as frequency components in the homogeneous solution. They can, however, be identified as resonances in response to the probe beam. Thus, in the bare-atom approach, the solution of the system under strong optical excitations does not explicitly give a picture of the restructured energy levels (although the latter can be obtained from the homogeneous solution of the Liouville equation), but the mathematical formalism is simple and straightforward. We now use two- and three-level systems as examples to illustrate the general description.

The case of an effective two-level system pumped by a strong beam and probed by a weak beam was discussed in Section 13.3. For clarity, we reproduce the essential steps of the calculation here and expand on the discussion.[3] First, we find the nearly exact solution of the Liouville equation for a system of two levels $|1\rangle$ and $|2\rangle$ under the strong excitation of a pump field $E(\omega)$. The nonnegligible components of the density matrix are

$$\rho_{11}(0) - \rho_{22}(0) = \Delta\rho° \Big/ \left[1 + \frac{\Gamma^2 I/I_s}{(\omega - \omega_{21})^2 + \Gamma^2} \right]$$

and   (22.1)

$$\rho_{12}(-\omega) = \rho_{21}^*(\omega) = \frac{-p_{12}E(-\omega)[\rho_{11}(0) - \rho_{22}(0)]}{\hbar(\omega - \omega_{21} - i\Gamma)}$$

where $I/I_s = 4\Omega^2 T_1/\Gamma$, $I = c|E(\omega)|^2/2\pi n$, $\Omega^2 = |p_{12}E(\omega)|^2/\hbar^2$, and $p_{12} = \langle 1|er|2\rangle$. Next, in the presence of the probe field $E(\omega')$, we seek the solution ρ linear in $E(\omega')$. There exist three nearly resonant components of ρ linear in $E(\omega')$: $\rho_{21}(\omega')$, $\rho_{11}(\omega' - \omega) - \rho_{22}(\omega' - \omega)$, and $\rho_{12}(\omega' - 2\omega)$. To find absorption at ω', we need only know $\rho_{21}(\omega')$, but it can only be obtained by solving the linearly coupled equations for the three components

$$\hbar(\omega' - \omega_{21} + i\Gamma)\rho_{21}(\omega')$$
$$= -p_{21}E(\omega')[\rho_{11}(0) - \rho_{22}(0)]$$
$$\quad - p_{21}E(\omega)[\rho_{11}(\omega' - \omega) - \rho_{22}(\omega' - \omega)],$$

$$\hbar\left(\omega' - \omega + i\frac{1}{T_1}\right)[\rho_{11}(\omega' - \omega) - \rho_{22}(\omega' - \omega)]$$

(22.2)

$$= -p_{12}E^*(\omega)\rho_{21}(\omega') + 2p_{21}$$
$$\quad \times E(\omega')\rho_{12}(-\omega) + 2p_{21}E(\omega)\rho_{12}(\omega' - 2\omega),$$

$$\hbar(\omega' - 2\omega - \omega_{12} + i\Gamma)\rho_{12}(\omega' - 2\omega)$$
$$= p_{12}E^*(\omega)[\rho_{11}(\omega' - \omega) - \rho_{22}(\omega' - \omega)]$$

where $\rho_{11}(0) - \rho_{22}(0)$ and $\rho_{12}(-\omega)$ are known from (22.1). The solution, expressed in a form slightly different from that in (13.16), is

$$\rho_{21}(\omega') = -\frac{1}{\hbar} p_{21} E(\omega') [\rho_{11}(0) - \rho_{22}(0)]$$

$$\times \left\{ \left(\omega - \omega' + i\frac{1}{T_1} \right)(\omega' - 2\omega + \omega_{21} + i\Gamma) \right.$$

$$\left. - 2\Omega^2 \left[1 - \frac{\omega' - 2\omega + \omega_{21} + i\Gamma}{\omega - \omega_{21} - i\Gamma} \right] \right\} / D, \tag{22.3}$$

$$D = (\omega' - \omega_{21} + i\Gamma) \left[\left(\omega' - \omega + i\frac{1}{T_1} \right)(\omega' - 2\omega + \omega_{21} + i\Gamma) - 2\Omega^2 \right]$$

$$- 2\Omega^2 (\omega' - 2\omega + \omega_{21} + i\Gamma).$$

The absorption coefficient at ω' is then obtained from

$$\alpha(\omega') = \left(\frac{4\pi\omega'}{c} \right) \operatorname{Im} \chi(\omega')$$

with

$$\chi(\omega') = \frac{N p_{12} \rho_{21}(\omega')}{E(\omega')}. \tag{22.4}$$

The expression for D can be recast into the form

$$D = (\omega' - \omega) \left[(\omega' - \omega)^2 - (\omega_{21} - \omega)^2 - 4\Omega^2 - \Gamma^2 - \frac{2\Gamma}{T_1} \right]$$

$$+ i \left[2(\omega' - \omega)^2 - 4\Omega^2 \right] \Gamma + \frac{i \left[(\omega' - \omega_{21})(\omega' - 2\omega + \omega_{21}) - \Gamma^2 \right]}{T_1}. \tag{22.5}$$

It shows clearly that $\operatorname{Re} D$ has three zeroes corresponding to three resonances in the absorption spectrum

$$\omega' = \omega,$$
$$\omega' = \omega \pm \Delta, \tag{22.6}$$
$$\Delta = \left[(\omega - \omega_{21})^2 + 4\Omega^2 + \Gamma^2 + \frac{2\Gamma}{T_1} \right]^{1/2}.$$

An explicit calculation of the absorption coefficient in (22.4) would show that at $\omega' = \omega$, α nearly vanishes, while at $\omega' = \omega + \Delta$, α is positive if $\omega < \omega_{21}$ and negative if $\omega > \omega_{21}$, and at $\omega' = \omega - \Delta$, α is negative if $\omega < \omega_{21}$ and positive if $\omega > \omega_{21}$. The absorption spectrum therefore shows an absorption line and an emission line located symmetrically on the two sides of the pump frequency ω. This has been demonstrated experimentally, as we shall discuss in a later section. When $|\omega - \omega_{21}| \gg \Omega \gg \Gamma$, the emission line is very weak, and the absorption line appears at $\omega' \cong \omega_{21} - 2\Omega^2/(\omega - \omega_{21})$, which is just the position one would expect from the simple perturbation calculation of an optical Stark shift given in (5.17) for the case of a relatively weak interaction of light with matter.

One may recognize that Δ in (22.6) is just the Rabi frequency described in Chapter 21, although we have not included Γ and T_1 in (21.16). Thus, physically, the resonances at $\omega' = \omega \pm \Delta$ can be considered as the sidebands created by modulation of the Rabi precession on the central resonant component at $\omega' = \omega$.

We can extend the above discussion to a three-level system, seen in Fig. 22.1, in which the levels $|1\rangle$ and $|2\rangle$ are connected by the strong pump field $E(\omega)$ and the transition from $|2\rangle$ to $|3\rangle$ is probed by the weak field $E(\omega')$. This is a double resonance problem. Again, we first obtain the solution (22.1) by assuming the presence of only the strong pump field, and then find $\rho_{32}(\omega')$ linear in $E(\omega')$ from a set of linearly coupled equations. The result should show that $\rho_{32}(\omega')$ has a frequency denominator whose real part has two zeroes, indicating two resonances in the absorption spectrum. When the absorption coefficient $\alpha(\omega') \propto \mathrm{Im}[\,p_{23}\rho_{32}(\omega')/E(\omega')]$ is calculated, one would find α to be positive at both resonances. The absorption spectrum then consists of two absorption lines, separated by the Rabi frequency Δ as one would expect from the earlier description. This effect was first discovered by Autler and Townes[4] in a microwave-optical double resonance experiment. More recently, it has also been repeatedly demonstrated in optical-optical double resonance experiments. We again postpone the discussion of the experiments to a later section.

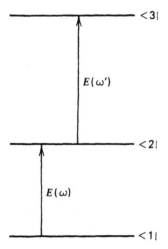

Fig. 22.1 A three-level system with two nearly resonant exciting fields.

What will happen if both $E(\omega)$ and $E(\omega')$ in the three-level case of Fig. 22.1 are strong? Clearly, the absorption or emission spectrum will be much more complicated. The result can be more easily visualized with the dressed-atom approach. We therefore postpone the discussion to the next section. Here we consider only the problem of how the steady-state population is redistributed in the three levels under the strong excitations. This is relevant not only in spectroscopy of strongly excited systems, but also in applications of optical pumping, such as isotope separation, multiphoton ionization, and studies of physics and chemistry of excited-state atoms. The calculation is in principle straightforward and has been discussed repeatedly in the literature.[3, 5, 6] Only the near-resonant terms in the density matrix formalism are kept and $\omega \sim \omega_{21}$ and $\omega' \sim \omega_{32}$ are assumed to be very different. The Liouville equation yields explicitly the following set of equations:[3]

$$\hbar\Delta_{21}\rho_{21}(\omega) = V_{21}(\omega)[\rho_{22}(0) - \rho_{11}(0)] + V_{23}(-\omega')\rho_{31}(\omega + \omega'),$$

$$\hbar\Delta_{32}\rho_{32}(\omega') = V_{32}(\omega')[\rho_{33}(0) - \rho_{22}(0)] - V_{12}(-\omega)\rho_{31}(\omega + \omega'),$$

$$\hbar\Delta_{31}\rho_{31}(\omega + \omega') = V_{32}(\omega')\rho_{21}(\omega) - V_{21}(\omega)\rho_{32}(\omega'),$$

$$0 = -(W_{21} + W_{31})\rho_{11} + W_{12}\rho_{22} + W_{13}\rho_{33}$$
$$+ V_{12}(-\omega)\rho_{21}(\omega) - V_{21}(\omega)\rho_{12}(-\omega), \quad (22.7)$$

$$0 = W_{31}\rho_{11} + W_{32}\rho_{22} - (W_{13} + W_{23})\rho_{33} + V_{32}(\omega')\rho_{23}(-\omega')$$
$$- V_{23}(-\omega')\rho_{32}(\omega'),$$

$$1 = \rho_{11} + \rho_{22} + \rho_{33}$$

where $\Delta_{21} = \omega - \omega_{21} + i\Gamma_{21}$, $\Delta_{32} = \omega' - \omega_{32} + i\Gamma_{32}$, $\Delta_{31} = \omega + \omega' - \omega_{31} + i\Gamma_{31}$, $V_{ij}(\omega_\alpha) = -p_{ij}E(\omega_\alpha)$, and the W_{ij} are relaxation rates from $|j\rangle$ to $|i\rangle$. We assume, for simplicity, that the population is conserved in the three levels. Equations (22.7) together with those for $\rho_{12}(-\omega)$, $\rho_{23}(-\omega')$, and $\rho_{13}(-\omega - \omega')$ form a set of nine linearly coupled equations. As expected, the formal expression of the solution is extremely complex and is not particularly illuminating. The results from numerical calculations may be more helpful.

Whitley and Stroud[6] conducted such a numerical calculation on a three-level system in which the relaxations are governed by spontaneous emission from $|3\rangle$ to $|2\rangle$ and from $|2\rangle$ to $|1\rangle$. Their results, shown in Fig. 22.2, indicate that if $\omega + \omega'$ is in exact resonance with ω_{31}, then in the limit of very strong fields, the population tends to become equalized between the ground and upper excited states, with a relatively small fraction occupying the lower excited state. This may be what one would expect when ω and ω' are detuned from ω_{21} and ω_{32}, respectively, since the process then becomes a direct two-photon absorption process. However, as seen in Fig. 22.2, the same is true even when ω and ω' are resonant with ω_{21} and ω_{32}, except that the population in the lower excited state is now somewhat higher. This result may be extended to the general case of an n-level system, and is most important in many applications of stepwise

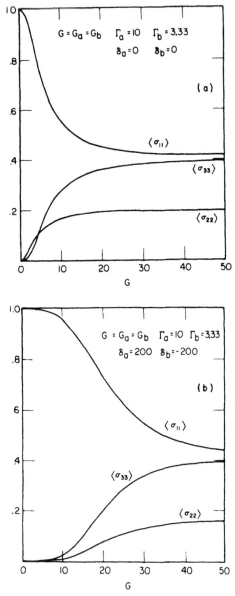

Fig. 22.2 Steady-state atomic population distribution in a three-level system as a function of the strength of the applied fields: (a) the two fields are exactly resonant with the two successive transitions, and (b) the two fields are detuned from resonances by equal magnitude but opposite sign. [After R. M. Whitley and C. R. Stroud, *Phys. Rev. A* **14**, 1498 (1976).]

multiphoton pumping. First, it is possible to pump nearly half of the population from the ground state to the final state. Second, the population in the intermediate states may appear depleted in the strong excitation limit. Third, a population inversion between the final state and the intermediate state can be established. An experimental demonstration of the effect, to be discussed in Section 22.4, was presented by Gray et al.[7]

We limited this discussion to the steady-state case, but the formalism can be readily extended to the time-dependent case.[6, 8] For example, one can use the time-dependent Liouville equation to discuss the time-dependent Autler-Townes effect.[9] One can also use it to find the development of the population redistribution in the three-level system under the resonant excitations of the two strong fields.[6] Transient effects in a multilevel system with strong resonant excitations should in general be a very interesting subject.[10] They are an extension of the coherent transient phenomena discussed in Chapter 21.

22.3   DRESSED-ATOM APPROACH

In the dressed-atom approach, the eigensolution of the combined system of atom and pump fields is sought first. The result yields a picture of the energy level structure of the "dressed" atom and hence a physical understanding of the absorption or emission spectrum. We use here a full quantum mechanical description to illustrate the approach.[2] To avoid excessive mathematical derivation, we consider only the qualitative or semiquantitative aspect of various problems in the following discussion.

Consider again the two-level system interacting with a strong monochromatic field. We first find the eigensolution of the combined atom-field system. Figure 22.3 shows that in the absence of the atom-field interaction with $\omega \sim \omega_{21}$, $|1, n\rangle$ and $|2, n\rangle$ are nearly degenerate with $|2, n - 1\rangle$ and $|1, n + 1\rangle$, respectively, where n indicates the number of photons in the field. With strong atom-field interaction, the degeneracies are lifted and the energy level structure appears as an infinite set of equally spaced doublets. The splitting $\hbar\Delta$ between the two states $\langle \alpha_n|$ and $\langle \beta_n|$ in a doublet can be obtained easily from degenerate perturbation theory knowing that the interaction connecting $\langle 1, n|$ and $\langle 2, n - 1|$ is

$$\langle 2, n - 1|\mathcal{H}_{int}|1, n\rangle = -p_{21}E(\omega) \tag{22.8}$$

with $c|E(\omega)|^2/2\pi = n\hbar\omega$ for $n \gg 1$. We find

$$\Delta = \left[(\omega - \omega_{21})^2 + 4\Omega^2\right]^{1/2}, \tag{22.9}$$

which is just the Rabi frequency. In terms of $\langle 1, n|$ and $\langle 2, n - 1|$, the two

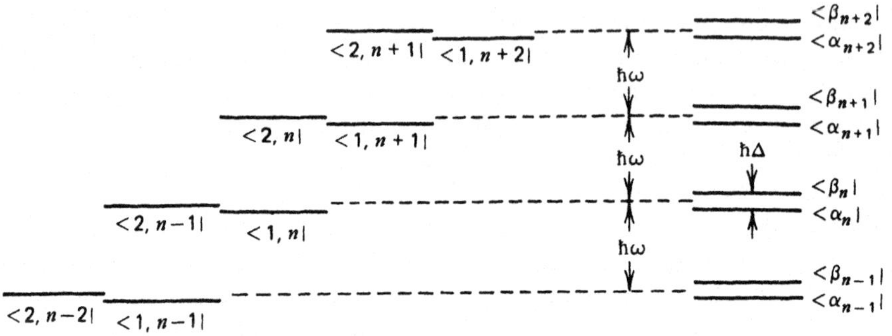

Fig. 22.3 Energy level structure of a two-level system in the dressed-atom picture.

eigenstates of the dressed atom are

$$|\alpha_n\rangle = \frac{2p_{12}E(\omega)^*}{\left[(\omega - \omega_{21} + \Delta)^2 + 4\Omega^2\right]^{1/2}} \, |1, n\rangle$$

$$+ \frac{\omega - \omega_{21} + \Delta}{\left[(\omega - \omega_{21} + \Delta)^2 + 4\Omega^2\right]^{1/2}} \, |2, n - 1\rangle$$

and   (22.10)

$$|\beta_n\rangle = \frac{\omega - \omega_{21} + \Delta}{\left[(\omega - \omega_{21} + \Delta)^2 + 4\Omega^2\right]^{1/2}} \, |1, n\rangle$$

$$+ \frac{2p_{21}E(\omega)}{\left[(\omega - \omega_{21} + \Delta)^2 + 4\Omega^2\right]^{1/2}} \, |2, n - 1\rangle.$$

From this eigensolution, one can then find the fluorescence and absorption spectra.[11] We need only consider transitions between two neighboring doublets. Because of mixing of wavefunctions, all transitions connecting the two pairs of states are allowed. As shown in Fig. 22.4, this leads to a fluorescence spectrum with three lines: the central component at ω and two side ones at $\omega \pm \Delta$. In the steady-state case, the detailed-balance condition usually requires that in equilibrium, the relaxation through fluorescence should obey the relation

$$\frac{\partial \rho_\alpha}{\partial t} = \Gamma_{\alpha\beta}\rho_\beta - \Gamma_{\beta\alpha}\rho_\alpha = 0$$

$$= \frac{\partial \rho_\beta}{\partial t}$$

(22.11)

where Γ_{ij} is the spontaneous transition rate from the state $|j, n\rangle$ to $|i, n - 1\rangle$,

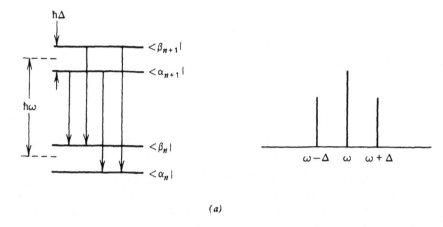

(a)

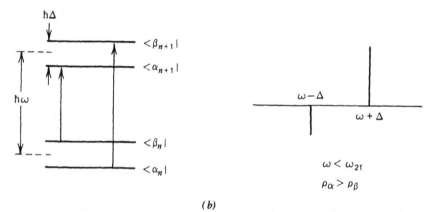

(b)

Fig. 22.4 (a) Fluorescent transitions and the corresponding fluorescent spectrum, and (b) absorptive transitions and the corresponding absorption spectrum in a two-level system under a strong optical excitation. The dressed-atom picture is used here to describe the energy level structure of the composite matter-field system.

and ρ_i is the steady-state population in $|i\rangle$. Equation (22.11) immediately leads to the conclusion that the two side components in the fluorescence spectrum are equal in intensity. The absorption spectrum should also have three components at ω and $\omega \pm \Delta$. However, as seen in Fig. 22.4, the central component comes from $|\alpha_n\rangle \rightarrow |\alpha_{n+1}\rangle$ and $|\beta_n\rangle \rightarrow |\beta_{n+1}\rangle$ transitions, and its intensity is zero because $\rho_{\alpha_n} = \rho_{\alpha_{n+1}}$ and $\rho_{\beta_n} = \rho_{\beta_{n+1}}$ (neglecting the very small difference in the probabilities of finding n and $n + 1$ photons in the strong field). The two side components have intensities

$$I(\omega + \Delta) = K\Gamma_{\alpha\beta}(\rho_\alpha - \rho_\beta)$$

and

$$I(\omega - \Delta) = K\Gamma_{\beta\alpha}(\rho_\beta - \rho_\alpha)$$

(22.12)

where K is a proportional constant. If $\omega < \omega_{21}$, then $\rho_\alpha > \rho_\beta$, and hence $I(\omega + \Delta) > 0$ and $I(\omega - \Delta) < 0$, corresponding to an absorption line and an amplification line, respectively. If $\omega > \omega_{21}$, then $\rho_\alpha < \rho_\beta$, and we have the reverse situation $I(\omega + \Delta) < 0$ and $I(\omega - \Delta) > 0$. The detailed-balance condition of (22.11) also leads to

$$I(\omega + \Delta) + I(\omega - \Delta) = K\left(\Gamma_{\beta\alpha}/\rho_\beta\right)\left(\rho_\alpha - \rho_\beta\right)^2 > 0, \qquad (22.13)$$

indicating that the absorption line is always more intense than the amplification line. A more detailed calculation taking into account the relaxations allows us to deduce also the lineshapes of the spectra.[2]

In the Autler–Townes double-resonance case, the transition from level $|2\rangle$ to a third level $|3\rangle$ is probed. Clearly, in the dressed-atom picture, having $|2, n\rangle$ mixed with $|1, n + 1\rangle$, the absorption spectrum from $|2\rangle$ should consist of two lines at $\omega_{32} \pm \Delta/2$. If $\omega < \omega_{21}$ so that $\rho_\alpha > \rho_\beta$, one finds $I(\omega_{32} + \Delta/2) > I(\omega_{32} - \Delta/2)$, and if $\omega > \omega_{21}$, the reverse is true.

The dressed-atom approach can be extended readily to the three-level system interacting with two strong resonant fields in Fig. 22.1. As seen in Fig. 22.5, the energy level structure of the dressed atom is an infinite set of triplets. It is easily seen that the fluorescence spectrum from $|3\rangle$ to $|2\rangle$ consists of a symmetric pattern of seven lines, with the central component at ω' and the side components at $\omega' \pm \Delta$, $\omega' \pm \Delta'$, and $\omega' \pm (\Delta + \Delta')$. The fluorescence spectrum from $|2\rangle$ to $|1\rangle$ is the same except that the central component is at ω. The absorption spectra from $|1\rangle$ to $|2\rangle$ and from $|2\rangle$ to $|3\rangle$ can also be calculated. Each should also have seven lines, but the central component again has zero intensity; three of the side lines are absorption lines and the other three are amplification lines. If the transition from one of the three levels to a fourth

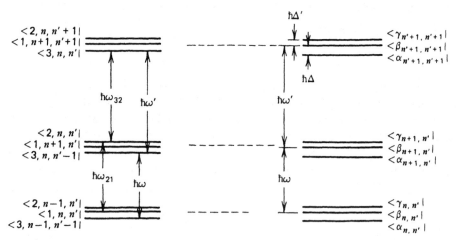

Fig. 22.5 Energy level structure of the composite system of a three-level atom strongly interacting with two nearly resonant exciting fields $E(\omega)$ and $E(\omega')$.

level of the atomic system is probed, then the absorption spectrum should consist of three lines. This is the Autler–Townes effect in a four-level system.

Another case of interest occurs when a single monochromatic field interacts strongly with a three-level system in which two levels, e.g., $|2\rangle$ and $|3\rangle$, are nearly degenerate with a frequency separation less than the Rabi frequency. In the dressed-atom picture, $\langle 1, n|$ is nearly degenerate with $\langle 2, n-1|$ and $\langle 3, n-1|$, and $\langle 2, n|$ and $\langle 3, n|$ are nearly degenerate with $\langle 1, n+1|$, so that the energy level diagram is also an infinite set of equally spaced triplets. The fluorescence spectrum should again be a symmetric pattern of seven lines centered at ω, while the corresponding absorption spectrum has the central component missing.

The foregoing examples show that the dressed-atom approach is indeed physically transparent, and is most helpful to the understanding of spectroscopy of atoms under strong resonant excitations. It can be generalized to an effective n-level system with the following general rules. Each level in the strong fields is split into n levels, all of which have the partial characters of the original n levels. Both the splittings and the partial characters of the split levels depend on the field-atom coupling strengths represented by the off-diagonal matrix elements of the perturbation Hamiltonian connecting the original n levels. They can be calculated by diagonalizing an $n \times n$ matrix following a degenerate perturbation calculation on the coupled field-atom system. The energy level diagram of the system now appears to have n sets of n levels, and the absorption and emission spectra can be directly deduced from transitions between the n sets of levels. We should remark that in general, strong coupling of two levels is not necessarily effected by a one-photon resonant excitation ($\omega \sim \omega_{ij}$). It can also be effected by fields connecting the levels via, for example, a two-photon transition ($2\omega \sim \omega_{ij}$). In this latter case, the effective Hamiltonian for two-photon transition [$\mathscr{H}'_{\mathrm{eff}} = -\mathbf{E}_1 \cdot \mathbf{M} \cdot \mathbf{E}_2 = -E_1 E_2 M$ with M defined in (12.1)] should be used in the degenerate perturbation calculation. However, the detailed calculation of a spectrum in the dressed-atom approach, taking into account the random relaxations, is actually as complicated as in the bare-atom approach.

22.4 EXPERIMENTAL DEMONSTRATION

The optical-field-induced line broadening, shifts, and splittings generally are known as the optical Stark effect. Similar effects have long been observed in microwave spectroscopy. With the advent of lasers, they can also be readily observed in the optical region. Laser-induced saturation with the resultant line broadening is of course the basis of saturation spectroscopy (Section 13.3). The optical Stark shift which occurs in the relatively weak atom-field interaction limit has also been extensively studied. With a high-intensity laser, even if the laser frequency is very far off resonance, the optical Stark shift is still observable. An example is presented in Fig. 22.6, where a shift of 0.12 cm^{-1} of

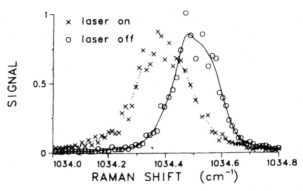

Fig. 22.6 The effects of high-intensity infrared radiation on the position and line shape of the $S(3)$ rotational transition of molecular hydrogen. Crosses represent data obtained by coherent anti-Stokes Raman spectroscopy in the presence of 1.06-μm radiation from a Q-switched Nd:YA1G laser; circles show data in the absence of this field. The solid and dashed lines are smooth curves drawn through the data points. (After Ref. 12.)

the $S(3)$ rotational transition (1034.5 cm^{-1}) in molecular hydrogen, induced by a Q-switched Nd:YAG laser pulse at 1.06 μm with a peak intensity of 8×10^{11} w/cm^2 is clearly seen.[12]

With strong atom-field interaction, usually achieved by resonant or near-resonant laser excitation, optical Stark splitting occurs. This has been studied in a large number of experiments. An atomic beam is often employed in the experiment to avoid complications due to Doppler broadening, to reduce the inhomogeneous linewidth, and to make the observation easier. In the study of the fluorescence spectrum induced by a strong resonant field, a single-frequency CW laser beam is used to excite an atomic beam, and the induced fluorescence is monitored and analyzed.[13] For example, in the experiment of Wu et al. (in Ref. 13), a circularly polarized CW dye laser with a linewidth less than 250 kHz was used to excite the $3^2S_{1/2}(F = 2, m_F = 2) \rightarrow 3^2P_{3/2}(F' = 3, m_{F'} = 3)$ transition of sodium in an orthogonally propagating atomic beam, which had the initial population in the $3^2S_{1/2}(F = 2, m_F = 2)$ ground state prepared by optical pumping. The fluorescence from the inverse transition emitted in the direction perpendicular to both beams was detected and analyzed by a Fabry–Perot interferometer. With sufficiently high laser intensity, three peaks in the spectrum could be readily observed, in agreement with the theoretical prediction. Figure 22.7 shows the observed spectra compared with theory for on- and off-resonance excitation with a peak laser intensity of 640 mW/cm^2, corresponding to an on-resonance Rabi frequency $\Omega = 78$ MHz. As predicted, the spectra are always centered on the exciting laser frequency, and the sideband separation 2Δ increased with the detuning $(\omega - \omega_{21})$ following $\Delta = [(\omega - \omega_{21})^2 + 4\Omega^2]$.

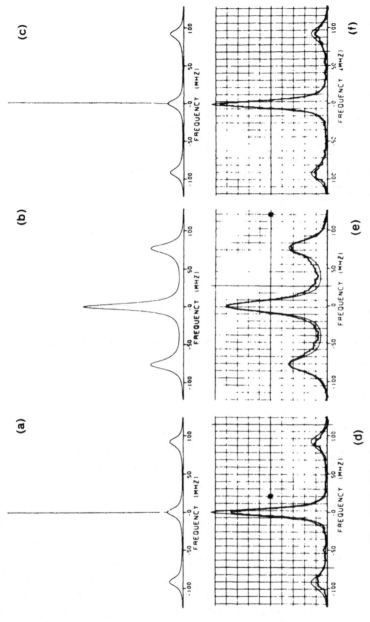

Fig. 22.7 Fluorescence spectra of the $3^2S_{1/2}(F = 2, m_F = 2) \leftarrow 3^2P_{3/2}(F = 3, m_F = 3)$ transition of sodium under a strong resonant excitation with a peak laser intensity of 640 mW/cm². Experimental spectra are in comparison with the theoretical spectra of (*a*), (*b*), and (*c*), convoluted with the instrumental lineshape (smooth curves) in (*d*), (*e*), and (*f*) for various amounts of detuning in excitation: (*a*) and (*d*): $\Delta\nu = -50$ MHz, (*b*) and (*e*): $\Delta\nu = 0$, (*c*) and (*f*): $\Delta\nu = +50$ MHz. [After R. E. Grove, F. Y. Wu, and S. Ezekiel, *Phys. Rev. A* **15**, 227 (1977).]

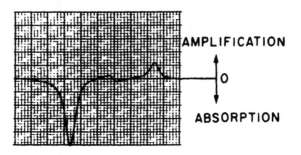

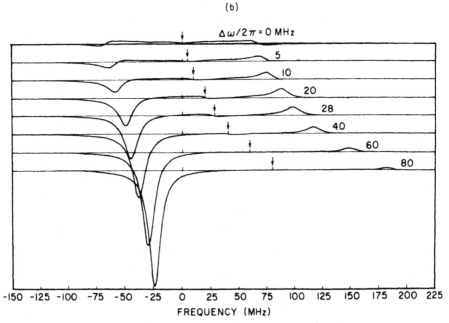

Fig. 22.8 (*a*) Measured absorption spectrum of the $3^2S_{1/2}(F = 2, m_F = 2) \rightarrow 3^2P_{3/2}(F = 3, m_F = 3)$ transition of sodium driven by a strong field detuned by 28 MHz above resonance. The field intensity is 560 mW/cm² corresponding to a Rabi frequency of 66 MHz. (*b*) Progression of theoretical line shapes with constant field strength (Rabi frequency = 66 MHz) and increasing detuning $\Delta\omega/2\pi = 0, 5, 10, 20, 28, 40, 60,$ and 80 MHz. Origin of horizontal axis is the transition resonance frequency. Arrows indicate frequency of driving field. (After Ref. 14.)

The absorption spectrum of the $3^2S_{1/2}(F = 2, m_F = 2) \rightarrow 3^2P_{3/2}(F' = 3, m_{F'} = 3)$ transition under strong laser excitation can also be probed if a second tunable dye laser is available. A representative piece of an experimental result of Wu et al.[14] is shown in Fig. 22.8. It is seen that here, as the Rabi frequency is much larger than the linewidth, the spectrum exhibits two sidebands separated by 2Δ. The central component at ω is indeed absent. The component on the high-frequency side is an amplification line, and the one on the low-frequency side is an absorption line. The latter is more intense than the former. These observations are all in good agreement with the theoretical predictions discussed in the previous sections.

In the Autler–Townes effect, the transition from one of the two levels connected by the strong field to a third level is probed, and two absorption lines are expected in the spectrum. An example of the experimental demonstra-

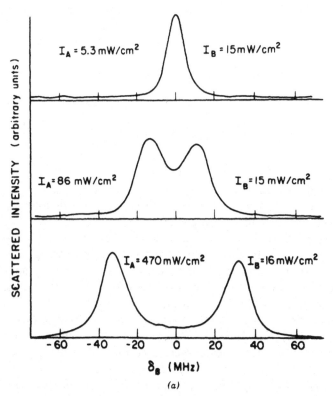

Fig. 22.9 (a) Autler–Townes absorption doublet observed in sodium when laser A is exactly on resonance with the $3^2S_{1/2}(F = 2, m_F = 2) \rightarrow 3^2P_{3/2}(F = 3, m_F = 3)$ transition and laser B scans over the $3^2P_{3/2}(F = 3, m_F = 3) \rightarrow 4^2D_{5/2}(F = 4, m_F = 4)$ transition. The splitting increases as the intensity of laser A increases. δ_B is the detuning of laser B from resonance. (b) Autler–Townes doublet when laser A is off resonance. As δ_A, the detuning of laser A from resonance changes, the spectra become asymmetric. (After Ref. 15.)

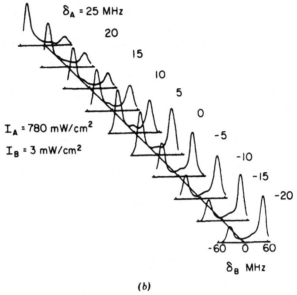

(b)

Fig. 22.9 (*Continued*).

tion[15] is given in Fig. 22.9. The first laser excites the $3^2S_{1/2}(F = 2, m_F = 2) \rightarrow 3^2P_{3/2}(F' = 3, m_{F'} = 3)$ transition of the sodium atom, and the second laser probes the absorption from $3^2P_{3/2}(F' = 3, m_{F'} = 3)$ to $4^2D_{5/2}(F'' = 4, m_{F''} = 4)$. The observed absorption spectrum shows two lines. It is symmetric with on-resonance pump excitation, and asymmetric with off-resonance excitation. The separation Δ between the two peaks increases with the detuning $\omega - \omega_{21}$ following (22.9). These results are in good agreement with theory.

When two lasers are used to excite a three-level system from $|1\rangle$ to $|2\rangle$ and from $|2\rangle$ to $|3\rangle$, respectively, the theory predicts that in the strong interaction limit, the two-photon resonant pumping tends to distribute the population evenly between the initial and final states $|1\rangle$ and $|3\rangle$, with little appearing in the intermediate state $|2\rangle$. This has been experimentally demonstrated by Gray et al.[7] The three-level system they chose was formed by the three levels, $3^2S_{1/2}(F = 1)$, $3^2P_{1/2}(F = 2)$, and $3^2S_{1/2}(F = 2)$, of atomic sodium. Two single-frequency CW dye laser beams were used to connect the two ground S states to the excited P states, as in Fig. 22.10, with the latter acting as the intermediate state. The population in the P state was monitored by the intensity of the fluorescence from that state. The result of an experiment in which the first laser was tuned to exact resonance with the $S_{1/2}(F = 1) \rightarrow P_{1/2}(F = 2)$ transition and the second laser was tuned through the two-photon resonance is shown in Fig. 22.10 in comparison with the theoretical calculation. Indeed, the population in the intermediate state dropped to nearly zero at exact two-photon resonance. The same happened even when the first laser was

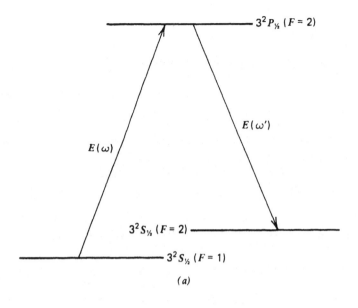

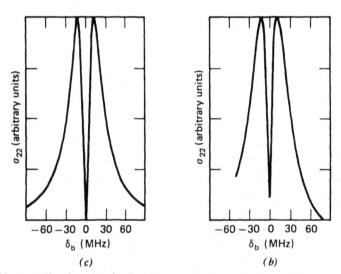

Fig. 22.10 (*a*) The three levels of sodium under the excitations of two nearly resonant strong fields. (*b*) Experimentally observed excited-state population σ_{22}. The fixed-frequency laser is at exact resonance, $\delta_a = 0$, with an intensity of 23 mW/cm². The second laser is detuned δ_b from exact resonance, with an intensity of 54 mW/cm². (*c*) Theoretical prediction of excited-state population σ_{22}. (After Ref. 7.)

detuned from resonance, except that a much higher pump power was needed to see the effect.

Experiments on strongly excited n-level systems with $n > 3$ are rare; but nonetheless, they could be interesting. For example, the absorption and emission spectra of a four-level system with two closely spaced intermediate levels, excited by two fields resonant with the ground-intermediate and inter-mediate-final transitions, respectively, could vary drastically with the field intensities. Experimental study on the transient effects in a strongly excited multilevel system has not yet been explored either. The time dependence of the absorption and emission spectra may also be interesting.

22.5 MULTIPHOTON EXCITATION AND IONIZATION

If the optical field is very intense, n-photon transitions with $n \gg 1$ can occur. We consider first transitions without intermediate resonances. In this case, the transition probability is very weak even with a strong excitation field and can be estimated from the nth order perturbation calculation. While the general expression of the transition probability is too cumbersome to be reproduced here, it can be symbolically written as

$$W^{(n)} \sim \frac{2\pi}{\hbar^2} \left| \frac{\mathscr{H}^n}{(\hbar\Delta\omega)^{n-1}} \right|^2 g^{(n)}(n\omega - \omega_{fg}) \qquad (22.14)$$

where $\mathscr{H}^n/(\hbar\Delta\omega)^{n-1}$ is actually a sum of many terms; in each term, $\mathscr{H}^n$ is a product of n off-diagonal matrix elements of the Hamiltonian $-e\mathbf{r}\cdot\mathbf{E}$, and $(\Delta\omega)^{n-1}$ is a product of $(n-1)$ factors of frequency detunings from resonances. The lineshape function $g^{(n)}(n\omega - \omega_{fg})$ has, in the ideal case, the simple Lorentzian form

$$g^{(n)}(n\omega - \omega_{fg}) = \frac{\Gamma/\pi}{(n\omega - \omega_{fg})^2 + \Gamma^2}. \qquad (22.15)$$

We have assumed here n photons of the same frequency. Generalization to n photons of different frequencies is straightforward with $n\omega$ replaced by $\omega_1 + \omega_2 + \cdots + \omega_n$.

Because $|\mathscr{H}/\hbar\Delta\omega| \sim |E/E_{at}| \ll 1$ far from resonance, knowing that the optical field E is much smaller than the atomic field E_{at} (see Section 1.3), the above perturbation calculation is justified. Recall, however, that an intense off-resonant light can still induce an optical Stark effect (see Section 22.2) proportional to the light intensity:

$$\delta\omega_{fg} = s_f - s_g,$$

$$s_f = \sum_l \frac{|\langle f|e\mathbf{r}\cdot\mathbf{E}|i\rangle|^2}{\hbar^2(|\omega_{f_l}| - \omega)} \frac{\omega_{f_l}}{|\omega_{f_l}|},$$

$$s_g = \sum_l \frac{|\langle g|e\mathbf{r}\cdot\mathbf{E}|i\rangle|^2}{\hbar^2(|\omega_{g_l}| - \omega)} \frac{\omega_{g_l}}{|\omega_{g_l}|}. \tag{22.16}$$

This causes a shift in the multiphoton resonant frequency so that $g^{(n)}(n\omega - \omega_{fg})$ in (22.15) should be replaced by $g^{(n)}(n\omega - \omega_{fg} - \delta\omega_{fg})$. Line broadening due to pump saturation is usually negligible for an n-photon transition with $n \gg 1$, but we can easily include it by replacing the linewidth Γ in $g^{(n)}$ by $\Gamma(1 + 2W^{(n)}T_1)^{1/2}$ where T_1 is the longitudinal relaxation time for the excitation. Thus, in the Lorentzian form, we have more exactly

$$g^{(n)}\left(n\omega - \omega_{fg} - \delta\omega_{fg}\right) = \frac{\Gamma\left(1 + 2W^{(n)}T_1\right)^{1/2}/\pi}{\left(n\omega - \omega_{fg} - \delta\omega_{fg}\right)^2 + \Gamma^2\left(1 + 2W^{(n)}T_1\right)}. \tag{22.17}$$

For $n = 2$ (two-photon transition), the shift in the resonant frequency is actually comparable to line broadening, both being proportional to the light intensity (they are neglected in the discussion of two-photon transition in Section 12.1). For $n > 2$, the shift is much larger than broadening. The former is measurable, but the latter often is not.

Experimentally, both saturation broadening[16] and light-induced line shifts[17] have been observed in two-photon transitions. The effect is larger if there exists a nearly resonant intermediate state. It is often said that an n-photon transition is characterized by an nth power dependence on the light intensity. This is, however, true only if both the lineshift and line broadening induced by the optical field are negligible, e.g., in the low intensity limit. For an n-photon transition with $n \gg 1$ between two discrete states, the lineshift may be appreciable when the signal reaches the detectable level. The nth power dependence on the intensity, I^n, then would never be observed. In experimental investigation, multiphoton transitions between discrete states with $n > 3$ have seldom been studied. This is presumably because the high laser intensity required for the observation also tends to ionize the atoms or molecules, and the subsequent effects induced by ionization tends to confuse the results. However, multiphoton ionization can also be considered a multiphoton transition, but the transition is between a discrete state and a continuum. In the latter case, the light-induced shift in the resonant frequency is immaterial in the presence of the continuum and while saturation broadening is certainly negligible. The I^n dependence of the signal can therefore be expected. Indeed, using a picosecond high-power Nd:glass laser, Manus and co-workers[18, 19] observed the n-photon ionization process with an I^n dependence in many atomic systems; for example, $n = 11$ was observed in the process $Xe \xrightarrow{nh\omega} Xe^+ + e^-$. The situation is, however, very different if there is a discrete m-photon intermediate resonant level in the n-photon ionization. As shown in Fig. 22.11,[20] the 4-photon ionization of Cs does have an I^4 dependence if 3ω is sufficiently far away from ω_{6S-6F}, but with $3\omega \sim \omega_{6S-6F}$, it is very different from I^4. Suppose the ionization signal S depends on I as I^K. Then, from the slope of $\log(S)$ versus $\log(I)$, we can find K. Figure 22.11 shows that as 3ω scans over ω_{6S-6F}, K varies drastically from 4 to 30 to 1 and finally back to 4. This result can be easily explained: since there is a light-induced shift on ω_{6S-6F}, the frequency denominator in the transition probability now also depends on I.[21] Qualitatively, when $3\omega > \omega_{6S-6F}$, an increase of I shifts ω_{6S-6F} closer to resonance and the ionization signal increases more rapidly than I^4, as indicated by a $K > 4$. For $3\omega < \omega_{6S-6F}$, an increase of I shifts ω_{6S-6F} farther away from resonance, and the ionization signal increases less rapidly than I^4 as indicated by a $K < 4$. The asymmetry of the curve in Fig. 22.11 can also be understood as due to interference of two ionization channels, one being a direct 4-photon ionization path and the other being a 4-photon ionization via the 3-photon intermediate resonance at ω_{6S-6F}.

Suppose from energy considerations that only N photons of the same frequency are required to ionize an atom, but one may find in an actual

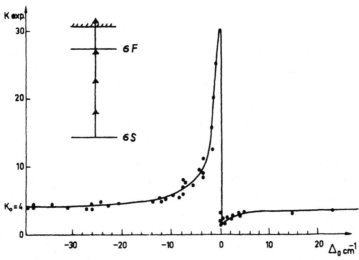

Fig. 22.11 Four-photon ionization of cesium with an intermediate resonant step: variation of the experimental order of nonlinearity K_{exp} as a function of the detuning from the three-photon transition $6S \rightarrow 6F$. (After Ref. 20.)

experiment that each atom is ionized after an absorption of more than N photons, depending on the laser intensity. This happens when the up-excitation rate in the continuum above the ionization level is larger than the ionization rate, and is evidenced by the appearance of higher-energy electrons released in the ionization process.[22] When the excitation rate is very high, it is even possible to excite an atom over the second, third, and nth ionization thresholds. Indeed, Manus and co-workers[23] demonstrated experimentally that Kr^{2+} can result from Kr via absorption of 33 photons from a 50-psec Nd:YAG laser pulse, and even Kr^{3+} and Kr^{4+} appear in the product.

Among the possible applications of multiphoton ionization are the detection of rare species (of atoms and molecules) (Section 19.3), isotope separation (Section 24.2), and generation of spin-polarized electrons. It has been suggested that multiphoton ionization of unpolarized atoms with circularly polarized light can produce highly polarized electrons.[24] Both the angular distributions and the polarizations of electrons produced in two-photon and multiphoton ionization of atoms have been measured.[25] An electron polarization as high as 0.8 has been observed in the two-photon ionization of Cs.

REFERENCES

1   See, for example, I. M. Beterov and V. P. Chebotaev, *Prog. Quant. Electron.* 3, 1 (1974), and references therein.

2   See, for example, C. Cohen-Tannoudji and S. Reynaud, *J. Phys. B* 10, 345, 365, 2311 (1977).

3   N. Bloembergen and Y. R. Shen, *Phys. Rev.* 133, A37 (1964).

4   S. H. Autler and C. H. Townes, *Phys. Rev.* 100, 703 (1955).

5   S. Yatsiv, *Phys. Rev.* 113, 1538 (1959); L R. Wilcox and W. E. Lamb, *Phys. Rev.* 119, 1915 (1960); M. S. Feld and A. Javan, *Phys. Rev.* 177, 540 (1969).

6   R. M. Whitley and C. R. Stroud, *Phys. Rev. A* 14, 1498 (1976); B. W. Shore and J. R. Ackerhalt, *Phys. Rev. A* 15, 1640 (1977); J. R. Ackerhalt, J. H. Eberly, and B. W. Shore, *Phys. Rev. A* 19, 248 (1979).

7   H. R. Gray, R. M. Whitley, and C. R. Stroud, *Opt. Lett.* 3, 218 (1978).

8   R. G. Brewer and E. L. Hahn, *Phys. Rev. A* 11, 1641 (1975); D. Grischkowsky, M. M. T. Loy, and P F. Liao, *Phys. Rev. A* 12, 2514 (1975).

9   K. I Osman and S. Swain, *Phys. Rev. A* 25, 3187 (1982).

10 See, for example, T. M Mossberg and S. R. Hartmann, *Phys. Rev. A* 23, 1271 (1981).

11 B R. Mollow, *Phys. Rev.* 188, 1969 (1969); *A* 5, 2217 (1972); *A* 12, 1919 (1975).

12 L A. Rahn, R. L. Farrow, M. L. Koszykowski, and P. L. Mattern, *Phys. Rev. Lett.* 45, 620 (1980).

13 F. Schuda, C. R. Stroud, and M. Hercher, *J. Phys. B* 7, L198 (1974); H. Walther, in S. Haroche, J. C. Pebay-Peyroula, T. W. Hansch, and S. H. Harris, eds., *Laser Spectroscopy II* (Springer-Verlag, Berlin, 1975); p. 358; F. Y. Wu, R. E. Grove, and S. Ezekiel, *Phys. Rev. Lett.* 35, 1426 (1975); R. E. Grove, F. Y. Wu, and S. Ezekiel, *Phys. Rev. A* 15, 227 (1977).

14 F. Y. Wu, S. Ezekiel, M. Ducloy, and B. R. Mollow, *Phys. Rev Lett* 38, 1077 (1977).

15 H. R. Gray and C. R. Stroud, *Opt. Commun.* 25, 359 (1978).

16 C. C. Wang and L. I. Davis, *Phys. Rev. Lett.* **35**, 650 (1975); J. F. Ward and A. V. Smith, *Phys. Rev. Lett.* **35**, 653 (1975).

17 P. F. Liao and J. E. Bjorkholm, *Phys. Rev. Lett.* **34**, 1540 (1975); M. M. T. Loy, *Phys. Rev. Lett.* **36**, 1454 (1976).

18 P. Agostini, G. Barjot, G. Mainfray, C. Manus, and J. Thebault, *IEEE J. Quant. Electron.* **6**, 782 (1970); M. LuVan, G. Mainfray, C. Manus, and I. Tugov, *Phys. Rev. A* **7**, 91 (1973); C. LeCompte, G. Mainfray, C. Manus, and S. Sanchez, *Phys. Rev. A* **11**, 1009 (1975).

19 L. A. Lompre, G. Mainfray, C. Manus, and T. Thebault, *J. Phys.* **39**, 610 (1978).

20 J. Morellec, D. Normand, and G. Petite, *Phys. Rev. A* **14**, 300 (1976).

21 C. S. Chang and P. Stehle, *Phys. Rev. Lett.* **30**, 1283 (1973); Y. Gontier and M. Trahin, *Phys. Rev. A* **7**, 1899 (1973).

22 P. Agostini, F. Fabre, G. Mainfray, G. Petite, and N. K. Rahman, *Phys. Rev. Lett.* **42**, 1127 (1979); P. Kruit, J. Kimman, and M. J. Van Der Wiel, *J. Phys. B* **14**, L597 (1981).

23 A. L'Huillier, L. A. Lompre, G. Mainfray, and C. Manus, *Phys. Rev. Lett.* **48**, 1814 (1982).

24 P. Lambropoulos, *Phys. Rev. Lett.* **30**, 413 (1973)..

25 M. J. Van der Wiel and E. H. A. Granneman, in J. H. Eberly and P. Lambropoulos, eds., *Multiphoton Processes* (Wiley, New York, 1978), p. 199; T. Hellmuth, G. Leuchs, S. J. Smith, and H. Walther, in W. O. N. Buimaraes, C. T. Lin, and A. Mooradian, eds., *Lasers and Applications* (Springer-Verlag, Berlin, 1981), p. 194.

BIBLIOGRAPHY

Balian, R., S. Haroche, and S. Liberman, eds., *Frontiers in Laser Spectroscopy* (North-Holland Publishing Co., Amsterdam, 1977).

Beterov, I. M., and V. P. Chabotayev, *Prog. Quant. Electron.* **3**, 1 (1974).

Delone, N. B., and V. P. Krainov, *Sov. Phys. Uspekhi* **21**, 309 (1978).

Eberly, J., and P. Lambropoulos, eds., *Multiphoton Processes* (Wiley, New York, 1978).

Letokhov, V. S., and V. P. Chabotayev, *Nonlinear Laser Spectroscopy* (Springer-Verlag, Berlin, 1977).

23

Infrared Multiphoton Excitation and Dissociation of Molecules

A most exciting late discovery in nonlinear optics is the phenomenon of infrared multiphoton excitation and dissociation of molecules. It was found that a molecule could be highly excited and eventually dissociated through frequency-selective absorption of tens of photons from an infrared laser pulse not much more intense than 10 MW/cm^2 and a few J/cm^2 per pulse. This was unexpected, as one would think that such a multiphoton excitation should have required a much more intense laser field. The process is of great fundamental importance and has far-reaching scientific and practical significance: the possibility of depositing a few electron volts of photon energy through vibration-rotational excitation in a molecule is extremely interesting for laser chemistry because it can drastically affect the chemical reaction involving the molecule. Being frequency-selective, infrared multiphoton dissociation is a viable method for isotope separation. Bond-selective (or mode-selective) multiphoton dissociation, if possible, could lead to a revolution in chemical synthesis. In this chapter, many important aspects of infrared multiphoton excitation and dissociation are discussed, with particular emphasis on the physical understanding of the process.

23.1 EARLY INVESTIGATIONS

Infrared multiphoton excitation (MPE) and dissociation (MPD) of single molecules were accidentally discovered by Isenor and Richardson[1] in 1971. In their experiment, when a high-power CO_2 laser pulse was focused into a resonantly absorbing gas medium (NH_3, CF_2Cl_2, etc.), visible luminescence appeared from the focal region, even when the laser intensity was well below the optical breakdown threshold. They identified the luminescence as emission from electronically excited dissociation products. A more careful study showed that the luminescence had an instantaneous part followed by a delayed part.[2] The instantaneous luminescence was believed to come from collisionless unimolecular dissociation and the delayed luminescence from collision-induced dissociation. Energy considerations indicated that the collisionless dissociation must have resulted from MPE of single molecules through the vibration-rotational ladder.

The resonant nature of MPE suggested the possibility of using MPD for isotope separation. In 1974, Ambartzumian et al. indeed found that MPD was isotopically selective.[3] This immediately aroused the interest of many researchers around the world, as it was envisioned that MPD used for isotope separation could be economically advantageous. Subsequent intensive studies showed that MPD was a general process applicable to a large number of molecules including SF_6, MoF_6, BCl_3, OsO_4, CF_2Cl_2. Through selective MPD in an isotopic mixture, appreciable isotope enrichment could be obtained. Consider the case of SF_6 as an example.[4] The natural abundance of $^{32}SF_6$ is 95% and that of $^{34}SF_6$ is 4.2%. This is reflected in the relative strength of the infrared absorption peaks of $^{32}SF_6$ and $^{34}SF_6$ shown in Fig. 23.1 for a gas mixture of 0.18 torr of SF_6 and 2 torr of H_2. Ambartzumian et al.[4] found that after the selective multiphoton excitation of $^{32}SF_6$ in the mixture by 2000 CO_2 laser pulses of 2 J per pulse, the $^{32}SF_6$ concentration was greatly reduced. This was seen from the resulting absorption spectrum, also shown in Fig. 23.1. The enrichment factor, defined by

$$\kappa = \frac{[^{34}S]}{[^{32}S]} \bigg/ \frac{[^{34}S]_0}{[^{32}S]_0}$$

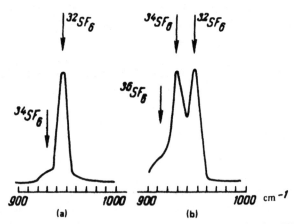

Fig. 23.1 IR absorption spectrum of the ν_3 vibrational mode in SF_6: (a) natural mixture, and (b) mixture enriched by infrared multiphoton molecular dissociation. [After R. V. Ambartzumian, Yu. A. Gorokov, V. S. Letokhov, and G. N. Makarov, JETP Lett. 21, 171 (1975).]

with [ ]$_0$ and [ ] denoting the initial and final concentrations, respectively, was 2800. The observed κ could be much higher if it were not for the fact that molecular collisions and chemical reactions had scrambled the isotopes.

The yields of MPD of SF_6 and other molecules were found to be very high, reaching a few tens of a percent if the laser fluence (energy per unit area) was above ~ 10 J/cm^2. That the process could be so efficient was a mystery. Clearly, the infrared MPE must be a resonant or near-resonant stepwise process; otherwise, any appreciable absorption of tens of photons would require a laser intensity much higher than 10 GW/cm^2. The resonant stepwise process would be possible if the absorbing molecule behaved as an harmonic oscillator in resonance with the incoming laser field. However, a molecular vibration is generally anharmonic. Because of the anharmonicity, a laser excitation resonant with the $v = 0$ to $v = 1$ transition is soon out of step with the vibrational ladder, as seen in Fig. 23.2. The stepwise resonant MPE would seem impossible.

It was pointed out by Ambartzumian et al.[5] and by Larsen and Bloembergen[6] that for the lower vibrational transitions in polyatomic molecules, the vibrational anharmonicity could be nearly compensated by the rotational energy, so that the laser field could remain in near resonance with the rotation-vibrational transitions between the vibrational excited states. Then, in a polyatomic molecule, the density of states increases rapidly with energy because of the large number of rotational and vibrational modes and forms a quasi-continuum.[2,7] Once the molecule is excited into the quasi-continuum, further multiphoton excitation up the quasi-continuum to and beyond the dissociation threshold is certainly stepwise resonant.

Aside from isotope separation, a considerable amount of interest and excitement was also generated with the suggestion that MPE could be a novel

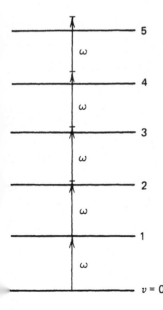

Fig. 23.2 Schematic diagram showing a multiphoton excitation getting out of step with an anharmonic vibrational ladder.

method for energizing molecules. It was hoped that by depositing energy into selected vibrational modes, molecules would be dissociated through certain dissociation channels different from those of thermal decomposition.[8] Initial experimental results appeared to support this hypothesis: the first reported analysis of the primary products of MPD of SF_6 indicated that the molecule dissociated into SF_4 and F_2 by passing over the lower dissociation channel with fragments SF_5 and F;[9] the results of MPD experiments on $CFCl_3$ were interpreted to evidence dissociation through a higher energy channel into $CFCl$ and Cl_2 rather than through the lowest energy channel $CFCl_2$ and Cl.[10] These results, however, were not substantiated by the subsequent MPD experiments with molecular beams.[11,12]

Many other interesting questions came up during the course of investigation. First, how does MPE modify the internal energy distribution in a molecule? Will the energy deposited into a molecule through excitation of a particular vibrational mode remain in that mode, or will it quickly randomize into many modes? In this respect, if the energy could be kept in the selected vibrational mode, then bond-selective (or mode-selective) MPD would be possible. Second, how many photons are actually absorbed before a molecule dissociates, and what limits the number of photons absorbed? Third, what are the dissociation products? As seen in Fig. 23.3, a polyatomic molecule such as SF_6 or CCl_3F has many dissociation channels with different dissociation energies. In MPD, would a molecule prefer to dissociate through the lowest energy channel, or could it selectively dissociate through a channel at a higher energy? In the latter case, one would have bond-selective dissociation. Fourth, what is the dynamics of MPD? After dissociation, how much excess energy

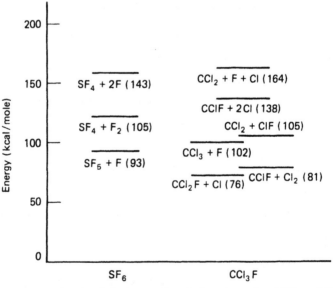

Fig. 23.3 Energy levels of the lower dissociation channels of SF_6 and CCl_3F.

appears as kinetic energy of the fragments and how much as internal energy in the fragments? How does the internal energy distribute in the fragments?

In the following section, a more detailed physical description of MPE providing some qualitative answers to these questions are presented. Section 23.3 offers a simple theoretical model to account for the many experimental observations.

23.2 PHYSICAL DESCRIPTION

Figure 23.4 is a schematic of a typical rotation-vibrational energy level diagram of a polyatomic molecule. It can be arbitrarily divided into three regions: the lower-energy discrete levels, the higher-energy quasi-continuum, and the true continuum above the lowest dissociation level. As mentioned earlier, to excite a significant fraction of the population into the continuum, the MPE process must be stepwise resonant or near-resonant. It is also necessary that the density of states be significantly higher at higher energies in order to have up-excitation dominate over downward stimulated emission. This is easy to understand because the transition rate is proportional to the density of states of the final state. (Molecules prefer to be excited into a level with less occupied states.)

Both of these conditions could indeed be satisfied in MPE of a polyatomic molecule, as we shall see in the following discussion, where the MPE process through the three regions of the energy-level diagram is described separately.

MPE Through Discrete Levels

In the discrete-level region, if excitation is restricted to pure vibrational transitions, then because of vibrational anharmonicity, stepwise resonant MPE is clearly impossible. However, it becomes possible when the complexity of the energy levels of a polyatomic molecule and the effects of intense laser excitation are taken into consideration. First, since the actual transitions can be vibration-rotational with allowed changes of quantum numbers $\Delta V = 1$ and $\Delta J = 0, \pm 1$, the anharmonic energy shift ΔE_{anh} between the $V \to (V + 1)$ and $(V + 1) \to (V + 2)$ transitions could be nearly compensated by the addition or removal of rotational energy through $|\Delta J| = 1$.[5,6] For example, the transitions $(V = 0, J) \to (V = 1, J - 1) \to (V = 2, J) \to (V = 3, J + 1)$ for a certain range of J values in a polyatomic molecule could have nearly equal transition frequencies. Then the anharmonic coupling between degenerate vibrational modes can induce anharmonic splittings in some overtone and combination levels which can also compensate the anharmonic shift over perhaps a wider range of initial J values.[13] Finally, the forbidden transitions, although much weaker, could still participate in MPE to make stepwise resonant transitions possible.[14] Since high laser intensities are used in the experiments, even the forbidden transition probability of MPE could be high. The high laser intensi-

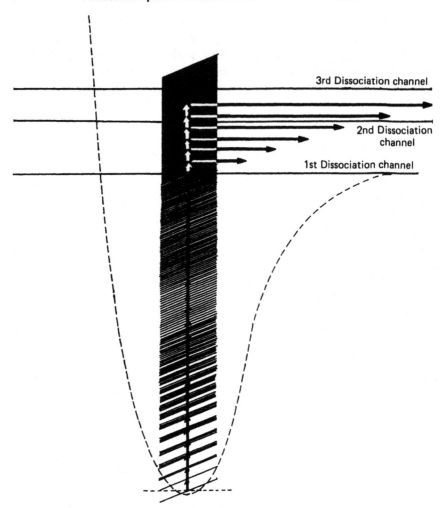

Fig. 23.4 Schematic energy level diagram of a polyatomic molecule. Multiphoton excitation in the various regions is denoted by the vertical arrows and dissociation by the horizontal arrows.

ties could also induce power broadening of the levels, which would further reduce the frequency mismatch in MPE. Thus one can expect that, in many polyatomic molecules, efficient stepwise resonant or near-resonant MPE from $V = 0$ (or 1 for thermally excited, vibrationally hot molecules) to $V = 3$–6 is possible. As we shall see later, at $V = 3$–6 the energy levels of many polyatomic molecules have already formed a quasi-continuum, assuming the vibrational frequency to be ~ 1000 cm^{-1}. This means that a significant fraction of the molecules can indeed be excited to the quasi-continuum.

The foregoing discussion shows that MPE is more efficient for molecules initially in certain (V, J) states satisfying the stepwise near-resonant condition. Consequently, at a given laser intensity, only a certain fraction of the molecules can be excited to the quasi-continuum, with the rest remaining in discrete levels.[15] This fraction increases with increasing laser intensity, approaching 1 at an intensity $\gtrsim 10^8-10^9$ W/cm^2 in many cases. It determines the maximum yield of MPD at a prescribed laser intensity because only molecules in the quasi-continuum can later be excited through the quasi-continuum to the true continuum above the dissociation threshold.

MPE Through the Quasi-Continuum

An n-atom molecule with $n \geq 3$ has $s = 3n - 6$ vibrational modes. As a result, the number of combination and overtone states increases rapidly with energy, more so in molecules with larger n. The levels soon become so dense that they practically form a continuum. This is known as the quasi-continuum. Superimposed on the vibrational levels are the rotational levels, which further increase the density of states.

Several approximate formulas have been proposed to calculate the density of vibrational states of a polyatomic molecule.[16] Among them, the Whitten–Rabinovitch approximation has been widely accepted. It gives the following expression for the density of vibrational states:

$$g(E) = \frac{(E + aE_0)^{s-1}}{(s - 1)! \prod\limits_{i=1}^{s} \hbar\omega_i} \left[1 - \beta \frac{dW}{d\eta} \right] \tag{23.1}$$

where E_0 is the total zero-point vibrational energy of the molecule, E is the energy measured from E_0, ω_i is the frequency of the ith vibrational mode, s is the number of vibrational modes, $\beta = (s - 1)\langle\omega^2\rangle/s\langle\omega\rangle^2$ with $\langle\omega\rangle$ and $\langle\omega^2\rangle$ being the mean and mean square vibrational frequencies, respectively, $a = 1 - \beta W(\eta)$, $\eta = E/E_0$, and

$$W(\eta) = (5.00\eta - 2.73\eta^{1/2} + 3.51)^{-1} \quad \text{for } 0.1 < \eta < 1.0$$

$$= \exp(-2.419\eta^{1/4}) \quad \text{for } 1.0 < \eta < 8.0.$$

Figure 23.5 shows the densities of states $g(E)$ versus E for a number of molecules. It is seen that $g(E)$ indeed increases much more rapidly with E in larger molecules. In SF$_6$, for example, $g(E)$ is more than 100 states/cm^{-1} at $E = 4000$ cm^{-1} and is close to 1000 states/cm^{-1} at $E = 5000$ cm^{-1}. The density of states is much higher when rotational states are taken into account.[16] Thus in every practical sense we can say that an SF$_6$ molecule is in quasi-continuum if it is vibrationally excited to $E \geq 4000$ cm^{-1}. This can be achieved in

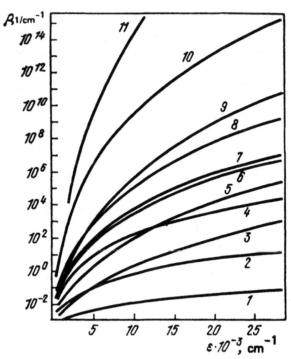

Fig. 23.5 Densities of states as a function of energy for different molecules: (1) D_2O; (2) OCS; (3) CH_3F; (4) BCl_3; (5) C_2H_4; (6) CF_3I; (7) OsO_4; (8) C_2F_3Cl; (9) SF_6; (10) UF_6; (11) S_2F_{10}. [After V. S. Letokhov and A. A. Makarov, *Uspekhi* **24**, 366 (1982).]

SF_6, for example, by MPE of the ν_3 mode ($\omega \sim 950$ cm^{-1}) from $V = 0$ or 1 to $V \geq 4$ using a pulsed CO_2 laser.

We should note that even in the quasi-continuum, the absorption spectrum may show a structure characteristic of the various vibrational modes. This depends on the mode–mode coupling. Anharmonic coupling in general causes significant red shifts and huge broadenings of the rotation-vibrational transitions. With the molecules in quasi-continuum, the absorption spectrum should exhibit very broad, red-shifted bands. In some special cases, however, a particular vibrational mode may couple only weakly with other modes for reasons such as large mismatches in the mode frequencies. Then, even in the quasi-continuum, the absorption spectrum may show relatively narrow peaks characteristic of the mode.[17]

In all practical cases we know, we can regard MPE in the quasi-continuum as stepwise resonant. The rapid increase of the density of states with the excitation energy also guarantees a net up-excitation of the population in the quasi-continuum. In exciting a molecule under the collisionless condition, each step is a one-photon transition with negligible relaxation (as the spontaneous emission is weak compared to stimulated emission and absorption with the laser intensities used in MPE). Therefore, the overall MPE in the quasi-continuum by an infrared laser pulse is proportional to the laser fluence (time integration of the laser intensity). In other words, the energy deposited into the molecule is directly proportional to the laser energy available.

If the coupling between modes is strong, the energy absorbed by the molecule in the quasi-continuum is expected to be distributed in all modes. Then MPE may be considered a laser heating process, in which the molecule as a whole acts as a small thermal reservoir to be heated up via successive absorption of photons at a selected wavelength.[18] However, it is different from the usual heating process in the sense that the excited population distribution is not thermal.[19] It is of course also possible that the energy deposited in the molecule may not get randomized into all the modes.

MPE and MPD in the True Continuum

With enough laser fluence (~ 1 J/cm^2 for SF$_6$, CF$_3$Cl, etc.), the MPE process can excite a molecule through the quasi-continuum into the true continuum above the dissociation threshold. Once in the true continuum, the molecule should dissociate, but the dissociation is in competition with the continuing stepwise resonant up-excitation.[19] Being successive one-photon transitions, the up-excitation has a rate proportional to the laser intensity but it does not strongly depend on the excitation energy in the molecule. On the other hand, the dissociation rate, which is nearly zero at the dissociation threshold, increases very rapidly with the excess energy above the threshold. As a result, the up-excitation in the continuum is soon limited by the dissociation. The final level a molecule can be excited to in the continuum should increase with increasing laser intensity. It determines the excess energy the fragments will carry when the molecule dissociates.

In real cases we should also consider the population distribution in the excited states.[19] The MPE process with a sufficiently strong laser fluence can create a population distribution that is partly in the quasi-continuum and partly in the true continuum. The high-energy tail of the distribution is usually truncated by rapid dissociation overcoming up-excitation. With a higher laser fluence, a larger fraction of the population can be excited into the continuum. With a higher laser intensity, on the other hand, the high-energy tail of the distribution is expected to extend higher up. (We should, of course, also remember that a higher laser intensity can drive a larger fraction of the molecules from the discrete levels to the quasi-continuum, as discussed earlier.) In dissociation, the mean excess energy appears as the mean energy carried away by the fragments.

It is possible that with sufficiently high laser intensity and fluence, part of the molecules can be excited beyond the second and third dissociation levels. In that case, the molecules can dissociate simultaneously through several channels, yielding different dissociation products.[20] The dissociation rate through each channel depends on the density of states, as described in Section 23.4.

Upon dissociation, the excess energy goes into the translational and rotation-vibrational degrees of freedom of the fragments. How the energy is distributed should depend on the nature of the bond that is broken and on the detailed dynamics of the bond breakage. In most cases, because of the large number of internal degrees of freedom in the molecular fragments, a major portion of the excess energy would appear as the internal energy in the fragments.[20]

In the following sections, a simple model is used to substantiate the above qualitative description. How experimental results support the predictions is also discussed.

23.3 A SIMPLE MODEL OF INFRARED MULTIPHOTON EXCITATION AND DISSOCIATION

We consider MPE of a polyatomic molecule by an infrared monochromatic field as a stepwise resonant excitation process over a set of equally spaced discrete levels,[19,21] shown in Fig. 23.6. The following assumptions are used in the model:

1 The lowest level in Fig. 23.6 is at the bottom of the quasi-continuum. The initial population in the ground levels is pumped into this level via MPE over the discrete-level region.

2 The degeneracy of each level is given by the density of states of the molecule according to (23.1).

3 Rate equations can be used to describe transitions.

4 A molecule excited over the dissociation energy level will dissociate with a rate depending on the excitation energy.

The thermal distribution of the initial population and the effect of coherent excitation are neglected in the model. The former is not important if we are interested only in the kinetics of MPE and MPD, while the latter is expected to be negligible because of the very short dephasing times of the highly degenerate levels.

Thus the rate equations governing MPE and MPD can be written as

$$\frac{dN_m}{dt} = \frac{I(t)}{\hbar\omega}\left[\sigma_{m-1}N_{m-1} + \frac{g_m}{g_{m+1}}\sigma_m N_{m+1} - \frac{g_{m-1}}{g_m}\sigma_{m-1}N_m - \sigma_m N_m\right]$$
$$- K_m N_m \qquad \text{for } m > 1$$

and   (23.2)

$$\frac{dN_1}{dt} = \frac{I(t)}{\hbar\omega}\left[\frac{g_1}{g_2}\sigma_1 N_2 - \sigma_1 N_1\right] - \frac{dN_0}{dt}.$$

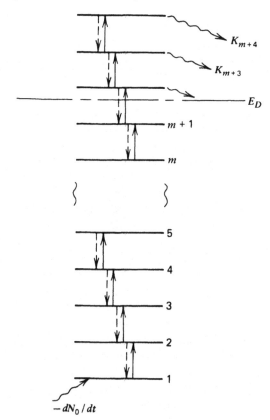

Fig. 23.6 Schematic diagram showing multiphoton excitation and dissociation of a simple model system with equally spaced energy levels.

Here N_m is the normalized population in the mth level with energy $E_m = E_1 + (m - 1)\hbar\omega$, g_m is the density of states at E_m, σ_m is the absorption cross section for the $m \rightarrow (m + 1)$ transition ($g_m\sigma_m = g_{m+1}\sigma_{m+1}$ from detailed balancing), K_m is the dissociation rate constant, which is nonzero only if E_m is larger than the dissociation energy E_D(or $E_D + E_B$ if the dissociation has also an exit barrier E_B to overcome), N_0 is the population in the ground levels, and $-dN_0/dt$ describes the rate of population increase in level 1 resulting from MPE over the discrete levels. Both absorption and stimulated emission are taken into account in (23.2). For a polyatomic molecule, the absorption cross sections σ_m are generally difficult to estimate because of lack of information about the excited vibrational states of a molecule. However, we expect that the anharmonic coupling between modes would cause σ_m to decrease with increasing m. In the present model, we then simply assume $\sigma_m = \sigma_0\exp(-\beta m)$ with the constants σ_0 and β to be determined by a fit to the experimental results. We also assume, for simplicity, $dN_0/dt = N_0\delta(t)$, so that immediately after $t = 0$, an initial population N_0 would appear in level 1. A more reasonable form of

$N_0(t)$ depending on the laser intensity $I(t)$ can be used in a more sophisticated calculation.[21]

The dissociation rate constant K_m as a function of E_m can be calculated using the RRKM model.[16] The basic assumption of the model is that the excitation energy in a molecule is distributed randomly in all modes; in other words, there is equal probability of finding the molecules in all the degenerate states. Let $g(E)dE$ be the total number of states between E and $E + dE$. If $E > E_D$, then part of these states should be the dissociation states in which a molecule would dissociate. The dissociation rate at E is proportional to the fraction of dissociation states in $g(E)dE$. We need, however, to define the dissociation states in order to find their number. We realize that in a dissociation process, a bond must be broken along a certain reaction coordinate. Consider, for example, the dissociation of $CF_3I \rightarrow CF_3 + I$. The reaction coordinate is along the CI bond, since, during dissociation, I moves away from C along that coordinate. If R, the distance between I and C, is within a critical range $d \pm \Delta$, and if at the same time, the momentum p_R is larger than zero, indicating a continuing separation of I from CF_3, then the molecule will dissociate. Therefore, a dissociation state is characterized by $E > E_D$, $(d - \Delta) \le R \le (d + \Delta)$, and $p_R > 0$. The corresponding translational energy of the dissociating fragments in the critical region is $\mathscr{E} = \frac{1}{2}p_R^2/\mu$, where μ is the reduced mass. If the exit barrier for dissociation is zero, then $\mathscr{E}$ is also the translational energy of the resulting dissociation products. Let the number of dissociation states between E and $E + dE$ with a translational energy between $\mathscr{E}$ and $\mathscr{E} + d\mathscr{E}$ be $D(E, \mathscr{E})dE\, d\mathscr{E}$. Then the probability that a molecule at E dissociates with a translational energy between $\mathscr{E}$ and $\mathscr{E} + d\mathscr{E}$ is $D(E, \mathscr{E})d\mathscr{E}/g(E)$. Knowing that the time for the dissociating system to cross the critical region is $2\mu\Delta/p_R$, one finds the dissociation rate

$$k(E, \mathscr{E})d\mathscr{E} = (p_R/2\mu\Delta)D(E, \mathscr{E})d\mathscr{E}/g(E). \tag{23.3}$$

The total dissociation rate for a molecule at E is

$$K(E) = \int_0^{E - E_D} k(E, \mathscr{E})d\mathscr{E} \tag{23.4}$$

and the dissociation lifetime is $\tau(E) = 1/K(E)$. In the case where the dissociation has no exit barrier, one can also obtain the translational energy distribution of the dissociation products from

$$P(E, \mathscr{E}) = k(E, \mathscr{E})/K(E). \tag{23.5}$$

The density of dissociation states $D(E, \mathscr{E})d\mathscr{E}$ is the product of two parts: $g^*(E - \mathscr{E})$, which is the density of states for the molecule at $(E - \mathscr{E})$ in all degrees of freedom other than the reaction coordinate, and $\nu(\mathscr{E})d\mathscr{E}$, which is the number of states between $\mathscr{E}$ and $\mathscr{E} + d\mathscr{E}$ associated with the reaction coordinate. We can treat the molecule in the reaction coordinate as a particle of mass μ in a one-dimensional box of length 2Δ. The total number of states between 0 and $\mathscr{E}$ with $p_R > 0$ is $(2\mu\Delta^2\mathscr{E}/\pi^2\hbar^2)^{1/2}$, and therefore

$$\nu(\mathscr{E})d\mathscr{E} = \left(\frac{\mu\Delta^2}{2\pi^2\hbar^2\mathscr{E}}\right)^{1/2} d\mathscr{E}$$

$$= \left(\frac{\mu\Delta}{\pi\hbar p_R}\right)d\mathscr{E} \tag{23.6}$$

$$D(E, \mathscr{E})d\mathscr{E} = \left(\frac{\mu\Delta}{\pi\hbar p_R}\right)g^*(E - \mathscr{E})d\mathscr{E}.$$

Insertion of $D(E, \mathscr{E})$ into (23.3) yields

$$k(E, \mathscr{E})d\mathscr{E} = \frac{g^*(E - \mathscr{E})d\mathscr{E}}{2\pi\hbar g(E)}. \tag{23.7}$$

Note that this final expression of $k(E, \mathscr{E})$ is independent of the dimension of the critical region we assumed. Therefore, using (23.1) to calculate $g(E)$ and $g^*(E - \mathscr{E})$ for a given molecule, we should be able to find the values for $k(E, \mathscr{E})$, $K(E)$, $\tau(E)$, and $P(E, \mathscr{E})$. With $K_m \equiv K(E_m)$ known, the rate equations of (23.2) can then be solved.

We use MPE and MPD of SF_6 by infrared laser excitation of the ν_3 mode (948 cm^{-1}) as an example.[19] The frequencies of the 15 vibrational modes of SF_6 are taken as 774(1), 642(2), 948(3), and 481(9) cm^{-1}, where the numbers in parentheses denote the degeneracies of the modes and 481 cm^{-1} is the harmonic mean frequency of all the bending modes. The MPD of $SF_6 \rightarrow SF_5$ + F has the reaction coordinate along an SF bond. We assume that in the critical configuration for dissociation, only three modes are affected by the change: one 948 cm^{-1} stretching mode disappears, and two of the 481 cm^{-1} bending modes are softened to $481\exp(-d/r_0)$ with $d = 3.7$ Å and the equilibrium bond distance $r_0 = 1.56$ Å. The quasi-continuum is assumed to begin at $E = 11$ kCal/mole and the dissociation energy of SF_6 is known to be 93 kCal/mole. Then, from (23.1), the densities of states and K_m can be calculated. The absorption cross section is taken as $\sigma_m = (8 \times 10^{-19})\exp(-0.042\ m)$ cm^2, which roughly reproduces the experimental results of multiphoton absorption in SF_6.[22] Finally, we assume the laser excitation to be a pulse of constant intensity I_0 and pulse duration T_p. All the coefficients in (23.2) are thus specified, and the rate equations can be solved numerically. The results are shown in Figs. 23.7 to 23.11.

Figure 23.7 describes MPE in the quasi-continuum of SF_6 by a resonant laser beam of 200 MW/cm^2. It is seen that the laser excitation broadens the

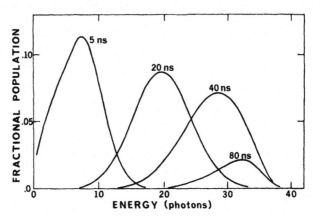

Fig. 23.7 Calculated population distribution at various times produced by a rectangular laser pulse excitation of 200 MW/cm^2. [After Aa. S. Sudbø, P. A. Schulz, Y. T. Lee, and Y. R. Shen, in *Proceedings of the First International School on Laser Applications to Atoms, Molecules, and Nuclear Physics*, Vilnius, USSR (1978), p. 338.]

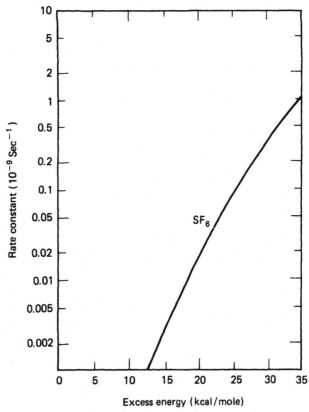

Fig. 23.8 Dissociation rate of SF$_6$ as a function of the excess energy above dissociation. (After Sudbø et al., cited in Fig. 23.7.)

population distribution with time and drives it up to higher levels; the average number of photons absorbed per molecule increases accordingly. After ~ 20 nsec, the high-energy tail of the population distribution starts to have an appreciable fraction above the dissociation threshold, indicating that dissociation has happened. The laser excitation continues to drive the population distribution up, but the action soon is limited by the increasing rate of dissociation. As shown in Fig. 23.8, the dissociation rate increases very rapidly with the excess energy, and hence dissociation would effectively deplete all the populations excited beyond certain levels. This is evidenced by the more abrupt cutoff on the high-energy side of the population distribution and by the net decrease of population at longer times in Fig. 23.7. If a 100-nsec, 20-J/cm² laser pulse is used for excitation, most of the molecules appear to have dissociated during the pulse with an excess of energy of 6–11 $\hbar\omega$ (16–30 kCal/mole), as seen in Fig. 23.9. A very small fraction of the molecules in lower levels will dissociate after the laser pulse is over because of the low dissociation rates. The excess energy is distributed mainly in the internal degrees of freedom of the fragment SF_5. As shown by $P(E, \mathscr{E})$ versus $\mathscr{E}$ in Fig. 23.10, the average translational energy in the fragments is only a few kilocalories per mole.

The calculation also answers the question on how laser intensity and fluence affect the population distribution above the dissociation threshold and the spread of excess energy with which SF_6 dissociates. Figure 23.11 shows the spread of excess energy in MPE of SF_6 using a 7.5-J/cm² laser pulse at two

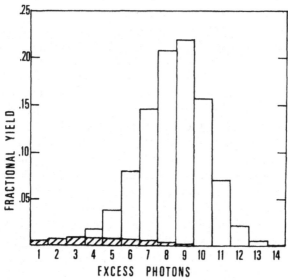

Fig. 23.9 Calculated dissociation yields from various levels above the dissociation energy during the laser pulse (unshaded region) and after the laser pulse (hatched region) for a 100-nsec, 200-MW/cm² laser pulse excitation. (After Sudbø et al., cited in Fig. 23.7.)

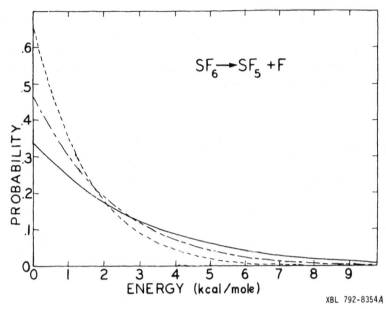

$SF_6 \rightarrow SF_5 + F$

XBL 792-8354A

Fig. 23.10 Calculated center-of-mass translational energy distribution in the fragments from the multiphoton dissociation of SF_6, assuming an excess energy of 5 (---), 8 (—·—), and 12 (——) kcal/mole. (After Ref. 12.)

different pulse durations: 60 and 0.6 nsec full widths at half-maxima. Because the up-excitation rate is much faster with the shorter and more intense pulse, the average excess energy is higher. However, for the 0.6-nsec pulse, only the fraction of molecules with excess energy larger than $13\hbar\omega$ can dissociate during the pulse. Their excess energy is limited by the balance of up-excitation with dissociation. Those molecules dissociating after the pulse is over have their excess energy limited by the laser fluence in the stepwise resonant pumping. With the 60-nsec pulse, the fraction with excess energy more than $7\hbar\omega$ can dissociate during the pulse. According to the calculation, if the laser fluence is sufficiently low (< 5 J/cm² in the present case), most of the molecules would dissociate after the laser pulse is over, irrespective of the laser pulsewidth (≤ 1 μsec). This is because energy-wise, the laser excitation cannot drive the molecules too far above the dissociation level. Then it is the laser fluence rather than intensity that determines the average excess energy. On the other hand, if, for a given pulsewidth, the laser fluence is sufficiently high such that most of the molecules are pumped to sufficiently high excited states, and are expected to dissociate during the laser pulse, the laser intensity should then determine the average excess energy. For a very short laser pulse, it is easily seen that the average level of excitation of the molecules should be limited by fluence consideration.

The calculation can, of course, be modified to take into account more realistic situations. For example, a molecule can have a number of dissociation channels at different energies. If the laser fluence and intensity are sufficiently

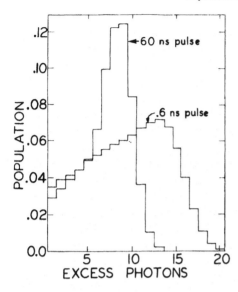

Fig. 23.11 Dissociation yield of SF_6 as a function of the number of excess photons for 0.6- and 60-nsec laser pulses, each at an energy fluence of 7.5 J/cm^2. (After Ref. 12.)

high, a significant fraction of the molecules can be excited beyond the higher dissociation channels; then, simultaneous dissociations through several channels can occur. This situation can be included in the calculation by adding terms in the rate equations of (23.2) describing dissociation through higher channels. The corresponding dissociation rate constants can be calculated using (23.4) for the different channels. In another case, the molecular fragment from MPD, such as SF_5, may contain so much internal energy that it also appears in the quasi-continuum. If most of the molecules dissociate before the pulse is over, then the fragments can be further excited by the laser via MPE through the quasi-continuum and into the true continuum above the dissociation threshold. This secondary MPD process can also be included in the calculation by incorporating in (23.2) another set of rate equations describing the populations of the fragments in different levels. Section 23.4 shows how the calculations sketched here can give a fair description of the experimental observations.

23.4 EXPERIMENTAL RESULTS

Earlier observations on MPE and MPD described in Section 23.2 were later substantiated by experiments more carefully designed to study particular features of the processes. In an experiment designed to show the two distinct stages of MPE, one over the discrete levels and one through the quasi-continuum, two infrared laser pulses at different frequencies were used.[23] One was nearly resonant with a fundamental vibrational frequency. It had sufficiently high intensity to excite the molecules over the discrete levels but not enough fluence to dissociate them. The other was far from resonance with any fundamental vibrational frequency and also was not intense enough to excite the molecules over the discrete states. It had, however, the appropriate frequency and enough fluence to be able to excite the molecules through the quasi-continuum to dissociation. Thus either pulse alone could not dissociate the molecules, whereas the two pulses together could. Tuning of the frequencies of the two lasers in the MPD experiment showed a sharp resonant structure with respect to the first laser excitation, and a strongly red-shifted, nearly featureless, broad spectrum with respect to the second laser excitation (Fig. 23.12).[24] They are apparently characteristics of MPE over the discrete levels and the quasi-continuum, respectively. The observed sharp resonant feature of the first laser excitation suggests that it is more advantageous to use the two-laser excitation scheme in isotope separation by MPD, especially for heavy isotopes in molecules with small isotope shift.[24]

The fraction of molecules excited to the quasi-continuum should depend on the exciting laser intensity, which can be measured by spontaneous Raman scattering since molecules in the quasi-continuum have a smaller vibrational Raman shift than those in the discrete levels.[25] This is shown in Fig. 23.13. It is seen in Fig. 23.13b that for the same laser pulseshape, the fraction of molecules in the quasi-continuum increases with the laser fluence, and hence with the

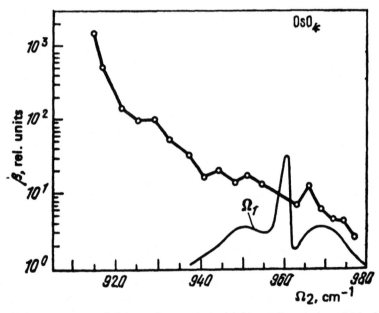

Fig. 23.12 Dependence of the two-frequency multiphoton dissociation yield of OsO_4 on the frequency Ω_2 of the second laser. Other parameters of the pulses were fixed: $\Omega_1 = 954.5$ cm^{-1}, $\Phi_1 = 0.24$ J/cm^2, $\Phi_2 = 0.22$ J/cm^2. The linear absorption spectrum of OsO_4 is represented by the curve denoted by Ω_1. [After R. V. Ambartzumian, V. S. Letokhov, G. N. Makarov, and A. A. Puretskii, *Opt. Commun.* **25**, 69 (1978).]

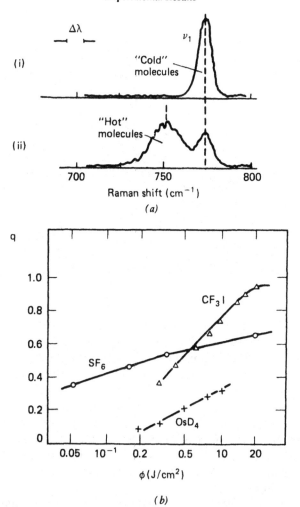

Fig. 23.13 (a) Raman spectrum of (i) unexcited SF_6 molecules, and (ii) SF_6 molecules excited by a CO_2 laser pulse of 0.5 J/cm^2 fluence. $\Delta\lambda$ indicates the instrumental linewidth. [After V. N. Bagratashvili, Yu. G. Vainer, V. S. Dolzhikov, V. S. Letokhov, A. A. Makarov, L. P. Malyavkin, E. A. Ryabov, and E. G. Sil'kis, *Opt. Lett.* **6**, 148 (1981).] (b) Dependence of the relative fraction q of the molecules excited into the quasi-continuum on the CO_2 laser fluence in SF_6, OsO_4, and CF_3I. [After V. S. Letokhov and A. A. Makarov, *Uspekhi* **24**, 366 (1982).]

laser intensity. At very high intensities, all the molecules can be excited into the quasi-continuum.

The MPE of a molecule can be monitored by the absorption measurement,[22,26] which yields the average number of photons absorbed per molecule, $\langle n \rangle$. Figure 23.14 shows the dependence of $\langle n \rangle$ on laser fluence Φ for a number of different molecules. For small molecules such as OCS and D_2O, the

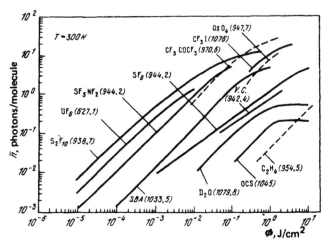

Fig. 23.14 Average number of infrared photons absorbed versus exciting laser fluence for a number of polyatomic molecules. [After V. S. Letokhov and A. A. Makarov, *Uspekhi* **24**, 366 (1982).]

absorption is at first linear and then quickly saturates with $\langle n \rangle < 1$ at higher fluences. This is expected since the quasi-continuum is too high to be easily reached by MPE in these molecules. They therefore behave more like a two-level system. For the larger polyatomic molecules, MPE is characterized by $\langle n \rangle$ increasing monotonically with Φ. If most molecules have already been excited to the quasi-continuum, then the absorption reflecting the stepwise one-photon excitations should show $\langle n \rangle$ increasing linearly with Φ until dissociation sets in. This happens when the exciting laser pulse is so intense that it can excite essentially all the molecules over the discrete levels.[22] With dissociation depleting the absorbing molecules, the dependence of $\langle n \rangle$ on Φ finally becomes sublinear.

The average excitation level above the dissociation threshold that a molecule can be excited to depends in general on both the laser fluence and the laser intensity. It can be determined in a crossed laser and molecular beam experiment in cases where the dissociation has no exit barrier, as in an atomic elimination process.[27] The laser beam dissociates the molecules in the molecular beam. The fragments produced can be identified, and their angular and velocity distributions can be measured by a rotatable mass spectrometer. From the angular and velocity distributions, the translational energy distribution of the fragments can be derived and compared with that calculated from the RRKM calculation. The fit then allows an estimate of the average excess energy. Figure 23.15 shows an example of the fit, from which it was concluded that the average excess energy carried by the fragments of SF_6 was $\sim 7\hbar\omega$ and that most of it appeared in the internal degrees of freedom of SF_5. That the experimental data could not be fit by a calculation assuming only a few vibrational modes participating in the randomization of the excitation energy

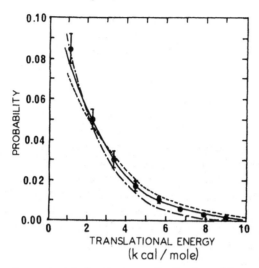

Fig. 23.15 Center-of-mass translational energy distribution of fragments in $SF_6 \rightarrow SF_5$ + F. Experimental data points obtained with ~ 6 J/cm² laser pulses are denoted by ●. Curves are calculated from the RRKM theory assuming a molecular excitation of $E = E_D + n\hbar\omega$ with $n = 7$ (—·—), $n = 9$ (——), and $n = 11$ (---) where E_D is the dissociation energy and $\hbar\omega$ is the CO_2 laser photon energy.

suggested the validity of the RRKM model; that is, the excitation energy must have been randomized in nearly all the modes. It has also been found experimentally that the average excess energy indeed increases with the laser intensity and fluence, as the theory predicted.[21]

Molecular dissociation may have an exit energy barrier. This is the case of bond rupture with molecular elimination. During the dissociation process, when the fragments have crossed the critical region and are on their way to infinite separation, the barrier energy is partly converted to translational energy and partly to internal energy in the fragments. Consequently, a relatively larger fraction of the excess energy would appear as translational energy. This was confirmed by experiment.[28]

The internal energy distribution in the fragments was measured by the laser-induced fluorescence technique (Section 19.2) in a few cases.[29] It was found that the energy distributions in the rotational and vibrational degrees of freedom of the fragments were Boltzmann-like and could be described by effective rotational and vibrational temperatures, T_R and T_V, respectively. Both T_R and T_V appeared to be higher than the approximate equivalent temperature describing the translational energy distribution, indicating that thermal equilibrium had not been established among the different degrees of freedom.

A complex molecule often has more than one low-lying dissociation level (or channel) over which the molecule can be excited through MPE. Then, in MPD, simultaneous dissociation over several competing channels can be observed. Their relative probabilities depend on the excited population distribution and the dissociation rates through the different channels. The latter can be estimated from the RRKM model. Depending on the densities of states, the dissociation rate through the lower channel may increase much more slowly with the excess energy than the one through the higher channel. As a result, it can happen that above a certain excitation energy, dissociation through the higher channel would dominate. Competition of two dissociation channels has actually been observed in the crossed laser-molecular beam experiment.[30] At low laser fluence, the lower channel dominated, but with increasing fluence, the upper channel contributed more and more to the products.

Secondary dissociation has also been observed in the MPD of some molecules.[21,27] In the SF_6 case, for example, when the dissociation was barely detectable, only the SF_5 and F were found in the products. Then, with higher laser fluence, SF_4 began to show up, yet no F_2 could be detected. The angular distribution of SF_4 was appreciably broader than that of SF_5. These results together with other measurements concluded that SF_4 was the secondary dissociation product from SF_5.

Many other experiments on MPD of polyatomic molecules, conducted either in a gas cell or in a molecular beam, have also contributed to the understanding of the problem. They are described in the review articles listed in the Bibliography.

23.5 ENERGY RANDOMIZATION IN A MOLECULE

Whether the excitation energy can be quickly randomized in a molecule is a very important question from both scientific and practical points of view. On the scientific side, energy randomization is the basic assumption in the theory of unimolecular dissociation and it would forbid mode-selective chemical reaction. On the practical side, energy nonrandomization would make MPE significantly different from thermal excitation and could open a new branch in chemical synthesis.

Two different pictures can be used to describe the energy randomization process. In the first picture, a molecule is treated as a bunch of modes or oscillators coupled by anharmonic forces. The pulsed laser energy is deposited into the molecule through selective excitation of a particular mode. Then, because of mode–mode coupling, energy is randomized into all modes in a certain characteristic time τ. Obviously, stronger mode–mode coupling should lead to shorter τ. In the second picture, the real eigenstates of the molecule are considered. Because of anharmonic coupling, the modes are so mixed that none of the eigenstates can be regarded as a pure mode state. When a coherent laser pulse of frequency ω and pulsewidth T_p is used to excite the molecule, different states in the frequency range between $\omega - \frac{1}{2}T_p^{-1}$ and $\omega + \frac{1}{2}T_p^{-1}$ are coherently excited to different extents. The coherent beat of the excitations causes the populations in these excited states to vary with time. If the number of these states is large, then interference of the beats makes the population in each state appear to relax toward an equilibrium value in a characteristic time τ. Again, if anharmonic coupling is larger, more excited states should participate effectively in the coherent beating and τ becomes shorter.

Experiments on MPE and MPD have so far yielded no firm evidence that excitation energy in the upper quasi-continuum or true continuum is not randomized during the laser pulse. The crossed laser-molecular beam experiment described earlier, for example, showed farily good agreement with the statistical RRKM model, which is based on the assumption of energy randomization, in the predictions of the dominating dissociation channels, and the translational energy distribution of the fragments. Other experimental results, such as the average number of photons absorbed per dissociating molecule, are also consistent with the RRKM model. It is possible that some particular modes of the molecule may not participate fully in the energy randomization because of their weak coupling to the other modes as a result of large frequency mismatch, for example. With this in mind, MPD of some molecules via MPE of the C–H bond has been studied.[31] The C–H vibrational frequency at ~ 3000 cm^{-1} was far from the frequencies of the other vibrational modes of the molecules, which were all around or below 1000 cm^{-1}. Yet the results obtained were still consistent with the RRKM model.

To see the effect of energy nonrandomization in MPE during a laser pulse, it is clear that we need a very short pulse and excitation of a mode more or less isolated from other modes. The characteristic time τ of energy randomization in the quasi-continuum is believed to be of the order of picoseconds or less. Therefore, picosecond- or subpicosecond-laser pulses would be needed to see energy nonrandomization. However, to excite the molecules to high energy states or into the dissociation continuum, the laser pulse must also have enough fluence. This would then make the laser peak intensity so high that many dissociation and ionization channels could open at the same time, thus confusing the results. Consequently, energy nonrandomization and hence selective bond breaking would be difficult to realize.

It might be possible that in a molecule with weakly connected, spatially separated groups, energy randomization during a certain time period is limited to the group that has been excited. In this case, MPD would break the weakest bond in the group, which is not necessarily the weakest of the whole molecule. Then MPD of the molecules could appear significantly different from thermal dissociation.

23.6   AN ANALOG MODEL OF MULTIPHOTON DISSOCIATION

We now discuss an analog that can help visualize many important aspects of MPD.[20] Consider a trough with many compartments, as in Fig. 23.16. There are small holes in the partition walls, so that as the trough is filled, water

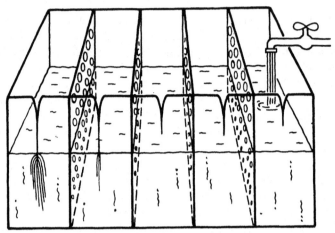

Fig. 23.16 An analog model for multiphoton excitation and dissociation of poly-atomic molecules.

pouring into one compartment will flow into the other compartments through the holes. With the trough acting as a molecule, the compartments as vibrational modes, the holes as mode–mode coupling, and water as excitation energy, this picture can be used to illustrate MPD of a molecule fairly well. The flow of water between compartments corresponds to energy flow between modes, and the overflow of water through the V-shaped openings in the containing wall corresponds to molecular dissociation through various channels. The hole size increases with height to indicate the stronger mode–mode coupling at higher energy. The V-shape of the openings is designed to make the overflow rate increase rapidly as the water level moves up.

A number of conclusions can be readily drawn from Fig. 23.16:

1. The equilibration of water in various compartments depends on the filling rate of the water and the size of the holes. Larger holes and slower filling rate help the equilibration. This is analogous to the energy randomization process in the excitation of a molecule.

2. If the water filling is sufficiently slow and the equilibration sufficiently fast, so that at all times water in all compartments is at the same level, then it will finally rise above the lowest V-shaped opening and leak through that opening. The final water level is determined by the balance between the filling and leaking rates. This is analogous to MPE of a molecule through the quasi-continuum into the true continuum and its dissociation through the lowest channel with an average excess energy.

3. If the filling rate is sufficiently fast, but still slow enough for equilibration to be established among all the compartments, the final water level may rise above the second or even third lowest opening. Then water can simultaneously leak through more than one opening. This corresponds to MPD of a molecule through several competing channels as a consequence of MPE by an intense laser pulse.

4. If the amount of water to be poured into the trough is small, the level water can reach may be limited by the amount of water available instead of the filling rate. This corresponds to the limitation of excitation of a molecule in MPE by the laser fluence rather than intensity.

5. Only if the filling rate is much faster than the rate of equilibration between compartments can water leak through the V-shaped opening in the same compartment that is being filled. This is equivalent to the possibility of bond-selective MPD by a very intense laser pulse with up-excitation much faster than energy randomization.

This model is admittedly very crude and cannot describe all the characteristic features of MPE and MPD of a molecule. For example, it does not give a proper description of MPE over the discrete levels. However, it does provide a simple, easily understandable physical picture that one can use to visualize the complicated MPE process through the quasi-continuum and the resultant MPD process.

23.7 SUMMARY AND FUTURE WORK

The present understanding of MPD of a polyatomic molecule can be summarized as follows:

1. MPE of a molecule over the discrete levels is a near-resonant, multiphoton absorption process.

2. Multiphoton up-excitation of all the molecules in the quasi-continuum is possible because of stepwise resonances and the rapidly increasing density of states with increase of excitation energy.

3. At least near and above the dissociation level, the excitation energy is expected to be randomized in a large number of modes within a few picoseconds. Then the resulting MPD, governed by statistical mechanics, is not very different from dissociation by thermal excitation.

4. Bond-selective MPD is not likely unless an extremely short but energetic laser pulse is used.

5. The average excitation level from which the molecules dissociate in MPD depends in general on both the laser intensity and the laser fluence. For a very short laser pulse, essentially none of the molecules dissociates before the pulse is over; the final excitation level is then determined only by the laser fluence. For a sufficiently long pulse with high enough fluence, the final excitation level is reached at the balance of up-excitation and dissociation and is therefore determined by the laser intensity.

6. MPD through competing dissociation channels and MPD with successive secondary dissociation can occur.

7. A heavier, more complex molecule with more degrees of freedom and lower vibrational frequencies has a dissociation rate increasing more slowly with the excitation energy. Consequently, the same laser pulse can excite such molecules to higher levels. In dissociation, their fragments should carry more excess energy, with a large fraction in the internal degrees of freedom. The highly excited fragments are also more likely to undergo a secondary MPD.

This suggests that one can predict fairly well the MPD of a not too complicated polyatomic molecule. Qualitatively, a larger, heavier, and more complex molecule has a quasi-continuum begin at lower energy and a dissociation rate increase more slowly with excess energy. Such a molecule requires a less intense laser pulse to excite it over the discrete levels. With sufficient laser intensity and fluence, it can be driven to a level high above the dissociation threshold. The excess energy carried away by the fragments in the subsequent dissociation is therefore large. The fragments having most of the excess energy in internal degrees of freedom may appear highly excited in the quasi-continuum. They can be further excited by the laser radiation through the quasi-continuum into dissociation. This description can apply, for example, to the case of MPD of UF_6.[31] Simultaneous dissociation through several channels competing in the statistical sense could also occur.

Although MPE and MPD of a polyatomic molecule are now fairly well understood, a number of relevant basic questions have not been answered. First, a quantitative understanding of MPE over the discrete levels is still lacking. This is mainly because of lack of spectroscopic information about polyatomic molecules. Few data are available on the higher rotation-vibrational states; even the linear absorption spectra of polyatomic molecules generally have not been interpreted. Only if MPE over the discrete levels is quantitatively understood can the dependence of the fraction of molecules excited into the quasi-continuum on laser intensity be predicted. Second, whether the coherent effect of MPE over the discrete levels is important or not is not yet resolved, although it is fairly certain that for a one-step MPE into the quasi-continuum, the effective dephasing time should be of the order of tens of picosecond or less. Third, the dynamics of energy transfer between modes is still an open problem. How the various modes come in at different excitation levels to participate in the energy randomization process and how fast the energy is redistributed among them are most important for the understanding of the structure of the quasi-continuum and MPE through the quasi-continuum. Other related questions to be answered are: What is the population distribution in the quasi-continuum and how does it vary during up-excitation?

Does the absorption spectrum of a molecule in the quasi-continuum show fine resolved structure,[32] and how does it change with the excitation level? Fourth, information about the detailed excess energy distribution in the fragments is scarce, but is crucial for a better understanding of unimolecular dissociation. It tells us not only which state the molecule dissociates from but also the detailed dynamics of how the fragments separate from each other in their path crossing the critical region of dissociation. Fifth, in special situations, bond-selective dissociation may still be a possibility. How would the dissociation pattern change if an energetic picosecond laser pulse is used for MPD? Can one find some molecules possessing highly isolated modes, through which the excitation would yield MPD products very different from thermal dissociation?

Applications of infrared MPE and MPD of molecules are many. As already mentioned, it can be used for isotope separation and for the study of unimolecular dissociation and laser-induced chemical reaction. In other applications, it has been suggested as a means to purify a material through MPD of impurities and to generate a large number of desired radicals for spectroscopic study or for chemical synthesis.

Multiphoton excitation of a molecule via electronic transitions is also possible. Two- or three-photon excitation is commonly used to selectively ionize a molecule, but little is known about excitation by more than three or four photons. The much shorter sponanteous emission lifetimes of the electronic excited states is probably the main reason for the limitation. However, with the now available intense ultrashort laser pulses it is conceivable that a molecule can be excited to a very high level in the ionization continuum via an n-photon absorption process with n much larger than 3 or 4. The theory of such an MPE process is expected to be difficult because of our lack of information about electronic excited states.

REFERENCES

1 N. R. Isenor and M. C. Richardson, *Appl. Phys. Lett.* **18**, 224 (1971).

2 N. R. Isenor, V. Merchant, R. S. Hallsworth, and M. C. Richardson, *Can J. Phys.* **51**, 1281 (1973); V. S. Letokhov, E. A. Ryabov, and O. A. Tumanov, *Opt. Commun.* **5**, 168 (1972); *JETP* **36,** 1069 (1973).

3 R. V. Ambartzumian, V. S. Letokhov, E. A. Ryabov, and N. V. Chekalin, *JETP Lett.* **20**, 273 (1974).

4 R. V. Ambartzumian, Yu. A. Gorokhov, V. S. Letokhov, and G. N. Makarov, *JETP Lett.* **21**, 171 (1975); J. L. Lyman, R. J. Jensen, J. Ring, C. P. Robinson, and S. D. Rockwood, *Appl. Phys. Lett.* **27**, 87 (1975).

5 R. V. Ambartzumian, Yu. A Gorokhov, V. S. Letokhov, G. N. Makarov, and A. A. Puretskii, *JETP Lett.* **23**, 22 (1976).

6 D. M. Larsen and N. Bloembergen, *Opt. Commun.* **17**, 254 (1976); D. M. Larsen, *Opt. Commun.* **19**, 404 (1976).

7 N. Bloembergen, *Opt. Commun.* **15**, 416 (1975); V. S. Letokhov and A. A. Makarov, *Opt Commun.* **17**, 250 (1976).

8   *Chemical Engineering News* **54** (47), 18 (1976).

9   K. L. Kompa, in A. Mooradian, T. Jaeger, and P. Stokseth, eds., *Tunable Lasers and Applications* (Springer-Verlag, Berlin, 1976), p. 177.

10 D. F. Dever and E. Grunwald, *J. Am. Chem. Soc.* **98**, 5055 (1976).

11 M. J. Coggiola, P. A. Schulz, Y. T. Lee, and Y. R. Shen, *Phys. Rev. Lett.* **38**, 17 (1977).

12 E. R. Grant, M. J. Coggiola, Y. T. Lee, P. A. Schulz, Aa. S. Sudbø, and Y. R. Shen, *Chem. Phys. Lett.* **52**, 595 (1977); Aa. S. Sudbø, P. A. Schulz, E. R. Grant, Y. R. Shen, and Y. T. Lee, *J. Chem. Phys.* **68**, 1306 (1978).

13 C. D. Cantrell and H. W. Galbraith, *Opt. Commun.* **18**, 513 (1976).

14 L. N. Knyazev, V. S. Letokhov, and V. V. Lobko, *Opt. Commun.* **25**, 337 (1978); I. N. Knyazev and V. V. Lobko, *JETP* **50**, 412 (1979).

15 R V. Ambartzumian, G. N. Makarov, and A. A. Puretskii, *JETP Lett.* **28**, 228 (1978); V. N. Bagratashvili, V. S. Dolzhikov, and V. S. Letokhov, *JETP* **49**, 8 (1979)

16 P. J. Robinson and K. A. Holbrook, *Unimolecular Reactions* (Wiley, New York, 1972).

17 E. Borsella, R. Fantoni, A. Giardini-Guidoni, and C. D. Cantrell, *Chem. Phys. Lett.* **87**, 284 (1982).

18 J. G. Black, E. Yablonovitch, N. Bloembergen, and S. Mukamel, *Phys. Rev. Lett.* **38**, 1130 (1977).

19 E. R. Grant, P. A. Schulz, Aa. S. Sudbø, Y. R. Shen, and Y. T. Lee, *Phys. Rev. Lett.* **40**, 115 (1978).

20 See, for example, Y. T. Lee and Y. R. Shen, *Physics Today* **33**, No. 11, 52 (1980).

21 P. A. Schulz, Aa. S. Sudbø, E. R. Grant, Y. R. Shen, and Y. T. Lee, *J. Chem. Phys.* **72**, 4985 (1980).

22 J. G. Black, P. Kolodner, M. J. Schulz, E. Yablonovitch, and N. Bloembergen, *Phys. Rev. A* **19**, 704 (1979).

23 R. V. Ambartzumian, Yu. A. Gorokhov, V. S. Letokhov, G. N. Makarov, A. A. Puretskii, and N P. Furzikov, *JETP Lett.* **23**, 194 (1976); R. V. Ambartzumian, N. P. Furzikov, Yu. A. Gorokhov, V. S. Letokhov, G. N. Makarov, and A. A. Puretskii, *Opt. Commun.* **18**, 517 (1976).

24 R. V. Ambartzumian, V. S. Letokhov, G. N. Makarov, and A. A. Puretskii, *Opt. Commun.* **25**, 69 (1978).

25 V. N. Bagratashvili, Yu. G. Vainer, V. S. Dolzhikov, S. F. Kol'yakov, A. A. Makarov, L. P Malyavkin, E. A. Rayabov, E. G. Sil'kis, and V. D. Titov, *JETP Lett* **30**, 471 (1979).

26 R. V Ambartzumian, Yu. A. Gorokhov, V. S. Letokhov, and G. N. Makarov, *JETP* **42**, 993 (1975).

27 Aa. S. Sudbø, P. A. Schulz, E R. Grant, Y. R. Shen, and Y. T. Lee, *J. Chem. Phys.* **70**, 912 (1979).

28 Aa. S. Sudbø, P. A. Schulz, Y. R. Shen, and Y. T. Lee, *J. Chem. Phys.* **69**, 2312 (1978)

29 D. S King and J. D. Stephenson, *Chem. Phys. Lett.* **51**, 48 (1977); J. D. Stephenson and D. S. King, *J. Chem. Phys.* **69**, 1485 (1978).

30 D. Krajnovitch, F. Huisken, Z. Zhang, Y. R. Shen, and Y. T. Lee, *J. Chem. Phys.* **77**, 5977 (1982); F. Huisken, D. Krajnovitch, Z. Zhang, Y. R. Shen, and Y. T. Lee, *J. Chem. Phys.* **78**, 3806 (1982).

31 J A. Horsley, P. Rabinowitz, A. Stein, D. M. Cox, R. O. Brickman, and A. Kaldor, *IEEE J. Quant. Electron.* **QE-16**, 412 (1980); P. Rabinowitz, A. Kaldor, A. Gnauk, R. L. Woodin, and J. S. Gethner, *Opt. Lett.* **7**, 212 (1982).

32 E. Borsella, R Fantoni, A. Giardini-Guidoni, and C. D. Cantrell, *Chem. Phys. Lett.* **87**, 284 (1982).

BIBLIOGRAPHY

Ambartzumian, R. V., and V. S. Letokhov, in C. B. Moore, ed., *Chemical and Biochemical Applications of Lasers*, vol. 3 (Academic Press, New York, 1977), p. 167.

Ashfold, M. N. R., and G. Hancock, Gas Kinetics and Energy Transfer (Chem. Soc. Spec. Per. Report), **4** (1980).

Bloembergen, N., and E. Yablonovitch, *Physics Today* **31**, No. 5, 23 (1978).

Lee, Y. T., and Y. R. Shen, *Physics Today* **33**, No. 11, 52 (1980).

Letokhov, V. S., *Uspekhi* **21**, 405 (1978).

Letokhov, V. S., and A. A. Makarov, *Uspekhi* **24**, 366 (1982).

Oref, I., and B. S. Rabinovitch, *Acc. Chem. Res.* **12**, 166 (1979).

Schulz, P. A., Aa. S. Sudbø, D. J. Krajnovitch, H. S. Kwok, Y. R. Shen, and Y. T. Lee, *Ann Rev Phys Chem.* **30**, 379 (1979)

24

Laser Isotope Separation

The advent of tunable lasers opens the prospect of frequency-selective excitation of matter. This is of fundamental importance in photochemistry, since the selective excitation could change the properties of atoms and molecules drastically. Of special interest in this respect is the use of tunable lasers for isotope separation. Since the existing isotope separation methods have various disadvantages, one hopes to develop new methods that could be cheaper, simpler to set up, and less power-consuming. Thus laser isotope separation has become a subject of immense research activity. It shows that lasers and quantum electronics can even have a major influence on nuclear power technology. In this chapter, we discuss briefly the ideas and practices of the various laser isotope separation methods.

24.1   GENERAL DESCRIPTION

The idea of using selective excitation for isotope separation is simple and straightforward. The optical spectra of isotopic atoms or molecules are generally the same except for the existence of small shifts in the positions of the spectral lines known as isotope shifts. If the isotope shifts of some spectral lines can be clearly resolved, then selective excitation of the desired isotopic atoms or molecules is possible. The excited atoms or molecules are expected to have very different physical and chemical properties from the unexcited ones. They can therefore be separated from the other isotopic atoms or molecules which have not been excited.

The difference in the number of neutrons in different isotopes is of course responsible for the isotope shifts. The electronic energy levels are shifted via the electron-nucleus interaction by the changes in nuclear mass, nuclear volume and hence the nuclear charge distribution, and nuclear spin angular momentum.[1] For light elements, the mass effect is important, since it changes significantly the reduced mass for the electron motion about the nucleus. It leads to an isotope shift $\Delta\omega_0 \sim (m\Delta M/M^2)\omega_0$, which is of the order of 1 cm^{-1} for Li^6 and Li^7, where m is the electron mass, M is the nuclear mass, and ΔM is the isotopic mass difference. For heavier elements, the isotope shifts are dominated by the change in the interaction of the electron angular momentum with the nuclear spin (the hyperfine interaction) and the change in the nuclear volume (the volume shift). They are of the order of a few tenths of a cm^{-1} or smaller. Figure 24.1 shows how the atomic isotope shift varies with the neutron number. In molecular spectra, the isotope shifts of vibrational and rotational levels are superimposed on the electronic energy shifts.[2] They are governed mainly by the mass effect: the vibrational level spacing is inversely proportional to the square root of the vibrational reduced mass and the rotational level spacing is inversely proportional to the moment of inertia. Except for some small molecules, however, the complexity of the very closely spaced rovibronic levels often forbids the description of isotope shifts of individual lines. Instead, one considers the isotope shifts of the rotation-vibrational bands. In many cases, even for heavier molecules, the isotope shift of a narrow band can be comparable to or larger than the half-bandwidth, which is of the order of a few cm^{-1} at room temperature.

The small isotope shifts necessitate the use of a narrow-band light source for selective excitation. Tunable lasers with their high monochromaticity are therefore ideal for such an application. In addition, the high intensity of the laser radiation greatly facilitates the selection process. Actually, the first

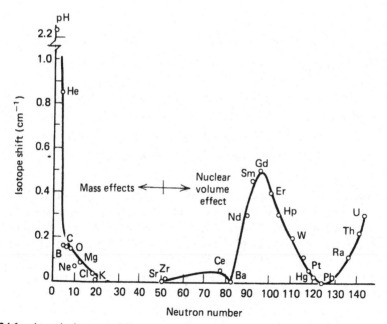

Fig. 24.1 Atomic isotope shifts versus the neutron number. [After R. C. Stern and B. B. Snavely, *Ann. N.Y. Acad. Sci.* **267**, 71 (1976).]

attempt at isotope separation by Hartley et al.[3] using optical selective excitation was made in 1922, long before the laser was discovered. The first successful experiment was the separation of Cl isotopes by Kuhn and Martin[4] in 1932, using the 2861.8-Å line of an aluminum spark to excite $CoCl_2^{35}$. Later, photochemical separation of Hg isotopes was also demonstrated.[5] The discovery of lasers prompted Tiffany et al.[6] to try isotope separation of Br by selective laser excitation. Their experiment failed because of isotopic scrambling in the subsequent photochemical reactions, but it revived the interest in isotope separation by optical means.

The basic requirements in laser isotope separation are summarized as follows:

1   The isotopic atoms or molecules should have an absorption spectrum with well-resolved isotope shifts.

2   The laser source should be sufficiently monochromatic and tunable to be able to selectively excite single isotopic species.

3   There should exist a physical or chemical process which can rapidly separate the excited and the unexcited atoms or molecules before the excited ones decay away or transfer their excitation energy to atoms or molecules of the undesired species.

The following changes in atomic or molecular properties can be induced by laser excitation:[7]

1   The chemical reactivity of atoms or molecules may increase.

2   The energy required to ionize an excited atom or molecules is less.

3   An excited atom or molecule may have a higher polarizability and a larger cross section of interaction with other particles or external fields.

4   Resonant excitation changes the trajectory of excited atoms or molecules.

5   The energy required to dissociate an excited molecule is less.

6   An excited molecule may predissociate or isomerize.

These changes allow us to devise effective physical or chemical methods to separate the excited atoms or molecules from the unexcited ones. Referring to the property changes listed above, the various methods are classified as photochemical reaction for (1), photoionization for (2), photophysical reaction for (3), photodeflection for (4), photodissociation for (5), and photopredissociation or photoisomerization for (6). In the two charts below, we summarize the more frequently considered schemes of laser isotope separation following, respectively, atomic and molecular excitations. We use A, B, C, AB, and so on to denote atoms or molecules.

Laser Isotope Separation Based on Atomic Excitation

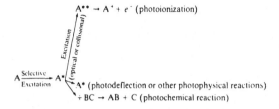

Laser Isotope Separation Based on Molecular Excitation

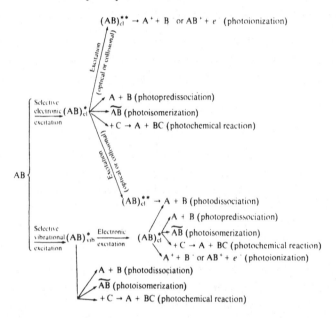

The relative isotope enrichment factor is defined as

$$K(I/J) = \frac{[I]/[J]}{[I]_0/[J]_0}, \tag{24.1}$$

where $[I]/[J]$ is the concentration ratio of the two isotopic species, and $[I]_0/[J]_0$ is the initial value of the concentration ratio. To have a large K value, a number of precautions must be taken. First, the selective excitation should have excitation energy much larger than kT to be distinguished from the nonselective thermal excitation. Next, the separated isotopic species should be removed before collisions scramble the isotopes. Finally, other isotope scrambling processes such as secondary chemical reactions should be avoided.

In the following sections we discuss the photophysical and photochemical methods of laser isotope separation with some illustrative examples.

24.2　PHOTOPHYSICAL METHODS

Photoionization

Laser isotope separation by photoionization stems from the fact that an atom can be selectively photoionized through multistep selective excitation,[8] as already discussed in some detail in Sections 18.3 and 19.3. With the available lasers, this is almost a universal method applicable to all elements in the periodic table except a very few. The actual ionization process can occur through direct excitation into the ionization continuum ($\sigma_i \sim 10^{-18}$ cm^2), or through excitation into an autoionization state ($\sigma_i \sim 10^{-15}$ cm^2), or through excitation into a Rydberg state followed by dc-field-induced ionization ($\sigma_i \sim 10^{-14}$ cm^2). The latter two have a much higher excitation probability because of the larger transition probability into a discrete final state, but they require an additional narrowband tunable laser. The selectively excited isotope can in principle also be ionized by collision with a partner.

As charged particles, the selectively ionized atoms can be physically separated from the neutral atoms by an applied electric field. In colliding with an atom, however, an ion usually has a large charge-transfer cross section. It is likely to lose its charge to the neutral atom in the collision, and subsequently the isotopic selectivity is lost. Such collisions thus should be avoided in the separation process. This can be best achieved by processing the materials to be isotopically enriched in an atomic beam. Even so, charge-transfer collisions are still the limiting factor on the maximum density the atomic beam can have.

As an example, we consider the case of uranium isotope separation. The arrangement of the first experiment is seen in Fig. 24.2.[9] An atomic beam of the ^{235}U–^{238}U isotopic mixture with a density of $\sim 5 \times 10^{10}$ atoms/cm^3 was generated from an oven heated to 2100°C. A CW dye laser with a 50-MHz bandwidth was used to selectively excite either ^{235}U or ^{238}U in the beam with a transition at ~ 5915 Å, at which the isotope shift between ^{235}U and ^{238}U was ~ 8 GHz. Then the uv radiation from a mercury arc lamp was used to ionize the excited atoms. To avoid direct ionization of the ground-state atoms, radiation with wavelengths shorter than 2100 Å had been filtered out. The resulting ionized atoms were transported out of the main beam by the deflection plates, sent through a quadrupole ion mass filter, and finally collected by a collector biased at -3100 v. In the initial attempt, the observed enrichment factor K was already ~ 100. In another experiment,[10] in which a pulsed dye laser was used for selective excitation and a pulsed N$_2$ laser for ionization, $K(^{235}$U$/^{238}$U) was about 140. It is also possible to selectively ionize the uranium atoms by successive resonant excitations over discrete states before the final ionization step. This requires more than one tunable laser system for the resonant excitations but should greatly improve the selectivity. With the uranium atoms excited to a high Rydberg state, an external dc electric field can then be used to efficiently ionize the atoms. The large-scale

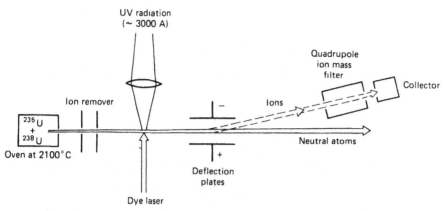

Fig. 24.2 Experimental arrangement of uranium isotope separation by the two-step photoionization method.

prototype system for uranium isotope separation using copper-laser-pumped dye lasers is presently being developed at Lawrence Livermore Lab.

Laser isotope separation of other isotopes, such as Na, K, Ca, Rb, rare earth atoms, and transuranium elements, using the stepwise photoionization method have also been demonstrated.[11] In general, this is a relatively simple and economical method for separation of small quantities of isotopes of essentially all elements.

Photodeflection

As discussed in Section 20.1, laser excitation followed by fluorescence can impart a net momentum to an atom or molecule. In absorbing a photon from a laser beam, the atom or molecule receives a momentum $\hbar k$, but, in emission, because the fluorescent photon appears in all directions with equal probability, the atom or molecule (assuming unpolarized) loses no momentum on the average. Thus if 10^3 visible photons are absorbed and emitted during the course of interaction of the atom or molecule with the laser field, a net momentum of $10^3 \hbar k$ is transferred to the atom or molecule. This is often sufficient to change appreciably the trajectory of the atom or molecule if the atomic or molecular mass is not exceptionally large.

We consider here the experimental demonstration of photodeflection of a ^{138}Ba beam as an example.[12] As shown in Fig. 24.3, the Ba atomic beam from an oven heated to 800 K had a mean velocity $\langle v \rangle \simeq 4 \times 10^4$ cm/sec and a mean transverse velocity $\langle v_T \rangle \simeq 80$ cm/sec corresponding to a beam spread of 2 mrad. A CW dye laser with a spectral width of 10 MHz was used to selectively excite the $6s^2(^1S_0) \rightarrow 6s6p(^1P_1)$ transition of ^{138}Ba in a transverse direction. Since $\hbar k/M \simeq 0.8$ cm/sec in this case, it would take the absorption

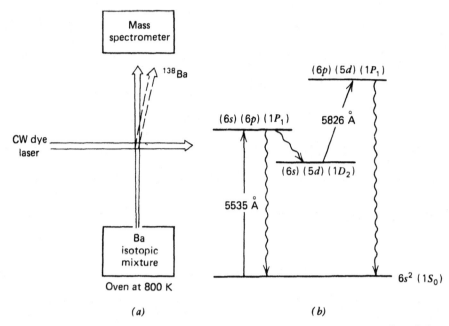

Fig. 24.3 Schematic diagrams of barium isotope separation by resonant photodeflection. (*a*) Experimental arrangement; (*b*) relevant energy levels of ^{138}Ba.

of ~ 100 photons to deflect the atoms away from the mean beam. The lifetime of the $6s^2 \to (6s)(6p)$ transition is ~ 8 nsec. With a laser intensity (~ 10 W/cm^2) sufficient to saturate the transition and an effective atom-field interaction length of 1 mm, a total of 300 photons could be absorbed and then emitted by each ^{138}Ba atom crossing the laser beam. Photodeflection of the selectively excited ^{138}Ba away from the main beam should therefore be possible and was actually demonstrated.[12] In the Ba case, there is a 4% probability that the excited atom would decay into the $(6s)(5d)(^1D_2)$ metastable state. The overall probability that a Ba atom may end up in the metastable state after n photon collisions is $\sum_n (24/25)^{n-1}(1/25)$, which is appreciable for large n. Since atoms in the metastable state can no longer absorb photons from the exciting laser beam and get deflected, the accumulation of population in the metastable state must be avoided. One possible way to quench the population in $(6s)(5d)$ is to use another laser to excite it to $(6p)(5d)$, from which it can decay rapidly to the ground state.

Another method to avoid metastable trapping is to use a coherent π pulse to excite the selected atoms to the excited state via adiabatic following.[13] During the process, a photon momentum $\hbar\mathbf{k}$ is transferred to the atom. The laser pulse is then reflected back from a mirror and is used to stimulate emission from the excited atom. The stimulated photon emission, being directional, also transfers a photon momentum $-\hbar(-\mathbf{k}) = \hbar\mathbf{k}$ to the atom. Thus, with the laser pulse going back and forth between mirrors, the resonantly excited and deexcited atoms can be effectively deflected with little loss of photons. The experimental demonstration of this method has not yet been reported.

The main difficulty of the photodeflection method is that to obtain a large isotope enrichment factor, a long atom-field interaction length is required. For heavier atoms, the requirement is more stringent.

Other Physical Methods

Several other physical methods for laser isotope separation have been proposed, but no successful experiment has yet been reported.

An excited atom or molecule has a different polarizability than one in the ground state. In an electric field gradient, therefore, the forces acting on the excited and unexcited atoms or molecules are different, and they can, in principle, be used to separate the two species. The sticking coefficients for the adsorption of the excited and unexcited atoms or molecules on a substrate can also be very different. By flowing the selectively excited isotopic mixture over a substrate, one may achieve separation of isotopes through the difference in adsorption, if admixing from secondary processes is negligible.

24.3 PHOTOCHEMICAL METHODS

Photochemical Reaction

An excited atom or molecule generally has a stronger chemical reactivity than the ground-state one and can be separated from the latter via an appropriate chemical reaction scheme. Although electronic excitation often is more effective in this respect, vibrational excitation, in the case of molecules, can also be operative. Two examples are given below to illustrate the isotopically selective photochemistry.

In a mixture of $ortho$-I_2 and $para$-I_2 with 2-hexane (X), $ortho$-I_2 was excited with the 5145 Å Ar$^+$ laser line.[14] After one hour of irradiation with a 0.2-W laser, the density of $ortho$-I_2 was found to have reduced to ~ 5%. This was believed to be due to the fact that the electronically excited $ortho$-I_2 could react with the 2-hexane, while the unexcited $para$-I_2 could not:

$$ortho\text{-}I_2 \overset{\hbar\omega}{\rightarrow} ortho\text{-}I_2^* + X \rightarrow XI_2.$$

In another example, a gas mixture of $1:1:1$ $H_3COH:D_3COD:Br_2$ was irradiated by a CW HF laser at ~ 2.7 μm.[15] The OH vibration in H_3COH was excited. The vibrationally excited H_3COH then reacted strongly with Br_2 to form 2HBr and H_2CO:

$$H_3COH^* + Br_2 \rightarrow 2HBr + H_2CO.$$

As a result, after the irradiation of the mixture by a 90-W laser for 60 sec, the

$H_3COH : D_3COD$ ratio reduced to less than $1 : 19$. The results have, however, been criticized, and need further confirmation.

The chemical reactions in these examples are simple. They yield stable primary products which can be easily separated from the parent molecules. This is, however, not the case in general. A chain reaction often occurs following selective excitation of the atoms or molecules. Isotopic scrambling could result from the uncontrolled secondary reactions.

One-Step Photopredissociation

Unimolecular dissociation is a special class of chemical reaction that can be induced by laser excitation. If the excitation is isotopically selective, the process can be used for isotope separation.

Some molecules have sharp ro-vibronic states superimposed on an electronic excited state, which is degenerate with the dissociation continuum of another excited electronic state (Fig. 24.4). These are predissociation states. They are

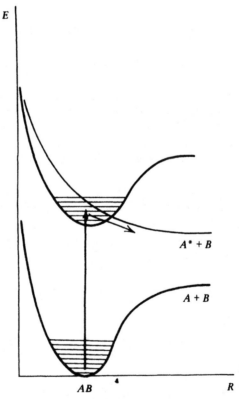

Fig. 24.4 Schematic energy level diagram of a molecule having predissociation states degenerate with the continuum of a dissociating electronic state.

similar in nature as the autoionization states to the ionization continuum. A molecule excited to a predissociation state can relax to the continuum and dissociate. If the sharp predissociation states have resolvable isotope shifts, selective excitation can lead to the predissociation of a selected isotopic species.[16] The dissociation products may be stable or unstable. In the latter case, a chemical scavenging process must be devised in order to remove the specific isotopic species before appreciable isotopic scrambling occurs. Secondary reactions that would scramble the isotopes should also be avoided. The advantage of the one-step photopredissociation method is its simplicity. Only a single, relatively low-power laser is needed for one-photon selective excitation.

As an example, consider the case of isotope enrichment of deuterium by photopredissociation of formaldehyde H_2CO.[16] The same process can be used for isotope enrichment of C and O. In the predissociation of H_2CO, the following reaction dominates:

$$H_2CO \xrightarrow{\hbar\omega} H_2CO^* \rightarrow H_2 + CO.$$

Both H_2 and CO are stable and can be readily removed. An enrichment factor $K(D/H)$ of 14 has been obtained from irradiation of a natural mixture of H_2CO and HDCO by a CW He-Cd laser at 3250.3 Å.

Two-Step Photodissociation and Photopredissociation

Molecular dissociation can be initiated by exciting a molecule either to a predissociation state or to the dissociation continuum. For isotopically selective dissociation, a two-step excitation scheme can be used: the first step selectively excites the desired isotopic species, and the second step further excites the molecules to a dissociation or predissociation state. Because the predissociation state may also exhibit an isotope shift, the two-step photopredissociation case is generally more selective than the one-step case, although two tunable lasers of high enough intensities for stepwise excitations will then be needed. The selective excitation can be either electronic or vibrational. The latter often is used, since for most molecules the isotope shifts are better resolved in the vibrational levels. Again, the dissociation products should be scavenged out before secondary reactions or other processes scramble the isotopic species.

The first demonstration of this method is used as an example here.[17] A $1:1$ mixture of $^{14}NH_3$ and $^{15}NH_3$ at a pressure of 10-20 mm Hg with 250 mm Hg of buffer gas (Xe or Ne) was irradiated simultaneously by a pulsed CO_2 laser at 10.6 μm and a uv spark. The CO_2 laser selectively excited the ν_2 vibrational mode of $^{15}NH_3$ in the ground state, $(\tilde{X}, v = 0) \rightarrow (\tilde{X}, v = 1)$, followed by the uv excitation to the predissociation state $(\tilde{A}, v' = 0)$. The ensuing sequence of chemical reactions was believed to be

$$^{15}NH_3^* \rightarrow {}^{15}NH_2 + H$$

$$^{15}NH_2 + {}^{15}NH_2 \rightarrow {}^{15}N_2H_4$$

$$^{15}N_2H_4 + H \rightarrow {}^{15}N_2H_3 + H_2$$

$$2^{15}N_2H_3 \rightarrow 2^{15}NH_3 + {}^{15}N_2.$$

It is seen that the secondary reactions here do not involve the unexcited NH_3 molecules. They are therefore the isotopically selective reactions. The resulting product N_2 should be enriched in ^{15}N. Indeed, the observed enrichment coefficient $K(^{15}N/^{14}N)$ in N_2 was about 2.5–6.

In another example, a pulsed CO_2 laser was used to selectively excite the ν_3 vibrational mode of $^{11}BCl_3$ in a mixture of $^{10}BCl_3$ and $^{11}BCL_3$.[18] The uv radiation from a Xe flashlamp then excited $^{11}BCl_3^*$ further into the continuum of a dissociating electronic state. The dissociation products were removed by using O_2 as the scavenger. A 10% enrichment of ^{10}B versus ^{11}B in the remaining mixture of BCl_3 was found in the experiment. The two-step photodissociation method has been considered for large-scale uranium isotope separation in the United States.[19]

Infrared Multiphoton Dissociation

Infrared multiphoton dissociation of molecules was discussed at great length in Chapter 23. Since transitions over the low-lying discrete levels are isotopically selective, the process can also be used for isotope separation. It has the advantage that only medium-power pulsed infrared lasers are required. Uranium isotope separation by infrared multiphoton dissociation of UF_6 was demonstrated by Rabinowitz et al.[20]

24.4   CONCLUDING REMARKS

The propelling force in laser isotope separation research is the energy crisis. For consumption in nuclear power plants, large quantities of D and ^{238}U are needed. They must be produced from separation of their natural isotopic mixtures. In comparison with existing isotope separation methods, laser schemes could yield more isotope enrichment per stage, consume less power, and save in capital investment. The requirement on the laser is, however, fairly stringent. Since the future development of laser technology is unpredictable, it is difficult to know for certain how profitable future large-scale laser isotope separation can be. Comparison of different laser isotope separation schemes is also a complicated matter, particularly because many details of these laser separation schemes are yet to be worked out. For example, in the photoionization method, even though selective photoionization of a certain isotopic species may not be a problem, extraction and collection of the resulting ion plasma may eventually limit the output. In the case where the desired amount of specific isotopes is small, laser isotope separation schemes can be of great advantage because of the relatively cheap capital investment. In other cases where existing separation methods are not very satisfactory because of their limited selectivity, laser schemes can be most useful. This applies to the separation of less abundant isotopes and of transuranium elements. To a large extent, the general success of laser isotope separation depends on advances in laser technology.

REFERENCES

1 See, for example, I I. Sobelman, *Introduction to the Theory of Atomic Spectra* (Pergamon Press, Oxford, 1973).

2 See, for example, G. Herzberg, *Spectra of Diatomic Molecules* (Van Nostrand, New York, 1950), p. 658.

3 H. Hartley, A. O. Ponder, E J. Bowen, and T. R. Merton, *Phil. Mag.* **43**, 430 (1922).

4 W. Kuhn and H. Martin, *Naturwiss.* **20**, 772 (1932); *Z. Phys. Chem. Abt.* **B21**, 93 (1933).

5 K. Zuber, *Nature* **136**, 796 (1935); *Helv. Phys. Acta* **8**, 487 (1935); **9**, 285 (1936).

6 W. B. Tiffany, N. W. Moos, and A. L. Schawlow, *Science* **157**, 40 (1967); W. B. Tiffany, *J. Chem Phys.* **48**, 3019 (1968).

7 V. S. Letokhov, *Ann. Rev. Phys. Chem.* **28**, 133 (1977).

8 R. V. Ambartzumian, V. P. Kalinin, and V. S. Letokhov, *JETP Lett.* **13**, 217 (1971).

9 S. A. Tuccio, J. W. Durbin, O G. Peterson, and B. B. Snavely, *IEEE J. Quant. Electron.* **QE-10**, 790 (1974).

10 G. S Janes, I. Itzkan, C. T. Pike, R. H. Levy, and L. Levin, *IEEE J. Quant. Electron.* **QE-11**, 101D (1974).

11 See the references in Ref. 7.

12 A. F. Burnhardt, D. E. Duerre, J. R. Simpson, and L. L. Wood, *Appl. Phys. Lett.* **25**, 617 (1974).

13 I. Nebenzahl and A. Szöke, *Appl. Phys Lett.* **25**, 327 (1974).

14 V. S. Letokhov and V. A. Semishen, *Sov. Phys. Doklady* **20**, 423 (1975); *Spectrosc Lett.* **8**, 263 (1975).

15 S. W. Mayer, M. A. Kwok, R. W. F. Gross, and D. J. Spencer, *Appl. Phys. Lett.* **17**, 516 (1970).

16 E. S. Yeung and C. B. Moore, *Appl. Phys. Lett.* **21**, 109 (1972); *J. Chem. Phys.* **58**, 3988 (1973).

17 R. V. Ambartzumian, V. S. Letokhov, G. N. Makarov, and A. A. Puretskii, *JETP Lett.* **17**, 63 (1973).

18 S. Rockwood and S. W. Rabideau, *IEEE J. Quant. Electron.* **QE-10**, 789 (1974).

19 S. D. Rockwood, in A. Mooradian, T. Jaeger, and P. Stokseth, eds., *Tunable Lasers and Applications* (Springer-Verlag, Berlin, 1976), p 140.

20 P. Rabinowitz, A. Kaldor, A. Gnauck, R. L. Woodin, and J. S. Gethner, *Opt. Lett.* **7**, 212 (1982).

BIBLIOGRAPHY

Aldridge, J. P., J. H. Birely, C. D. Cantrell, and D. C. Cartwright, in M. O. Scully and C. T. Walker, eds., *Laser Photochemistry, Tunable Lasers, and Other Topics* (Addison-Wesley, Reading, MA, 1976), p. 57.

Ambartzumian, R. V., and V. S. Letokhov, in C. B. Moore, ed., *Chemical and Biochemical Applications of Lasers*, Vol. 3 (Academic Press, New York, 1977), p. 166.

Cantrell, C. D., S. M. Freund, and J. L. Lyman, in M. L. Stitch, ed., *Laser Handbook*, Vol. 3 (North-Holland Publishing Co., Amsterdam, 1979).

Letokhov, V. S., *Science* **180**, 451 (1973).

Letokhov, V. S., *Ann. Rev. Phys. Chem.* **28**, 133 (1977).

Letokhov, V. S., and C. B. Moore, *Sov. J. Quant. Electron.* **6**, 129, 259 (1976).

Letokhov, V. S., and C. B. Moore, in C. B. Moore, ed., *Chemical and Biochemical Applications of Lasers*, Vol. 3 (Academic Press, New York, 1977), p. 1.

Moore, C. B., *Acc. Chem. Res.* **6**, 323 (1973).

25

Surface Nonlinear Optics

Research in nonlinear optics has also been extended from the bulk to the surface. This is most interesting and exciting since it allows the field of nonlinear optics to be in close contact with the growing field of surface science. While a full exploration of this new area has not yet been carried out, a few initial attempts have proven to be very successful. As in the bulk case, research in surface nonlinear optics has proceeded along two lines: first, to obtain a better understanding of the nonlinear optical effects occurring at surfaces or interfaces, and second, to look into the possibility of applying surface nonlinear optics to surface and interfacial studies. The latter can lead to the development of new surface probes very different from, but complementary to, the conventional ones.

25.1 GENERAL DESCRIPTION

A question that often arises in the discussion of nonlinear optical effects is how the boundary surfaces of the media affect the results. In Section 6.4 we saw that a boundary surface not only modifies the transmitted sum-frequency generation but also yields a reflected sum-frequency wave. In the derivation there, however, we neglected the fact that the surface atomic or molecular layers generally have very different optical properties from the bulk. This is a good approximation in many cases, yet if we are interested in problems more specific to the surface, we must recognize the distinction between the surface microscopic layer and the bulk.

In the usual simplified approach, the surface microscopic layer is assumed to have a characteristic thickness with optical constants different from the bulk. Two such surface layers normally exist at an interface, as illustrated in Fig. 25.1. Because the linear transmission and reflection of light at the boundary surface usually are dominated by the bulk properties, the surface layers have little effect on the linear wave propagation. As far as nonlinear

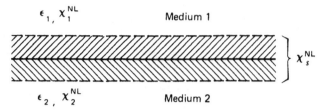

Fig. 25.1 A model of an interface between two media.

optical effects are concerned, we can assume that the surface layers have the same linear refractive indices as the adjoining bulk media, but their nonlinear optical susceptibilities are different from the bulk. Unlike the linear case, the surface layers can strongly affect the nonlinear optical output in some cases. It is often convenient to characterize the nonlinear optical response of the surface layers by a surface nonlinear susceptibility χ_S^{NL} instead of the usual volume nonlinear susceptibility χ_V^{NL}. The two are related by

$$\chi_S^{NL} = \int \chi_V^{NL} dz. \tag{25.1}$$

The integration here runs across the surface layers along the surface normal. The surface contribution to the overall nonlinear optical signal becomes nonnegligible if $|\chi_S^{NL}|$ is comparable to $|\chi^{NL} l_{eff}|$ from the bulk, where l_{eff} is the effective length of the nonlinear optical interaction in the bulk. With some modification, the discussion here can also apply to the case of a thin-film layer at an interface.

Let N_S and N be the surface density (per unit area) and the bulk density (per unit volume) of atoms or molecules, respectively. In order to have $|\chi_S^{NL}|$ comparable to $|\chi^{Nl} l_{eff}|_{bulk}$, one must have $|\chi_S^{NL}|/N_S$ larger than $|\chi^{NL}|/N$, and the surface layer thickness smaller than l_{eff} by no more than a few orders of magnitude. The former is possible if χ_S^{NL} is resonantly enhanced but χ^{NL} of the bulk is not. It is also possible if, as a result of their different symmetry properties, $|\chi_S^{NL}|$ is allowed and $|\chi^{NL}|$ forbidden for a certain nonlinear optical process, or for a certain combination of input and output polarizations, For example, second-order processes are forbidden in centrosymmetric media, but they are always allowed at the surface layers. The effective interaction length l_{eff}, on the other hand, can also be limited in a number of ways. For mixing processes with a phase mismatch Δk, $l_{eff} \sim \pi/|\Delta k|$ can be limited by deliberately choosing a maximum $|\Delta k|$. In an absorbing medium, l_{eff} is limited by the attenuation length. If total reflection of the pump field occurs at the boundary, l_{eff} is limited by the penetration depth of the field. Finally, surface electromagnetic waves can be propagated on a boundary surface in some cases, and l_{eff} corresponds to the penetration depth of the surface wave into the bulk.

Nonlinear optics involving surface waves is an interesting subject in its own right. First, since the surface wave is confined to a thin layer of the order of a wavelength at the boundary, its propagation characteristic is surface-specific. This means that it can be very sensitive to small perturbations on the surface. Then, if a large fraction of the incoming laser energy can be coupled into a surface wave, the field intensity of the surface wave can be very high. Consequently, nonlinear optical effects arising from the surface wave interaction can be readily observable. With these in mind, one may wonder if nonlinear optical effects involving surface waves can even be sensitive enough to be used as surface probes. In the following section we dwell on this question after a general discussion of the surface wave interaction. We then discuss in a subsequent section how a second-order nonlinear process, being surface-specific by symmetry, can be an effective tool for surface studies even without the help of surface wave interaction.

25.2 NONLINEAR OPTICS WITH SURFACE ELECTROMAGNETIC WAVES

Surface Electromagnetic Waves

The term surface electromagnetic waves here refers to em waves propagating along an interface between two media with their amplitudes decaying exponentially away from the interface. Sometimes they are known as surface polaritons. The existence of surface em waves was predicted by Sommerfeld as early as 1909.[1] They appear in a variety of circumstances. The ground wave propagation on earth is just one example. Here we are concerned with surface em waves on condensed matter.

Let us consider the simple case of a plane interface between two semi-infinite cubic or isotropic media (Fig. 25.2). In this case, the surface em wave

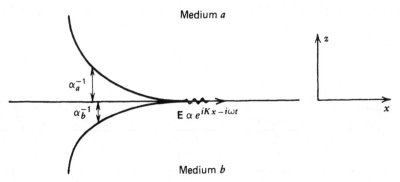

Fig. 25.2 Schematic describing a surface electromagnetic wave propagating on the interface between media a and b. The field penetration depths into the two media are α_a^{-1} and α_b^{-1}, respectively.

must be transverse magnetic (TM). This is seen as follows. We start by
assuming that the surface em wave would exist for both TE and TM polarizations. In the TM case, the surface wave propagating along $\hat{x}$, in the coordinate
system of Fig. 25.2, can be described by

$$\begin{aligned}
\mathbf{E} &= \left(\hat{x}\mathcal{E}_{ax} + \hat{z}\mathcal{E}_{az}\right)e^{iKx - \alpha_a z - i\omega t} \quad \text{for } z > 0 \\
&= \left(\hat{x}\mathcal{E}_{bx} + \hat{z}\mathcal{E}_{bz}\right)e^{iKx + \alpha_b z - i\omega t} \quad \text{for } z < 0.
\end{aligned} \tag{25.2}$$

Note that in order to satisfy the wave equation, K and α must be related by

$$\begin{aligned}
K^2 - \alpha_a^2 &= (\omega/c)^2 \varepsilon_a \quad \text{in medium } a(z > 0) \\
K^2 - \alpha_b^2 &= (\omega/c)^2 \varepsilon_b \quad \text{in medium } b(z < 0).
\end{aligned} \tag{25.3}$$

To match the boundary conditions at $z = 0$, we must have

$$\mathcal{E}_{ax} = \mathcal{E}_{bx} \quad \text{and} \quad \varepsilon_a \mathcal{E}_{az} = \varepsilon_b \mathcal{E}_{bz}. \tag{25.4}$$

Since $\nabla \cdot \mathbf{E} = 0$ in both media, (25.4) can be transformed into

$$\mathcal{E}_{ax} = \mathcal{E}_{bx}$$

and $\hspace{9cm}$ (25.5)

$$\varepsilon_a(iK/\alpha_a)\mathcal{E}_{ax} = \varepsilon_b(-iK/\alpha_b)\mathcal{E}_{bx}.$$

For $\mathcal{E}_{ax}$ and $\mathcal{E}_{bx}$ to be nonvanishing, the determinant of this set of coupled
algebraic equations should vanish. This leads to

$$\varepsilon_a \alpha_b + \varepsilon_b \alpha_a = 0, \tag{25.6}$$

which, with the help of (25.3), yields the dispersion relation for the surface
wave

$$K^2 = \left(\frac{\omega}{c}\right)^2 \frac{\varepsilon_a \varepsilon_b}{\varepsilon_a + \varepsilon_b}. \tag{25.7}$$

For the surface wave to exist, α_a and α_b must be positive and real (assuming
for the moment $\text{Im}\,\varepsilon = 0$), and hence $K^2 > (\omega/c)^2 \varepsilon_a$, $(\omega/c)^2 \varepsilon_b$. As seen from
(25.7), this can be true only if $\varepsilon_a < 0$ and $|\varepsilon_a| > \varepsilon_b$, or $\varepsilon_b < 0$ and $|\varepsilon_b| > \varepsilon_a$. In
other words, one of the two media must have a negative dielectric constant.
There actually exist a number of such media in nature; crystals with either
phonon or exciton restrahlung bands are good examples. The more commonly
used media for propagation of surface em waves are, however, the metals.
Below the plasma frequency, the dielectric constant of a metal is always

negative. Surface em waves on a metal surface are often known as surface plasmon waves. The preceding calculation and discussion can be repeated for the TE wave (**E** along $\hat{y}$), but it is easy to show that in this case no dispersion relation for surface em waves can result. This indicates that no TE waves can propagate as surface em waves.

Equation (25.3) shows that the wavevector K of the surface em wave is always larger than that of the bulk em wave. This is seen explicitly in Fig. 25.3, where the dispersion curves of the bulk and surface waves appear to intersect only at the point where $\varepsilon_b(\omega) \to -\infty$. Because of the wavevector mismatch, it is not possible for a bulk em wave to excite a surface em wave at the interface between two semi-infinite media. Conversely, a surface wave is not capable of radiating either. One can, however, borrow the various excitation schemes used in integrated optics for coupling light in and out of waveguides. We consider here the prism coupling method. Either the Otto configuration,[2] shown in Fig. 25.4a, or the Kretschmann configuration,[3] in Fig. 25.4b, can be effective. The refractive index of the prism must be large enough so that by adjusting the incident angle, the incoming wave through the prism can have a wavevector component along the surface equal to the wavevector of the surface wave. Then, if the film thickness in Fig. 25.4a, or the air gap in Fig. 25.4b is properly chosen, the excitation of the surface em wave can be efficient. An example is given in Fig. 25.5, where it is shown that using the Kretschmann geometry, nearly 100% of the incoming light can be coupled into the surface mode.

Strictly speaking, the prism would modify the dispersion curve of the surface wave. In practice, however, the change often is not appreciable. A more rigorous description of the surface wave excitation by the prism coupling method follows the analysis of a light wave incident on a thin film sandwiched between two semi-infinite media.[4] In the Kretschmann geometry of Fig. 25.4b, for example, the incoming TM wave $\mathbf{E}_0 = (\hat{x} - \hat{z}k_x/k_{0z})\mathscr{E}_{0x}\exp(ik_x x + ik_{0z}z$

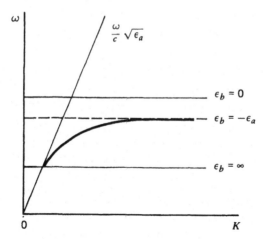

Fig. 25.3 Dispersion curve of a surface electromagnetic wave following (25.7).

$$\epsilon_2' < 0, \epsilon_0' > \epsilon_1' > 0 \qquad\qquad\qquad \epsilon_1' < 0, \epsilon_0' > \epsilon_2' > 0$$

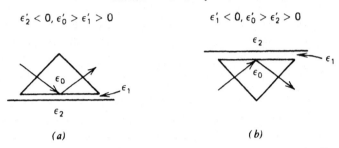

(a)   *(b)*

Fig. 25.4 (*a*) Otto configuration and (*b*) Kretschmann configuration for linear excitation of surface polaritons.

$- i\omega t)$ in the prism yields the following transmitted and reflected waves:

$$\mathbf{E}_T = (\hat{x} - \hat{z}k_x/k_{2z})\mathscr{E}_{Tx}\exp(ik_x x + ik_{2z}z - i\omega t),$$

$$\mathbf{E}_R = (\hat{x} + \hat{z}k_x/k_{0z})\mathscr{E}_{Rx}\exp(ik_x x - ik_{0z}z - i\omega t) \tag{25.8}$$

with

$$\mathscr{E}_{Tx} = \left[t_{01}t_{12}\exp(ik_{1z}d)\right]\left(k_{2z}^2\epsilon_0/k_{0z}^2\epsilon_2\right)\mathscr{E}_{0x}/D,$$

$$\mathscr{E}_{Rx} = \left[r_{01} + r_{12}\exp(i2k_{1z}d)\right]\mathscr{E}_{0x}/D, \tag{25.9}$$

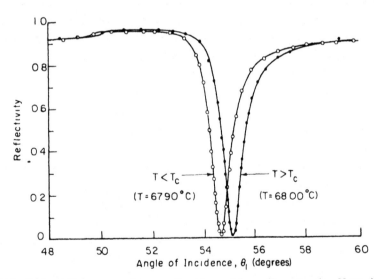

Fig. 25.5 Reflectivity curves from a liquid crystal medium in the Kretschmann geometry versus the angle of incidence θ_1 at $T < T_c$ and $T > T_c$ where T_c is the isotropic-mesomorphic transition temperature. The solid curves are theoretical curves obtained by nonlinear least square fitting. [After N. M. Chao, K. C. Chu, and Y. R. Shen, *Molec. Cryst. Liq. Cryst.* **67**, 261 (1981).]

and

$$D = 1 + r_{01}r_{12}\exp(i2k_{1z}d)$$

where r_{01}, r_{12}, t_{01}, and t_{12} are the Fresnel coefficient

$$r_{ij} = (\epsilon_j k_{iz} - \epsilon_i k_{jz})/(\epsilon_j k_{iz} + \epsilon_i k_{jz})$$
$$t_{ij} = 2\sqrt{\epsilon_i \epsilon_j}\, k_{iz}/(\epsilon_j k_{iz} + \epsilon_i k_{jz})$$

(25.10)

and d is the film thickness. The subindices 0, 1, and 2 refer to the prism, the film, and the dielectric medium above the film, respectively. Equation (25.9) shows that the solution is in resonance when $D = 0$. This is possible only if k_{1z} is imaginary or $\epsilon_1 < 0$ (assuming $\mathrm{Im}\,\epsilon_1 = 0$), suggesting that the resonance here corresponds to a surface mode. In fact, it can be shown that $D = 0$ can be rewritten in the form

$$\tanh\beta_1 d - \frac{q_0 q_1 + q_1 q_2}{q_1^2 + q_0 q_2} = 0,$$

(25.11)

which is the dispersion relation for surface waves on the thin film. Here, $\beta_1 \equiv ik_{1z}$ and $q_i = k_{iz}/\epsilon_i = [(\omega/c)^2\epsilon_i - k_x^2]^{1/2}/\epsilon_i$. In the limit of $\beta_1 d \gg 1$, (25.11) reduces to

$$\left[k_x^2 - \left(\frac{\omega}{c}\right)^2 \frac{\epsilon_0 \epsilon_1}{\epsilon_0 + \epsilon_1}\right]\left[k_x^2 - \left(\frac{\omega}{c}\right)^2 \frac{\epsilon_1 \epsilon_2}{\epsilon_1 + \epsilon_2}\right] = 0.$$

(25.12)

We then recognize that the two factors in (25.12) give exactly the dispersion relations for the two surface em waves at the interfaces between media 0 and 1 and between 1 and 2, respectively. This is what one expects physically when the film with $\epsilon_1 < 0$ is so thick that waves on the two sides are unable to communicate.

In the practical case, the dielectric constant of the film is a complex quantity, so that the resonant denominator in (25.9) is always finite, and (25.9) can be used to find the intensity of the excited surface wave. The example in Fig. 25.5 corresponds to a silver film ~ 500 Å thick. This film thickness is near optimum for coupling of an incident wave into a surface wave, but it is also large enough to have the dispersion relation of (25.11) fairly well approximated by (25.12). The field amplitude of the surface wave is then greatly enhanced from that of the incoming wave. The enhancement is limited by the loss in the film, described by the imaginary part of K_x. Among all metals, silver films appear to be best for surface wave propagation in the visible because of their low loss.

The field distribution in the film can also be found from the solution of the wave equation. For $0 \le z \le d$, we have

$$E_1 = \{\hat{x}[A \cos k_{1z}z + B \sin k_{1z}z] + \hat{z}(ik_x/k_{1z})[-A \sin k_{1z}z + B \cos k_{1z}z]\}$$
$$\times \exp(ik_x x - i\omega t) \tag{25.13}$$

with

$$A = (1 + r_{01})[1 + r_{12}\exp(i2k_{1z}d)]\mathscr{E}_{0x}/D$$

and

$$B = (1 - r_{01})[1 - r_{12}\exp(i2k_{1z}d)](i\varepsilon_0 k_{1z}/\varepsilon_1 k_{0z})\mathscr{E}_{0x}/D.$$

Surface waves can exist in general in a multilayer system. Their dispersion relation can be derived in a way similar to the solution of linear wave propagation.[4]

Nonlinear Optical Interaction Involving Surface Electromagnetic Waves

Nonlinear optics involving surface em waves can be easily understood if we realize that surface waves are nothing but propagating waves of a special class of modes.[4] The general theory of nonlinear optics developed in earlier chapters can be applied here with little modification. We consider only optical mixing in this section.

The wave equation governing the optical mixing process is

$$\left[\nabla \times (\nabla \times) - (\omega_s^2/c^2)\varepsilon(z)\right]\mathbf{E}(\omega_s) = (4\pi\omega_s^2/c^2)\mathbf{P}^{\mathrm{NL}}(\omega_s) \tag{25.14}$$

where ε is assumed to be scalar. The output field $\mathbf{E}(\omega_s)$ can be either a surface or a bulk wave, and so are the pump fields inducing $\mathbf{P}^{\mathrm{NL}}(\omega_s)$. The procedure for finding the solution of (25.14) is the same as that outlined in Section 6.4. The pump waves are initially specified, and hence the expression for $\mathbf{P}^{\mathrm{NL}}(\omega_s)$ is known. Pump depletion is negligible in the present case. Therefore, the homogeneous and particular solutions of (25.14) can be obtained in a straight-forward way. Finally, the amplitudes of the homogeneous waves and the wavevectors of all waves can be fixed by the boundary conditions.

As an illustration, we consider the case of nonlinear wave interaction at the plane boundary between two semi-infinite media 1 ($z < 0$) and 2 ($z > 0$). For simplicity, we assume that $\mathbf{P}^{\mathrm{NL}}(\omega_s) \approx 0$ in medium 1 and $\mathbf{P}^{\mathrm{NL}}(\omega_s) = \hat{x}P_x^{\mathrm{NL}}(\omega_s) + \hat{z}P_z^{\mathrm{NL}}(\omega_s) \propto \exp(i\mathbf{k}_{2s}\cdot\mathbf{r} - i\omega_s t)$ in medium 2. The solution of (25.14) takes the form

$$\mathbf{E}_1(\omega_s) = \left(\hat{x} + \hat{z}\frac{k_x}{k_{1z}}\right)\mathscr{E}_{\mathrm{R}x}e^{i(k_x x - k_{1z}z - \omega t)} \qquad \text{for } z < 0$$

$$\mathbf{E}_2(\omega_s) = \left(\hat{x} - \hat{z}\frac{k_x}{k_{\Gamma z}}\right)\mathscr{E}_{\mathrm{T}x}e^{i(k_x x + k_{2z}z - \omega t)} + \hat{x}\left(\gamma_{xx}P_x^{\mathrm{NL}} + \gamma_{xz}P_z^{\mathrm{NL}}\right)$$

$$+ \hat{z}\left(\gamma_{zx}P_x^{\mathrm{NL}} + \gamma_{zz}P_z^{\mathrm{NL}}\right) \qquad \text{for } z > 0 \tag{25.15}$$

where

$$\gamma_{xx} = \frac{4\pi k_{2z}^2}{\varepsilon_2\left(k_{2s}^2 - k_2^2\right)},$$

$$\gamma_{xz} = \gamma_{zx} = \frac{4\pi k_x k_{2s,z}}{\varepsilon_2\left(k_{2s}^2 - k_2^2\right)},$$

$$\gamma_{zz} = \frac{4\pi\left(k_{2s,z}^2 - k_2^2\right)}{\varepsilon_2\left(k_{2s}^2 - k_2^2\right)},$$

and

$$k_{2s,x} = k_x.$$

The amplitudes, $\mathscr{E}_{Rx}$ and $\mathscr{E}_{Tx}$, of the homogeneous waves are obtained by matching of the fields at $z = 0$:

$$\mathscr{E}_{Rx} - \mathscr{E}_{Tx} = \gamma_{xx}\mathscr{P}_x^{NL} + \gamma_{xz}\mathscr{P}_z^{NL}$$

$$\varepsilon_1\left(\frac{k_x}{k_{1z}}\right)\mathscr{E}_{Rx} + \varepsilon_2\left(\frac{k_x}{k_{2z}}\right)\mathscr{E}_{Tx} = \varepsilon_2\left(\gamma_{zx}\mathscr{P}_x^{NL} + \gamma_{zz}\mathscr{P}_z^{NL}\right)$$

Here, $\mathscr{P}^{NL}$ is the amplitude of $\mathbf{P}^{NL}$. The preceding set of equations yields

$$\mathscr{E}_{Tx} = \frac{\varepsilon_2\left(k_{1z}k_{2z}/k_x\right)\left(\gamma_{zx}\mathscr{P}_x + \gamma_{zz}\mathscr{P}_z\right) - \varepsilon_1 k_{2z}\left(\gamma_{xx}\mathscr{P}_x^{NL} + \gamma_{xz}\mathscr{P}_z^{NL}\right)}{D'},$$

$$\mathscr{E}_{Rx} = \frac{\varepsilon_2 k_{1z}\left(\gamma_{xx}\mathscr{P}_x^{NL} + \gamma_{xz}\mathscr{P}_z^{NL}\right) + \varepsilon_2\left(k_{1z}k_{2z}/k_x\right)\left(\gamma_{zx}\mathscr{P}_x^{NL} + \gamma_{zz}\mathscr{P}_z^{NL}\right)}{D'},$$

$$D' = \varepsilon_2 k_{1z} + \varepsilon_1 k_{2z} \tag{25.16}$$

$$= \frac{\varepsilon_2 - \varepsilon_1}{\left(\varepsilon_2 + \varepsilon_1\right)\left(\varepsilon_1 k_{2z} - \varepsilon_2 k_{1z}\right)}\left[k_x^2 - \frac{\omega^2}{c^2}\frac{\varepsilon_1\varepsilon_2}{\varepsilon_1 + \varepsilon_2}\right].$$

Therefore, the output generated by $\mathbf{P}^{NL}$ is completely determined.

We discuss here the generation of a bulk wave by nonlinear interaction of surface waves or surface and bulk waves, and then the generation of a surface wave by interaction of bulk waves, or surface and bulk waves, or all surface waves. In the first case, because the pump fields involve surface waves, only the boundary media within the penetration depth of the surface waves contribute effectively to the nonlinear polarization $\mathbf{P}^{NL}$. Therefore, the process should be fairly surface-specific. As an example, we consider here second-harmonic generation by a surface em wave at the boundary between a metal and a nonlinear dielectric. We assume that the metal nonlinearity is negligible. The nonlinear polarization induced in the dielectric medium ($z > 0$) by a surface wave $\mathbf{E}(\omega) = (\hat{x}\mathscr{E}_x + \hat{z}\mathscr{E}_z)\exp(ik_x x - \beta z)$ is

$$\mathbf{P}^{(2)}(2\omega) = \mathbf{\chi}^{(2)} : (\hat{x}\mathscr{E}_x + \hat{z}\mathscr{E}_z)^2 \exp(i2k_x x - 2\beta z - i2\omega t), \quad (25.17)$$

which we assume to have only $\hat{x}$ and $\hat{z}$ components. With $\mathbf{P}^{NL}$ given, the second harmonic wave generated in the dielectric is explicitly given by $\mathbf{E}_2(\omega_s = 2\omega)$ in (25.15) with media 1 and 2 referring to the metal and the dielectric, respectively. Since $k_x(2\omega) = k_{sx}(2\omega) = 2k_x(\omega)$, the homogeneous part of the solution appears as a well-collimated propagating bulk wave when $|k_2(2\omega)| = |(2\omega/c)\sqrt{\varepsilon(2\omega)}| > |2k_x(\omega)|$. The same output appears as a surface wave if $|k_2(2\omega)| < |2k_x(\omega)|$.

Second-harmonic generation by a surface em wave at a metal–dielectric interface is easily observable. The Kretschmann geometry can be used to launch the surface wave at the metal–dielectric interface. The surface wave intensity is much higher than the incoming beam intensity if a sizable fraction of the input is coupled into the surface mode and the surface wave attenuation is small. Since the second-harmonic output depends quadratically on the fundamental input intensity, it becomes easily detectable, even though the interaction region is limited to a boundary layer of the order of a wavelength, and by the attenuation length of the surface wave ($\leq 10\,\mu$m in the visible). The process was first demonstrated by Simon et al.[5] A ruby laser with 1 MW in power, 20 nsec in pulsewidth, and 1 cm^2 in cross section at a quartz–silver interface could yield a second-harmonic signal of $\sim 10^4$ photons/pulse.[6] In fact, even second-harmonic generation by surface waves at an air–silver interface was easily observed.[5] In this latter case, the nonlinearity of silver was responsible for the nonlinear process. The output from the prism side in all these cases was found to be highly collimated with the predicted wavevector $\mathbf{k}(2\omega) = 2k_x(\omega)\hat{x} + [(4\omega^2/c^2)\varepsilon(2\omega) - (2k_x)^2]^{\frac{1}{2}}\hat{z}$ in the prism. When two counterpropagating surface em waves of the same frequency were used in second-harmonic generation, an output appeared along the surface normal, as required by the wavevector relation $k_{sx}(2\omega) = k_x(\omega) - k_x(\omega) = 0$. This peculiar effect resulting from the surface boundary condition is a special feature characteristic of surface or guided wave interaction.

While surface waves can interact to generate bulk waves, bulk waves can also interact at an interface to generate surface waves.[7] The latter is possible if the sum of the pump wavevectors has a component along the interface equal to the wavevector of the generated surface wave. The solution given in (25.15) and (25.16) is still applicable here. With $k_{sx}(\omega_s) = k_x(\omega_s) > k(\omega_s)$, the homogeneous part of the solution in (25.15) appears as a surface wave, which, according to (25.16), is resonantly excited by the nonlinear wave mixing via $P^{NL}(\omega_s, \mathbf{k}_s)$ when

$$\mathrm{Re}[k_x^2 - K^2] \simeq 0 \quad (25.18)$$

where $K^2 = (K' + iK'')^2 = (\omega_s^2/c^2)\varepsilon_1\varepsilon_2/(\varepsilon_1 + \varepsilon_2)$ is the dispersion relation for the surface em waves. The surface wave generated here with a wavevector $k_x(\omega_s) = k_{sx}(\omega_s)$ (not necessarily equal to K') actually corresponds to a driven wave.[7] In general, there should also exist at the interface a free surface wave with a wavevector $K'(\omega_s)$. Its amplitude is determined by the condition of field continuity in the interfacial plane. In the infinite plane wave approximation, however, the free wave can be neglected.

To illustrate the feasibility of nonlinear excitation of surface em waves, we use second-harmonic excitation of surface exciton-polaritons at a ZnO–liquid He interface as an example.[8] The experimental setup is seen in Fig. 25.6. The fundamental wavevector component $k_x(\omega)$ along the interface can be varied by varying the incident angle of the input laser beam. The generated surface exciton-polariton at the second-harmonic frequency can be detected either by the prism coupling method or simply through surface roughness scattering. From (25.16) and (25.18), we expect that the output signal should exhibit the resonant structure described by

$$S \propto \left[\left(2k_x(\omega) - K' \right)^2 + K''^2 \right]^{-1} \qquad (25.19)$$

as $2k_x(\omega)$ scans over $K'(2\omega)$. This was actually observed in the experiment, as shown in Fig. 25.7 at four different frequencies. From the positions and widths of the resonant peaks at different frequencies, $K'(2\omega)$ and $K''(2\omega)$ could be deduced. They were compared with the theoretical dispersion curve in Fig. 25.8 calculated from a dispersion relation somewhat different from (25.7) because of anisotropy of ZnO. In another example, difference-frequency generation of surface phonon-polaritons at an air–GaP interface was demonstrated.[9] The measured dispersion curve was also in good agreement with the theoretical prediction. We note that unlike the linear excitation of surface waves, the nonlinear excitation method has the advantage that it can be used to excite and

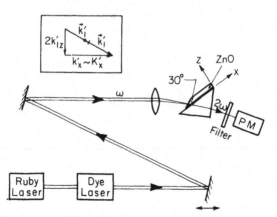

Fig. 25.6 Experimental setup for observing second-harmonic generation of surface exciton-polariton on ZnO. The inset shows the wavevector relation. (After Ref. 8.)

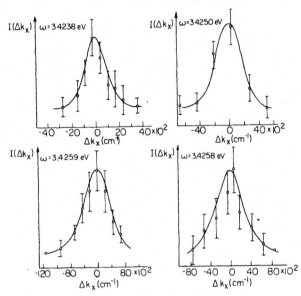

Fig. 25.7 Examples of experimental results of $I(2\omega, \Delta k_x)$ versus Δk_λ in the ZnO case. The solid curves are Lorentzian used to fit the data. (After Ref. 8.)

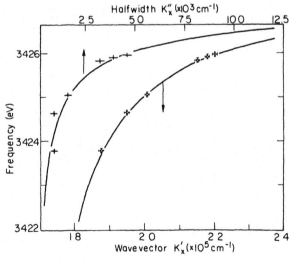

Fig. 25.8 Dispersion and damping characteristics of surface exciton-polaritons on ZnO. The solid curves are theoretical curves. (After Ref. 8.)

study surface polaritons at the interface of two semi-infinite media, or at an interface of a multilayer medium that cannot be reached by linear excitation.

Surface wave excitation by nonlinear mixing of surface waves is also possible. We consider here the problem of surface coherent anti-Stokes Raman scattering (CARS) at a metal–dielectric interface.[10] The process is interesting for many reasons:

1   It is an all-surface wave process.
2   It is phase-matchable in the surface plane.
3   It can be used to study Raman resonances of the dielectric medium.
4   The small field penetration depth (of the order of a wavelength) renders the process surface-sensitive.
5   The high surface wave intensity allows such a third-order surface process to be easily detectable.

We can again use the solution given by (25.15) and (25.16) to describe the anti-Stokes output. For simplicity, the nonlinear polarization is assumed to be dominated by that in the dielectric:

$$\mathbf{P}^{(3)}(\omega_a) = \chi^{(3)}(\omega_a = 2\omega_1 - \omega_2) : \mathbf{E}_1(\omega_1)\mathbf{E}(\omega_1)\mathbf{E}_2^*(\omega_2) \qquad (25.20)$$

where both $\mathbf{E}_1$ and $\mathbf{E}_2$ are surface waves with wavevectors $\mathbf{k}_{1,\parallel}(\omega_1)$ and $\mathbf{k}_{2,\parallel}(\omega_2)$, respectively, and the nonlinear susceptibility $\chi^{(3)}$ can be decomposed into a resonant and a nonresonant term

$$\chi^{(3)} = \chi_R^{(3)} + \chi_{NR}^{(3)} \qquad (25.21)$$

with

$$\chi_R^{(3)} = A/[\omega_1 - \omega_2 - \omega_{ex}) + i\Gamma].$$

Then (25.15) and (25.16) show that the anti-Stokes output is a surface wave if $|\mathbf{k}_{s,\parallel}(\omega_a)| = |2\mathbf{k}_{1,\parallel} - \mathbf{k}_{2,\parallel}|$ is larger than $(\omega_a/c)\sqrt{\varepsilon_1}$. This surface wave is resonantly excited when $\mathbf{k}_{s,\parallel}(\omega_a) \simeq \mathbf{K}'(\omega_a)$. Since the output is proportional to $|P^{(3)}(\omega_a)|^2$ and hence $|\chi^{(3)}|^2$, it is also resonantly enhanced when $(\omega_1 - \omega_2)$ approaches ω_{ex}.

Surface CARS has been demonstrated using the experimental arrangement in Fig. 25.9.[10] The Kretschmann geometry was adopted to launch the pump surface waves at a silver–benzene interface. The wavevectors of the pump waves could be adjusted to satisfy the surface phase-matching condition $2\mathbf{k}_{1,\parallel} - \mathbf{k}_{2,\parallel} = \mathbf{k}_{a,\parallel}(\omega_a) = \mathbf{K}'(\omega_a)$. The anti-Stokes signal coupled out by the prism appeared as a highly collimated beam along $\mathbf{k}_a(\omega_a)$. The result as a function of $(\omega_1 - \omega_2)$ is shown in Fig. 25.10, in comparison with a theoretical plot of $|\chi^{(3)}(\omega_1 - \omega_2)|^2$ for benzene. The resonant peak here corresponds to

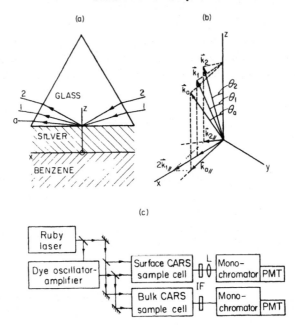

Fig. 25.9 Experimental setup for surface CARS measurements: (*a*) the prism-metal-liquid assembly; (*b*) wavevectors in the glass prism with their components in the *xy*-plane phase matched; (*c*) block diagram of the experimental arrangement. (After Ref. 10.)

the 992-cm^{-1} breathing mode of benzene. The signal level was fairly high. With the two input pulses having 0.5 and 5 mJ, respectively, in a 30-nsec pulsewidth and in a 0.5-cm^2 beam cross section at the interface, the output at the resonant peak was 2×10^5 photons/pulse, which was readily detectable. This shows that the process can be a valuable spectroscopic technique for probing thin films, overlayers, and perhaps even adsorbed molecules. The highly directional output allows the detection of CARS even in the presence of a strong luminescence background. Moreover, the attenuation length, $1/K''$, of the surface waves is of the order of 10 μm in the visible, indicating that the surface CARS signal will not be hurt very much by absorption in the dielectric medium as long as the absorption length is longer than $1/K''$. Thus surface CARS can be useful in spectroscopic studies of absorbing and luminescent materials.

The signal strength of surface CARS depends on the input pump fields as

$$S \propto I_1^2(\omega_1) I_2(\omega_2) A T \tag{25.22}$$

where $I(\omega_1)$, $I(\omega_2)$ are the pump intensities, A the beam cross section, and T the pulsewidth. It is obvious that to increase S we should increase I. However,

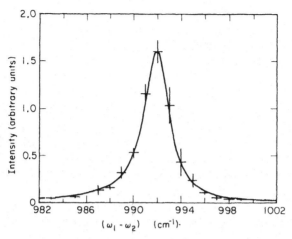

Fig. 25.10   Anti-Stokes output in surface CARS versus $\omega_1 - \omega_2$ around the 992 cm^{-1} Raman resonance of benzene. The solid curve is a theoretical curve.

the maximum laser intensity on the surface often is limited by optical damage. It happens that the damage may have a fluence (energy/unit area) threshold rather than an intensity threshold. Then, according to (25.22), shorter pump pulses should yield much stronger signals below the damage threshold. As an example, if we use 10-psec input pulses with 10 μJ/pulse instead of the nanosecond pulses in the above-mentioned experiment on the silver–benzene interface, we can obtain a signal of ~ 10^{11} photons/pulse from a focal spot of 0.15 mm^2 at the interface. Realizing that the signal comes essentially from a benzene boundary layer ~ 1000 Å thick, the preceding estimate suggests that surface CARS with picosecond pulses can have a sensitivity of detecting a submonolayer or molecules at the interface. Unfortunately, in such circumstances $\chi^{(3)}$ from the metal may dominate $\chi^{(3)}$ from the molecules. Some sort of background suppression techniques must be invented before surface CARS can be used for spectroscopic studies of monolayer adsorbates.

25.3   NONLINEAR OPTICAL EFFECTS AS SURFACE PROBES

The possibility of applying lasers to surface studies has opened up a new area of research in surface science. Laser annealing, for example, has aroused a great deal of interest for both scientific and technical reasons. Lasers have also been used to probe molecule–surface interaction by detecting and analyzing molecules desorbed or scattered from surfaces,[11] and to yield vibrational spectra of adsorbed molecules by laser desorption or photoacoustic spectroscopy.[12] In this section we discuss the problem of exploiting nonlinear optical effects, particularly second-harmonic generation (SHG), for surface studies.[13] Unlike the conventional surface probes, which rely on emission, absorption, or scattering of massive particles, nonlinear optical techniques are applicable to interfaces between two dense media and therefore may offer some unique and intriguing possibilities.

We saw in the previous section that surface CARS can have a submonolayer sensitivity and may become a useful surface spectroscopic tool. Another coherent Raman technique that can be used for monolayer spectroscopy is the stimulated Raman gain spectroscopy discussed in Section 10.6. Heritage and Allara[14] showed that by using CW mode-locked lasers as the pump and probe, the sensitivity of the technique could be greatly improved and a Raman gain smaller than 10^{-9} could be measured. Since a 100-W pump laser focused to a spot of 10 μm on a monolayer of molecules with a Raman cross section of 10^{-29} cm^2/sterad can yield a Raman gain of $\sim 2 \times 10^{-8}$, the technique should have a sensitivity high enough for monolayer spectroscopy. Indeed, Heritage and Allara were able to obtain a monolayer spectrum of p-nitrobenzoic acid adsorbed on sapphire, as shown in Fig. 25.11. The technique, however, required two extremely stable dye lasers and a low-noise detection system to observe the weak Raman gain. Small residual absorption and thermal fluctuations in the substrate could easily mask the spectrum.

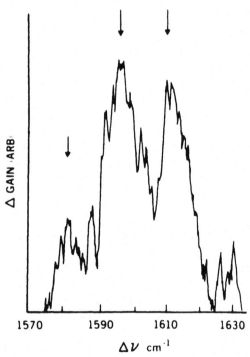

Fig. 25.11 Raman spectrum of a monolayer of p-nitrobenzoic acid (PNBA) on a thin film of aluminum oxide supported by sodium fluoride obtained by stimulated Raman gain spectroscopy. Three Raman peaks are marked. (After Ref. 14.)

Among the few nonlinear optical techniques that have been considered for surface studies, the second-harmonic and sum-frequency generation processes are probably most attractive. They are particularly useful for probing interfaces between media with inversion symmetry for the following reason. These processes are forbidden by symmetry in media with an inversion center but are allowed on surfaces because of the lack of inversion symmetry of the surface layers. They are therefore highly surface-specific and can be used as surface probes. In comparison with surface CARS and stimulated Raman gain techniques, SHG or sum-frequency generation has the advantages of being much simpler in the experimental arrangement and much stronger in the signal output.

The theory of SHG or sum-frequency generation from an interface modeled after Fig. 25.1 follows closely the theory discussed in Section 6.4, except that we now have a multilayer nonlinear medium. The general solution of this problem was worked out by Bloembergen and Pershan.[15] Here we consider the case of SHG from the interface between two isotropic media shown in Fig. 25.12. The linear dielectric constant of the surface layer is taken to be the same as that of the medium at $z > 0$ for simplicity. The second-order nonlinearity of the surface layer is described by the surface nonlinear susceptibility $\chi_S^{(2)}$; those of the bulk media vanish in the electric dipole approximation, but become finite if the electric quadrupole and magnetic dipole contributions are taken into account. The electric quadrupole and magnetic dipole contributions

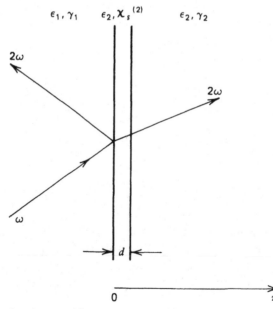

Fig. 25.12 Sketch of second-harmonic generation from an interface between two isotropic media. The interfacial layer of thickness d is specified by a linear dielectric constant ϵ_2 and a second-order surface nonlinear susceptibility $\chi_S^{(2)}$.

originate from the nonlocal response of the media to the field; they are proportional to the first spatial derivative of the field. By symmetry, the induced second-order polarization in an isotropic medium can be written as

$$
\begin{aligned}
\mathbf{P}^{(2)}(2\omega) &= \chi^{(2)} : \mathbf{E}(\omega)\nabla\mathbf{E}(\omega) \\
&= \alpha[\mathbf{E}(\omega) \cdot \nabla]\mathbf{E}(\omega) + \beta\mathbf{E}(\omega)[\nabla \cdot \mathbf{E}(\omega)] \\
&\quad + i(2\omega/c)\gamma[\mathbf{E}(\omega) \times \mathbf{B}(\omega)].
\end{aligned}
\tag{25.23}
$$

The first two terms on the right are of the electric quadrupole character and the last term is magnetic dipole.

With the nonlinear polarizations of both the surface layer and the bulk media specified, the solution of the wave equation for SHG from the boundary surface is fairly straightforward, although it is rather tedious. Two approaches can be taken. One uses the solution in Ref. 15 for a three-layer system; the middle layer corresponding to the surface layer has a thickness d approaching zero. The other assumes no clear boundary between the surface layer and medium 2 ($z > 0$), but uses a combined $\mathbf{P}^{(2)}(2\omega)$ to describe the nonlinear polarization induced in the surface layer and the substrate:

$$
\begin{aligned}
\mathbf{P}^{(2)}_{\text{eff}}(2\omega) &= \alpha_2[\mathbf{E} \cdot \nabla]\mathbf{E} + \beta_2\mathbf{E}[\nabla \cdot \mathbf{E}] + (i2\omega/c)\gamma_2(\mathbf{E} \times \mathbf{B}) \\
&\quad + \chi_S^{(2)}\delta(z) : \mathbf{E}\mathbf{E}
\end{aligned}
\tag{25.24}
$$

where $\delta(z)$ is a δ-function at $z = 0^+$. The solution of the problem is then the same as the one described in Section 6.4 for SHG from a single interface, except that the presence of the dipole layer, $\chi_S^{(2)}\delta(z) : \mathbf{E}\mathbf{E}$ at $z = 0^+$, changes the boundary conditions of the fields to[16]

$$
\Delta E_x = -\frac{4\pi}{\epsilon_2}\frac{\partial}{\partial x}P_{sz}, \qquad \Delta B_x = -4\pi i\left(\frac{\omega}{c}\right)P_{sy},
$$

$$
\Delta E_y = -\frac{4\pi}{\epsilon_2}\frac{\partial}{\partial y}P_{sz}, \qquad \Delta B_y = 4\pi i\left(\frac{\omega}{c}\right)P_{sx},
\tag{25.25}
$$

$$
\Delta(\epsilon E_z) = -4\pi\left(\frac{\partial}{\partial x}P_{sx} + \frac{\partial}{\partial y}P_{sy}\right), \qquad \Delta B_z = 0
$$

with $\Delta E \equiv E(z = 0^+) - E(z = 0^-)$ and $\mathbf{P}_s = \chi_S^{(2)} : \mathbf{E}\mathbf{E}$. Both approaches yield the same result. In the present case, where a single fundamental beam is assumed, the α and β terms in (25.24) should vanish, while $\chi_S^{(2)}$ for an isotropic surface layer has only the following nonzero elements: $\chi_{S,zzz}^{(2)}$, $\chi_{S,zii}^{(2)} = \chi_{S,izi}^{(2)}$, and $\chi_{S,izz}^{(2)}$ ($i = x, y$). The result is then greatly simplified. For the second-harmonic output in the reflected direction, for example, the intensity has the expression[16]

$$
I(2\omega) = \frac{32\pi^3\omega^2\sec^2\theta_{2\omega}}{c^3\epsilon_1(\omega)\epsilon_1^{1/2}(2\omega)}\left|\mathbf{e}_{2\omega} \cdot \chi_{S,\text{eff}}^{(2)} : \mathbf{e}_\omega\mathbf{e}_\omega\right|^2 I^2(\omega)
\tag{25.26}
$$

where $e_\Omega = L(\Omega) \cdot \hat{e}_\Omega$, with $\hat{e}_\Omega$ denoting the unit polarization vector at frequency Ω in medium 2 ($z > 0$); L is the Fresnel factor for the field, which has the diagonal elements

$$L_{xx} = \frac{2\varepsilon_1 k_{2z}}{\varepsilon_2 k_{1z} + \varepsilon_1 k_{2z}},$$

$$L_{yy} = \frac{2k_{1z}}{k_{1z} + k_{2z}}, \tag{25.27}$$

$$L_{zz} = \frac{2\varepsilon_1 k_{1z}}{\varepsilon_2 k_{1z} + \varepsilon_1 k_{2z}},$$

and the effective surface nonlinear susceptibility $\chi^{(2)}_{S,\text{eff}}$ is defined as

$$\left[\chi^{(2)}_{S,\text{eff}}\right]_{z\shortparallel} = \chi^{(2)}_{S,z\shortparallel} + \frac{\varepsilon_2(2\omega)}{\varepsilon_1(2\omega)}\gamma_1 - \gamma_2,$$

$$\left[\chi^{(2)}_{S,\text{eff}}\right]_{zzz} = \chi^{(2)}_{S,zzz} + \frac{\varepsilon_2(2\omega)\varepsilon_2^2(\omega)}{\varepsilon_1(2\omega)\varepsilon_1^2(\omega)}\gamma_1 - \gamma_2, \tag{25.28}$$

$$\left[\chi^{(2)}_{S,\text{eff}}\right]_{\shortparallel z\shortparallel} = \chi^{(2)}_{S,\shortparallel z\shortparallel}.$$

The above solution shows that the s-polarized second-harmonic output, proportional to $|\chi^{(2)}_{S,\shortparallel z\shortparallel}|^2$, is generated entirely from the surface layer, while the p-polarized output, having contributions from all the $\chi^{(2)}_{S,\text{eff}}$ elements, is generated from both the surface layer and the bulk. Here, however, only the magnetic dipole part appears in the bulk contribution. Generally, with two input laser beams, or in crystalline bulk media having inversion symmetry, the electric quadrupole part can also appear as a bulk contribution. To estimate the relative importance of the bulk versus the surface in SHG, we note that the electric quadrupole and magnetic dipole part of a susceptibility is usually about ka times smaller than the allowed electric dipole part, where a is the size of atoms or unit cells. Therefore, since $\chi_S^{(2)}$ for the surface layer is always electric dipole allowed, we have $|\chi_S^{(2)}/\chi_{\text{vol}}^{(2)}| \sim d/ka$, with d being the surface layer thickness. For SHG by reflection from a surface, it was mentioned in Section 6.4 that the bulk contribution comes essentially from a layer of $\lambda/2\pi$ thick near the surface. Then the relative contribution of surface to bulk in SHG is about $|\chi_S^{(2)}/\chi_{\text{vol}}^{(2)}(\lambda/2\pi)|^2 \sim (d/a)^2$, which is larger than or of the order of unity. This ratio can be further enhanced through resonances or polarization selection if the structure of $\chi_S^{(2)}$ is sufficiently different from $\chi_{\text{vol}}^{(2)}$. The foregoing discussion indicates that surface SHG may be used for the probing of interfaces between two centrosymmetric media. Yet we must still show that the signal strength of SHG from a surface monolayer is strong enough for detection. Assuming $|\chi_S^{(2)}| \sim 10^{-15}$ esu for the surface monolayer, and a pump laser pulse with a pulsewidth of 10 nsec and an energy of 20 mJ/pulse at 1.06 μm focused to 0.2 cm^2, we find, from (25.26), a second-harmonic output of ~ 10^4 photons/pulse. Such a signal should be readily detectable.

Surface SHG is clearly a viable method for studying adsorbates at interfaces. In this case, the surface susceptibility can be written as

$$\chi_S^{(2)} = \chi_{SA}^{(2)} + \chi_{SS}^{(2)} \tag{25.29}$$

where $\chi_{SA}^{(2)}$ denotes the part arising from adsorbed atoms or molecules, and $\chi_{SS}^{(2)}$ from the surface layers of the adjoining media. If $|\chi_{SA}^{(2)}| \gg |\chi_{SS}^{(2)}|$, or by some means, the contribution of $\chi_{SS}^{(2)}$ can be suppressed or subtracted, then surface SHG can be used to probe the adsorption. This happens, for example, with molecular adsorbates having a large second-order nonlinearity so that $\chi^{(2)}$ from the molecules dominates in $\chi_S^{(2)}$. We assume here that this is the case and consider the various possible applications of surface SHG to studies of surface adsorbates.

The experimental arrangement of surface SHG is fairly simple. As seen in Fig. 25.13, it basically involves the direction of a laser beam onto a sample and

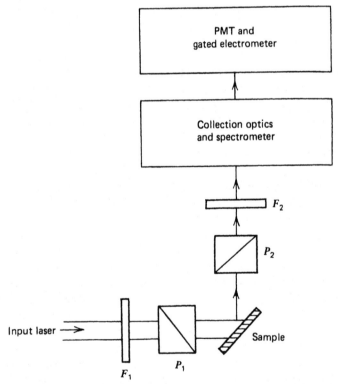

Fig. 25.13 Experimental schematic for second-harmonic generation by reflection from a sample.

the collection of second-harmonic output from the interface by an appropriate detection system. The polarizer and the analyzer allow the selection of polarizations of the input and output fields. Because pulsed lasers are used, time-resolved *in situ* probing of adsorbates is possible.

The high sensitivity of the surface SHG to adsorbates is clearly demonstrated in Fig. 25.14, in which the variation of SHG from a silver electrode during an oxidation-reduction cycle in a 0.1 M KCl electrolyte is presented.[17] The signal rises sharply as AgCl begins to form on the electrode and drops precipitously when the last layers of AgCl are reduced. From the measured amount of charge transfer at the electrode at various times, we can arrive at the conclusion that the better part of the abrupt change in the SH signal corresponds to a deposition or removal of a single adsorbed layer. The bulk layers

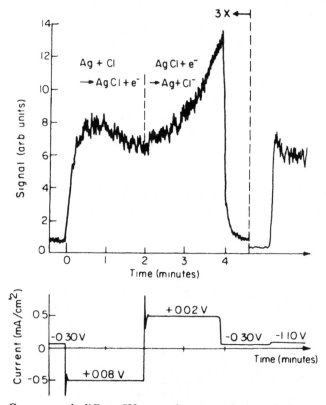

Fig. 25.14 Current and diffuse SH as a function of time during and after an electrolytic cycle. The voltages listed in the lower curve are V_{Ag} with respect to a standard electrode in the cell. Pyridine (.05 M) was added to the 0.1 M KCl solution following the completion of the electrolytic cycle. With V_{Ag} adjusted to -1.1 V, a sudden rise of the SH signal was observed, corresponding to the adsorption of pyridine molecules on the Ag electrode. (After Ref. 17.)

of AgCl do not appear to contribute appreciably to the SHG. With 0.05 M pyridine in the electrolyte and with a sufficiently negative bias on the Ag electrode, a monolayer or submonolayer of pyridine molecules can be adsorbed on the electrode. This is also manifested by a dramatic increase of the SHG in Fig. 25.14. The curve in the figure was obtained with a Q-switched YAG laser of only 0.2 mJ/pulse in 0.2 cm^2, and the SH signal from the adsorbed pyridine was already as high as 8×10^5 photons/pulse. This signal strength is several orders of magnitude larger than the one estimated from (25.26). It is the result of local-field enhancement on the rough surface structure of the Ag electrode. Because of the local plasmon resonance at the surface and the pointing rod effect, the local fields on the tips of the local Ag structures can be much larger than the incoming field and can lead to an enhancement of $\sim 10^4$ in the SH output.[18] The signal level given above indicates that even without surface enhancement, as in the case of a smooth surface, SHG from a monolayer of adsorbates like AgCl and pyridine should be easily detectable using a laser of ~ 10 mJ/pulse.

If a tunable pump laser is used in the surface SHG, then spectroscopic data of adsorbates can be obtained from the resonant feature of the signal when either ω or 2ω hits a transition. This is illustrated in Fig. 25.15, in which the

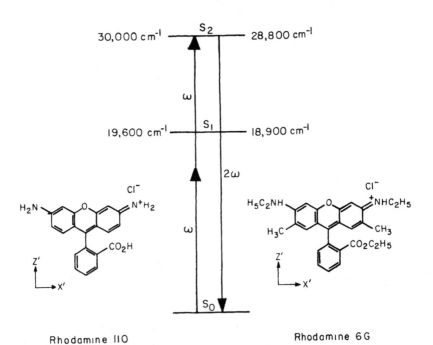

Fig. 25.15 (a) Energy level diagrams for rhodamine 110 and rhodamine 6G dissolved in ethanol. (b) Normalized SH intensity for p-polarized excitation of half-monolayer samples of rhodamine 110 and rhodamine 6G on fused silica as a function of the SH wavelength in the region of the $S_0 \rightarrow S_2$ transition. (After Ref. 19.)

SH signal from a half monolayer ($\sim 5 \times 10^3$ molecules/cm^2) of rhodamine 6G and rhodamine 110 on a smooth fused quartz plate versus 2ω is plotted.[19] The two peaks correspond to the 2ω resonant excitation of the $S_0 \to S_2$ transition in the two dyes. The resonantly enhanced signal was very strong; with 10-nsec, 1-mJ pump pulses focused to 10^{-3} cm^2 on the sample, $\sim 10^4$ photons/pulse were generated. This was several orders of magnitude stronger than the SH signal from the quartz substrate. It therefore shows that studies of adsorbates at much lower coverage than a monolayer should not pose major difficulties.

The submonolayer sensitivity of SHG allows us to measure the adsorption isotherm.[20] An example is given in Fig. 25.16, where the adsorption isotherm of p-nitrobenzoic acid (PNBA) on fused quartz immersed in an ethanol : PNBA solution is shown. From the adsorption isotherm, a free energy of adsorption for PNBA at the ethanol–quartz interface can be deduced. The result here also provides an example that surface SHG can be used to probe adsorbates at interfaces between two dense media.

With input and output beam polarizations and geometry properly chosen, the various elements of $\chi_S^{(2)}$ can be selectively measured by the surface SHG. The symmetry of $\chi_S^{(2)}$ in the $(\hat{x}-\hat{y})$ plane is a reflection of the symmetry of the

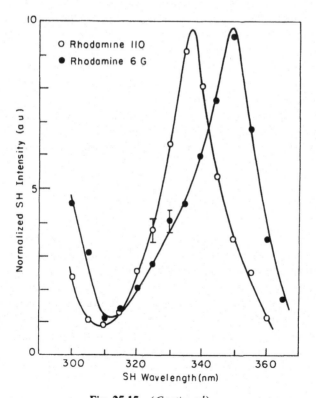

Fig. 25.15 (*Continued*).

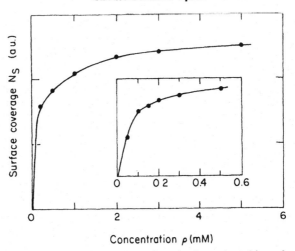

Concentration p (mM)

Fig. 25.16 Isotherm for the adsorption of p-nitrobenzoic acid to fused silica from ethanolic solution, measured by the surface second-harmonic generation technique. (After Ref. 20.)

average molecular arrangement at the interface. Therefore, surface SHG can also be used to probe the structural symmetry of the molecular monolayer adsorbed at an interface. For example, it was found that the surface SHG signal from the adsorbed monolayer of dye or PNBA molecules on a fused quartz substrate remained unchanged when the substrate was rotated around its surface normal; this indicated that the molecules were randomly or isotropically distributed on the substrate.[19, 20]

The surface susceptibility tensor $\chi_S^{(2)}$ also reflects the average orientation of the adsorbates in the following sense.[20] If the local-field correction is neglected, then $\chi_S^{(2)}$ and the second-order molecular polarizability $\alpha^{(2)}$ are related by the equation

$$\chi^{(2)}_{S,ijk} = N_S \langle T^{\lambda\mu\nu}_{ijk} \rangle \alpha^{(2)}_{\lambda\mu\nu}. \tag{25.30}$$

Here, N_S is the surface density of the adsorbates, and T_{ijk} represents the coordinate transformation between the molecular (ξ, η, ζ) system and the lab (x, y, z) system. The average of $T^{\lambda\mu\nu}_{ijk}$ over the molecular orientations, denoted by $\langle T^{\lambda\mu\nu}_{ijk} \rangle$, is then a description of the average orientation of the adsorbates. To find $\langle T^{\lambda\mu\nu}_{ijk} \rangle$, we need to know both $\chi^{(2)}_{S,ijk}$ and $\alpha^{(2)}_{\lambda\mu\nu}$ in general. While $\chi^{(2)}_{S,ijk}$ can be measured by the surface SHG, $\alpha^{(2)}_{\lambda\mu\nu}$ are unfortunately difficult to obtain. This makes the determination of average orientation difficult. In some cases, however, measurements of only the ratios of various elements of $\chi^{(2)}_{S,ijk}$ can already yield some information about the orientation of the adsorbates. This is illustrated by the example of PNBA adsorbed on fused quartz below.[20]

The nonlinearity of PNBA molecules is dominated by a signal element $\alpha^{(2)}_{\zeta\zeta\zeta}$ in $\boldsymbol{\alpha}^{(2)}$, and the molecular distribution on fused quartz is isotropic. These molecules can therefore be considered as long rods defined by the ζ-axis; their orientation at the interface is specified by the angle θ between $\hat{z}$ and $\hat{\zeta}$. The ratio of two independent elements of $\boldsymbol{\chi}_S^{(2)}$ can then give the weighted average of θ, for example,

$$\frac{\chi^{(2)}_{S,zzz}}{\chi^{(2)}_{S,zxx}} = \frac{N_S\langle\cos^3\theta\rangle\alpha^{(2)}_{\zeta\zeta\zeta}}{\frac{1}{2}N_S\langle\cos\theta\sin^2\theta\rangle\alpha^{(2)}_{\zeta\zeta\zeta}} = \frac{2\langle\cos^3\theta\rangle}{\langle\cos\theta\sin^2\theta\rangle}. \tag{25.31}$$

The average orientation of PNBA on fused quartz has been determined in this fashion at solid–air and solid–ethanol interfaces. Assuming a sharply peaked orientation distribution, θ was found to be $\sim 40°$ in ethanol and $\sim 70°$ in air.

The surface SHG technique can certainly be extended to sum- and difference-frequency generation. With a tunable infrared radiation and a visible laser as the pumps, the sum-frequency generation should permit the study of vibrational transitions of adsorbates and facilitate the determination of molecular orientation. If tunable picosecond lasers are available, then the technique can also be used to study the dynamic properties of adsorbates at interfaces in the picosecond time regime.

In comparison with the conventional surface probes, the second-order nonlinear optical processes have a number of important advantages: the experimental setup is relatively simple, *in situ* probing of interfaces between two dense media is possible, studies of dynamic properties of interfaces with a subpicosecond time resolution can become a reality, and information about molecular arrangement and molecular orientation at interfaces may be obtained in a fairly straightforward way. The technique, however, is still in the developmental stage. How well it can apply to various substrates and adsorbates is yet to be tested. In many ways, the limitation of the technique will also depend on our theoretical understanding of linear and nonlinear optical properties of interfaces.

REFERENCES

1 A. Sommerfeld, *Ann. Phys.* **28**, 665 (1909).

2 A Otto, *Z. Phys.* **216**, 398 (1968).

3 E. Kretschmann, *Z. Phys.* **241**, 313 (1971).

4 See, for example, Y. R. Shen and F. DeMartini, in V. M. Agranovich and D. L. Mills, eds., *Surface Polaritons* (North-Holland Publishing Co., Amsterdam, 1982), p. 629.

5 H. J. Simon, D. E. Mitchell, and J. G. Watson, *Phys. Rev. Lett.* **33**, 1531 (1974); H. J. Simon, R.E. Benner, and J.G. Rako, *Opt. Commun.* **23**, 245 (1977).

6 C. K. Chen, A. R. B. de Castro, and Y. R. Shen, *Opt. Lett.* **4**, 393 (1979).

7 F. DeMartini and Y. R. Shen, *Phys. Rev. Lett.* **36**, 216 (1976).

8 F. DeMartini, M. Colocci, S. E. Kohn, and Y. R. Shen, *Phys. Rev. Lett.* **38**, 1223 (1977).

9   F. DeMartini, F. G. Giuliani, M. Mataloni, E. Palange, and Y. R. Shen, *Phys Rev. Lett* **37**, 440 (1976).

10   C K. Chen, A R. B. de Castro, Y. R. Shen, and F. DeMartini, *Phys. Rev. Lett.* **43**, 946 (1979).

11   See, for example, M. Asscher, W. L. Guthrie, T. H. Lin, and G A. Somorjai, *Phys. Rev. Lett.* **49**, 76 (1982); H. Zacharias, M. M. T. Loy, and P. A. Roland, *Phys. Rev. Lett.* **49**, 1790 (1982).

12   T. J. Chuang and H. K. Seki, *Phys. Rev. Lett.* **49**, 382 (1982); F. Träger, H. Coufal, and T. J. Chuang, *Phys. Rev. Lett.* **49**, 1720 (1982).

13   C. K. Chen, T. F. Heinz, D. Ricard, and Y. R. Shen, in C. B. Collins, ed., *Proceedings of the International Conference on Lasers '81* (STS Press, McLean, VA, 1982), p. 61; H. W. K. Tom, T. F. Heinz, and Y. R. Shen, in A. Zewail, ed., *Proceedings of the International Conference on Photochemistry and Photobiology* (Harwood Academic Publishing, Chur, Switzerland, 1983).

14   J. P. Heritage and D. L. Allara, *Chem. Phys. Lett.* **74**, 507 (1980).

15   N. Bloembergen and P. S. Pershan, *Phys. Rev.* **128**, 606 (1962).

16   Thanks to T. F. Heinz for providing the derivations.

17   C. K. Chen, T. F. Heinz, D. Ricard, and Y. R. Shen, *Phys. Rev. Lett.* **46**, 1010 (1981).

18   C. K. Chen, A. R. B. de Castro, and Y. R. Shen, *Phys. Rev. Lett.* **46**, 145 (1981).

19   T. F. Heinz, C. K. Chen, D. Ricard, and Y. R. Shen, *Phys. Rev. Lett.* **48**, 478 (1982).

20   T. F. Heinz, H. W. K. Tom, and Y. R. Shen, *Phys. Rev. A28*, 1883 (1983).

BIBLIOGRAPHY

Agranovich, V. M., and D. L. Mill, eds., *Surface Polaritons*, (North-Holland Publishing Co., Amsterdam, 1982).

Heinz, T. F., H. W. K. Tom, and Y. R. Shen, *Laser Focus* **19**, No. 5, 101 (1983).

26

Nonlinear Optics
in Optical Waveguides

In the development of fiber and integrated optics for communications and data processing, nonlinear optical effects in waveguides have played a unique role. On the one hand, such effects impose a limit on the power that can be transmitted through an optical fiber or waveguide. On the other hand, they have led to the construction of novel optical devices potentially useful not only in optical signal processing but also in other applications. The characteristic features of nonlinear optics in waveguides are the high field intensities resulting from the beam confinement and the long interaction length achievable in low-loss fibers or waveguides. Both are important factors in the buildup of nonlinear wave interaction, making nonlinear optical effects in waveguides easily observable even with CW lasers. This chapter outlines the general theory of wave interaction in optical waveguides, and briefly describes the experimental observations. Pulse propagation in a long fiber is discussed as a special topic of interest.

26.1 GENERAL THEORY

We consider first linear wave propagation in a waveguide.[1] A guided wave is generally defined as a propagating wave that is confined in the transverse dimensions. The surface em wave discussed in Chapter 25 is an example. Waves propagating in thin films and fibers are other examples. In all cases, the wave in a specific waveguide mode can be described by the field

$$\mathbf{E}^{(i)} = \left[\frac{A^{(i)} \mathbf{F}^{(i)}(\rho)}{\sqrt{D^{(i)}}} \right] \exp(iK^{(i)}z - i\omega t)$$

with

$$D^{(i)} = \int d^2\rho \, \mathbf{F}^{(i)}(\rho) \cdot \mathbf{F}^{(i)}(\rho). \qquad (26.1)$$

Here the direction of wave propagation is chosen as $\hat{z}$, the superindex i denotes the waveguide mode, $\mathbf{F}^{(i)}(\vec{\rho})$ is the normalized field distribution of the ith mode in the transverse plane, $A^{(i)}$ is the amplitude of the wave, and $K^{(i)}$ is the wavevector. Both $K^{(i)}$ and $\mathbf{F}^{(i)}(\rho)$ can be determined by solving the wave equation with the proper boundary conditions in the transverse directions. The function $\mathbf{F}^{(i)}(\rho)$ describes the confinement of the field in the transverse plane. For a waveguide with a close boundary in the transverse plane, two indices are generally needed to specify a waveguide mode. If the waveguide is open in one dimension, one index is sufficient to specify a mode. For the special case of surface em waves propagating along an interface between two semi-infinite media, no index is needed since for a given ω only one surface wave mode can exist (Section 25.1).

The theory of wave interaction in a waveguide is essentially the same as that developed in earlier chapters for plane waves in a bulk medium. The difference is merely in the fact that guided waves now play the role of plane waves. The equation is, of course, the same in all cases:

$$\nabla \times (\nabla \times \mathbf{E}) - \frac{\omega^2 \varepsilon}{c^2} \mathbf{E} = \frac{4\pi\omega^2}{c^2} \mathbf{P}^{\mathrm{NL}} \qquad (26.2)$$

where, for the nth order nonlinear process,

$$\begin{aligned}\mathbf{P}^{\mathrm{NL}}(\omega) = \mathbf{P}^{(n)}(\omega) = \boldsymbol{\chi}^{(n)}(\omega = \omega_1 + \omega_2 + \cdots + \omega_n) \\ : \mathbf{E}_1(\omega_1)\mathbf{E}_2(\omega_2)\ldots\mathbf{E}_n(\omega_n).\end{aligned} \qquad (26.3)$$

For the waveguide case, we notice that $\mathbf{E}^{(i)}$ in (26.1) with a constant $A^{(i)}$ is a homogeneous solution of (26.2). In the presence of $\mathbf{P}^{\mathrm{NL}}$, the amplitude $A^{(i)}$ is expected to change with the propagation distance z. Then, in the slowly varying amplitude approximation (see Section 3.3), (26.2) can be transformed into a first-order differential equation for $A^{(i)}(z)$:

$$\left[\frac{\mathbf{F}^{(i)}(\rho)}{\sqrt{D^{(i)}}}\right] \frac{\partial}{\partial z} A^{(i)} = \frac{i2\pi\omega^2}{K^{(i)}c^2} \mathbf{P}^{\mathrm{NL}}\exp(-iK^{(i)}z + i\omega t). \qquad (26.4)$$

Multiplication of both sides by $\mathbf{F}^{(i)}(\rho)$, followed by an integration over ρ, yields[2]

$$\frac{\partial}{\partial z} A^{(i)} = \frac{i2\pi\omega^2}{K^{(i)}c^2} \frac{1}{\sqrt{D^{(i)}}} \int d^2\rho \, \mathbf{F}^{(i)}(\rho) \cdot \mathbf{P}^{\mathrm{NL}}\exp(-iK^{(i)}z + i\omega t). \qquad (26.5)$$

Equation (26.5) can now be used to describe how the guided wave varies with z under the action of the nonlinear polarization $\mathbf{P}^{NL}$.

For illustration, we consider here three different cases: optical mixing, parametric amplification, and stimulated Raman scattering. In the first case, let us use four-wave mixing as an example. For simplicity, we assume that the pump depletion is negligible. The nonlinear polarization induced by the three pump waves, each in a specific waveguide mode, is

$$
\mathbf{P}^{(3)}(\omega) = \boldsymbol{\chi}^{(3)}(-\omega = \omega_1 + \omega_2 + \omega_3) : \mathbf{E}^{(1)}(\omega_1)\mathbf{E}^{(2)}(\omega_2)\mathbf{E}^{(3)}(\omega_3)
$$

$$
= \left\{ \frac{\boldsymbol{\chi}^{(3)} : A^{(1)}\mathbf{F}^{(1)}(\rho)A^{(2)}\mathbf{F}^{(2)}(\rho)A^{(3)}\mathbf{F}^{(3)}(\rho)}{[D^{(1)}D^{(2)}D^{(3)}]^{1/2}} \right\} \tag{26.6}
$$

$$
\times \exp\left[i(K^{(1)} + K^{(2)} + K^{(3)})z - i\omega t \right].
$$

Equation (26.5) for the output field in the fth waveguide mode can then be written as

$$
\frac{\partial}{\partial z}A^{(f)} = \frac{i2\pi\omega^2}{K^{(f)}c^2}\langle f123\rangle A^{(1)}A^{(2)}A^{(3)}\exp(i\,\Delta K\,z) \tag{26.7}
$$

where

$$
\langle f123\rangle \equiv \int \frac{d^2\rho\,\mathbf{F}^{(f)}(\rho)\cdot\boldsymbol{\chi}^{(3)} : \mathbf{F}^{(1)}(\rho)\mathbf{F}^{(2)}(\rho)\mathbf{F}^{(3)}(\rho)}{[D^{(f)}D^{(1)}D^{(2)}D^{(3)}]^{1/2}},
$$

$$
\Delta K \equiv K^{(1)} + K^{(2)} + K^{(3)} - K^{(f)},
$$

and $\boldsymbol{\chi}^{(3)}$, in general, is a function of ρ. The preceding equation can be readily solved to yield $A^{(f)}$ and hence, an output power in the fth mode at $z = l$,

$$
\mathscr{P}^{(f)} = \frac{cn^{(f)}}{2\pi}|A^{(f)}|^2
$$

$$
= \frac{2\pi\omega^2}{cn^{(f)}}|\langle f123\rangle|^2|A^{(1)}A^{(2)}A^{(3)}|^2\frac{\sin^2(\Delta K\,l/2)}{(\Delta K\,l/2)^2}l^2 \tag{26.8}
$$

assuming $n^{(f)} = K^{(f)}c/\omega$, and $A^{(f)} = 0$ at $z = 0$. This result is very similar to that of the plane wave case (see Section 14.2). In fact, if the pump beam intensities, the beam cross sections, the nonlinearity of the medium, and the length were assumed to be the same in the two cases, the output powers would be comparable. In practice, however, because of the beam confinement, a waveguide allows a much longer wave interaction length, hence resulting in much stronger nonlinear optical effects.

The long interaction length also makes the four-wave parametric amplification process possible in a waveguide. We consider here the amplification of the signal and idler waves at ω_s and ω_i by the parametric pumping of $2\omega_p = \omega_s + \omega_i$, assuming each wave in a definite waveguide mode. Using (26.5), we find the following coupled equations for the amplitudes of the coupled waves:[2]

$$\frac{\partial}{\partial z}A^{(p)} = \frac{i2\pi\omega_p^2}{K^{(p)}c^2}\langle pppp\rangle|A^{(p)}|^2 A^{(p)},$$

$$\frac{\partial}{\partial z}A^{(s)} = \frac{i2\pi\omega_s^2}{K^{(s)}c^2}\left[\langle sspp\rangle A^{(s)}|A^{(p)}|^2 + \langle sppi\rangle A^{(p)}A^{(p)}A^{(i)*}e^{i\Delta K z}\right],\quad (26.9)$$

$$\frac{\partial}{\partial z}A^{(i)*} = \frac{-i2\pi\omega_i^2}{K^{(i)}c^2}\left[\langle iipp\rangle^* A^{(i)*}|A^{(p)}|^2 + \langle ipps\rangle^* A^{(p)*}A^{(p)*}A^{(s)}e^{-i\Delta K z}\right]$$

where

$$\langle klmn\rangle = \int \frac{d^2\rho\, \mathbf{F}^{(k)}\cdot\boldsymbol{\chi}^{(3)}:\mathbf{F}^{(l)}\mathbf{F}^{(m)}\mathbf{F}^{(n)}}{[D^{(k)}D^{(l)}D^{(m)}D^{(n)}]^{1/2}}$$

and $\Delta K = 2K_p - K_s - K_i$. In the equation of $A^{(p)}$, we neglected the pump depletion effect, but included the term corresponding to an effective refractive index change induced by the pump field. With $|A^{(p)}|^2$ taken to be a constant, the pump amplitude has the form

$$A^{(p)}(z) = A^{(p)}(0)\exp(i\,\delta K_p z) \quad (26.10)$$

with $\delta K_p = (2\pi\omega_p^2/K^{(p)}c^2)\langle pppp\rangle|A^{(p)}|^2$. Then, by the substitution of $\mathscr{A}^{(s)} = A^{(s)}\exp(-i\,\delta K_s z)$ and $\mathscr{A}^{(i)} = A^{(i)}\exp(-i\,\delta K_i z)$, with $\delta K_s = ((2\pi\omega_s^2)/K^{(s)}c^2)\langle sspp\rangle \times |A^{(p)}|^2$ and $\delta K_i = (2\pi\omega_i^2/K^{(i)}c^2)\langle iipp\rangle|A^{(p)}|^2$, the set of equations for $A^{(s)}$ and $A^{(i)}$ in (26.9) becomes

$$\frac{\partial}{\partial z}\mathscr{A}^{(s)} = i\frac{2\pi\omega_s^2}{K^{(s)}c^2}\langle sppi\rangle\left[A^{(p)}(0)\right]^2\mathscr{A}^{(i)*}e^{i\gamma z},$$

$$\frac{\partial}{\partial z}\mathscr{A}^{(i)*} = -i\frac{2\pi\omega_i^2}{K^{(i)}c^2}\langle ipps\rangle^*\left[A^{(p)}(0)^*\right]^2\mathscr{A}^{(s)}e^{-i\gamma z} \quad (26.11)$$

where $\gamma = \Delta K + 2\delta K_p - \delta K_s - \delta K_i^*$. Except for the coupling coefficients, (26.11) is the same as the equation obtained in Section 9.1 for parametric amplification. The solution has the form

$$\mathscr{A}^{(s)}(z) = \left(C_{1s}e^{gz} + C_{2s}e^{-gz}\right)e^{i(\gamma/2)z},$$

$$\mathscr{A}^{(i)*}(z) = \left(C_{1i}e^{gz} + C_{2i}e^{-gz}\right)e^{-i(\gamma/2)z} \quad (26.12)$$

where

$$g = \left[\frac{4\pi^2\omega_s^2\omega_i^2}{K^{(s)}K^{(i)}c^4}|\langle sppi\rangle|^2|A^{(p)}(0)|^4 - (\gamma/2)^2\right]^{1/2}$$

and the C can be easily obtained in terms of $\mathscr{A}^{(s)}(0)$ and $\mathscr{A}^{(\iota)}(0)^*$. Note that if $|\mathscr{A}^{(\iota)}(z)|^2 \simeq |\mathscr{A}^{(\iota)}(0)|^2$, and $\mathscr{A}^{(s)}(0) = 0$, the result here should reduce to that of four-wave mixing discussed earlier. The parametric gain is a maximum when $\gamma = 0$. For $\mathrm{Re}(gz) \gg 1$, both $\mathscr{A}^{(s)}(z)$ and $\mathscr{A}^{(\iota)}(z)$ grow exponentially.

The foregoing derivation for parametric amplification applies equally well to the case of coupled Stokes–anti-Stokes generation in stimulated Raman scattering if $\omega_p - \omega_s = \omega_t - \omega_p$ is near a Raman resonance. The nonlinear susceptibility $\chi^{(3)}$ is then a complex quantity. The Stokes and anti-Stokes waves are effectively decoupled if the overall phase mismatch γ is large. In that case, the Stokes amplitude in the waveguide is simply described by

$$\frac{\partial}{\partial z} A^{(s)} = \frac{i2\pi\omega_s^2}{K^{(s)}c^2} \langle sspp \rangle |A^{(p)}|^2 A^{(s)}. \tag{26.13}$$

With negligible pump depletion, (26.13) yields

$$|A^{(s)}(z)|^2 = |A^{(s)}(0)|^2 \exp(G_R z) \tag{26.14}$$

where the stimulated Raman gain is

$$G_R = -\frac{4\pi\omega_s^2}{K^{(s)}c^2} \mathrm{Im}\langle sspp \rangle |A^{(p)}|^2. \tag{26.15}$$

As expected, this result is the same as the one given in (10.13) for the plane wave case with $\mathrm{Im}\langle sspp \rangle |A^{(p)}|^2$ replacing $\mathrm{Im}\,\chi^{(3)}_{R2}|E_1|^2$.

These examples show that most nonlinear optical processes can be expected in a waveguide, with a theoretical description not much different from that of the plane wave case. The phase mismatch ΔK for a particular process depends on the dispersion of the waveguide modes. For a single-mode waveguide, it is difficult, in general, to have ΔK approach zero. For a multimode waveguide, however, there exists the flexibility in appropriating various waves to various modes, and then it is possible to make ΔK nearly vanishing. In principle, a coherent length larger than a few meters would be achievable. In practice, because of structural imperfection, ΔK could vary along a waveguide, thus limiting the effective coherent length.

26.2 EXPERIMENTAL STUDIES

Nonlinear optical effects have been studied in both thin-film waveguides and optical fibers. Thin-film waveguides are the key element in integrated optics. By epitaxial growth, it is possible to construct a thin-film waveguide out of a crystalline medium with no inversion center. In such a waveguide, second-order nonlinear processes are allowed and can be easily observed. Second-harmonic generation has actually been demonstrated in various crystalline film structures.[3]

Note, however, that although a guided wave is generally confined to the core region of a waveguide, the tail portion of its transverse distribution does penetrate into the boundary media over a distance of the order of a wavelength. Therefore, if the boundary media are nonlinear, they can also contribute to the nonlinear optical process. In the theoretical description, as illustrated by the examples in the preceding section, this is taken into account by the spatial variation of the nonlinear susceptibility in the transverse dimensions. A second-order nonlinear optical process, therefore, can still be appreciable if the waveguide film is centrosymmetric but the substrate is not. Indeed, second-harmonic generation has been observed in such a waveguide structure.[4]

Phase matching often is important for optimizing the efficiency of a nonlinear optical process. The wavevector of a wave in a waveguide mode generally depends on the dimensions of the waveguide and on the refractive indices of the waveguide and the surrounding materials. For a thin-film waveguide, it is possible to vary the relative magnitudes of $K(\omega)$ and $K(2\omega)$ by adjusting the film thickness or by immersing the film waveguide in a liquid and adjusting the refractive index of the liquid. Phase matching, $K(2\omega) = 2K(\omega)$, of second-harmonic generation can in fact be achieved this way, as has been demonstrated experimentally.[3, 4] With phase matching, even CW second-harmonic generation in the uv has been observed with an input laser power of only ~ 0.5 W[4].

There are, however, some difficulties which prevent the thin-film waveguide from being a practical second-harmonic generator. First, the film thickness can hardly be made uniform. As a result, the phase-matching condition cannot be satisfied over the entire length of the waveguide. To have a coherent length larger than 1 mm, the variation of the film thickness should be less than a few percent.[5] Surface imperfection of the waveguide then can result in strong attenuation of the guided waves and limit the propagation distance.[4] Even if better waveguides can be designed and fabricated, there still exist the difficulties of efficiently coupling the waves in and out of the waveguide, and avoiding high-power laser damage of the waveguide.

Note, that the nonlinear output generated by $\mathbf{P}^{NL}$ need not always be a guided wave, although our discussion thus far has been limited to the guided wave case. More generally, the output can also be a bulk wave propagating into a medium adjoining the waveguide. The boundary condition requires the wavevector components of $\mathbf{P}^{NL}$ and the bulk wave along the boundary surface be matched; consequently, the propagation direction of the bulk wave is specified. This is very similar to the case of bulk wave generation by surface wave mixing discussed in Section 25.2. Experimentally, second-harmonic generation of bulk waves by guided waves in a thin-film waveguide can be easily demonstrated.[5]

While thin-film waveguides are often used for second-harmonic or second-order sum- and difference-frequency generation, optical fibers are more suitable for other types of nonlinear optical processes. The latter are usually made of glassy materials possessing an inversion symmetry. Therefore, the lowest-order nonlinear processes allowed in such a medium are the third-order processes. However, unlike the thin-film waveguides, optical fibers are structurally more perfect. They can have an attenuation constant as low as 0.2 dB/km and a coherent length for phase-matched wave interaction longer than a few meters. Because of the very long interaction length, third-order nonlinear optical effects can be readily observed in a fiber even with CW lasers. These include stimulated Raman scattering,[6, 7] stimulated Brillouin scattering,[8, 9] four-wave mixing,[10] four-wave parametric amplification,[11] optical Kerr effect,[12] and self-phase modulation.[13,14]

Stimulated Raman scattering in optical fibers is a subject that has been extensively studied. Both ordinary glass and liquid-core fibers have been used. The latter allows selection of an appropriate liquid as the Raman scatterer in the fiber[6] with a relatively large peak Raman cross section. The Raman shift can be varied by varying the liquid medium. The former has a much lower Raman cross section per unit frequency, but it has the advantage of having a very broad Raman spectrum,[15] as shown in Fig. 26.1. This allows tuning of stimulated Raman output over a broad range of frequencies. With the help of an optical cavity (Fig. 26.2), tunable Raman oscillation can be achieved.[16] In the case of Fig. 26.2 where a 100-m fused silicate fiber with a 17 dB/km loss was pumped by a CW argon laser, the Raman oscillator was found to be tunable over 80 Å. The tuning range could be larger if pulsed lasers were used. Higher-order Stokes radiation may also appear at the output of a fiber Raman oscillator. In the example of Fig. 26.2, four orders of Stokes radiation were observed. They could be used to extend the output tuning range over 350 Å in the visible.

The Stokes conversion efficiency of a fiber Raman oscillator can be higher than 20%.[17] Therefore, as a tunable light source, the fiber Raman oscillator is an attractive alternative to CW dye lasers. Yet it has the disadvantage of

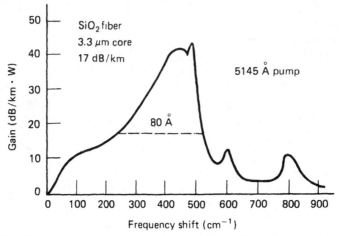

Fig. 26.1 Raman gain curve for a fused silica fiber. (After Ref. 7.)

having a relatively large output linewidth (≥ 10 GHz), which cannot be narrowed by the insertion of an etalon in the cavity.

As mentioned in Section 11.1, the steady-state Brillouin gain in a condensed medium is generally larger than the Raman gain. One would expect backward stimulated Brillouin scattering to appear before stimulated Raman scattering in an optical fiber, at least in the CW case. It would then be difficult to obtain Raman oscillation in a fiber without a Brillouin output. However, because the Raman gain spectrum of a fiber is much broader than the Brillouin gain spectrum, a laser beam with a linewidth larger than the Brillouin spectral width can efficiently pump the Raman oscillator but not the Brillouin oscillator. In that case, the threshold for Raman oscillation can appear lower than that for Brillouin oscillation. With a sufficiently narrow linewidth, however, Brillouin oscillation will indeed dominate. Figure 26.3 describes the arrangement of a Brillouin ring oscillator.[18] A single-mode CW Ar$^+$ laser, with a linewidth less than the 150-MHz Brillouin width of silica, was used to pump the ring oscillator consisting of a single-mode silica fiber of 2.4 μm in diameter and 9.5 m in length. The oscillator had a pump threshold of 25.0 mW and an output of 20 mW at a pump input of 750 mW. A conversion efficiency of $\sim 20\%$ could be obtained if the mirror reflectivities of the cavity were optimized.

More efficient pumping of a Brillouin oscillator has been achieved by a two-mirror straight Brillouin cavity in which the input mirror is also the output mirror of the Ar$^+$ pump laser.[19] At high pump powers, higher-order Stokes

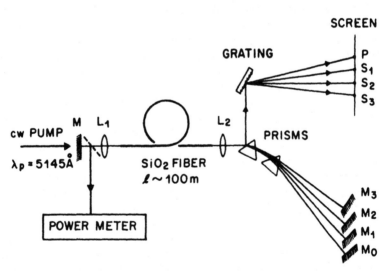

Fig. 26.2 Schematic of the experimental configuration used for the multiresonant fiber oscillator. L_1 and L_2 are AR-coated 20X microscope objectives; M is the output mirror of the argon ion laser and common mirror of the Raman oscillator; M_0, M_1, M_2 and M_3 are used to reflect the pump, first, second, and third Stokes; P, S_1, S_2, and S_3 are the diffracted spots for the pump, first, second, and third Stokes, respectively. (After Ref. 16.)

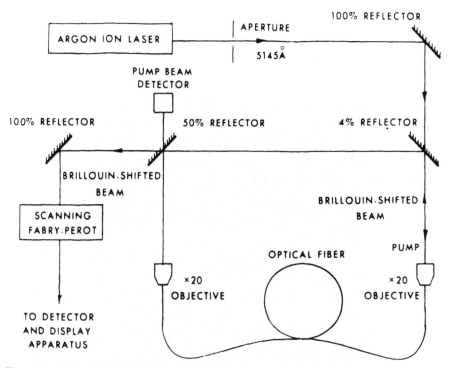

Fig. 26.3 Schematic of Brillouin ring oscillator apparatus. The ring consists of the optical fiber together with the beam paths defined by the 4% and 50% partial reflectors. (After Ref. 18.)

components would show up, as illustrated in Fig. 26.4. Up to 21 orders of Stokes lines have actually been observed, with an overall Stokes shift of 714 GHz in frequency. Anti-Stokes lines can also be generated. Phase locking of the many Stokes and anti-Stokes lines can yield an output in the form of mode-locked pulses.[20]

The anti-Stokes generation is only a special case of four-wave mixing with $\omega_a = 2\omega_l - \omega_s$. In general, $(\omega_l - \omega_s)$ does not have to be in resonance with a transition in the fiber medium. For a small frequency difference between ω_l and ω_s, phase matching is nearly fulfilled if all waves are propagating in the same direction. The coherent length can be larger than 2 km in a silica fiber with $\omega_l - \omega_s = 1$ cm^{-1}. For larger $(\omega_l - \omega_s)$, however, the color dispersion of the medium becomes increasingly significant. Consequently, phase matching of four-wave mixing in a fiber can be realized only if the color dispersion can be canceled by the modal dispersion of the fiber. This is possible at discrete frequencies when a multimode fiber is used. In practice, an optical fiber is often not perfect; the effective coherent length for four-wave mixing under the optimum phase-matching condition is usually limited by imperfection, such as variation of the fiber diameter along the length. By having the pump in the

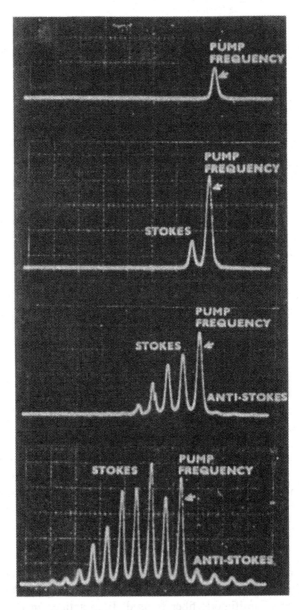

Fig. 26.4 Output spectra from the multiple Stokes-Brillouin oscillator. The spectra are shown for a range of pump powers from 100 nW (top trace) to 3.0 W (bottom trace). The spectral lines are separated by 34 GHz. (After Ref. 19.)

same two modes as the Stokes and anti-Stokes, one can show that the phase-matching condition becomes less sensitive to the variation of the fiber diameter.[11, 21] The effective coherent length can then be as long as 10 m at $\omega_l - \omega_s = 3000$ cm^{-1}. With such a long interaction length, even parametric four-wave amplification starting from noise becomes possible. It has been observed in a silica fiber using a 5320-Å pump pulse with a peak power no more than a few hundred watts.[7]

Some third-order nonlinear optical effects originate from the optical-field-induced refractive index change Δn. In a glass medium, Δn is usually small, with $n_2 = 2\Delta n/|E|^2$ of the order of 10^{-13} esu. Yet a guided wave propagating through a long fiber can still have a large accumulated phase change resulting from Δn. As seen in (26.10), the phase change is given by

$$\Delta\phi(l) = \delta K_p l..$$ (26.16)

For an order-of-magnitude estimate, we can approximate (26.16) by

$$\Delta\phi(l) \sim (\omega/c)\Delta n\, l.$$ (26.17)

If a 100-mW green beam is propagated through a fiber 3.5 μm in diameter and 100 m in length, one would find $\Delta\phi(l) \sim 1$ radian. This indicates that the optical Kerr effect, that is, birefringence induced by a linearly polarized pump beam, should be easily observable in a fiber. The ordinary fiber is, unfortunately, not perfect: an input beam would have its polarization state change continuously along the fiber. This makes the optical Kerr effect difficult to observe. The effect is, however, observable in a birefringent fiber, in which an input pump beam with a polarization along the birefringent axes can retain its polarization in the propagation. The additional birefringence induced by the pump field can then be detected by the induced polarization change on a probe beam initially polarized away from the birefringent axes. Optical Kerr effect in a birefringent fiber has been demonstrated,[12, 22] and has been suggested as a means for optical switching and pulse shaping.[22]

If an intense light pulse is propagated through a fiber, the field-induced phase change on the pulse should be time-dependent. This means that the transmitted light should now experience a self-phase modulation described by $\Delta\phi(t) = \delta K_p(t)l \propto |A^{(p)}(t)|^2$, according to (26.10). The situation here is very similar to that of self-phase modulation of light in a trapped filament discussed in Section 17.7. An immediate consequence of the phase modulation is a spectral broadening on the transmitted light. As shown in Section 17.7, the broadened spectrum is expected to have a semiperiodic structure. The number of peaks in the broadened spectrum is determined by the integer that is closest to, but smaller than, $\Delta\phi_{max}/2\pi$, and the farthest peaks on the two sides have their frequency shifts given by $|\partial\phi/\partial t|_{max}$. This has been demonstrated in an experiment in which mode-locked Ar$^+$ laser pulses of $\sim$ 150-psec pulsewidth were propagated through a silica fiber of a few μm in diameter and $\sim$ 100 m in length.[14] The spectral broadenings observed with various input power levels are shown in Fig. 26.5. They agree well with the theoretical spectra expected from the predicted self-phase modulation. As discussed briefly in Section 17.7, a self-phase-modulated pulse can be significantly compressed when it is sent through a suitable dispersive delay line, such as a grating pair. The principle behind the pulse compression is as follows. In Fig. 26.6, it is seen that the frequency modulation on the middle section of the self-phase-modulated light pulse can be approximated by a linear shift $\Delta\omega \propto (t - t_0)$; the leading part of

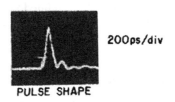

200ps/div

PULSE SHAPE

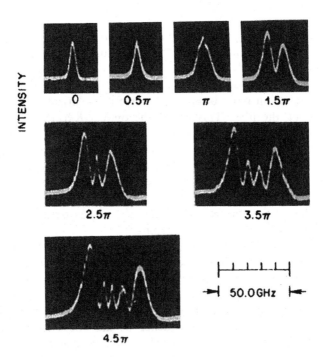

FREQUENCY SPECTRUM

Fig. 26.5 Photographs of input pulse shape and the output spectrum from a 3.35-μm silica-core fiber. Spectra are labeled by the maximum phase shift, which is proportional to peak power. (After Ref. 14.)

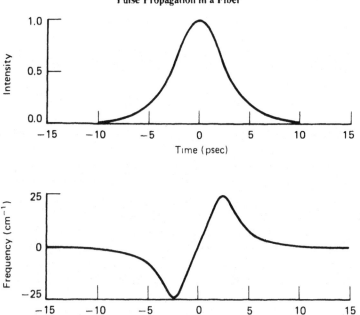

Fig. 26.6 A 6-psec pulse experiencing a self-phase modulation proportional to the instantaneous pulse intensity. (*a*) Pulseshape of the 6-psec pulse. (*b*) Frequency modulation proportional to the derivative of the pulse shape. (After Ref. 28.)

the pulse is at lower frequency. If, in passing through a dispersive delay line, the low-frequency part is delayed and the high-frequency part advanced by an appropriate amount, then the pulse should appear significantly compressed.[23] As a self-phase modulator, a single-mode optical fiber has the advantage of imposing a uniform phase modulation on a guided wave over its entire transverse profile. In addition, the induced Δn in glass is of electronic origin and should respond instantaneously to even subpicosecond pulses. Therefore, together with a dispersive delay line, an optical fiber can be an effective compressor for short pulses.[24] Actually, for picosecond and subpicosecond pulse propagation in a fiber, because of the appreciable spectral width of the pulse, color dispersion of the fiber alone can already cause the pulse to deform significantly. The interplay between the color dispersion and the nonlinearity in the fiber can lead to very interesting results: a short pulse propagating through a fiber can be appreciably broadened or compressed depending on the pulse intensity and frequency. This is the subject we discuss next.

26.3   PULSE PROPAGATION IN A FIBER

Pulse propagation in a fiber is a subject of great importance in many respects. From the basic point of view, study of the problem can yield information about pulse propagation in a nonlinear medium over a long distance that cannot be obtained otherwise. Theoretically, the problem is interesting because it is governed by a nonlinear wave equation that belongs to the same class of partial differential equations as the nonlinear Schrödinger equation and the Landau–Ginzberg equation. From the practical point of view, pulse broadening is an important factor that may limit the data transmission rate through a fiber.

We consider here only the simple case of pulse propagation in a single-mode fiber. The modal dispersion (which is important for a multimode fiber) is absent in this case, and we only have to consider the effects of color dispersion and field-induced refractive index change on the pulse propagation. Color dispersion leads to a group velocity dispersion, $\partial v_g / \partial \omega = -v_g^2 \partial^2 K / \partial \omega^2$, such that even in the linear case, a pulse propagating over a distance is expected to experience a pulse broadening. This is, of course, a well-known phenomenon. With the presence of a field-induced Δn, the situation is more complicated. Depending on the circumstances, a pulse propagating in a fiber can undergo broadening, shrinkage, deformation, or even splitting into multiple pulses.

The formal description of pulse propagation in a single-mode fiber is governed by a nonlinear wave equation in which the nonlinear term arises from the field-induced Δn. With the field $E^{(i)}$ given by (26.1) and using the slowly varying amplitude approximation (see Section 3.5), the nonlinear wave equation can be transformed into the amplitude equation[25]

$$\left(\frac{\partial}{\partial z} + \frac{1}{v_g} \frac{\partial}{\partial t} \right) A^{(i)}(z, t) = -i \frac{\partial v_g^{-1}}{2 \partial \omega} \frac{\partial^2}{\partial t^2} A^{(i)} + i K_2 |A^{(i)}|^2 A^{(i)} \quad (26.18)$$

where $K_2 = (2\pi \omega_p^2 / K^{(i)} c^2) \langle iiii \rangle$ and we assume $\Delta n = (K_2 c / \omega) |A^{(i)}|^2$. The first term on the right of (26.18) comes from the group velocity dispersion, while the second term comes from the field-induced Δn. By the following change of variables

$$s = T^{-1} \left(t - \frac{z}{v_g} \right), \qquad \xi = \left| \frac{\partial v_g^{-1}}{\partial \omega} \right| T^{-2} z,$$

and

$$a = T \left(\frac{K_2}{|\frac{1}{2} \partial v_g^{-1} / \partial \omega|} \right)^{1/2} A^{(i)}, \qquad (26.19)$$

(26.18) can be reduced to the dimensionless form[25]

$$-i \frac{\partial a}{\partial \xi} = \frac{1}{2} \left(\frac{\partial v_g}{\partial \omega} \Big/ \left| \frac{\partial v_g}{\partial \omega} \right| \right) \frac{\partial^2 a}{\partial s^2} + |a|^2 a. \qquad (26.20)$$

Here, in (26.19), T is a measure of the input pulsewidth. If $\partial v_g / \partial \omega > 0$, (26.20) is in the same form as the nonlinear Schrödinger equation. It is also of the same type of partial differential equations that govern pulse formation and propagation in a wide variety of physics problems.[26] Although the general solution of the equation is not available, the particular solution has been found and is sketched briefly below. Reference 26 provides more detail.

We consider first, in physical terms, the combined action of a group velocity dispersion $\partial v_g/\partial\omega > 0$ and a field-induced Δn on the pulse deformation. As we described in Section 26.2, the field-induced Δn imposes a frequency modulation on the propagating pulse. The leading part of the pulse appears to have a lower frequency than the lagging part. This is sketched in Fig. 26.6. If the pulse experiences at the same time a group velocity dispersion $\partial v_g/\partial\omega > 0$, the leading part of the pulse will travel more slowly than the lagging part, and the pulse will shrink. This pulse narrowing effect is opposite to the pulse broadening effect from the group velocity dispersion alone. The narrowing action arising from Δn is expected to increase as the pulse intensity increases. First, the pulse broadening is reduced from the linear case. Then, if the pulse has the right amplitude and the right shape, the narrowing action may just balance the broadening action, and the pulse can propagate without any change in shape. Such a pulse is often known as a fundamental soliton, which appears quite generally in nonlinear pulse propagation in a wide variety of physical cases.[26] At even higher pulse amplitudes, the pulse narrowing action can overcome the broadening action, and the pulse may shrink. In propagating along the fiber, the pulse shape may change continuously, undergoing repeated narrowing and expansion before arriving at a stable form.

This physical picture is supported by the detailed solution of (26.18) with $\partial v_g/\partial\omega > 0$.[26] In addition, the calculation shows that the fundamental soliton solution has a hyperbolic secant pulse shape, $a = \mathrm{sech}(s)$, with a definite pulse area $\mathcal{A}_0$. Here a pulse area is defined by $\mathcal{A} = \int_{-\infty}^{\infty} A^{(1)}\, dt$. If the input pulse has a pulse shape $a = N\,\mathrm{sech}(t/T)$, with N being an integer larger than 1 ($\mathcal{A} = N\mathcal{A}_0$), then the solution of (26.20) is periodic in ξ with a period $\xi_0 = \pi/2$. For $N = 2$, the pulse shrinks to a minimum width at $\xi = \frac{1}{2}\xi_0$ and then expands to the original width at $\xi = \xi_0$. For $N = 3$, it shrinks to a minimum width at $\xi = \frac{1}{4}\xi_0$; as it expands, it splits into two pulses of equal strength at $\xi = \frac{1}{2}\xi_0$. Finally, the two pulses merge back into the original pulse at $\xi = \xi_0$. The soliton solutions with $N = 1, 2, 3$ are illustrated in Fig. 26.7. A more detailed pulse evolution of the $N = 3$ soliton case is shown in Fig. 26.8. For $N = 4$, the pulse undergoes a threefold splitting at $\xi = \frac{1}{2}\xi_0$. The theoretical calculation also shows that an input pulse with the wrong amplitude and shape may evolve, over a long distance, into a pulse variation which is the same as that given by an input pulse $a = N\,\mathrm{sech}(t/T)$. For example, if $\mathcal{A}_0/2 \leq \mathcal{A} \leq 3\mathcal{A}_0/2$, the input pulse should evolve into an exact fundamental soliton. This recalls the picture of self-induced transparency discussed in Section 21.6. Indeed, the pulse developed in self-induced transparency can also be identified as a soliton.

We can use the theory to estimate the length of the fiber and power of the laser pulse needed for observing the soliton effect. From (26.19) and $a =$

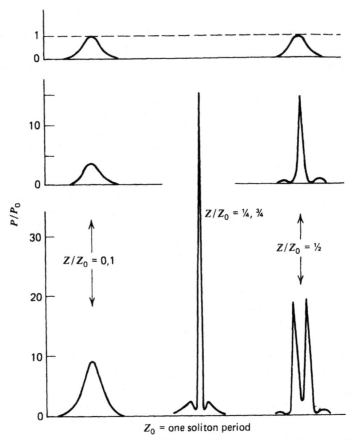

Fig. 26.7 Theoretical behavior of solitons as they propagate down a fiber. The fundamental soliton (top) propagates without change in shape or amplitude; higher-order solitons, $N = 2$ (middle) and $N = 3$ (bottom), exhibit more complex behavior, undergoing sequences of narrowing and splitting. [After L. F. Mollenauer, R. H. Stolen, and J. P. Gordon, *Phys. Rev. Lett.* **45**, 1095 (1980).]

sech(s), we find, in terms of the lab parameters, the soliton period to be

$$z_0 = \frac{\pi T^2}{2} \left| \frac{\partial v_g^{-1}}{\partial \omega} \right|^{-1} \tag{26.21}$$

and the peak intensity of the fundamental soliton to be

$$I_0 = \frac{n_0 c}{16 K_2 z_0}. \tag{26.22}$$

The group velocity dispersion of silica fibers, $\partial v_g / \partial \omega$, usually changes from negative to positive at $\lambda \simeq 1.3$ μm. If we choose $\lambda = 1.55$ μm at which

$\partial v_g^{-1}/\partial\lambda = (2\pi c/\lambda^2)\,\partial v_g^{-1}/\partial\omega \simeq -16(\mathrm{psec/nm})/\mathrm{km}$, and $T = 4$ psec corresponding to an input pulse of 7 psec full width at half maximum, we find $z_0 = 1260$ m. The linear and nonlinear refractive indices for the fiber are $n_0 = 1.45$ and $n_2 = 1.1 \times 10^{-13}$ esu $(K_2 \equiv \frac{1}{2}\omega n_2/c)$. Then (26.22) yields $I_0 = 1 \times 10^6$ W/cm^2, and for an effective beam cross section of 10^{-6} cm^2 in the fiber, a critical peak power of 1 W for the fundamental soliton. Such a picosecond pulse is readily obtainable from an infrared picosecond laser system.

The experiment has actually been carried out by Mollenauer et al. using a picosecond mode-locked color center laser operating at 1.55 μm.[25] The silica fiber used in the first experiment was only 700 m long, but it had a corresponding ξ larger than $\pi/4$, and therefore the pulse narrowing and pulse splitting phenomena could still be observed. The autocorrelation traces, $\int_{-\infty}^{\infty} I(t)I(t + \tau)\,dt$ versus τ, for the transmitted pulses through the fiber were measured for various input peak powers. The results are presented in Fig. 26.9. At low input power, $P = 0.3$ W, the output pulse is clearly broader than the input pulse. At $P = 1.2$ W, the output pulse is about the same as the input pulse, suggesting that the critical power for the fundamental soliton has just been reached. This agrees with the predicted value of $P_0 = 1$ W. At $P = 5$ W ($\sim N^2 P_0$ with $N = 2$) the output pulsewidth reduces to nearly a minimum with a full width of ~ 2 psec. Then, at $P = 11.4$ W ($\sim 3^2 P_0$) and $P = 22.5$ W ($\sim 4^2 P_0$), the autocorrelation traces exhibit three and five peaks, respectively, indicating that the pulse has split into two and three in the respective cases. With a longer

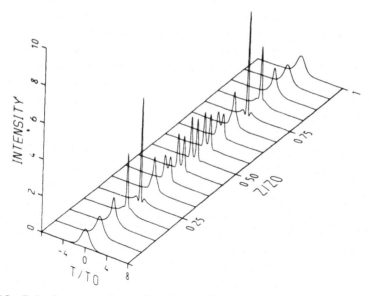

Fig. 26.8 Pulseshape at various points along a fiber for the $N = 3$ soliton. [After R. H. Stolen, L. F. Mollenauer, and W. J. Tomlinson, *Opt. Lett.* **8**, 186 (1983).]

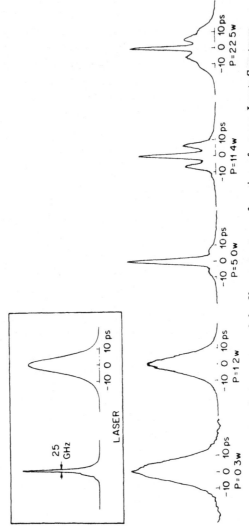

Fig. 26.9 Autocorrelation traces of the fiber output as a function of power. Inset: Spectrum and autocorrelation trace of the direct laser output. The various curves have been roughly normalized to a common height. (After Ref. 25.)

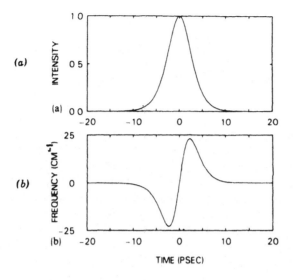

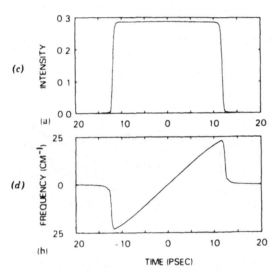

Fig. 26.10 (*a*) A 6-psec input pulse; (*b*) frequency modulation on the pulse resulting from self-phase modulation due to the optical Kerr effect; (*c*) a square pulse resulting from the combined action of the group velocity dispersion, $\partial v_g/\partial \omega < 0$, and the optical Kerr effect; (*d*) frequency modulation on the square pulse. (After Ref. 28.)

fiber ($\xi \sim \xi_0 = \pi/2$), the return of the transmitted soliton pulse shape to the input form has also been observed.[25]

The study of solitons is a field in which theoretical calculations abound, but quantitative experiments are rare. Optical fibers provide an ideal opportunity for soliton studies. Here, in a fiber, both theoretical and experimental details for soliton propagation can be worked out. Soliton–soliton interaction and soliton reflection and transmission at a boundary can also be investigated. They are among the problems that have fascinated many researchers in the field. The pulse narrowing effect of soliton propagation can be used in practice to compress picosecond pulses. As seen in Fig. 26.7, for the $N = 3$ soliton, the output pulse at $\xi = \pi/8$ and $3\pi/8$ can have its pulsewidth reduced by an order of magnitude in comparison with the input pulse. Experimentally, using this method, a 30-fold compression of a picosecond pulse with a high soliton number, $N \geq 10$, has actually been demonstrated.[27]

The nonlinear pulse propagation in an optical fiber behaves very differently if $\partial v_g/\partial \omega < 0$.[28] Figure 26.10$a$ and b reproduces the sketch in Fig. 26.6 for a short input pulse and the frequency modulation on it arising from the field-induced Δn. Since the higher frequency part travels more slowly than the lower frequency part, the group velocity dispersion in the present case tends to stretch the pulse, flatten the central peak, and sharpen the leading and lagging edges. At appropriate pulse intensities or fiber lengths, the pulse can be

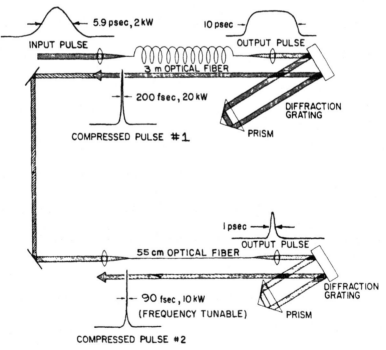

Fig. 26.11 Schematic diagram of a two-stage optical pulse compressor. (After Ref. 31.)

deformed into a nearly square pulse with more than 90% of the pulse assuming a linear frequency modulation, as in Fig. 26.10c and d. Such a pulse, in passing through a properly adjusted linearly dispersive delay line, such as a grating pair, can be greatly compressed, with more than 90% of the energy contained in the narrow peak. This physical picture has been substantiated by the numerical solution of (26.18).

Experiments have indeed demonstrated that this is a viable method for short pulse compression. In the initial trial, 10-W, 5-psec pulses passing through a 70-m silica fiber together with a dispersive delay line were compressed to 1.5 psec.[24] In other experiments, 7-kW, 90-fsec pulses were compressed to 30 fsec using a 15-cm silica fiber,[29] and 1-kW, 5.4-psec pulses were compressed to 450 fsec using a 30-m fiber.[30] With two stages of pulse compression, a total compression factor of 65 has been achieved.[31] The experimental arrangement

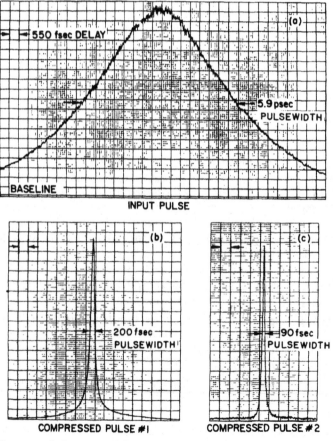

Fig. 26.12 Autocorrelation traces of (a) input pulse; (b) compressed pulse from the first stage of the optical pulse compressor, and (c) compressed pulse from the second stage of the optical pulse compressor. (After Ref. 31.)

of the last experiment is shown in Fig. 26.11. The prism-grating pairs act as the dispersive delay lines. In the first stage, 2-kW, 5.9-psec pulses from a dye laser system were compressed to 20 kW and 200 fsec. In the second stage, the pulses were further compressed to 10 kW and 90 fsec. The measured autocorrelation traces of the input and compressed pulses are seen in Fig. 26.12. This pulse compression scheme has the advantage of being simple and applicable to tunable picosecond dye laser systems. It is therefore expected to be an extremely useful tool in extending picosecond optical studies to the subpicosecond regime.

REFERENCES

1   See, for example, D. Marcuse, *Theory of Dielectric Optical Waveguides* (Academic Press, New York, 1974).

2   R. H. Stolen and J. E. Bjorkholm, *IEEE J. Quant. Electron.* **QE-18**, 1062 (1982).

3   D. B Anderson and J. T. Boyd, *Appl. Phys. Lett.* **19**, 266 (1971); S. Zemon, R. R. Alfano, S L. Shapiro, and E. Conwell, *Appl. Phys. Lett.* **21**, 327 (1972); J. P. van der Ziel, R. C. Miller, R. A. Logan, W. A. Nordland, and R. M. Mikulyak, *Appl. Phys. Lett.* **25**, 238 (1974).

4   W. K. Burns and A. B. Lee, *Appl. Phys. Lett.* **24**, 222 (1974); B. U. Chen, C. L. Tang, and J. M. Telle, *Appl. Phys. Lett.* **25**, 495 (1974).

5   P. K. Tien, *Appl. Opt.* **10**, 2395 (1971).

6   E. P. Ippen, *Appl. Phys. Lett.* **16**, 303 (1970).

7   R. H. Stolen, in D. B. Ostrowsky, ed., *Fibers and Integrated Optics* (Plenum, New York, 1979), p. 21, and references therein.

8   E P Ippen and R. H. Stolen, *Appl. Phys. Lett.* **21**, 539 (1972).

9   K. O. Hill, B. S. Kawasaki, D. C. Johnson, and Y. Fujii, in B. Bendow and S S. Mitra, eds., *Fiber Optics—Advances in Research and Development* (Plenum, New York, 1979), p. 211, and references therein.

10   R. H. Stolen, J. E. Bjorkholm, and A. Ashkin, *Appl. Phys. Lett.* **24**, 308 (1974).

11   R. H. Stolen, *IEEE J. Quant. Electron.* **QE-11**, 100 (1975); C. Lin and M. A. Bösch, *Appl Phys. Lett.* **38**, 479 (1981).

12   R. H. Stolen and A. Ashkin, *Appl. Phys. Lett.* **22**, 294 (1973).

13   E. P. Ippen, C. V. Shank, and T. K. Gustafson, *Appl. Phys. Lett.* **24**, 190 (1974).

14   R H. Stolen and C. Lin, *Phys. Rev. A* **17**, 1448 (1978).

15   R. H. Stolen, E. P. Ippen, and A. R. Tynes, *Appl. Phys. Lett.* **20**, 62 (1972); R. H. Stolen and E. P. Ippen, *Appl. Phys. Lett.* **22**, 276 (1973).

16   R. K Jain, C. Lin, R. H. Stolen, and A. Ashkin, *Appl. Phys. Lett.* **31**, 89 (1977).

17   R. H. Stolen, *Appl. Phys. Lett.* **26**, 163 (1975).

18   K. O. Hill, B. S. Kawasaki, and D. C. Johnson, *Appl. Phys. Lett.* **28**, 608 (1976).

19   K. O. Hill, D. C. Johnson, and B. S. Kawasaki, *Appl. Phys. Lett.* **29**, 185 (1976).

20   B S. Kawasaki, D. C. Johnson, Y. Fujii, and K. O. Hill, *Appl. Phys. Lett.* **32**, 429 (1978).

21   R. H. Stolen and W. N. Liebolt, *Appl Opt.* **15**, 239 (1976).

22   J. M. Dziedzic, R. H. Stolen, and A. Ashkin, *Appl. Opt.* **20**, 1403 (1981).

23   J. R. Klauder, A. C. Price, S. Darlington, and W. J. Albersheim, *Bell System Tech. J.* **39**, 745 (1960); E. B. Treacy, *Phys. Lett.* **28A**, 34 (1968).

24 H Nakatsuka and D. Grischkowsky, *Opt. Lett.* **6**, 13 (1981); H. Nakatsuka, D. Grischkowsky, and A. C. Balant, *Phys. Rev. Lett.* **47**, 910 (1981).

25 L. F. Mollenauer, R. H. Stolen, and J. P. Gordon, *Phys. Rev. Lett.* **45**, 1095 (1980); R. H. Stolen, L. F. Mollenauer, and W. J. Tomlinson, *Opt. Lett.* **8**, 186 (1983); L. F. Mollenauer and R. H. Stolen, *Laser Focus* **18**, No. 4, 193 (1982).

26 J. Satsuma and N. Yajima, *Prog. Theoret. Phys. Supp.* **55**, 284 (1974); V. E. Zakharov and A. B. Shabat, *JETP* **37**, 823 (1973); see also the R. K. Bullough and P. J. Caudrey, eds., *Solitons* (Springer-Verlag, Berlin, 1980).

27 L. F. Mollenauer, R. H. Stolen, J. P. Gordon, and W. J. Tomlinson, *Opt. Lett.* **8**, 289 (1983).

28 D. Grischkowsky and A. C. Balant, *Appl. Phys. Lett.* **41**, 1 (1982).

29 C. V. Shank, R. L. Fork, R. Yen, R. H. Stolen, and W J. Tomlinson, *Appl. Phys. Lett.* **40**, 761 (1982).

30 B. Nikolaus and D. Grischkowsky, *Appl. Phys. Lett.* **42**, 1 (1983).

31 B. Nikolaus and D. Grischkowsky, *Appl. Phys. Lett.* **43**, 228 (1983).

27

Optical Breakdown

Optical breakdown here refers to the catastrophic evolution of damage inflicted in a transparent medium by a strong laser field. It results from avalanche ionization initiated by a laser and is different from the laser-induced thermal breakdown which arises from direct laser heating. The process is critical in high-power laser technology as it limits the amount of laser power that can be transmitted through a medium. Generally speaking, optical breakdown is a consequence of the rapid energy deposition into a medium by a laser. The very strong optical excitation together with the subsequent highly complex plasma formation due to avalanche ionization makes a rigorous solution of the problem extremely difficult. This chapter is meant to give only a short introduction to the subject. Emphasis is on physical understanding.

27.1 GENERAL DESCRIPTION

A laser field as strong as the Coulomb field in an atom can certainly rip the valence electrons away from an atom. We already saw in Section 22.5 that multiphoton ionization of nearly all atoms and molecules can readily take place in a laser beam with an intensity 10^{11} W/cm^2 or higher. This, however, is a process involving only single atoms or molecules; ionization of one atom does not affect that of the others. It happens when the gas pressure is low (e.g., $\lesssim 10^{-3}$ torr) and the laser pulse is short ($\lesssim 10^{-8}$ sec) in comparison with the time interval between atomic collisions ($\gtrsim 10^{-6}$ sec). The electron mean free path ($\gtrsim 10$ cm) should also be much longer than the dimensions of the focal volume ($\lesssim 10^{-2}$ cm), so that the secondary effect of ionization by electron collisions with atoms can be neglected. Under such conditions, multiphoton ionization is the only operating mechanism for ionization of a gas medium.

Optical breakdown, however, refers to an ionization process with a subsequent plasma formation in a relatively dense gas medium or in condensed matter. It commonly occurs when a Q-switched laser pulse of $\gtrsim 0.1$–1 J is focused into a medium. This phenomenon was discovered as early as 1962, right after the laser was invented.[1] Because of its importance in high-power laser applications, it has attracted the attention of many researchers ever since. Optical breakdown is generally signified by a visible flash or spark in the medium, resulting from the formation of a plasma. In gases, the spark can extend over a long distance. A record spark length of 60 m was reported by a Soviet group using a 160-J pulsed Nd: glass laser beam weakly focused in air.[2] In condensed matter, plasma formation usually leads to the appearance of a damage spot (or series of spots) in solid, or the appearance of a cavity (or cavities) in liquid.

Plasma formation in optical breakdown is the result of an electron avalanche process. It starts with a small number of free electrons (or quasi-free electrons in solid) floating in space. These electrons may initially happen to be there, or they may be generated by laser-induced (multiphoton) ionization. Electron avalanche ionization can develop if the electrons can gain energy from the laser field, since they can then attain enough energy to ionize an atom in collision, and repetitions of the process can lead to a rapid multiplication of electrons. From the requirement of energy and momentum conservation, an electron can absorb a photon from a laser field only if it is colliding with an atom or ion. This is just the inverse of the so-called bremsstrahlung process and is known as inverse bremsstrahlung.[3] Apparently such a process is effective only in a relatively dense medium in which electrons can have frequent collisions with atoms. From inverse bremsstrahlung, the electrons can absorb photon by photon and gain enough energy to allow electron impact ionization to occur. Cascade ionization or electron avalanche follows, with the resultant formation of a plasma.[3] As soon as the level of ionization becomes appreciable, the incoming light can be readily absorbed by electrons via free–free transitions in the field of ions. This causes intense heating of the electron plasma and a consequent rapid hydrodynamic expansion of the plasma in the form of a shock wave.[4] The final result is the appearance of a spark in the gas case and visible damage in the condensed matter case.

In many respects optical breakdown is similar to dc or microwave break-down.[5] Both arise from electron avalanche induced by a field although the creation of initial electrons and the detailed dynamics of the two processes are different. Optical breakdown is, however, categorically different from laser-induced thermal breakdown. In the latter case, the medium absorbs energy from the laser beam and quickly converts it into heat. The resultant temperature rise may then cause impact ionization in the gas case and eventually lead to plasma formation. In the solid case, laser heating can effect melting of the solid and is the basis of laser annealing. With excess heating, vaporization of the matter can result. A dense plasma can be formed at the solid surface, leading to a fireball emitting visible, uv, and even x-ray radiation.[6] Laser-induced thermal breakdown is an important process in laser science and technology but is not a subject of discussion of this chapter. Here we are concerned only with optical breakdown.

Optical breakdown in gases and in condensed matter have many features in common. Theoretically, both result from electron avalanche ionization and involve the absorption of more than one (sometimes many) photons per electron created. Experimentally, they also run into similar problems. In both cases the early experiments encountered a great deal of difficulty in establishing reproducible breakdown thresholds for various materials. The irreproducibility is often caused by the following factors. First, being a highly nonlinear phenomenon, optical breakdown has a threshold depending critically on laser intensity variations. To find reproducible results, single-mode (both longitudinal and transverse) lasers with well-defined intensity profiles must be used. Second, the threshold can be significantly lowered if there are absorbing particles, or impurities with low ionization energies, present in the transparent medium, because they would then provide the primary electrons in the avalanche ionization. Only in a pure material can we expect to observe an intrinsic breakdown threshold. Third, self-focusing (see Chapter 17) of the laser beam leads to an apparent breakdown threshold which is significantly lower than the true breakdown threshold. It is important to avoid self-focusing if one intends to measure the intrinsic breakdown threshold.

Gases and solids are nonetheless very different, and should exhibit very different breakdown characteristics. In the following sections we discuss optical breakdown in gases and in solids separately. There is no discussion of optical breakdown in liquids, since little is known about the subject. The material presented here was taken from the review articles listed in the bibliography. Research on optical breakdown is still going strong, although basic understanding of the process was mostly achieved before 1975.

27.2   OPTICAL BREAKDOWN IN GASES

Initiation of optical breakdown in a gas relies on two steps: creation of the prime electrons and development of the avalanche ionization process.[3] In the absence of free electrons initially, the prime electrons in the laser focal volume can be created only by multiphoton ionization of atoms or molecules. This requires a very strong laser intensity if the number n in the n-photon ionization process is high. If absorbing submicroscopic particles or impurity atoms or molecules are present, then the laser intensity required to create the first few electrons would be much lower. The prime electrons control the initiation of avalanche ionization. However, the development of electron avalanche ionization comes from an interplay between gain and loss of electrons and electron energy. First, the ionization rate is directly proportional to the rate of net electron energy gain, which is the difference between energy gain by the electrons via inverse bremsstrahlung and energy loss by the electrons due to collisions. Then there is loss of electrons due to diffusion out of the interaction region or binding to atoms and ions. Avalanche ionization can lead to optical breakdown only if the net rate of electron multiplication is so fast that it is above a threshold value necessary for plasma formation during the laser pulse. Since the rate of electron energy gain is proportional to the laser intensity, while the loss is more or less independent of the field, one expects to find for a given laser pulsewidth a threshold laser intensity for avalanche ionization.

Quantification of this qualitative description of optical breakdown is difficult, especially if the quantum nature of avalanche ionization is to be considered. Here we limit ourselves to a classical description using a very simple model.[3] We assume that creation of prime electrons and avalanche multiplication of electrons are two successive independent processes. If the prime electrons are created by one-step n-photon ionization, then they have a density

$$\rho_0 \cong AI^n \qquad (27.1)$$

where I is the laser intensity, and A is proportional to the laser pulsewidth. With absorbing particles or easily ionizable impurities present in the gas medium, A would be much larger and n smaller. It is also possible to use preionization to provide the prime electrons; in that case, we would have a prescribed value for ρ_0.

We now assume that the electron multiplication process would start with an initial value of electron density ρ_0. Let the ionization rate be η and the electron loss rate be g. The resultant electron multiplication rate is

$$\frac{d\rho}{dt} = (\eta - g)\rho \qquad (27.2)$$

and hence we have

$$\rho(t) = \rho_0 \exp[(\eta - g)t]. \qquad (27.3)$$

For optical breakdown, ρ must reach a critical value, $\rho_{cr} (\sim 10^8/\text{cm}^3)$, signifying the initial stage of plasma formation during the laser pulse. If the laser pulsewidth is τ_p, the optical breakdown threshold is characterized by a threshold ionization rate η_{cr}. Note that η should be proportional to the laser intensity I, and g independent of I; consequently, η_{cr} means that there is a threshold intensity I_{cr} for breakdown. From (27.3), we have

$$\eta_{cr} = g + \tau_p^{-1} \log_e\left(\frac{\rho_{cr}}{\rho_0}\right). \qquad (27.4)$$

In order to relate η_{cr} to I_{cr}, we use a classical free-electron model. The rate of energy gain of an electron is given by

$$\frac{d\mathscr{E}}{dt} = \frac{e^2|E|^2\tau}{m(1 + \omega^2\tau^2)} \qquad (27.5)$$

where E is the optical field at frequency ω, τ is the momentum transfer collision time, and e and m are the electron charge and mass, respectively. If the ionization energy of the atoms or molecules is $\mathscr{E}_I$, then the ionization rate is

$$\eta = \frac{e^2 |E|^2 \tau}{m \mathscr{E}_I (1 + \omega^2 \tau^2)}. \tag{27.6}$$

From (27.4) to (27.6), we find

$$I_{cr} = \frac{mc\mathscr{E}_I (1 + \omega^2 \tau^2)}{2\pi e^2 \tau} \left[g + \frac{1}{\tau_p} \log_e \left(\frac{\rho_{cr}}{\rho_0} \right) \right]. \tag{27.7}$$

The last equation allows us to see explicitly how the breakdown threshold depends on the various physical parameters.

Before we discuss the implications of (27.7), we should realize that the equation provides only a crude description of the process. In a more general sense, Yablonovitch suggested the use of a similarity principle to describe optical breakdown.[7] It states that η/p, with p being the gas pressure, satisfies a scaling law of the form

$$\eta/p = f(x) \tag{27.8}$$

where $x = |E|/p(1 + \omega^2\tau^2)^{1/2}$. Since $\tau \propto 1/p$, we notice that (27.6) actually has the form of (27.8), with $f(x) \propto x^2$. The experimental result on helium, however, seems to confirm this similarity principle with an $f(x)$ function much steeper than x^2. The preceding theoretical description is actually a simple extension of the classical theory of microwave cascade ionization.[5] One may question the validity of such a description considering that it requires the amount of energy acquired between collisions, $\tau \, \partial \mathscr{E}/\partial t$, to be much larger than the photon energy $\hbar\omega$. In the case of optical breakdown, we usually have $\tau \, \partial \mathscr{E}/\partial t \sim 0.01$ eV, while $\hbar\omega$ is around 0.1–1 eV; therefore, the quantum effect is clearly important. An analysis of the problem using the quantum Boltzmann transport equation to describe the dynamic electron distribution in the cascade process, however, shows that (27.6) or, more correctly, (27.8) roughly holds.[8]

We now use (27.7) to discuss how the threshold intensity for optical breakdown varies with different parameters. We consider first the effect of prime electrons. To initiate avalanche ionization, we need at least one prime electron in the laser focal volume. This means that we must have $\rho_0 \gtrsim \rho_{min} = 1/V_f$, where V_f is the focal volume. For $V_f \sim 10^{-7}$ cm^3, the corresponding ρ_{min} is 10^7/cm^3. If $\rho_0 \ll \rho_{min}$, the chance of finding an electron in V_f is very small, and the avalanche process is not likely to occur. If $\rho_0 \gg \rho_{min}$, then the reverse is true. Let $I_m \tau_p$ be the laser fluence required to generate ρ_{min}. We can conclude that if I_m is larger than I_{cr} of (27.7) with $\rho_0 = \rho_{min}$, the threshold of optical breakdown is determined by I_m. On the other hand, if $I_{cr} \gg I_m$, then the breakdown threshold is given by I_{cr}. As mentioned earlier, I_m can be very large if ρ_{min} is created by multiphoton ionization. However, I_m can be drastically lowered if absorbing particles or easily ionizable impurities are present in the gas. In experiments designed to eliminate the effect of generation of prime electrons, the gas medium can be preionized to give a prescribed ρ_0 much larger than ρ_{min}.[9] In that case, the subsequent optical breakdown in the gas is solely controlled by the development of avalanche ionization. Experiments have indeed shown that when a transparent gas with an ionization energy $\mathscr{E}_I$ much larger than $\hbar\omega$ is purified, the optical breakdown threshold becomes much higher; it also becomes more irreproducible since the initiation of the avalanche process is more strongly affected by the statistics of finding the prime electrons in the focal volume.[9] The latter is also true when the focal volume is reduced.

We now assume $I_{cr} \gg I_m$, so that the breakdown is controlled by the avalanche process. From (27.7), it is seen that I_{cr} should still depend on ρ_0. This is actually demonstrated by the experimental results shown in Fig. 27.1,

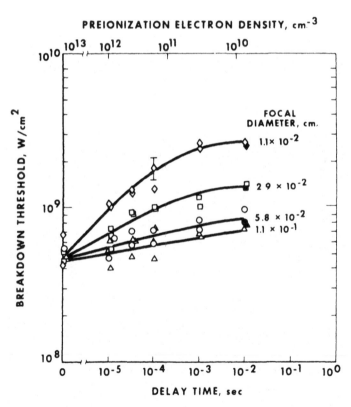

Fig. 27.1 Breakdown threshold in helium versus electron density prescribed by preionization. The solid points correspond to no preionization. The data were obtained using a CO_2 TEA laser at 10.6 μm with four different focusing parameters. (After Ref. 9.)

where ρ_0 was prescribed by preionization. The breakdown threshold I_{cr} appears to decrease as ρ_0 increases. The results in Fig. 27.1 also depend on the focal diameter. For a smaller focal diameter, corresponding to a smaller focal volume, the electron loss due to diffusion out of the focal volume during the laser pulse is expected to be more. Then, according to (27.7), the loss rate g is larger, and hence the breakdown threshold should be higher. At high ρ_0, the electron diffusion loss is less important because the diffusion process becomes more ambipolar and less rapid. As a result, the dependence of I_{cr} on the focal diameter is no longer so obvious.

Equation (27.7) also predicts the dependence of I_{cr} on the gas pressure p, since $\tau \propto 1/p$. We have $I_{cr} \propto 1/p$ when $\omega^2\tau^2 \gg 1$ at low pressures, and $I_{cr} \propto p$ when $\omega^2\tau^2 \ll 1$ at high pressures. This prediction was qualitatively confirmed by the experimental results given in Fig. 27.2a.[10] For comparison, Fig. 27.2b shows the microwave breakdown thresholds as a function of pressure for the various gases.[5] The curves in the two figures have the same qualitative behavior except that the minimum breakdown threshold in the

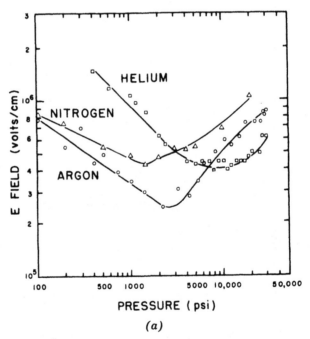

PRESSURE (psi)

(a)

Fig. 27.2 (a) Optical breakdown threshold versus pressure in Ar, He, and N_2. A 50-nsec ruby laser pulse focused to a diameter of 100 μm was used. (After Ref. 10.) (b) Microwave breakdown threshold versus pressure in air, N_2, and O_2. Microwave frequency used was 0.994 GHz, and the diffusion length of discharge volume was 1.51 cm. [After A. D. MacDonald, *Microwave Breakdown in Gases* (Wiley, New York, 1966).]

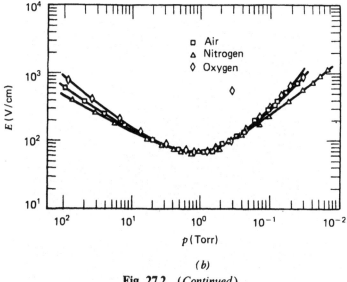

(b)

Fig. 27.2 (*Continued*).

microwave case is at a much lower pressure. This can be easily understood from (27.7) knowing that the minimum of I_{cr} versus p should appear at $\omega\tau \sim 1$, and τ is proportional to p^{-1}.

After the initial breakdown, the electrons in the plasma can readily absorb more energy from the laser field in the presence of ions. The rapidly heated plasma soon leads to the formation of an expanding shock wave and the simultaneous appearance of a spark.[3] Then the incoming laser beam is intercepted and preferentially absorbed by the shock wavefront propagating toward the laser. As a result, the laser energy is continuously fed into the shock wavefront, and the spark appears to propagate toward the laser.[11] With sufficient laser energy in the pulse, the spark can propagate over a very long distance.[2] A more quantitative description of how a laser beam heats up a plasma and gives rise to a long propagating spark is certainly very difficult. This is a problem of great importance in the field of laser interaction with plasmas,[12] but it is outside the scope of this book. Readers are referred to the bibliography and the references therein.

Optical breakdown in gases has found applications in a number of areas. The rapid formation of plasma and the subsequent blocking of the incoming laser light by the plasma (see Fig. 27.3) can act as a fast optical switch. Laser-induced plasma via optical breakdown is clearly a means to generate a high-temperature, dense plasma. Such a plasma can be used as a light source of very high brightness. The possibility of sustaining a laser-induced plasma by a CW laser and the more imaginative applications of the optical discharge have been reviewed by Raizer (see bibliography).

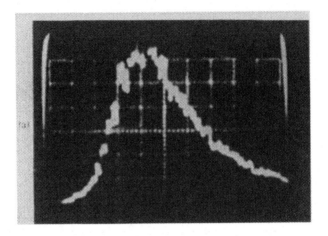

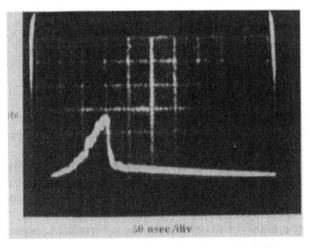

Fig. 27.3 Oscilloscope traces of (*a*) an input CO_2 laser pulse 200 nsec pulsewidth (FWHM), and (*b*) the output pulse after optical breakdown in air. [After D. C. Smith, *Appl. Phys. Lett.* **19**, 405 (1971).]

27.3 OPTICAL BREAKDOWN IN SOLIDS

Although optical breakdown in gases and in solids were discovered at the same time,[1] optical breakdown in solids was not well understood until much later because of experimental difficulties intrinsic to solids. Unlike the gas case, optical breakdown in a solid leaves permanent damage in the solid. Then the measurement is nonrepeatable, unless one can find a large piece of solid with extremely high uniformity or many pieces of identical quality. The quality control of the solid sample is actually a major problem in the optical break-down experiment. Absorbing inclusions in a solid can drastically lower the apparent breakdown threshold, since local heating at the inclusions can readily lead to thermal breakdown in the solid. Only if these absorbing inclusions are avoided or eliminated can we expect to find reproducible optical breakdown thresholds. It has been found that normal optical breakdown leads to a funnel-shaped damage track while inclusion breakdown gives a spherical damage spot. Therefore, by examining the damage tracks, one may distinguish the two breakdown mechanisms.[13] Another experimental difficulty is the effect of self-focusing, which occurs more readily in a transparent solid than in a transparent gas. With self-focusing, the observed breakdown threshold would be determined by the self-focusing threshold (see Section 17.5). The difficulty, however, can be more or less eliminated by using a tightly focused laser beam such that the effect of self-focusing becomes negligible.[13] We consider here only the case of optical breakdown in a pure transparent solid induced by a tightly focused single-mode laser pulse.

The physical mechanism governing optical breakdown in solids is basically the same as in the gas case.[14] The conduction electrons here play the role of free electrons, and excitation of valence electrons to the conduction bands is equivalent to ionization of atoms in a gas. Again, the laser-induced avalanche ionization process in a solid should start from a few prime conduction electrons in the laser focal volume.[15] In the present case, the prime electrons could be created by thermal excitation of electrons out of the donor levels. Except for ultrapure crystals, the electron density in the conduction bands can easily be $10^8/cm^3$ at room temperature. Then, in a focal volume of 10^{-7} cm^3, the average number of prime electrons is certainly more than 1, and the development of avalanche ionization is clearly possible. The avalanche process is again governed by (27.2). Following the same classical model used to derive the ionization rate η in (27.6), we arrive at the expression for the breakdown threshold I_{cr} in (27.7). Therefore, at least the qualitative behavior of avalanche ionization in solids should be similar to that in gases.

As in the gas case, (27.7) suggests that the optical breakdown threshold is directly connected to the dc breakdown threshold by the relationship

$$I_{cr} = \frac{c}{2\pi} \frac{\varepsilon(0)}{\varepsilon^{1/2}(\omega)} |E_{dc}|_{cr}^2 (1 + \omega^2 \tau^2). \tag{27.9}$$

In solids, the collision lifetime τ is estimated to be $\sim 10^{-15}$ sec.[16] Equation (27.9) predicts that for $\omega \lesssim \tau^{-1}$, the threshold is nearly independent of ω.[17] Experimentally, the observed thresholds for alkali halides do seem to remain roughly the same from dc to $\lambda = 1$ μm and show a slight decrease at higher λ.[13,17,18] The dependence of the breakdown threshold on the laser pulsewidth τ_p is also specified in (27.7). If the loss rate g is negligible, I_{cr} is inversely proportional to τ_p and the breakdown process should have a fluence (energy/cm^2) threshold rather than an intensity threshold. If g dominates in (27.7), then the breakdown should have an intensity threshold. Experimental results on NaCl with a 1.06-μm laser light show that the breakdown threshold field changes from 2×10^6 v/cm for $\tau_p = 10^{-8}$ sec to 2×10^7 v/cm for $\tau_p = 10^{-11}$ sec.[19] This indicates that it is neither strictly intensity dependent nor strictly fluence dependent.

Optical breakdown in solids is also characterized by the rapid buildup of the plasma density. This is experimentally evidenced by the sudden cutoff in the transmitted laser intensity, as seen in Fig. 27.4. The oscilloscope traces in Fig. 27.4 also reveal the statistical nature of the breakdown process. The four traces were taken with input laser pulses of the same shape and magnitude. In the top three, breakdown occurred at slightly different times, and in the bottom trace, no breakdown occurred at all. The statistical fluctuations came in because the number of prime electrons in the focal region was, after all, quite small. In more careful studies, one should use a probability distribution to characterize the breakdown threshold.[20] It is customary to define the breakdown threshold as the value at which the breakdown occurs in 50% of the pulses.

We assumed here that multiphoton excitation of electrons to the conduction bands is negligible. This is certainly true at room temperature if the laser photon energy $\hbar\omega$ is much smaller than the energy gap of the solid, because even at the breakdown intensity, the number of electrons excited by multipho-

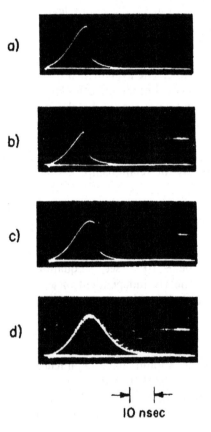

a)

b)

c)

d)

10 nsec

Fig. 27.4 Oscilloscope traces describing the transmitted TEM_{00} ruby laser pulse through a NaCl crystal. The laser pulse with an energy of 0.3 mJ was focused into the crystal by a lens of 14-mm focal length. (a) Optical breakdown occurs at the peak of the pulse. (b) Breakdown occurs before the peak at the energy $\mathscr{E}_{break} = 0.896 \, \mathscr{E}_{peak}$. (c) Breakdown occurs after the peak at $\mathscr{E}_{break} = 0.954 \, \mathscr{E}_{peak}$. (d) Three consecutive pulses without breakdown. [After D. W. Fradin, E. Yablonovitch, and M. Bass, *Appl. Opt.* **12**, 700 (1973).]

ton excitation is much smaller than the number of prime electrons created by thermal excitation. If, however, $\hbar\omega$ is comparable with the energy gap, then the multiphoton excitation process can become so important that it may even appear to be the dominant mechanism in determining the breakdown threshold.[17] Equation (27.2) should then be modified to take the form

$$\frac{\partial \rho}{\partial t} = (\eta - g)\rho + \left(\frac{\partial \rho}{\partial t}\right)_{\text{M}} \qquad (27.10)$$

where $(\partial\rho/\partial t)_{\text{M}}$ denotes the rate of increase of the conduction electron density due to multiphoton excitation. One expects that as ω increases, $(\partial\rho/\partial t)_{\text{M}}$ becomes increasingly important, and can eventually dominate the initial buildup of the electron avalanche process. This results in a transition from avalanche ionization to multiphoton excitation as the leading mechanism controlling the breakdown threshold.

Optical breakdown can also occur on solid surfaces. The physical process should be the same as in the bulk, and one would expect the same breakdown threshold for the surface as for the bulk. Experimentally, however, it was found that the surface breakdown threshold was usually much lower. In most cases, this was due to contamination of the surface by absorbing dust particles. For a clean surface, the breakdown threshold could be lower because of the existence of scratches and pores on the surface.[21] It is well known that the local field around a local surface structure with a sharp curvature can be significantly higher than the average field in the bulk. Consequently, optical breakdown is more likely to occur at such local structures, leading to an apparently lower breakdown threshold for the surface. The surface imperfections can be removed by superpolishing. It has been demonstrated that a superpolished surface can indeed have a breakdown threshold approaching that of the bulk.[22] Another method of eliminate the high field strength at local surface structure is to construct a surface layer with a graded refractive index.[23] The breakdown threshold for such a surface also approaches that of the bulk.

The relevance of optical breakdown in solids to high-power lasers and applications is obvious. Laser damage limits the maximum laser power one can hope to obtain from a high-power laser system. It also limits the laser power one can transmit through windows, lenses, and other optical components. The subject is of such technical importance that there have been annual conferences on it since 1970. The readers should consult the conference proceedings for details and advances in this field.[24]

REFERENCES

1 P. D. Maker, R. W. Terhune, and C. M. Savage, in P. Grivet and N. Bloembergen, eds., *Proc. Third International Conference of Quantum Electronics* (Paris, Dunod, 1964), p. 1559.

2 V. N. Parfenov, L. N. Pakhomov, V. Yu Petrun'kin, and V. A. Podlevskii, *Sov. Tech. Phys. Lett.* **2**, 286 (1976).

3   Ya. B. Zel'dovich and Yu. P. Raizer, *JETP* **20**, 772 (1965); Yu. P. Raizer, *Uspekhi* **8**, 650 (1965).

4   Yu. P. Raizer, *JETP* **21**, 1009 (1965).

5   See, for example, A. D. MacDonald, *Microwave Breakdown in Gases* (Wiley, New York, 1966); H. Raether, *Electron Avalanches and Breakdown in Gases* (Butterworth, Washington D.C., 1964).

6   R. V. Ambartzumian, N. G. Basov, V. A. Boiko, V. S. Zuev, O. N. Krokhin, P. G. Kryukov, Yu. V. Senat-skii, and Yu. Yu. Stoilov, *JETP* **21**, 1061 (1965); W. I. Linlor, *Appl. Phys. Lett.* **3**, 210 (1963).

7   E. Yablonovitch, *Appl. Phys. Lett.* **23**, 121 (1973).

8   N. M. Kroll and K. M. Watson, *Phys. Rev. A* **5**, 1883 (1972).

9   R. T. Brown and D. C. Smith, *Appl. Phys. Lett.* **22**, 245 (1973).

10 D. H. Gill and A. A. Dougal, *Phys. Rev. Lett.* **15**, 845 (1965).

11 S. A. Ramsden and W. E. R. Davis, *Phys. Rev. Lett.* **13**, 227 (1964).

12 H. J. Schwarz and H. Hora, eds., *Laser Interaction and Related Plasma Phenomena*, 5 vols. (Plenum, New York, 1971, 1973, 1975, 1977, 1979); C. Yamanaka and H. J. Schwarz, eds., *Laser Interaction with Matter* (Japan Society for Promotion of Science, Tokyo, 1973).

13 E. Yablonovitch, *Appl. Phys. Lett.* **19**, 495 (1971).

14 A. Wasserman, *Appl. Phys. Lett.* **10**, 132 (1967); G. M. Zverev, T. N. Mikhailova, V. A. Pashkov, and N. M. Solov'eva, *JETP* **26**, 1053 (1968).

15 E. Yablonovitch and N. Bloembergen, *Phys. Rev. Lett.* **29**, 907 (1972).

16 F. Seitz, *Phys. Rev.* **76**, 1376 (1949).

17 N. Bloembergen, *IEEE J. Quant. Electron.* **QE-10**, 375 (1974).

18 D. W. Fradin, E. Yablonovitch, and M. Bass, *Appl. Opt.* **12**, 700 (1973); D. W. Fradin and M. Bass, *Appl. Phys. Lett.* **22**, 206 (1973).

19 D. W. Fradin, N. Bloembergen, and J. P. Letellier, *Appl. Phys. Lett.* **22**, 635 (1973).

20 M. Bass and H. H. Barrett, *Appl. Opt.* **12**, 690 (1973); *IEEE J. Quant. Electron.* **QE-8**, 338 (1972); M. Bass and D. W. Fradin, *IEEE J. Quant. Electron.* **QE-9**, 890 (1973).

21 N. Bloembergen, *Appl. Opt.* **12**, 661 (1973).

22 C. R. Giuliano, *Appl. Phys. Lett.* **21**, 39 (1972); D. W. Fradin and M. Bass, *Appl. Phys. Lett.* **22**, 157 (1973).

23 W. H. Lowdermilk and D. Milam, *Appl. Phys. Lett.* **36**, 891 (1980); W. H. Lowdermilk, in *Handbook of Laser Science and Technology*, vol. 3: *Optical Materials* (CRC Press, Boca Raton, Fla.), in press.

24 *Symposium on Laser Damage and Optical Materials*, Boulder, Colo. (NBS Special Publications, 1970–1980).

BIBLIOGRAPHY

Bloembergen, N., *IEEE J. Quant. Electron.* **QE-10**, 375 (1974).

DeMichelis, C., *IEEE J. Quant. Electron.* **QE-5**, 188 (1969).

Grey-Morgan, C., *Rep. Prog. Phys.* **38**, 621 (1975).

Krokhin, O. N., in F. T. Arecchi and E. O. Schulz-Dubois, eds., *Laser Handbook* (North Holland Publishing Co., Amsterdam, 1972), p. 1371.

Raizer, Yu. P., *Uspekhi* **8**, 650 (1965).

Raizer, Yu. P., *Uspekhi* **23**, 789 (1981).

28

Nonlinear Optical Effects in Plasmas

Plasma is a highly nonlinear optical medium. With high-power pulsed lasers, nonlinear optical effects in plasmas are easily observable. They are in fact hardly avoidable in laser heating of plasmas and in laser-induced fusion work. A thorough understanding of such effects is therefore necessary for progress in these areas. Nonlinear interaction of light in a plasma is also a very interesting subject in its own right. As a charged fluid, a plasma is readily perturbed by external fields. Its extremely strong and complex response to intense laser fields can yield many fascinating nonlinear optical phenomena, but quantitative interpretations of these phenomena appear to be difficult. Here we restrict ourselves to a basic formulation of the theory and a brief description of some experimental observations.

28.1 THEORETICAL DESCRIPTION

That a plasma can be a highly nonlinear optical medium was known early in the development of nonlinear optics. Various nonlinear optical effects in plasmas, such as harmonic generation, parametric amplification, and stimulated Raman scattering, were predicted in the 1960s.[1] Experimentally, however, they were not studied with any concerted effort until the 1970s. The immense research activities in this area started only when the importance of laser-induced fusion in future technology was recognized. Understanding nonlinear optical effects in plasmas is essential since they have direct influence on laser heating of plasmas. In most cases, one is concerned with a gas plasma created by focusing of a laser pulse on a solid target. While the plasma is being created, the same laser pulse also induces the nonlinear optical effects. The high complexity of the laser-induced plasma growth process and the strong nonlinearity of the expanding plasma make the problem very difficult to analyze. We consider here only the basic theory of nonlinear laser interaction with plasmas. For this purpose, we assume a fully ionized stationary plasma with a given density profile, under the excitation of quasi-monochromatic infinite-plane waves.

The basic formalism for laser interaction with a plasma was described in Section 1.4 in connection with second-harmonic generation from a free-electron plasma. Here we simply expand the formalism to include plasma damping, and apply it to the case of a two-component (electron-ion) plasma. Let the electron and ion densities in the plasma be

$$N_e(\mathbf{r}, t) = N_{e0} + \rho_e \quad \text{and} \quad N_i(\mathbf{r}, t) = N_{i0} + \rho_i \qquad (28.1)$$

where ρ_e and ρ_i are the induced changes in the electron and ion densities away from their unperturbed values N_{e0} and N_{i0}. The subindices e and i refer to electrons and ions, respectively. The dynamic equations governing the plasma are: (1) the continuity equations for electrons and ions

$$\begin{aligned}
\frac{\partial \rho_e}{\partial t} + \nabla \cdot (N_e \mathbf{v}_e) + \nu_e \rho_e &= 0, \\
\frac{\partial \rho_i}{\partial t} + \nabla \cdot (N_i \mathbf{v}_i) + \nu_i \rho_i &= 0
\end{aligned} \qquad (28.2)$$

where $\mathbf{v}$ is the velocity and ν is the effective collision frequency responsible for damping, and (2) the equations of motion for electrons and ions

$$\begin{aligned}
\frac{\partial \mathbf{v}_e}{\partial t} + (\mathbf{v}_e \cdot \nabla)\mathbf{v}_e &= -\frac{1}{m_e N_e}\left(\frac{\partial p_e}{\partial \rho_e}\right)\nabla \rho_e + \frac{q_e}{m_e}\left(\mathbf{E} + \frac{1}{c}\mathbf{v}_e \times \mathbf{B}\right), \\
\frac{\partial \mathbf{v}_i}{\partial t} + (\mathbf{v}_i \cdot \nabla)\mathbf{v}_i &= -\frac{1}{m_i N_i}\left(\frac{\partial p_i}{\partial \rho_i}\right)\nabla \rho_i + \frac{q_i}{m_i}\left(\mathbf{E} + \frac{1}{c}\mathbf{v}_i \times \mathbf{B}\right)
\end{aligned} \qquad (28.3)$$

where p is the pressure, m is the mass, and q is the charge. The quantity $\partial p / \partial \rho$ can be related to the temperature of the plasma at equilibrium by $\partial p / \partial \rho = \gamma k_B T$, with γ being the adiabatic exponent and k_B the Boltzmann constant. As to the fields $\mathbf{E}$ and $\mathbf{B}$, they are governed by the Maxwell equations driven by the charge density

$$\rho_Q = q_e \rho_e + q_i \rho_i \qquad (28.4)$$

and the current density

$$\mathbf{J} = N_e q_e \mathbf{v}_e + N_i q_i \mathbf{v}_i. \qquad (28.5)$$

The foregoing set of coupled equations (28.2)–(28.5), together with the Maxwell equations, formally describes all the possible optical effects in the idealized plasma. In practice, for relatively high-frequency optical fields, the $(q_i/m_i)[\mathbf{E} + \mathbf{v}_i \times \mathbf{B}/c]$ term can be neglected because of the heavy ion mass. The plasma is then driven only by the electric and Lorentz forces on electrons. The response of ions to the field is coupled to that of electrons mainly through interaction with the field via the Gauss law

$$\nabla \cdot \mathbf{E} = 4\pi(\rho_e q_e + \rho_i q_i). \tag{28.6}$$

Optical nonlinearities of a plasma arise from the $(\mathbf{v} \cdot \nabla)\mathbf{v}$ terms and the Lorentz force on electrons in (28.3).

Second harmonic generation from a plasma has already been discussed in Section 1.4. At optical frequencies, the ionic contribution to the nonlinearity is negligible. The second-order current density responsible for the second-harmonic generation can be derived from (28.2) and (28.3) by iterative expansion. It takes the form

$$\mathbf{J}^{(2)}(2\omega) = \frac{iq_e^2}{m_e^2 \omega^3}\left\{\tfrac{1}{4}N_{e0}\nabla(\mathbf{E}_1 \cdot \mathbf{E}_1) + \frac{(\nabla N_{e0} \cdot \mathbf{E}_1)\mathbf{E}_1}{1 - \omega_{e0}^2/\omega^2}\right\} \tag{28.7}$$

where $\omega_{e0} = (4\pi N_{e0}q_e^2/m_e)^{1/2}$, $\mathbf{E}_1$ is the fundamental field, and ν_e has been neglected. As pointed out in Section 1.4, for a single incoming laser beam, $\mathbf{J}^{(2)}(2\omega)$ cannot radiate in the bulk of a uniform plasma, but it is responsible for the second-harmonic generation at the surface of a uniform plasma or in the bulk of a nonuniform plasma. Extension of the calculation to sum- and difference-frequency generation and higher-order mixing processes is straightforward but tedious. The principle is, however, very much the same: the nonlinear optical effects arise from the nonlinear response of individual particles to the applied fields.

Plasma waves exist in a plasma. They are collective excitations of electrons and ions in a plasma. Nonlinear optical effects can also result from coupling of light with plasma waves. The plasma wave equations for a two-component uniform plasma can be derived from (28.2) and (28.3) together with (28.6). We find, by eliminating $\partial\mathbf{v}/\partial t$ in the equations,

$$\left[\frac{\partial^2}{\partial t^2} - \frac{1}{m_e}\left(\frac{\partial p_e}{\partial \rho_e}\right)\nabla^2 + \nu_e\frac{\partial}{\partial t}\right]\rho_e = -\omega_{e0}^2\left(\rho_e - \left|\frac{q_i}{q_e}\right|\rho_i\right) + \nabla \cdot \mathbf{F}_e,$$

$$\left[\frac{\partial^2}{\partial t^2} - \frac{1}{m_i}\left(\frac{\partial p_i}{\partial \rho_i}\right)\nabla^2 + \nu_i\frac{\partial}{\partial t}\right]\rho_i = \omega_{i0}^2\left|\frac{q_e}{q_i}\right|\left(\rho_e - \left|\frac{q_i}{q_e}\right|\rho_i\right),$$

and $\tag{28.8}$

$$\mathbf{F}_e = \frac{q_e}{m_e}\rho_e\mathbf{E} + N_e(\mathbf{v}_e \cdot \nabla)\mathbf{v}_e - \mathbf{v}_e\frac{\partial\rho_e}{\partial t}$$

$$+ N_e\left(\frac{q_e}{m_e c}\right)\mathbf{v}_e \times \mathbf{B}$$

where $\omega_{i0}^2 = 4\pi N_{i0} q_i^2 / m_i$. We neglect the $\nabla \cdot \mathbf{F}_i$ term in the equation for ρ_i because of the heavy ion mass and the slow ion velocity. As shown in (28.8), the two wave equations for ρ_e and ρ_i are not independent but are linearly couped. The dispersion curve of the plasma waves is obtained from the resonant structure of the coupled wave equations. It has an optical branch and an acoustic branch.[2] Since $m_i \gg m_e$, the following approximations can be used in describing the two branches.[3] For the optical modes, only electrons in the plasma can effectively respond to the rapid oscillation. The density variation of the ions is negligible. Therefore, the corresponding plasma wave equation, from (28.8), is simply

$$\left[\frac{\partial^2}{\partial t^2} - \frac{1}{m_e} \left(\frac{\partial p_e}{\partial \rho_e} \right) \nabla^2 + \omega_{e0}^2 + \nu_e \frac{\partial}{\partial t} \right] \rho_e = \nabla \cdot \mathbf{F}_e. \qquad (28.9)$$

The dispersion relation for the optical plasma waves is

$$\omega_{pe}^2 = \omega_{e0}^2 + \frac{1}{m_e} \left(\frac{\partial p_e}{\partial \rho_e} \right) k^2. \qquad (28.10)$$

The quantity ω_{pe} is usually known as the electron plasma resonance frequency. For acoustic modes, the charge neutrality condition $\rho_e q_e + \rho_i q_i = 0$ is approximately satisfied, because electrons can easily follow the slow motion of ions in the plasma. The plasma wave equation is derived by first combining the two wave equations in (28.8) with the elimination of the $(\rho_e - |q_i/q_e|\rho_i)$ terms, and then replacing ρ_e by $|q_i/q_e|\rho_i$. Neglecting $\partial^2 \rho_e / \partial t^2$ in comparison with $(m_i/m_e)\partial^2 \rho_i / \partial t^2$, we find

$$\left(\frac{\partial^2}{\partial t^2} - V_a^2 \nabla^2 + \nu_a \frac{\partial}{\partial t} \right) \rho_i = \left(\frac{m_e}{m_i} \right) \nabla \cdot \mathbf{F}_e,$$

$$V_a^2 = \frac{1}{m_i} \left(\frac{\partial p_i}{\partial \rho_i} \right) + \frac{1}{m_i} \left| \frac{q_i}{q_e} \right| \left(\frac{\partial p_e}{\partial \rho_e} \right), \qquad (28.11)$$

and

$$\nu_a = \nu_i + \left(\frac{m_e}{m_i} \right) \left| \frac{q_i}{q_e} \right| \nu_e.$$

The dispersion relation for the ion-acoustic plasma waves is

$$\omega_{pa}^2 = V_a^2 k^2 \qquad (28.12)$$

where ω_{pa} is the ion-acoustic plasma resonance frequency. In all the above wave equations, (28.8), (28.9), and (28.11), the optical fields act as the driving sources through $\nabla \cdot \mathbf{F}_e$. Since ρ_e and v_e in $\nabla \cdot \mathbf{F}_e$ are both induced by the optical fields [via (28.6) and (28.3)], the plasma waves appear to be driven nonlinearly by the fields. The optical waves propagating in the plasma, in turn, are influenced by the plasma variation as governed by the wave equation

$$\nabla \times (\nabla \times \mathbf{E}) + \frac{1}{c^2} \frac{\partial^2}{\partial t^2} \mathbf{E} = -\frac{4\pi}{c^2} \frac{\partial}{\partial t} \mathbf{J} \qquad (28.13)$$

with $\mathbf{J}$ given by (28.5), which depends on ρ and $\mathbf{v}$. Coupling of (28.13) with the plasma wave equations gives rise to a number of interesting nonlinear optical effects in plasmas, including stimulated Raman and Brillouin scattering processes and parametric instabilities. They are of practical importance in the consideration of laser heating of a plasma.

Consider first the stimulated Raman scattering process.[1] Here the Raman process refers to the scattering of light by the electron plasma resonance. As noted in Section 10.3, stimulated Raman scattering can be considered a parametric process resulting from coupling of a pump wave at ω_l, a Stokes wave at ω_s, and a material excitational wave at $\omega_l - \omega_s \simeq \omega_{ex}$. In the present case, the material excitational wave is the optical plasma wave with a resonant frequency $\omega_{ex} = \omega_{pe}$. The theory, therefore, closely follows the one developed in Section 10.3. First, we should write down explicitly the wave equations for the three waves. From (28.5) and (28.13), we find, for the pump wave, $\mathbf{E}(\omega_l) = \mathscr{E}_l \exp(i\mathbf{k}_l \cdot \mathbf{r} - i\omega_l t)$, and the Stokes wave, $\mathbf{E}(\omega_s) = \mathscr{E}_s \exp(i\mathbf{k}_s \cdot \mathbf{r} - i\omega_s t)$, in a uniform plasma, the following wave equations

$$\left(\nabla^2 + \frac{\omega_l^2 \varepsilon_l}{c^2} \right) \mathbf{E}(\omega_l) \cong -i\left(\frac{4\pi\omega_l}{c^2} \right) q_e \rho_e(\omega_l - \omega_s) \mathbf{v}_e(\omega_s)$$

$$\cong \left(\frac{4\pi\omega_l q_e^2}{m_e \omega_s c^2} \right) \rho_e(\omega_l - \omega_s) \mathbf{E}(\omega_s)$$

and   (28.14)

$$\left(\nabla^2 + \frac{\omega_s^2 \varepsilon_s}{c^2} \right) \mathbf{E}(\omega_s) \cong \left(\frac{4\pi\omega_s^2 q_e^2}{m_e \omega_l c^2} \right) \rho_e^*(\omega_l - \omega_s) \mathbf{E}(\omega_l).$$

Then, from (28.9), we can write the optical plasma wave equation as[3]

$$\left[-\frac{1}{m_e} \frac{\partial p_e}{\partial \rho_e} \nabla^2 + \omega_{e0}^2 - (\omega_l - \omega_s)^2 - i(\omega_l - \omega_s)\nu_e \right] \rho_e(\omega_l - \omega_s)$$

$$\cong \nabla \cdot \left[\frac{N_{e0} q_e (\mathbf{v}_e \times \mathbf{B})}{m_e c} \right] \qquad (28.15)$$

$$\cong \left[\frac{N_{e0} (\mathbf{k}_l - \mathbf{k}_s)^2 q_e^2}{m_e^2 \omega_l \omega_s} \right] E(\omega_l) E^*(\omega_s)$$

assuming, for simplicity, that $\mathbf{E}(\omega_l)$ and $\mathbf{E}(\omega_s)$ are linearly polarized in the same direction. The solution of such a set of coupled wave equations, (28.14) and (28.15), was illustrated many times in Chapters 9 to 11, and is not repeated here. We note that in the limit of negligible pump depletion, the Stokes intensity should grow exponentially. The maximum exponential gain is

$$G_{max} = \frac{4\pi N_{e0} q_e^4 (k_l + k_s)^2 \varepsilon_s^{1/2} |E(\omega_l)|^2}{m_e^3 c^3 k_l k_s \omega_l \omega_{e0} \nu_e},$$

(28.16)

which occurs when the Stokes scattering is in the backward direction and $(\omega_l - \omega_s)$ equals the electron plasma resonance frequency ω_{pe}. For a weakly damped plasma with $\omega_{e0}/\nu_e \sim 100$, the above result predicts a strong Stokes backscattering from the focal region of a 1-GW laser beam. Generation of Stokes and anti-Stokes radiation of many orders should also be possible in a plasma, assuming a sufficiently long and uniform interaction length. The calculation can be extended to a plasma with a density gradient and special boundary conditions.[4]

The electron plasma wave in the above stimulated scattering process can be replaced by an ion-acoustic plasma wave. We then have the stimulated Brillouin scattering process. The theory of stimulated Brillouin scattering is essentially the same as that of stimulated Raman scattering, except that instead of (28.9), we now have (28.11), which can be rewritten in the form

$$\left[-V_a^2 \nabla^2 - (\omega_l - \omega_s)^2 - i(\omega_l - \omega_s)\nu_a \right] \rho_e$$

$$= -\left(\frac{m_e q_i}{m_i |q_e|} \right) \left[\frac{N_{e0}(\mathbf{k}_l - \mathbf{k}_s)^2 q_e^2}{m_e^2 \omega_l \omega_s} \right] E(\omega_l) E^*(\omega_s).$$

(28.17)

In this case, because of the linear dispersion relation of (28.12) for the ion-acoustic plasma waves, the situation is very similar to that of an ordinary stimulated Brillouin scattering process discussed in Section 11.1

We have not included the Stokes attenuation in this calculation because we neglected the loss terms in the equations of motion in (28.3), so that the linear dielectric constant $\varepsilon(\omega)$ of the plasma appears as a real quantity. More generally, we should have, at optical frequencies,

$$\varepsilon(\omega) = 1 - \frac{\omega_{e0}^2}{\omega(\omega + i\Gamma_e)}$$

(28.18)

where Γ_e is the damping rate of the electron velocity. The dielectric constant is now a complex quantity. Only when the Stokes gain G is larger than the linear attenuation $(\omega_s/c\sqrt{\varepsilon(\omega_s)})\text{Im}[\varepsilon(\omega_s)]$ can the stimulated scattering process actually take place. We should also note that since light waves with $\omega < \omega_{e0}$ cannot propagate in a plasma, the pump wave in stimulated Raman scattering must have a frequency ω_l at least two times larger than ω_{e0}, since otherwise $\omega_s = \omega_l - \omega_{e0}$ would be smaller than ω_{e0}. There is no such restriction for stimulated Brillouin scattering because $\omega_a \ll \omega_l$.

In this discussion, stimulated Raman or Brillouin scattering was treated as a three-wave parametric process, in which the signal is an optical wave and the idler is a plasma wave. The process can certainly be generalized to the case where both the signal and the idler are plasma waves. This case is generally known as a plasma parametric decay process or parametric instability.[3,5] It usually has a threshold lower than stimulated Raman scattering in a plasma. We consider here the parametric process involving an electron plasma wave and an ion-acoustic plasma wave with $\omega_l = \omega_e + \omega_a$, $\mathbf{k}_l = \mathbf{k}_e + \mathbf{k}_a$, $\omega_e \simeq \omega_{pe}(\mathbf{k}_e) \gg \omega_a \simeq \omega_{pa}(\mathbf{k}_a)$, and $\mathbf{k}_e \simeq -\mathbf{k}_a$. In the limit of negligible pump depletion, it can be described by the coupled equations (28.9) and (28.11). With appropriate approximations, they reduce to the form[3]

$$\left[-\omega_e^2 + \omega_{pe}^2(\mathbf{k}_e \to -i\nabla) - i\omega_e\nu_e \right] \rho_e(\omega_e) = -i\left(\frac{q_e}{m_e}\right)\left|\frac{q_i}{q_e}\right|(\mathbf{k}_e \cdot \mathbf{E}_l)\rho_i^*(\omega_a)$$

and (28.19)

$$\left[-\omega_a^2 + \omega_{pa}^2(\mathbf{k}_a \to -i\nabla) + i\omega_a\nu_a \right] \rho_i^*(\omega_a) = i\frac{q_e}{m_i}(\mathbf{k}_a \cdot \mathbf{E}_l)^*\rho_e(\omega_e).$$

Following the same derivation used in Section 9.1 for parametric amplification, we can find from (28.19) that the threshold pump intensity for the parametric instability is

$$I_{\text{th}} = \frac{c}{2\pi}\sqrt{\epsilon_l}\,\frac{m_e}{q_e^2}\,\frac{\omega_{e0}^2}{\omega_e\omega_a}\nu_e\nu_a\left[\left(\frac{\partial p_e}{\partial \rho_e}\right) + \left|\frac{q_e}{q_i}\right|\left(\frac{\partial p_i}{\partial \rho_i}\right)\right]. \qquad (28.20)$$

Above the threshold, the pump wave can effectively feed energy into the plasma, and the plasma waves grow exponentially. It is also possible to have a parametric process generate two electron plasma waves. In this case, the two plasma wavevectors, $\mathbf{k}_{e1}$ and $\mathbf{k}_{e2}$, generally satisfy the relations $\mathbf{k}_{e1} + \mathbf{k}_{e2} = \mathbf{k}_l$ and $|k_{e1}| \simeq |k_{e2}| \gg |k_l|$, where $\mathbf{k}_l$ is the pump wavevector. Therefore, the frequencies of the two plasma waves should be nearly equal.

Four-wave mixing and other third-order nonlinear optical processes can also occur in a plasma. Some of them result from an optical-field-induced refractive index change in the plasma. A third-order iterative calculation (see Section 1.4) using (28.2–28.6) can lead to a formal expression of $\mathbf{J}^{(3)}(\omega)$ in the form $\mathbf{J}^{(3)}(\omega) = \sigma^{(3)}|E(\omega)|^2E(\omega)$. Since $\sigma^{(3)}(\omega) = -i\omega\chi^{(3)}(\omega)$, the refractive index $n(\omega) = n_0 + \Delta n$ has a field-induced term $\Delta n(\omega) = \chi^{(3)}(\omega)|E(\omega)|^2/2n_0$. More physically, the plasma considered as a uniform dielectric medium can have a refractive index change induced by a field intensity gradient via electrostriction (see Section 16.2). To minimize the free energy of the system, the electrons and ions will redistribute themselves so as to increase the refractive indices in regions of higher field intensities. Laser heating of the plasma can also produce a refractive index change via $(\partial n/\partial T)\Delta T$ with ΔT depending on the laser intensity. This laser-induced thermal effect reinforces the electrostrictive effect in a plasma since it tends to expand the plasma and therefore increase the refractive index in the higher-intensity region. (Note that at optical frequencies $n^2 \cong 1 - 4\pi N_{e0}q_e^2/m_e\omega^2$.) As a result of $\Delta n(|E|^2)$, self-focusing and self-phase modulation can occur in a plasma.[6] Degenerate four-wave mixing should also be possible. These processes, however, can be very complicated in a real plasma, which is often highly nonuniform, partially ionized, and strongly nonstationary.

28.2 EXPERIMENTAL STUDIES

The experimental situation of nonlinear optics in plasmas is far more complicated than the theoretical picture just presented. The main reason is that a real plasma is often far from the ideal plasma we assumed. For a thorough understanding of nonlinear laser interaction in a real plasma, it is necessary to know the detailed initial characteristics of the plasma: the degree of ionization, the electron and ion density distributions, the temperature distribution, the variation with time, and so on. The laser pulse used should also be well characterized in its pulse shape and intensity profile. Then the experimental results properly measured may be compared with the numerical solution of the appropriate set of coupled equations. Unfortunately, information about a real plasma is never so complete. This is particularly true for a laser-induced high-density plasma. We must therefore be satisfied with only a very qualitative description of the experimental observations.

Harmonic Generation

Optical second-harmonic generation from a plasma was first attempted using metals as the samples.[7,8] It was believed that a highly conductive metal might behave like a very dense electron plasma. The plasma is uniform and second-harmonic generation can occur only at the surface. The second-harmonic reflection from a metal surface could indeed be easily detected. However, whether the signal was dominated by free-electron or bound-electron contributions could not be determined.[8] The experiments could also be badly affected by surface contamination. For reliable and reproducible results, clean metal surfaces located in an ultrahigh vacuum chamber are clearly needed. The laser frequency should be well below the interband transitions to avoid the bound-electron effect.

    Second-harmonic radiation has also been detected in reflection from a laser-induced gas plasma.[9] The plasma was created by a laser pulse focused on a solid target. Being an expanding plasma, it was highly nonuniform. The second harmonic can then be generated in the bulk of the plasma. Quantitative analysis, however, requires a good characterization of the laser-induced plasma.[10]

Generation of higher harmonics in a laser-induced plasma has also been observed.[11] The most surprising result occurred when a high-energy CO_2 laser pulse was focused on a target (for example, a 100-J, 1-nsec pulse focused to $\sim 8 \times 10^{14}$ W/cm^2). Harmonics as high as the forty-sixth could be seen in the output, with little difference in their production efficiencies.[12] Some·examples are shown in Fig. 28.1. Clearly, this cannot be explained by the usual perturbation theory. Bezzerides et al. showed theoretically that when resonant absorption ($\omega \sim \omega_{e0}$) takes place in a steep plasma density profile, the plasma restoring force is strongly anharmonic.[13] It then gives rise to a strongly anharmonic oscillator system which radiates the high harmonics. The emission apparently originates from the critical surface·in the steep density profile. If the measurement has enough resolution in space and time, it can be used to locate the temporal position of the critical surface in an expanding plasma. There is no basic reason why this process cannot occur with a higher-frequency laser excitation. This then opens the possibility of producing intense vacuum uv to soft x-ray radiation by simply irradiating metal surfaces with intense laser light.

Stimulated Raman and Brillouin Scattering and Parametric Processes

Both stimulated Raman and stimulated Brillouin scattering have been observed in laser-induced plasmas with high-energy pulsed laser excitation ($\geq 10^{10}$ W/cm^2). Since the latter involves a less dispersive plasma wave than the former [V_a^2 in (28.11) is much smaller than $(\partial p_e/\partial \rho_e)/m_e$ in (28.9)], it can be better phase matched in the backward direction. Consequently, stimulated Brillouin scattering usually appears to have a lower threshold and can be more readily observed.[14] The backscattered Brillouin radiation is characterized by an initial red shift of a few angstrom units, which corresponds to the ion-acoustic plasma frequency, a linear polarization in the same direction as the incoming laser beam, a ray path retracing the incoming beam path, and an exponential rise followed by saturation in the backscattered energy above an input threshold energy (see Fig. 28.2). As much as several percent of the input energy can appear in the backward Brillouin output. The process is presumably responsible for the observed intensity-dependent reflection from a target. From the laser-plasma-heating or laser-fusion point of view, this is a detrimental process since it decreases the laser energy deposited into the plasma. To prevent this from happening, one can use a broadband laser source as the pump beam to raise the stimulated Brillouin threshold significantly.

With higher pump intensities, stimulated Raman scattering can also occur in a plasma.[15] It appears later than the Brillouin scattering,[16] and has a maximum gain in the backward direction. The output is characterized by a frequency shift proportional to the square root of the electron density ($\Delta \omega \simeq$

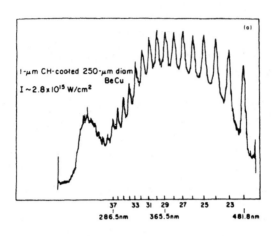

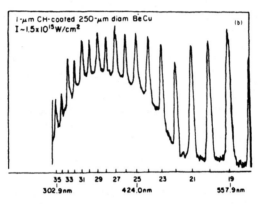

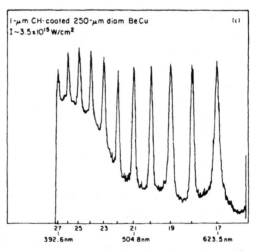

Fig. 28.1 Microdensitometer traces of the raw data taken with a 0.5-m vacuum spectrograph. The numbers on the abscissa indicate the harmonic order of the P20 transition in the 10-μm band. Three different shots are shown with the intensities on target indicated where the center wavelength of the spectra was shifted between shots. [After R. L. Carman, C. K. Rhodes, and R. F. Benjamin, *Phys. Rev. A* **24**, 2649 (1981).]

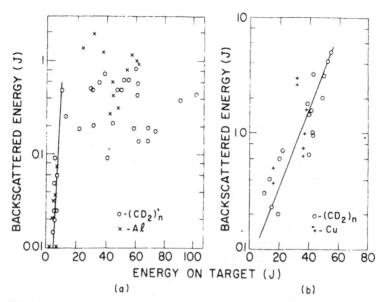

Fig. 28.2 Backscattered Brillouin energy versus the incident laser energy at 1.06 μm. (a) 900-psec laser pulses focused on $(CD_2)_n$ and Al targets by an f/14 lens. (b) 900-psec laser pulses focused on $(CD_2)_n$ and Cu targets by an f/1.9 lens. Saturation occurs only in case (a) after an initial exponential growth. [After B. H. Ripin, J. M. McMahon, E. A. McLean, W. M. Manheimer, and J. A. Stamper, *Phys. Rev. Lett.* **33**, 634 (1974).]

$\omega_{e0} = [4\pi N_{e0}q_e^2/m_e]^{1/2}$), and a sharp rise followed by saturation in the output energy above an input intensity threshold (see Fig. 28.3). In a magnetically confined plasma column with a CO_2 laser excitation, a Raman conversion efficiency as high as 0.7% has been reported.[16] The efficiency is much lower ($\sim 10^{-5}$) in a laser-induced plasma.[15] The saturation comes in mainly because of the nonlinear damping effect:[16] hot electrons are created in the plasma by heating via the stimulated Raman process. They are effective in increasing the Landau damping on the plasma oscillation. The balance between the increase of the Raman gain and the increase of the Landau damping with the pump intensity leads to the observed saturation. The spectrum of the Raman output is usually very broad, reflecting the variation of the electron density in the plasma.

Stimulated Raman scattering generates the electron plasma wave, which is then dissipated into heat in the plasma. This suggests that the process can be used for heating of a plasma. The efficiency is, however, very poor. One may increase the heating efficiency by using optical mixing to produce the electron plasma oscillation.[17] In that case, two input laser beams at ω_l and ω_s, with $\omega_l - \omega_s = \omega_{pe}$, are used to resonantly excite the plasma wave. If the generated plasma wave is probed by a third beam, we then have effectively a four-wave mixing process, the output of which should reflect the characteristics of the electron plasma. The same mixing process can be used to excite the ion-acoustic wave in a plasma if $\omega_l - \omega_s = \omega_{pa}$.[18]

Parametric instability, however, offers a more efficient way to heat a plasma, since the incoming photons participating in the process are completely converted into plasmons. This has been carefully studied using microwave pumping on a low-density cold plasma.[19] With laser pumping, parametric instability has also been observed.[9,20] It actually has a lower threshold than stimulated Raman scattering. While parametric instability generally involves the parametric photon decay into either two electron plasmons of the same frequency, or one electron and one ion-acoustic plasmon, the former seems to be more easily detectable. The generated electron plasma waves at $\omega_{pe} = \omega_l/2$ can radiate or coherently scatter the incoming beam at ω_l, leading to the appearance of $\omega_l/2$, $3\omega_l/2$, and possibly other higher subharmonics in the reflected light.[9,20] Parametric instability may produce highly energetic electrons in a laser-induced plasma. It may then cause preheating of the target core in laser-fusion experiments.

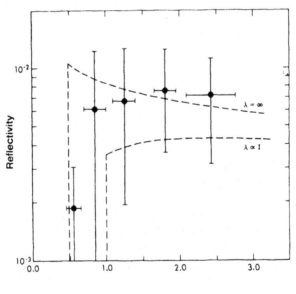

Intensity (10^{11} W / cm^2)

Fig. 28.3 Raman reflectivity as a function of incident laser intensity. The experiment was conducted using a long-pulse CO_2 laser beam focused by an f/15 mirror into a magnetically confined, laser-heated, hydrogen plasma column. The circles represent the average measured reflectivities from numerous laser shots. Error bars indicate the standard deviation. The dashed curves are theoretical curves with two different values of a parameter λ, taking into account nonlinear damping on the plasma wave. (After Ref. 16.)

Self-Focusing

Experiments on self-focusing due to intensity-dependent refractive index change in a plasma are usually difficult. The initially existing refractive index variation arising from the density distribution in a nonuniform plasma always tends to confuse the observed results. This is particularly true in a laser-induced plasma.

The earlier observations on self-focusing in plasmas were carried out with optical breakdown in air.[21] Time-integrated photographs of the laser focal volume showed the appearance of bright filaments at breakdown. These filaments were only ~ 5 μm in diameter, which was 10 times smaller than the focal diameter. They were formed by the very bright breakdown spots moving in the backward direction. This is characteristically the same as that happens with self-focusing in solids (see Section 17.5). Spectroscopic studies showed that the scattered radiation from the breakdown region was spectrally broadened, presumably because of self-phase modulation. The difficulty of such experiments was that one could not be sure if the optical nonlinearity of the plasma was the sole contributor to the refractive index variation. In many cases, it was likely that self-focusing was simply due to the plasma density gradient created in the dynamic formation of the plasma.[22]

The self-focusing effect can be enhanced by the excitation of electron plasma resonance via laser mixing.[23] The large pondermotive force of the excited plasma waves then effectively creates a depression of the plasma density on the beam axis, and the resultant refractive index change causes the laser beam to self-focus. With a CO_2 laser beam of only 5×10^9 W/cm^2 in intensity and a plasma density of 0.6% of the critical density ($N_{ec} = m_e \omega^2 / 4\pi q_e^2$), self-focusing was readily observed when the plasma wave was resonantly excited by optical mixing.[23]

Nonlinear optical process will undoubtedly occur in laser compression of a fusion target. They can predominantly affect the way energy is deposited onto the target. They can also play an important role in producing uv and x-ray radiation from a laser-induced plasma. The problems are, however, extremely complex. In generating and heating a plasma, the laser irradiation creates such chaos that all imaginable processes could take place. Difficulties in attaining a quantitative understanding of the problems may seem insurmountable. Yet the history of science has witnessed many such successes. Through perseverance and intelligence, one can always hope to find order in chaos. It is in fact the creation of order out of disorder that brings joy and excitement to devoted scientists.

REFERENCES

1 See, for example, N. Kroll, A. Ron, and N. Rostoker, *Phys. Rev. Lett.* **13**, 83 (1964); D. F. Dubois and M. V. Goldman, *Phys. Rev. Lett.* **14**, 544 (1965); S. S. Jha, *Phys. Rev.* **140**, A2020 (1965); N. Bloembergen and Y. Shen, *Phys. Rev.* **141**, 298 (1966).

2 L. Spitzer, *Physics of Fully Ionized Gases*, 2nd ed. (Interscience, New York, 1961).

3 R. E. Kidder, in P. Caldirole and H. Knoepfel, eds., *Physics of High-Energy Density*, Proceedings of the International School of Physics "Enrico Fermi," Course XLVIII, (Academic Press, New York, 1971), p. 306.

4 D. W. Forslund, J. M. Kindel, and E. L. Lindman, *Phys. Rev. Lett.* **30**, 739 (1973); A. A. Galeev, G. Laval, T. M. O'Neil, M. N. Rosenbluth, and R. Z. Sagdeev, *JETP Lett.* **17**, 48 (1973); D. Biskamp and H. Welter, *Phys. Rev. Lett.* **34**, 312 (1975).

5 T. P. Huges, in P. G. Harper and B. S. Wherrett, eds., *Nonlinear Optics* (Academic Press, New York, 1977), p. 365.

6 C. E. Max, *Phys. Fluids* **19**, 74 (1976).

7 F. Brown, R. E. Parks, and A. M. Sleeper, *Phys. Rev. Lett.* **14**, 1029 (1965).

8 N. Bloembergen, R. K. Chang, S. S. Jha, and C. H. Lee, *Phys. Rev.* **174**, 183 (1968).

9 J. L. Bobin, M. Decroisette, B. Meyer, and Y. Vitel, *Phys. Rev. Lett.* **30**, 594 (1973); P. Lee, D. V. Giovanelli, R. P. Godwin, and G. H. McCall, *Appl. Phys. Lett.* **24**, 406 (1974); K. Eidman and R. Sigel, *Phys. Rev. Lett.* **34**, 799 (1975).

10 G. Auer, K. Sauer, and K. Baumgärtel, *Phys. Rev. Lett.* **42**, 1744 (1979).

11 E. A. McLean, J. A. Stamper, B. H. Ripin, H. R. Griem, J. McMahon, and S. E. Bodner, *Appl. Phys. Lett.* **31**, 825 (1977).

12 N. H. Burnett, H. A. Baldes, M. C. Richardson, and G. D. Enright, *Appl. Phys. Lett.* **31**, 172 (1977); R. L. Carman, D. W. Forslund, and J. M. Kindel, *Phys. Rev. Lett.* **46**, 29 (1981); R. L. Carman, C. K. Rhodes, and R. F. Benjamin, *Phys. Rev. A* **24**, 2649 (1981).

13 B. Bezzerides, R. D. Jones, and D. W. Forslund, *Phys. Rev. Lett.* **49**, 202 (1982).

14 L. M. Goldman, J. Sources, and M. J. Lubin, *Phys. Rev. Lett.* **31**, 1184 (1973); C. Yamanaka, T. Yamanaka, T. Sasaki, J. Mizui, and H. B. Kang, *Phys. Rev. Lett.* **32**, 1038 (1974); B. H. Ripin, J. M. McMahon, E. A. McLean, W. M. Manheimer, and J. A. Stamper, *Phys. Rev. Lett.* **33**, 634 (1974).

15 R. G. Watt, R. D. Brooks, and Z. A. Pietrzyk, *Phys. Rev. Lett.* **41**, 170 (1978); K. Tanaka, L. M. Goldman, W. Seka, M. C. Richardson, J. M. Sources, and E. A. Williams, *Phys. Rev. Lett.* **48**, 1179 (1982); K. A. Nugent and B. Luther-Davies, *Phys. Rev. Lett.* **49**, 1943 (1982), and references therein.

16 A. A. Offenberger, R. Fedosejevs, W. Tighe, and W. Rozmus, *Phys. Rev. Lett.* **49**, 371 (1982).

17 B. I. Cohen, A. N. Kaufman, and K. M. Watson, *Phys. Rev. Lett.* **29**, 581 (1972); M. N. Rosenbluth and C. S. Liu, *Phys. Rev. Lett.* **29**, 701 (1972).

18 C. J. Pawley, H. E. Huey, and N. C. Luhmann, *Phys. Rev. Lett.* **49**, 877 (1982).

19 M. Porkolab, V. Arunasalam, and R. A. Ellis, *Phys. Rev. Lett.* **29**, 1438 (1972).

20 C. Yamanaka, T. Yamanaka, T. Sasaki, J. Mizui, and H. B. Kang, *Phys. Rev. Lett.* **32**, 1038 (1974); K. Tanaka, L. M. Goldman, W. Seka, M. C. Richardson, J. M. Sources, and E. A. Williams, *Phys. Rev. Lett.* **48**, 1179 (1982), and references therein.

21 V. V. Korobkin and A. J. Alcock, *Phys. Rev. Lett.* **21**, 1433 (1968); A. J. Alcock, in H. J. Schwarz and H. Hora, eds., *Laser Interaction and Related Plasma Phenomena*, vol. 2 (Plenum, New York, 1972), p. 155; C. Yamanaka, T. Yamanaka, T. Mizui, and N. Yamaguchi, *Phys. Rev. A* **11**, 2138 (1975).

22 L. C. Johnson and T. K. Chu, *Phys. Rev. Lett.* **32**, 517 (1974).

23 C. Joshi, C. E. Clayton, and F. F. Chen, *Phys. Rev. Lett.* **48**, 874 (1982).

BIBLIOGRAPHY

H. J. Schwarz and H. Hora, eds., *Laser Interaction and Related Plasma Phenomena*, 5 vols. (Plenum, New York, 1971–1979).

Index

Acoustic wave equation, 290
Adiabatic following, 399–401
Adiabatic inversion, 400
Anharmonic oscillator, 5–8, 439
Anomalies in stimulated Raman scattering, 159, 316–319
  anti-Stokes rings, 159, 319
  due to self-focusing, 159, 316–319
  effective Raman gain, 159, 316
  forward-backward asymmetry, 158–159, 318
Anti-Stokes Raman scattering:
  spontaneous, 168–169
  stimulated, 167–168
Applications of stimulated Raman scattering:
  high-resolution spectroscopy, 167–169
  tunable infrared sources, 164–167
  tunable UV sources, 167–169
Autler-Townes effect, 418, 425, 429–430
Approximation, slow varying amplitude, 47–49
Autoionization spectroscopy, 346–347

Backward parametric amplification and oscillation, 248–249
Bare atom, 415–421
  three-level system, 418–421
  two-level system, 416–418
Bistability, optical, 299–301
Blackbody radiation, 108
Bloch equation, 379–383
  pseudo-dipole in, 382
  in rotating frame, 382
Bond model, for calculations of susceptibilities, 29–31
  additivity rule, 30
  bond polarizabilities, 30
Breakdown:
  DC or microwave, 529
  dependence on gas pressure, 534
  laser-induced thermal, 529, 537
  optical, 528–540
  similarity principle for, 532
Brillouin doublet, 192
Brillouin scattering, stimulated, 187–192

Broadening of spectral line, 211–213
  collisional, 213
  Doppler, 212
  homogeneous, 212–213
  inhomogeneous, 212
  power, 213
  pressure, 213

Cherenkov radiation, 200, 314
Coherent anti-Stokes Raman scattering (CARS), 267–272
  applications, 270–271
  experimental arrangement, 271
  nonresonant background, 268–270
  signal strength, 270
  spectrum, 268–269
  surface, 491–493
Coherent length, 76
Coherent Raman spectroscopy, 267–275
  of excited states, 280–281
  pressure-induced, 281
Coherent Stokes Raman scattering, 275
Coherent transient effects, see Transient coherent optical effects
Compression of ultrashort pulses, 524–526
  experimental arrangement, 524
Cotton-Mouton effect, 55
  inverse, 60–66
Coupled-wave approach:
  boundary conditions, 50
  energy and momentum conservation, 44
  energy transfer among waves, 44
  phase matching condition, 44
  slowly varying amplitude approximation, 47–49
  see also Stimulated Raman scattering

Debye equation, 198, 298, 319
Debye relaxation, 198
Degenerate four-wave mixing, 249–251
  connection with halography, 250
  polarization dependence, 250–251
  symmetry consideration, 250

Density matrix formalism, 13–17, 415–419
  for transient four-wave mixing, 259–262,
      393–395
Density variation, laser-induced, 290
Detection of rare atoms and molecules,
      270–271, 275, 349–365
  applications, 363–364
  basic theory, 349–354
  conditions, 352
  demonstrations, 360–363
  experimental techniquès, 354–360
  laser-induced fluorescence, 354–358
  photoionization, 358–360
  requirement, 349–350
  signal-to-background ratio (discrimination
      factor), 352–354
Difference frequency generation, 108–116
  of far-infrared radiation, 110–116
  of tunable infrared radiation, 109
Doppler broadening, 212
Doppler-free spectroscopy:
  coherent transient spectroscopy, see Transient
      coherent optical effects
  four-wave mixing spectroscopy, 277–279
  multiphoton absorption and saturation
      spectroscopy, 240
  polarization spectroscopy, 232–235
  saturation spectroscopy, 216–229
  two-photon absorption spectroscopy, 229–232
Dressed atom, 220, 421–425
  absorption and fluorescence spectra, 423
  effective n-level system, 425
  three-level system, 424–425
  two-level system, 421–424
Dynamic (AC) Stark splitting, 220. See also
      Optical Stark effect

Electron multiplication, 531
Electrooptical effects, 53–54
  DC Kerr effect, 54
  Pockel's effect, 53
Energy randomization, 458–459
Energy relation for fields, 44–47
Exciton molecules, 208
Exciton polaritons, 207–208

Fabry-Perot interferometer, nonlinear, 299–301
Faraday effect, inverse, 61–66
Faraday rotation, 55
Far-infrared generation by difference-frequency
      mixing, 110–116, 176
  absorption, 112
  diffraction effect, 111–112

phase matching effect, 112
  reflection at boundary, 112
  total reflection, 112
Far-infrared generation by ultrashort pulses,
      113–116
  power spectrum, 114–115
Feynman diagram, double, 19–23, 259–261,
      393–397
Field energy:
  effective density, 59–60
  energy density, 46–47
  in nonlinear medium, 46–47
  time-average conservation relation, 47
Filament formation:
  in CW self-focusing, 323–324
  in quasi-steady-state self-focusing, 315
  in self-focusing in plasmas, 553
  in transient self-focusing, 320–321
Fluorescence, multiphoton-induced spectroscopy,
      337. See also Laser-induced fluorescence
Forced light scattering, 281–282
Four-wave mixing, 242–265
  backward parametric process, 248–249
  degenerate, 249–251
  general theory, 247–249
  nonlinear susceptibilities for, 242–247
  in optical fibers, 513
  with output in same mode as input field, 248
  phase conjugation, 251–254
  in plasma, 547
  pressure-induced extra·resonance (PIER), 281
  spectroscopy, 266–285. See also Four-wave
      mixing spectroscopy
  with three pump fields, 247–248
  transient, 259–264. See also Transient four-
      wave mixing
  tunable infrared and uv generation by,
      254–259
Four-wave mixing spectroscopy, 266–285
  coherent anti-Stokes Raman spectroscopy
      (CARS), 267–272
  coherent Raman spectroscopy, 267–275
  forced light scattering spectroscopy, 281–282
  general description, 266–267
  as high-resolution Doppler-free spectroscopy,
      277–279
  multiply resonant spectroscopy, 277–281
  polarization CARS, 272–275
  Raman-induced Kerr effect spectroscopy,
      275–277
  resonant characteristics in, 266–267
  transient, 283
Free electron gas, 8–11

Hamilton-Jacobi equation, 308
Harmonic generation, 86–116
  in plasmas, 548–550
High-resolution nonlinear optical spectroscopy, 211–241
High-resolution Raman spectroscopy, 167–169, 270
Hole burning, 218–222

Index ellipsoid, 54
Infrared multiphoton excitation and dissociation of molecules, 437–465
  early investigations, 437–441
  experimental results, 453–458
  models, 446–453, 459–461
  physical description, 441–446
Infrared-to-visible converter, 256
Inhomogeneous broadening, 211–212
Intensity-dependent ellipse rotation, 295–296
  experimental arrangement, 296
Interference between Raman and two-photon resonances, 275
Inverse Bremsstrahlung process, 529
Inverse Faraday and Cotton-Mouton effect, 60–66
Ionization:
  cascade or avalanche, 529, 532
  electron-impact, 529
  multiphoton, 337–339, 528–529, 533
  single-atom, 528
  threshold rate, 531
Isotope enrichment factor, 438, 469
Isotope separation, 374, 438, 466–478. *See also* Laser isotope separation
Isotope shifts, 466–467
  hyperfine interaction, 467
  mass effect, 466–467
  volume shift, 467

Jacobian, 104
Jacobi elliptical integral, 80

Lamb dip, 225–226
  inverted, 226–227
Lamb shift, 223–224
Laser fusion, 271, 550, 552, 278–279
Laser-induced fluorescence, 229, 354–358, 457
  coincidence counting technique, 356
  for detection of rare atoms and molecules, 354, 357
  experimental arrangement, 356–357
Laser isotope separation, 466–478
  of B, 476

  of Ba, 471–472
  based on atomic excitation, 469
  based on molecular excitation, 469
  basic requirement, 468
  general description, 466–469
  of hydrogen/deuterium, 473, 475
  by infrared multiphoton dissociation, 476
  by one-step photopredissociation, 474–475
  of ortho-I_2, 473
  by photochemical methods, 473–476
  by photochemical reaction, 473–474
  by photodeflection, 240, 471–473
  by photoionization, 470–471
  by photophysical methods, 470–473
  by two-step photodissociation and photopredissociation, 475–476
  of uranium, 462, 470–471, 476
Laser manipulation of particles, 366–378
Laser steering of atomic beams, 373–375
  deceleration of atoms, 374
  deflection of atomic beam, 374
  experimental arrangement, 374
  focusing of atomic beam, 374–375
  for isotope separation, 374
Liouville's equation, 14, 380, 416, 419
Liquid crystals, 293

Magneto-optical effects, 54–56
  Cotton-Mouton effect, 55
  Faraday effect, 55
  inverse effects, 60–66
Maker fringes, 99–100
Manley-Rowe relation, 78
Maxwell equations, 2–3
Maxwellian velocity distribution, 212
Measurement of nonlinear optical suceptibilities, 98–103
  interference method, 99–100
  phase factor, 99–100
  powder method, 102–103
Molecular redistribution, 95, 291, 293
Molecular reorientation, 195–198, 291–293
  by circularly polarized field, 292
  by linearly polarized field, 292
  in liquid crystalline media, 293
  optical-field-induced, 195–198, 291–293
Multichannel quantum defect theory, 340
Multiphoton dissociation, infrared:
  analog model, 459–461
  bond-selective (or mode selective), 440, 461–462
  competing dissociation channels, 458

Multiphoton dissociation, infrared: (*Continued*)
in continuum, 445–446
critical configuration, 448
dependence on laser intensity and fluence, 452
dissociation channels, 440, 445, 453, 458, 462
dissociation lifetime, 448
dissociation rate, 448
dissociation states, 448
dynamics, 440, 457
excess energy, 445, 452, 457
exit energy barrier, 448, 457
isotopically selective, 438
products, 440
secondary dissociation, 453, 458, 462
simple model, 446–453
summary, 461–463
two-laser scheme, 454
yields, 439
Multiphoton excitation, 354, 432–435, 539
infrared, 437–465
average excitation level, 456
coherent effect in, 462
through discrete levels, 441–443
induced luminescence, 437–438
isotopically selective, 438
as laser heating process, 445
limited by dissociation, 445
population distribution after, 449–451
through quasi-continuum, 443–445
stepwise resonant or near-resonant, 441–442
in true continuum, 445–446
Multiphoton-induced fluorescence, 337
Multiphoton ionization, 528–529, 533
spectroscopy, 337–339
Multiphoton spectroscopy, 334–348
Doppler-free, 336
experimental techniques, 337–339
general considerations, 334–336
Multiphoton transitions, 334–336
detection, 337–339
n-th order perturbation calculation, 335
population excitation, 334–335
(m + 1)-step n-photon transition, 336
single-step n-photon transition, 33
transition probability, 334
Multiply resonant four-wave mixing, 227–281
coherent Raman spectroscopy of excited states, 280–281
as Doppler-free spectroscopy, 227–279
doubly resonant case, 244–245, 278–279
measurement of longitudinal relaxation time, 279–280

triply resonant case, 245–246, 275
Multipoles, 3

Navier-Stokes equation, 192
Nonlinear optical effect in optical waveguides, 505–527
effective interaction length, 507, 511
four-wave mixing, 507, 513
four-wave parametric amplification, 507–509, 515
general theory, 505–509
in optical fibers, 510–517
optical field-induced refractive index change, 515, 518–519
optical Kerr effect, 515
phase matching condition, 510
second harmonic generation, 509, 510
self phase modulation, 515–517
spectral broadening, 515–516
stimulated Raman and Brillouin scattering, 507, 511–515
in thin-film waveguides, 509–510
wave interaction, 506
Nonlinear optical effect in plasmas, 541–554
basic formalism, 8–11, 542–548
coupling of light with plasma waves, 543–548
experimental studies, 548–553
four-wave mixing, 552
harmonic generation, 548–550
optical-field-induced refractive index change, 547–548, 553
optical nonlinearities, 8–10, 542–543
parametric instabilities, 547, 552
second harmonic generation, 10–11, 543, 548
self-focusing, 548, 553
self-phase modulation, 548, 553
stimulated Raman and Brillouin scattering, 545–547, 549, 551–552
Nonlinear optical effects involving surface electromagnetic waves, 481–493
experimental setup, 489, 492
generation of bulk wave, 487–488
generation of surface wave, 488–493
phase matching, 491
at plane boundary, 486–488
second harmonic generation, 488
surface CARS, 491–493
surface wave excitation by optical mixing, 488–493
Nonlinear optical effects as surface probes, 493–503
second harmonic generation, 495–503

stimulated Raman gain spectroscopy, 494
surface CARS, 494
Nonlinear optical susceptibilities, 13–41
bond-charge model, 33–34
bond model, 30–36
bulk, 480, 497
charge transfer model, 34
conventions on, 38–40
diagrammatic technique for calculation, 19–23
empirical pseudopotential calculation, 36
Kleinman's conjecture on, 26
LCAO methods, 37
local field correction, 23–25
measurement of, 98–103
microscopic expressions for, 17–19
Miller's coefficient, 37–38
for molecular crystals, 37
permutation symmetry, 25–26, 57
perturbation calculation, 16–19
phase factor, 99–100
practical calculations, 29–37
semiempirical Hartree-Fock method, 37
structural symmetry, 26–29
surface, 480, 497–498
tables, 27, 28, 101
Nonlinear Schrodinger's equations, 518

Optical breakdown, 528–540
in gases, 530–536
general description, 528–530
initiation, 530–531
self-focusing effect, 530, 537
in solids, 536–539
statistical nature, 533, 538
on surfaces, 539
threshold, 531–537
apparent, 530, 536
intrinsic, 530
Optical cooling of atoms, 375–376
Optical-field-induced refractive index change
(optical-field-induced birefringence),
286–302
in absorbing medium, 293
applications, 298–301
in atomic vapor, 289
electronic contribution, 287–289, 297
electrostriction, 290–291, 297
general form, 286–287
measurements of, 294–297
in mixture, 293
molecular reorientation and redistribution,
291–293, 298
in optical fibers, 515, 518–519
in photorefractive materials, 293

physical mechanisms, 287–294
in plasmas, 547–548, 553
population redistribution, 288
Raman-induced, 289–290
saturation effect, 288–289
transient effect, 297–298
Optical Kerr effect, 294–296, 518–519
experimental arrangement, 295
as four-wave mixing, 295
in optical fibers, 518–519
Optical levitation, 317–373
applications, 372–373
experimental setup, 372
Optical Ramsey fringes, 235–240
experimental arrangement, 235, 238–239
with two-photon transition in two-level
system, 237–240
Optical rectification, 57–66
inverse magneto-optical effect, 60–66
Optical Stark effect, 61, 414, 425
Optical trapping of atoms and ions, 376–377
Orientational distribution function, 197
Orientational order parameter, 197, 291

Parametric amplification, 117–119
vs. difference-frequency generation, 118
high-conversion limit, 119
signal and idler waves, 118
Parametric coupling between:
light and acoustic and entropy waves, 192
light and acoustic waves, 187
light and concentration variation, 199
light and material excitation, 148–151
light and molecular orientation, 198
Parametric fluorescence (or scattering), 134–135
Parametric four-wave mixing, 153
in optical fibers, 507–509, 515
Parametric instabilities, 547, 552
Parametric oscillator, 120–134
backward, 138–139
conversion efficiency, 124, 127
doubly resonant, 120–124
frequency tuning, 127–134
power output, 124, 127
singly resonant, 124–127
stability, 122–123, 124
threshold, 122, 125–126
Parametric superfluorescence, 136–138
bandwidth, 138
output power, 137
as tunable picosecond infrared source,
138
Phase conjugation, 251–254
applications, 254

Phase conjugation (*Continued*)
    correction of aberration, 251–253
    from stimulated light scattering, 254
Phase matching, Type I and Type II, 76
Photon echoes, 388–399
    backward and forward echoes, 398–399
    conditions for, 389, 396–398
    other echoes, 399
    multi-level system with multi-pulse
        excitations, 392, 399
    phase matching requirement, 391
    quantitative analysis, 391
Photoionization, 358–360
Plasma dispersion function, 278
Plasma formation, laser-induced, 528–531
    by electron avalanche, 529
    hydrodynamic expansion, 529
    primary electrons for, 530, 531, 537
    threshold laser intensity, 531–533
Plasma resonance, 11
Plasmas:
    laser-heating of, 541, 551–552
    laser-induced, 548, 552–553
    two-component (electron-ion), 542
Plasma waves, 543–547
    dispersion curve, 544
    electron (or optical), 544
    ion-acoustic, 544
    resonant frequencies, 544
Polariton dispersion curve, 170
Polarization CARS, 272–275
    experimental arrangement, 273
    as heterodyne technique, 274
    suppression of nonresonant background,
        273–274
Polarization labelling spectroscopy, 232–235
Polarization spectroscopy:
    CARS, 272–275
    high-resolution, 232–235
    labelling, 232–235
    Raman-induced Kerr effect, 275–277
Preionization, 531–533
Pressure-induced extra resonance in four-wave
    mixing (PIER-4), 281
Pulse propagation in fiber, 517–526
    effect of field-induced refractive index, 518
    effect of group velocity dispersion, 518
    formal description, 518–519
    pulse broadening, 519
    pulse compression, 523–526
    pulse narrowing, 519–522
    solitons, 519–524

Quantum beats, 213–216
Quasi-continuum states of molecule, 429,
    443–445, 453–456

Rabi frequency, 384, 418, 421
Radiation forces, 366–371
    dipole force, 369–370, 374, 376–377
    electrostrictive force, 188, 367
    radiation pressure force, 368–369
    scattering force, 370, 376–377
Raman gain and inverse Raman spectroscopy,
    167–169, 275
Raman-induced Kerr effect, 275–277, 289
    induced birefringence, 276, 289
    spectroscopy, 275–277
    suppression of nonresonant background,
        276–277
Raman scattering, stimulated, *see* Stimulated
    Raman scattering
Raman susceptibilities, 147
    microscopic expression, 147
    relation with:
        differential Raman cross-section, 147
        Raman gain, 147
        Raman transition probabilities, 147
Raman transitions between excited states,
    280–281
Ramsey fringes, 235–240
Rayleigh component, 192
Rayleigh scattering, stimulated, 194
    thermal, 192–195
Rayleigh-wing scattering, stimulated, 195–199
Relaxation, 15
    longitudinal (population), 15
    transverse (dephasing), 15
Relaxation times, 15, 181–184
RRKM model, 448–450, 459
Rydberg atoms, 340–346
    diamagnetic effect, 342–343
    exhibiting amplified spontaneous emission,
        345
    exhibiting maser action, 345–346
    experimental setup, 342
    fine structure splittings, 340, 341, 344
    ionization, 344
    linewidth *vs.* principal quantum number,
        341
    pressure shifts and pressure broadening, 341
    properties of, 341
    transition probabilities, 343
    two-photon Doppler-free spectra, 341
Rydberg spectrometer, 344

Saturation Raman spectroscopy, 275
Saturation spectroscopy, 216–229
  coherent effect, 220, 221–223
  copropagating pump and probe, 218–220
  counter-propagating pump and probe,
    220–227
  experimental arrangement, 223
  in multi-level system, 227–229
  saturation effect, 217
  two counter-propagating beams of equal
    intensity, 225–227
  in two-level system, 216–227
  weak probe and strong pump, 218–222
  weak saturation limit, 217, 221
Second-harmonic generation, 1, 86–93
  collinear phase matching, 88–89
  double refraction effect, 90–92
  effective length, 91
  efficiency, 88, 90, 92
  with focused Gaussian beam, 89–93
  with 90°-phase matching, 89, 90
  optimum focusing, 91–93
  output power, 86
  in plasma, 10–11, 543, 548
  with surface waves, 488
  theory, 86–89
  with ultrashort pulses, 103–106
  in waveguides, 509, 510
Second-harmonic generation as surface probe,
    495–503
  effective surface nonlinear susceptibility,
    496–497
  experimental arrangement, 498
  between media with inversion symmetry, 495
  sensitivity, 499–501
  signal strength, 497–498
  studying submonolayer adsorbates, 499–503
  surface vs. bulk contribution, 497
Self-defocusing, thermal, 330–331
Self-focusing, 303–333
  in amplifying media, 323
  CW, 323
  in gases, 323
  in liquids with colloidal suspension, 323
  physical description, 303–307
  in plasmas, 548, 553
  polarization dependence, 314
  quasi-steady-state, 313–319
    connection with anomalous stimulated
      Raman and Brillouin effects, 316–319
    critical beam size, 313–314
    filament diameter, 315, 319

  filament and moving focus, 314–319
    moving focal spot, 315–316
    transient dynamics in focal region, 319
    U-curve for, 315–318
  resulting in damage in solids, 322
  resulting in filaments, 314–315, 320, 323
  resulting from optical-field-induced refractive
    index, 303
  as result of stimulated Rayleigh scattering, 199
  vs. self-trapping, 308, 311–312
  in solids, 321–322, 530, 537
  vs. stimulated Raman scattering, 307,
    316–319
  theory, 307–313
  thermal, 323
  transient, 319–321
    as dynamic self-trapping, 320
    filament formation, 320
    formation of horn-shaped pulse, 320
    qualitative description, 319–321
    from quasi-steady-state to transient, 321
Self-induced transparency, 401–407
  basic idea, 401–402
  characteristics of, 406
  hyperbolic secant pulse, 404
  pendulum equation for, 403
  $2n\pi$ pulse, 401, 404–406
Self-phase modulation, 324–329
  effect on self-steepening, 330
  in optical waveguides, 515–517
  in plasmas, 548, 553
  in quasi-steady-state self-focusing, 325–327
  resulting in spectral broadening, 324–328
  in self-trapped filaments, 324–325
  in space, 328–329
  in transient self-focusing, 328
Self-steepening, 329–330
Self-trapping, 308, 311–312
Slowly varying amplitude approximation, 47–49
Snell's law, nonlinear, 50, 69
Solitons in optical fibers, 519–524
  experiment, 521–522
  fundamental soliton, 519
  intensity, 520
  interaction of, 524
  period, 519–520
  pulse narrowing and splitting, 520–524
  pulseshape, 519
  solution, 519–520
Spectral broadening, 324–328
  maximum, 327
  in optical waveguides, 515–517

Spectral broadening (*Continued*)
  from self-phase modulation, 324–328
  semi-periodic structure, 325–326
Spin-flip Raman scattering, 172–176
  cross-section, 173–174
  resonant enhancement, 173–174
  stimulated, 174–176
Stimulated Brillouin scattering, 187–192
  in optical fibers, 507, 511–515
  in plasmas, 545–547, 550–552
Stimulated Compton scattering, 200
Stimulated concentration scattering, 199–200
Stimulated light scattering, 187–201
  Brillouin scattering, 187–192
  Compton scattering, 200
  concentration scattering, 199–200
  Rayleigh-wing scattering, 195–199
  thermal Brillouin and Rayleigh scattering,
    192–195
Stimulated polariton scattering, 169–172
Stimulated Raman scattering, 141–186
  anomalies, 159, 307, 316–319
  applications, 164–169
  competition with stimulated Brillouin
    scattering, 317–318
  connection with self-focusing, 307, 316–319
  coupled-wave description, 146–151
  experimental observations, 159–169
  higher-order Raman effects, 156–157
  in Kerr liquids, 159
  in optical fibers, 507, 511–515
  as parametric process, 148–151
  in plasmas, 545–547, 550–552
  quantum theory, 143–146
  Stokes-anti-Stokes coupling, 152–156
  transient, 176–181
Stimulated Rayleigh-wing scattering, 195–199
Stimulated spin-flip Raman scattering, 172–176
  conversion efficiency, 175
  for infrared output, 176
  operation characteristics, 174–175
Stimulated thermal Brillouin and Rayleigh
    scattering, 192–195
  gain spectrum, 194–195
Strong interaction of light with atoms, 220,
    413–436
  bare-atom approach, 415–421
  dressed atom approach, 220, 421–425
  experimental demonstration, 425–432
  general description, 413–415
  multiphoton excitation and ionization,
    432–435
Sub-Doppler Raman spectra, 275

Sum-frequency generation, 67–85
  with boundary reflections, 73–76
  in bulk, 71–73
  coupled wave formalism, 68–70
  with focused beams, 84
  with high-conversion efficiency, 77–82
  limiting factors for high conversion efficiency,
    83–84
  Manley-Rowe relation, 78
  with 90°-phase matching, 84
  phase matching, 76
  practical example, 82–83
Superfluorescence, 407–410
  characteristic features, 408
  effective emission lifetime, 408
  initiation, 410
  quantum fluctuations, 410
Surface coherent anti-Stokes Raman scattering
    (surface CARS), 491–493
  phase matching condition, 491
  signal strength, 492–493
  as spectroscopic technique, 491–493
Surface electromagnetic waves (surface plasmon
    waves), 481–486
  dispersion relation, 482–483
  excitation schemes, 483
    Kretschmann configuration, 483
    Otto configuration, 483
  field enhancement, 485
  in multilayer system, 486
  see also Nonlinear optical effects involving
    surface electromagnetic waves
Surface exciton-polaritons, 489–490
Surface nonlinear optics, 479–504
  effective interaction length, 480
  general description, 479–481
  see also Nonlinear optical effects involving
    surface electromagnetic waves; Nonlinear
    optical effects as surface probes
Surface nonlinear susceptibility, 480, 495–498
  effective, 497
Surface photon-polaritons, 489
Susceptibilities, *see* Nonlinear optical
    susceptibilities

Third harmonic generation:
  in crystals, 93
  in gases, 93–98
Third-order nonlinear susceptibilities,
    242–247
  doubly resonant cases, 244–245
  local environment effect, 246–247
  resonant and nonresonant parts, 242–243

singly resonant cases, 243
triply resonant cases, 245–247
Time-dependent wave propagation, 50–51
Time-resolved infrared spectroscopy, 258
Transient coherent optical effects, 379–410
  adiabatic following, 399–401
  adiabatic inversion, 400
  in effective two-level system, 379–383
  experimental setup, 385
  free induction decay, 386–387
  photon echoes, 388–399
  self-induced transparency, 401–407
  superfluorescence, 407–410
  superradiation, 408
  transient four-wave mixing, 393–399
  transient nutation, 383–385
Transient four-wave mixing, 259–264, 283,
    393–399
  calculated by density matrix formalism,
    259–262
  coherent transient optical effects, 262,
    393–399
  forced light scattering, 281–283
  spectroscopy, 283
Transient n-wave mixing, 399
Tunable far-infrared radiation:
  by coherent Cherenkov radiation, 200
  by electron cyclotron motion, 200
  by stimulated Compton scattering, 200
Tunable infrared generation by four-wave
    mixing, 254–258
  collinear phase matching, 256
  as four-wave parametric amplification process,
    256
  limitation, 256–257
  resonant enhancement, 256
  tunable range, 256
Tunable infrared sources:
  in atomic vapor, 164

conversion efficiency, 164, 167
limitations, 164
in molecular gas, 164–167
by stimulated polariton scattering, 172
by stimulated Raman scattering, 164–167
tuning range, 164–167
Tunable uv generation by four-wave mixing,
    254–258
  in alkali earth vapor, 257
  conversion efficiency, 258
  detrimental effects, 258
  in molecular beam, 258
  in molecular gases, 258
  tuning range, 258
  vacuum, 258
Two-level system, 379, 416
Two-photon absorption, 202–209
  absorption coefficient, 203
  in atoms, 209
  coupled wave approach, 203
  detection schemes, 204–205
  experimental techniques, 204–205
  in molecular fluids and gases, 209
  in solids, 206–208
  spectroscopy, 206–209
  transition probability, 202
Two-photon Doppler-free absorption
    spectroscopy, 229–232
Two-photon resonant pumping in three-level
    system, 419–421, 430–432

Up-conversion process, 70

Wave equation, driven, 68
  homogeneous solution (free wave), 68
  particular solution (driven wave), 68
Whitten-Rabinovitch approximation for density
    of vibrational states, 443